D0793459

Alcohol in Western Society from Antiquity to 1800

Alcohol in Western Society from Antiquity to 1800

A Chronological History

Gregory A. Austin

with the staff of the
Southern California Research Institute

Foreword by Robin Room

ABC-CLIO

Santa Barbara, California
Oxford, England

This book is Smyth sewn and printed on acid-free paper to meet library standards.

Library of Congress Cataloging in Publication Data

Austin, Gregory A.
Alcohol in Western society from antiquity to 1800.

Bibliography: p.
Includes indexes.
1. Drinking of alcoholic beverages—History.
2. Alcoholism—History. 3. Alcoholic beverage industry—History. I. Title.

HV5020.A97 1985 394.1'3'09 84-21668

ISBN 0-87436-418-3

10 9 8 7 6 5 4 3 2

ABC-Clio, Inc.
2040 Alameda Padre Serra
P.O. Box 4397
Santa Barbara, CA 93140-4397

Clio Press Ltd.
55 St. Thomas Street
Oxford, OX1 1JG, England

Manufactured in the United States of America

To Dan J. Lettieri, without whose advice and support this research would never have been possible. We have been sustained by his belief in its value and by his patience during its long gestation.

Contents

Foreword

In 1887, after "twenty-two years of special attention to the various questions comprised within its scope," the temperance scholar Daniel Dorchester published his compilation of *The Liquor Problem in All Ages.*[1] About 130 pages were devoted to the topic of "drinks, drinkers, and abstainers in the olden and later times", that is, the world before the nineteenth century; most of the rest of the 728-page volume was devoted to "the temperance reformation." Until just a few years ago contemporary historical studies in the alcohol field showed a parallel imbalance—although American studies of the 1950s and 1960s implicity took a much more critical and indeed antagonistic approach to the temperance movement than had Dorchester. As Dannenbaum had documented, it is only after 1974 that published articles on state or national prohibition lost their dominance in American studies of alcohol history and that social historical studies came to the fore.[2]

Among historians, then, there has been an "explosion of interest" in recent years, and the range of interests has been broadened to encompass such issues as the cultural position of drinking and of alcoholic beverages, the relation of alcohol consumption to alcohol-related problems, the politics of alcohol control, and the place of alcohol issues in popular movements. A newsletter on *Alcohol in History*[3] now serves the growing constituency of those interested in the whole range of topics involving alcohol.

Meanwhile, researchers from other disciplines have also found themselves turning to history. The researchers involved in the International Study of Alcohol Control Experiences, attempting a "social history of control policy" in the past thirty years as a way of understanding widespread rises in alcohol consumption and problems, quickly found that their research questions required a much deeper look into the past.[4] The heavy attendance of social scientists at a recent Berkeley conference on the social history of alcohol shows that, for an increasing number of scholars in the alcohol field, history is too important to be left only to historians.

But those who wish to learn from history have not found the path easy. Original historical work on alcohol tends to deal with one country or culture at a time and to concentrate on the last two centuries. Those desiring a synoptic view, and those interested in drinking in earlier times, have very soon found themselves back in the hands of temperance movement antiquarians—whose approach was limited not only by the enthusiasms of the time but also by the fact that their pursuit of the topic was sporadic, serving in Dorchester's case "as a relaxation from the regular duties of the Christian ministry" (p. 3).

The present volume represents a long leap forward in making an understanding of alcohol history a manageable task. Few will be able to read the volume without discovering facets of alcohol history of which they had no previous inkling. Besides its status as a fascinating reference source, the volume is truly a *vade mecum* for social and policy researchers wishing to ransack the past for contemporary policy implications. In its arrangement, too, the volume is pathbreaking and thought-provoking: the very fact of bringing the disparate national stories together in a common chronology frame of reference makes it a fruitful source of speculation and hypotheses.

— ROBIN ROOM
Berkeley, California

1. Daniel Dorchester. *The Liquor Problem in All Ages.* New York and San Francisco: Phillips & Hunt; Cincinnati, Chicago and St. Louis: Cranston and Stowe, 1888 (revised from 1887 edition).

2. Jed Dannenbaum. "The Social History of Alcohol." *Drinking and Drug Practices Surveyor* 19 (April 1984), pp. 7–11.

3. Available from David Fahey, Department of History, Miami University, Oxford, Ohio 45056.

4. Klaus Mäkelä, Robin Room, Eric Single, Pekka Sulkunen, Brendan Walsh, et al. *Alcohol, Society, and the State: 1. A Comparative Study of Alcohol Control.* Toronto: Addiction Research Foundation, 1981.
Eric Single, Patricia Morgan, and Jan de Lint, eds. *Alcohol, Society, and the State: 2. The Social History of Control Policy in Seven Countries.* Toronto: Addiction Research Foundation, 1981.

Acknowledgments

Although the ultimate responsibility for this volume must rest with the compiler, many people participated in its research and compilation. I offer special thanks and recognition to those staff members and colleagues who helped conduct the research in various topical areas, especially Michael Prendergast, my associate at the Southern California Research Institute. Dr. Prendergast's thoughtful historical research and editorial suggestions substantially improved this volume. He helped research the areas of colonial America, England, Finland, Ireland, Russia, and Sweden. He also deserves credit for the preparation of the Subject Index. Dr. Harry Levine, Queens College, SUNY, provided much of the material on colonial America. Staff researchers included: Brian Bunnett (France), Bohdan Futala (Russia), J. Andrew Greig (Scotland), David Mayfield (Germany), and Boris Odynocki (Russia). The thoughtful comments of Peter Clark and Thomas Brennan helped our efforts in the final editing and revision of the manuscript. Hermann Beck and Lisa Roberts reviewed the text on early modern Europe. Both the content of this volume and my own understanding of the subject were greatly improved by the January 1984 Conference on the Social History of Alcohol sponsored by the Alcohol Research Group, Berkeley, California.

The following staff members assisted in locating source materials and preparing the manuscript: Beverly D. Barnett, Pat Benson, Nancy Fitzgerald, Barbara Garrison Hawkins, Rita Marks, and Mieko Morita.

This chronology was prepared as part of two larger projects in which a comprehensive data base on the world history of alcohol and drug use is being compiled. Funding for this research has been provided by the National Institute on Drug Abuse (Grant DA02203) and the National Institute on Alcohol Abuse and Alcoholism (Grant AA04570); the research is being conducted at the Southern California Research Institute, Los Angeles, California.

Introduction

The use of alcoholic beverages is as old as civilization itself, as are concerns over alcohol abuse. Until recently, however, alcohol history has largely been ignored as a field of research. Fortunately, the interest in this field is growing, an endeavor this volume seeks to encourage and aid. *Alcohol in Western Society from Antiquity to 1800* is a comprehensive overview of the history of alcohol use and controls in the Western world through the end of the eighteenth century. It is designed not only for alcohol researchers but for anyone interested in the role played by alcohol in Western society, in terms of both the benefits derived and the problems created. By studying the experiences of the past we may not only begin to better understand and appreciate the functions of drinking but also improve our ability to deal with this major social issue.

Subtitled "a chronological history," this book is essentially a hybrid between a chronology and a traditional monograph. It was designed to facilitate the identification and acquisition of historical data, and, particularly, to assist in the comparative analysis of similarities and differences in alcohol use among various times and countries. To these ends, the content is organized in three levels—first by date, second by country, and third by topic. This organization, as discussed below, allows the chronology to be useful to three types of readers: 1) those interested in studying alcohol use in specific periods; 2) those interested in specific countries; and 3) those interested in specific topics. Although organized chronologically, it is not merely a list of dates and events. For the most part, we have tried to let the information speak for itself, allowing readers to learn what has been said about a topic and to draw their own conclusions and interpretations. Yet as much as possible, within the confines of the volume's intent and organization, we tie the data together, discuss causal factors, and provide guidelines for assessing the data.

The book is perhaps best described as a compendium of current knowledge and a guide to what contemporaries and later commentators have said about drinking and the alcohol trade. In content, it is like a banquet of information from which readers can make their own selections. Much of the information is highly detailed and focused on epidemiological issues. Many of the entries are anecdotal, and though it deals with a serious social problem, the work is often entertaining and amusing. In addition to covering the subject of alcohol use, the text also touches on popular culture, recreation and leisure, diet, lifestyles, morals and manners, physical and mental health, attitudes toward work and poverty, class structure and conflicts, the growth of capitalism, urbanization, and industrialization. It is one of the fascinating and valuable features of alcohol history that it relates to so many varied aspects of life: political, economic, social, religious, medical, legal, and cultural.

This is actually a work-in-progress which we are continuing to expand and revise. The decision to publish the current version was not easily reached, primarily because there was much more information that could have been included. The more research we conducted, the more we uncovered new issues, problems, and subjects that warranted further research. Yet, as it stands, this volume is the most comprehensive, up-to-date general survey to current knowledge, and we felt that it was time to make the information available, with expectations for subsequent editions.

Content

The chronology was written in the present tense to improve readability and promote a sense of immediacy and participation in events. It should be pointed out, however, that this approach sometimes tends to obscure the complex relationship between contemporary perceptions and "reality." For example, because people blamed increased drunkenness on a particular development does not mean that drunkenness had actually increased or that it had increased for the reasons people identified. As the chronology demonstrates, many factors influence attitudes and concerns about drinking besides actual changes in consumption patterns or levels. It is also the nature of a reference book such as this that we were unable to confirm some statements. Documentation for all our information is provided, but in recognition of these considerations we issue this *caveat*: the reader should realize that an unspoken "it has been asserted that" should precede each statement.

The depth of coverage is not uniform across all countries and periods for several reasons. First, our primary focus was classical Greece and Rome, England, France, and, to a lesser extent, Germany. Second, there is more information available on some countries and topics than on others. Third, for this initial chronology, our research was limited largely to materials in English and French. In compiling the material, we first examined the major secondary sources; then began the long process of verifying data in primary sources and tracking down new material to answer questions or fill gaps. As many primary sources were examined as was feasible, given the book's large time-span and geographic scope.

Special attention was devoted to collecting information in the following areas: 1) the relationship of use to abuse; 2) factors that maintain uniform levels and patterns of use over time; 3) causes for changes in prevalence and patterns of use; 4) attitudes towards use; and 5) effects of popular attitudes and public policies on levels and patterns of use. Among the specific subjects covered by this chronology are the economic role of brewing, viticulture, and distilling; the functions served by alcoholic beverages and drinking outlets; the effects of the Protestant Reformation; the introduction of distillation and spirit drinking; the changes brought about by the commercialization of the alcohol trade in the early modern period; and the role of alcohol revenues in public financing. In addition to alcoholic beverages, the chronology surveys the history of coffee and tea in considerable detail, both because the introduction of these hot beverages affected alcohol consumption and because their history sheds light on the process by which an innovation in drug use is adopted and spreads.

Organization

Events are first grouped together in uniform chronological divisions. In the chapters on antiquity and the Middle Ages, the material is grouped in 25-year periods. For antiquity, there are often large gaps in the chronological divisions; for the Middle Ages, the divisions are more regular and each century begins on a new page. From the sixteenth through the eighteenth centuries, the chronology is divided into increments of ten years, with each decade beginning on a new page. Under each date, the text is subdivided according to national, ethnic, or religious groupings, as indicated by the alphabetically arranged headings. Most of the subdivisions are by country; the major exceptions are sections on Hebrews/Judaism and Christianity during antiquity. For simplicity, we will refer to these headings as "geographic" divisions or "countries." There are two exceptions to the alphabetical arrangement of "countries." Sections devoted to Europe in general appear first in each chronological division, preceding the sections for specific countries; sections on colonial America follow those devoted to European countries and are thus the last section of each division. Information about each country is grouped into topical paragraphs or entries, each preceded by a short topic descriptor identifying the type of information included in the entry (e.g., drinking practices, attitudes, price controls). To assist in locating information, running heads have been provided at the top of each page which give the date and country of the first geographic heading listed on the page, or the prior heading if no change occurs.

The chronological divisions are intended only as guideposts and should not necessarily be interpreted as the exact date of occurrence for any of the events discussed. Under each date, events are covered that may have occurred around that date or between it and the next chronological division. Summaries of major trends, such as developments that occurred over a long period of time, are generally found at the beginning of each century. Specific dates of events are indicated within the entry itself. Not all events could be easily assigned to a specific chronological division, especially when the information dealt with developments or practices that extended over long periods of time or had no clear point of origin. In some cases we simply had to made an arbitrary decision as to the best place to insert the information. Whenever possible, we linked related events together so that associations could be seen.

Referencing

Primary and secondary sources of information are indicated within the text of each entry. When information is taken from a secondary source, a short citation is given to the author, publication date, and page(s) from which the information was taken (e.g., Clark 1978:78). Full citations to these secondary sources are included in the list of References preceding the Index. Sources are listed alphabetically by author's last name; works by the same author are then arranged chronologically by date of publication.

A short citation system is also used for most of the primary materials. Classical Greek and Roman works are cited by author and document section rather than by publication date and page (e.g., Herotodus 4.23.4). For medieval and early modern sources, we provide in the short citation the date of original composition or publication as well as the date of the edition used. For example, the citation, (Fielding 1751/1871:341), means that the cited information came from page 341 of the 1871 edition of the work by Fielding listed in the References and originally published in 1751. In the list of References, both dates are again given, with the original date provided in parentheses. Generally, primary sources are listed in the References only when we examined an edition other than the original or when the document is cited more than once in the text. For other primary sources, author names, titles, and publication dates are given in the text. If no secondary source is cited, the information was derived directly from the primary source. We did not include in the text the titles of primary materials listed in the References unless it avoided confusion or the title communicated important information.

Index

Because of the unique nature of this reference tool, we designed a special indexing system to facilitate locating information of interest. This system is explained in more detail immediately preceding the indexes themselves. Briefly, three indexes are provided: Name and Anonymous Title Index, Peoples and Places Index, and Subject Index. The Name Index indentifies all historical individuals who appear in the text, including titles of anonymously published works. The Peoples and Places Index contains references to groups of people by their country or nation and to specific geographical locations such as cities and provinces. Here we have indexed all references to alcohol production, consumption, and trade by each country. This index can also be used to identify the major geographical sections in the text. The Subject Index refers to the topics discussed in each entry.

The unique feature of these indexes is that the locators consist of dates and geographic headings rather than page numbers. Following each index term, all the applicable country headings are listed alphabetically; each country heading is then followed by the relevant dates. Note, however, that whereas the text is organized first by date and then by country, in the indexes the country appears first, followed by the date. Furthermore, all country headings are arranged alphabetically, whereas in the text a section on Europe would precede, and a section on colonial America would follow, all the European countries.

The primary advantage of this system is that it provides both chronological and geographical subdivisions for each index term. In addition, the system often permits more precise access to information than a page reference, since the majority of geographical divisions under a given date are less than a page long. Many divisions are longer than a page, particularly for France and England, but the topic descriptors in the text should enable the reader to locate information quickly even in those cases.

Historical Overview

To conclude this introduction, I will briefly touch on some of the trends, themes, and major developments that appear in the text. Given the amount of material contained in it, such an overview must be very impressionistic. It is intended simply to set the stage and orient readers to the subject matter, to reduce the possibility of their becoming lost in a myriad of details, and to assist them in tying the strands of information together.

To begin on the most general level, during most of the period covered by the *Chronology*, "alcohol" meant beer and wine. Recreational spirit drinking did not begin until the sixteenth century and did not become a significant factor in most countries until the eighteenth century. Throughout the *Chronology*, we find drunkenness consistently attacked by moralists, statesmen, and physicians. Recognition of the potential danger of alcohol abuse is as old as alcohol use itself, and drinking early became subject to social regulation. However, both the level of use and the nature of the concerns about drinking have varied greatly across time and place. Use has not always lead to widespread abuse, and several factors besides changes in consumption levels have contributed to the perceptions that drinking poses a social problem. Social concerns have often focused not so much on inebriety as on particular drinking practices, drinking outlets, and drinkers.

Although concerns about chronic inebriety were voiced in every country and age, one of the most important characteristics of the entire period from antiquity through the eighteenth century is the esteem in which alcohol was held. This esteem was rooted in the many vital functions alcohol played in society. So essential were alcoholic beverages that their consumption in moderation was rarely questioned. First, beer and wine were the major thirst quenchers. Water supplies generally were uncertain and unhealthful; coffee, tea and cocoa were not introduced until the mid-seventeenth century, and it was only toward the end of the period surveyed that they began to become popularly consumed on a daily basis. Second, alcoholic beverages provided important nutritional supplements to the diet of most people. Along with bread, they were regarded as necessities of life and as important sources of nourishment for the body. Third, they were widely consumed for medicinal purposes, especially wine, both as menstruums for the administration of drugs and as therapeutic agents in their own right. Alcohol was believed to confer strength and aid digestion, and it was the most common anaesthesia available in a period in which almost everyone lived with some pain. Indeed, the belief in the therapeutic value of alcohol was so strong that some physicians recommended getting drunk periodically, and alcohol consumption for prophylactic purposes appears to have risen appreciably during major outbreaks of disease. Fourth, alcohol served as a work adjunct, supplying a sensation of well-being and relief from the fatigue of hard agricultural labor. Fifth, socializing over food and drink was one of the few means by which personal and social tensions could be released; few alternative means of entertainment existed. Thus, drinking was an important component of festivals, feasts, and community gatherings.

Turning to classical antiquity, the Greco-Romans lavished poetic praise on the glories of wine but strove to encourage moderation. They emphasized the physical and mental harm inflicted by inebriety and advocated that wine be diluted and drunk—in limited amounts—only during and after eating. However, excessive drinking still occurred. With the exception of the Spartans, the Greeks were often not as temperate as has sometimes been alleged, although they were more restrained than most of their contemporaries. The expansion of the cities and the growing popularity of Dionysian rituals prompted several efforts to control drunkenness during the seventh century B.C. Judging from the number of complaints and calls for temperance, excessive drinking was also common in Athens during the height of its glory in the fifth century B.C. One contemporary critic, Plato, argued that the state had the right and duty to protect citizens from the adverse effects of drinking and to impose severe restrictions over wine consumption.

While the early Romans were known for their sobriety (the legendary King Numa purportedly prohibited Roman women, young men, and slaves from drinking wine), many Romans during the late republic and early empire period engaged in a prolonged drinking binge as the wealth and power of the city grew and viticulture expanded. Although most descriptions of inebriety in classical literature refer to acute intoxication, during this period Pliny and Seneca both described chronic alcoholism, a phenomenon with which they were evidently quite familiar. Seneca distinguished between the occasional drunkard and the drunkard who has no control and is a slave to the habit. He regarded such uncontrollable drinking as a form of insanity, a voluntary madness.

Despite their concerns over abuses, the Greeks and Romans generally did not consider inebriety as great a social problem as it is regarded today. It was viewed as not significantly different from excessive eating, luxury, greed, or any other indulgence. There was no specific advocacy of the virtue of abstinence as an end in itself, and there was no legislation penalizing inebriety among the general population. Neither the Athenians in the fifth century B.C. nor the Romans in the early empire took legislative action to reduce the level of inebriety. Concerns about alcohol were muted by their great appreciation of wine's value to life. Furthermore, most drinking took place in the home, in conjunction with eating; the lower classes usually drank only vinegary, diluted wine; and for most people drunkenness was a periodic occurrence associated with celebrations or social gatherings. Habitual, chronic drunkenness was largely limited to the upper classes who had the leisure time and wealth to indulge in it. Thus, the changes in the level of inebriety during antiquity seem to have been most influenced by changes in the drinking habits of the upper classes and changes in societal wealth.

In the Judeo-Christian tradition, the same duality exists between the praise of alcohol's positive benefits and the condemnation of inebriety that characterized the classical tradition. In some respects, early Hebrews were more critical in their attitudes toward drinking than were the Greco-Romans. Inebriety is soundly condemned in the Old Testament. The Book of Deuteronomy is the only known legal code in the ancient Near East that stipulated punishment for drunkenness. Some hostility to wine drinking itself appeared in early Hebraic writings, and the Rechabites embraced abstinence in the ninth century B.C. But reflecting the esteem with which wine was held by the ancients, the Hebrew scriptures frequently praised wine with as much enthusiasm as the Greco-Romans. Nor did the Rechabites advocate abstinence as an ideal in itself; they embraced it as only one facet of a wider ascetic, conservative reaction to the growing sophistication of Hebraic culture. Furthermore, the hostility to wine drinking diminished as it was integrated into Jewish rituals after the end of the Babylonian Capitivity in the sixth century B.C., a development which appears in part to have been a successful attempt to moderate drinking practices.

Following Judaic tradition, early Christians emphasized that wine was one of God's creations and therefore inherently good; the problem lay in the drinker who abused it. Drinking was sanctioned by Christ, who incorporated symbolic wine drinking into the Eucharist. In their efforts to convert pagans and, later, barbarians, Church authorities even incorporated features of pagan festivals into Christian celebrations, including drinking practices. Yet temperance was a preeminent virtue manifested by Jesus, and Christian warnings against excessive drinking were far stronger than those of the pagans. The latter advocated temperance largely on the basis of the individual's physical well-being and, to a lesser extent, on the preservation of social order. Christians held that chronic drunkenness was a sin that endangered one's soul and, therefore, the chance for everlasting life. The drunkard lived a life of physical distress and spiritual deadness. As St. Paul asserted, drunkenness barred the gates of Heaven and desecrated the body. A Christian should not even associate with a drunkard. Almost all Church fathers maintained that while the use of wine was lawful, voluntary abstinence, except for reasons of health, was a praiseworthy way to remove temptation and safeguard the soul. It was this emphasis on chronic drunkenness as a sin that made early Christianity a moderating influence on drinking.

During the Middle Ages, as in classical antiquity, the drinking of fermented beverages served multiple purposes and found its way into almost every aspect of life. Tradition encouraged feasting, drinking, and other public rituals. Among the nobility, the holding of public banquets was intended as a sign of their generosity, a virtue they were expected to demonstrate as justification for their superior position. Among the villagers, festive drinking functioned as a mechanism of community self-consciousness.

Given the pervasiveness of alcohol throughout medieval society, together with the pervasiveness of Christianity, it is not surprising that complaints about inebriety abound in medieval literature. The ideal of a sober Christian society was not always embodied in reality. Even the religious festivities were often celebrated with considerable drinking, at times within the church itself, as at parish wakes. Another example of the blending of religion and drinking during the Middle Ages was the English church-ale, a drinking party at which the proceeds from selling ale went to support the church. How extensive drunkenness was in different periods and areas of medieval Europe cannot be determined precisely. As national groupings began to appear, distinct drinking reputations developed. Northern Europeans (the Germans, Flemish, Dutch, and, to a lesser extent, the English) acquired reputations as heavy drinkers; southern Europeans (the Spanish, French, and Italians) as lighter drinkers. Contemporaries attributed these distinctions to the differences between wine and beer drinking, to the effects of climate, and to the greater prevalence of convivial drinking customs that characterized the more Germanic north. But throughout Europe, lay inebriety was not a widespread concern, except as it affected individual salvation; it was not considered a crime. Although chronic inebriety was a sin, occasional inebriety was accepted as a natural aspect of life. One of the few examples of legislation against drunkenness was a decree by Archbishop Theodore of Canterbury in the seventh century ordering that anyone who drank to excess must do penance for fifteen days. Because of the importance of beer and wine to the diet, drink controls largely focused on protecting the drinker from unscrupulous sellers, maintaining a good supply and a fair price, and reducing the adverse consequences (such as public disorders) of too much drinking. As in antiquity, inebriety was largely associated with occasional festivities and with a few specific populations (nobles, students, and clerics) who had the wealth, free time, or access to supplies that enabled more regular indulgence. The level of drunkenness was also moderated by the influence of Christianity, the simplicity of life, the weakness of the beer and wine consumed, the local and small-scale nature of alcohol production, and the fact that most drinking occurred in the home or in conjunction with agricultural work. Inebriety does seem to have increased during the Late Middle Ages as living standards rose in the aftermath of the plague.

In the sixteenth and seventeenth centuries, per capita consumption reached prodigious proportions in almost every country of western Europe. For example, average Swedish beer consumption is estimated to have been forty times higher in the mid-sixteenth century than the mid-twentieth century. Complaints proliferated about rising levels of inebriety, particularly in the cities. The negative effects of drunkenness on society as well as on the individual were increasingly stressed, and greater restrictions to reduce drunkenness began to be imposed. This rise of inebriety as a social problem was most apparent in Germany: the sixteenth century is considered by all sources to be the greatest period of drunkenness in German history. Repeated ministerial warnings and legislative enactments failed to control it. By the end of the century, the problem began to draw the attention of other countries as well, especially the English. The Holy Roman Empire, France, and England established their first penalties for drunkenness in 1495, 1536, and 1606, respectively. Greater controls over drink outlets also were instigated, including the beginnings of licensing.

Several factors have been suggested as contributors to this rise in drinking. First, as the availability of fresh meat declined, salt and spices began to be added to meat and other foods to preserve them and make them more palatable, giving rise to greater thirst. Second, the major social, economic, political, and religious changes of the period may have undermined existing controls over drinking and fostered more drinking by generating greater insecurities. Third, the overall availability of alcohol increased with the beginning of the commercialization of the alcohol trade, improved brewing and distilling techniques, better transportation, the growth of towns, and the development of capitalism. Trade in alcohol provided a ready source of income as inflation increased prices and reduced the value of earnings and rents. Fourth, beverages generally became stronger as hops were added to beer, the recreational drinking of distilled spirits began, wines became fortified, and stronger southern wines became more available. Fifth, the number and functions of taverns increased in response to population growth, urbanization, community needs, and economic conditions. As taverns grew in numbers from the twelfth century onwards, concerns about them were frequently voiced. They were attacked by authorities because they fostered inebriety and because on Sundays and holy days people gathered there instead of at church. Secular authorities often viewed taverns as potential centers of criminality and social disorder. In the sixteenth century, these attacks noticeably increased. Sixth, the growth in the wealth of the upper and commercial sectors brought an expansion of the medieval tradition of conspicuous public consumption. Inebriety was still most closely associated with the economic ability to purchase alcohol; it was primarily a component of a more general propensity towards self-indulgence and thus most characteristic of the wealthy. In northern Europe, individual as well as national economic well-being was closely related to the level of wine drinking. The frequent association of drunkenness and wine drinking in sixteenth-century Germany is indicative of the extent of inebriety among the upper classes. Conversely, the relish with which German peasants attacked wine cellars during their revolt in the 1520s reflects how rarely wine was available to them. Among the general population, traditions plus the lack of consumer goods and advancement opportunities also encouraged a preference for expending surplus income on such leisure-time activities as drinking instead of working harder to acquire and save money. Finally, due to population growth and economic change, the numbers of the poor, unemployed, and uprooted were increasing. Many of these "masterless men" had nothing better to do than frequent taverns. The drinking problem seems to have been so acute in Germany largely because more of these conditions were prominent there than elsewhere.

Many of the changes occurring in society, as the medieval world was transformed into the early modern world, also contributed to more hostile attitudes in some circles toward drinking levels and practices that once had been tolerated. Indeed, it may be that changes in attitudes toward alcohol were as routed in concerns over social order, economic conditions, and individual salvation. These social concerns appear to have been as significant in motivating the attacks on drinking as were the changes in consumption levels or patterns. In a time of widening social cleavages and disorders, with bands of poor vagabonds descending on cities all over Europe, fears increased that alcohol and taverns would contribute to the disruption of the established order. This was particularly evident in England. As part of an effort to preserve the existing class distinctions that were giving way under the impact of the expanding commercial economy and the collapse of the hegemony of the warrior-knights, sumptuary legislation attempted to restrict the amounts of money different groups could expend on feasting. To come to grips with price inflation, more limited food supplies, and greater economic competition, the traditional emphasis on conspicuous public feasting came under attack by some in the commercial sector and by government financial authorities. The merits of hard work were also enshrined, and poverty and idleness, both of which were linked to drunkenness, were stigmatized more than ever before. Social utilitarians began attacking inebriety on the basis of the harm it inflicted on the economy and social order. Among the upper classes, there began a withdrawal from and desire to reform many popular drinking practices and other aspects of popular culture that were no longer considered respectable.

One of the most important influences on attitudes towards alcohol was the Protestant Reformation, which both reflected and contributed to contemporary concerns over drinking. The Reformation brought a renewed emphasis on the fundamental Christian opposition to alcohol abuse and a desire to end the close associations (such as drinking at church festivals) between the secular and the sacred within the medieval Church. John Calvin further emphasized the need to rely on civil authority to institute and enforce moral behavior among a community of saints. The radical Anabaptists so thoroughly deplored drunkenness that temperance became one of their distinguishing characteristics, and the Hutterian Brethren prohibited all involvement in the alcohol trade. Still, the differences in attitudes toward drinking between the medieval Catholic Church and Protestants have been exaggerated. All Protestant sects continued to praise moderate alcohol consumption. Luther enjoyed a good drink even though he repeatedly rebuked the inebriety of his fellow Germans, and Catholics attacked the prevalence of inebriety in sixteenth-century Germany as vigorously as did Protestants. Although the origins of modern prohibitionism have been traced to a combination of the Anabaptist moral code and Calvinist reliance on civil authority, neither the Anabaptists nor the Calvinists contemplated prohibition of alcohol use. There were restrictions which even Calvin's followers would not support, and efforts to reform the taverns of Geneva quickly failed. As for the later English Puritans, there is little evidence that England under the Protestant Commonwealth was significantly more sober than under the early Stuarts, or that the drink legislation was significantly harsher.

Although concerns over inebriety were increasing during the sixteenth and seventeenth centuries, for most people alcohol was one of the great pleasures (as well as necessities) of life, and occasional drunkenness was accepted as a natural occurrence. The attitude of many Europeans was probably expressed best by Montaigne, who regarded drunkenness as a "gross and brutish vice" that caused one to lose control over body and mind but was less hurtful and mischievous than many other vices that "clash more directly with society in general." Furthermore, the heightened concerns that characterized the sixteenth and early seventeenth centuries seem to dissipate during the second half of the seventeenth century. In the case of Germany, this decline is curious after all the unsuccessful efforts of governments and ministers to moderate drinking in the sixteenth century. It has been suggested that consumption was reduced by the general devastation, destruction of vineyards, and high alcohol taxes that resulted from the Thirty Years' War. In the case of England, inebriety undoubtedly was just as prevalent in the late-seventeenth century as in the beginning, but it was not considered as great a problem, probably because of the alleviation of many of the social problems that generated previous concerns. However, too little research has been conducted on this period to draw any definite conclusions.

In the eighteenth century, concerns again rose as inebriety became more regular among more people, reaching unprecedented heights. The upper classes and the towns continued to lead the way, but chronic inebriety was no longer primarily the prerogative of the upper-classes. The major development of the century was the expansion of drinking among the lower classes and into rural villages. It was most prevalent in England, but everywhere complaints about inebriety multiplied. This was the great century of convivial drinking in Scotland, and the drinking excesses of the small German courts were notorious. France by all accounts was more sober than England, as it had been since the Middle Ages. High wine taxes, the weakness of most of the wine consumed (which was often diluted), less urbanization, and greater poverty helped to moderate drinking levels there, but in France, too, rural drinking rose appreciably during the 1720s and 1730s, and in the second half of the century, Mirabeau and Mercier described squadrons of Parisians staggering back into the city after indulging in weekend drinking bouts in rural cabarets where they could avoid the high city taxes on wine.

The widespread prevalence of inebriety in the eighteenth century partly resulted from long-standing beliefs about the benefits of alcohol consumption. Drinkable water was still scarce, and it was in this century that the popularity of alcohol as a medicine and a source of health and strength reached its height. Liberal spirits rations were given to the military, a practice identified by contemporaries as contributing to the spread of spirit consumption. For many poor people, alcohol remained an important source of nutrition: in Poland, alcohol supplied the least expensive source of calories; in France, wine thefts increased in the winter when other food sources became scarce. For the majority of people, there were still few recreational activities besides drinking and few alternative social centers to the drink house. Indeed, at least in England, public houses became even more central to lower class life than before. Finally, periodic drunkenness was still not only accepted by many, but even considered a sign of virility and prestige.

At the same time, several new developments increased the availability of cheaper, stronger drinks. First, population growth and economic expansion fueled demand and production; rising living standards in many areas allowed more drink to be purchased. Second, stronger beverages became available. In England, this was the age of port, porter, and gin; throughout Europe, spirit consumption expanded. Third, the continued commercialization and incipient industrialization of the alcohol trade made it possible to increase production and hold down costs. Fourth, expansion of the alcohol industry was encouraged by the belief that it conferred important economic benefits: it utilized expanding agricultural surpluses, it employed laborers, and it provided needed government revenues.

Possibly the most important contributor to inebriety was a great expansion of spirit production and consumption, particularly such grain-based spirits as gin and whiskey. Although one could get just as drunk on wine and beer as on spirits, distillation made it possible to increase the potency of alcoholic beverages and freed people from dependency on viticulture and the difficulties of transporting beer and wine. By making it easier to get drunk, distillation greatly intensified the problem of alcohol use. The early history of distillation is obscure and the subject is of considerable dispute. The art of distilling wine into brandy was probably discovered, or at least first developed, in the medical schools of Italy in the twelfth and early-thirteenth centuries, and the new substance quickly became the heir to all the beliefs about the benefits of wine. Indeed, because brandy was more potent than wine, its therapeutic benefits were believed to be greater. Extravagantly praised as a universal panacea that induced a sense of vigor, youth, and buoyant health, the "spirit" of alcohol was called "water of life" *(aqua vitae).* There is general agreement that until at least the end of the fifteenth century, spirits were produced only in limited quantities and at considerable expense by monks, alchemists, and apothecaries who often carefully guarded the secrets of their recipes.

It was during the sixteenth century that knowledge of the techniques of wine distilling spread among the general populace and that recreational spirit drinking appeared. The earliest evidence comes from Germany. A poem in 1493 described the woes of brandy drinking, and an ordinance in Nuremberg in 1496 restricted the sale of brandy because of the "serious misconduct and disorder" daily resulting from its consumption. The extent of brandy drinking in sixteenth-century Germany is uncertain. The many complaints and ordinances directed against the new practice would indicate it promoted more inebriety. However, brandy drinking was not criticized any more than wine drinking. In other countries, the spread was much slower, delayed in part by the secrecy that at first surrounded the process. Until the mid-seventeenth century, in most areas of Europe, brandy was the only spirit known and, because of limited availability, its consumption was often confined to medicinal purposes. What recreational spirit consumption did occur was not considered a major concern. Though whiskey was established as the distinctive drink of Ireland and the Scottish Highlands by the seventeenth century, it was largely unknown outside of these areas, and even inside the most common drink was ale until the eighteenth century.

The Dutch were the primary popularizers of spirits in the seventeenth century. In the early century, as Europe's greatest wine tradesmen, they encouraged brandy distilling in the south of France to reduce the bulk of their wine cargoes and to help fortify and preserve their wines. The preference for stronger drinks in northern Europe also was influential in encouraging the brandy trade. In the second half of the century, the Dutch developed Europe's greatest distilling industry and helped popularize the drinking of grain spirits. Although the technique of grain distilling was known in Germany at least as early as the 1530s, uncertainties about limited grain supplies and the bad taste of most grain spirits generated considerable hostility toward them. Distilling in Holland began to thrive in the mid-seventeenth century following the opening up of the Baltic grain supplies and the discovery of a recipe for a more palatable grain spirit, which became known as "gin."

In the eighteenth century, distilling became easier and cheaper, and the availability and affordability of better-tasting spirits increased dramatically throughout Europe. At the end of the seventeenth century, the English government actively began to promote gin production to utilize surplus grain supplies. By the mid-1730s, there was an epidemic of gin abuse among London's lower classes, giving rise to what has been called the first drug crisis. When Parliament sought to end gin abuse through prohibitive taxes in 1736, riots, bootlegging, and smuggling ensued rendering the law impossible to enforce. Brandy and liqueur consumption also spread in France. In 1690, the English physician Martin Lister observed in Paris a noticeable increase in inebriety and obesity, which he correlated with the recent introduction of spirits. By the 1780s, Mercier complained that women who once scorned brandy now reeked of it, and he alluded to the soberer workers who drank wine, not spirits. In Scotland and Ireland, whiskey drinking became firmly established, and illicit distilling proliferated. Indeed, distilling had become an economic necessity for many tenant farmers. In America, the colonies were inundated with rum, prompting an unsuccessful attempt by Oglethorpe in the 1730s to ban rum imports and production in Georgia. High taxes on traditional fermented beverages may have also contributed to increased spirit consumption in Scotland, England, and France, although evidence for the latter is more tenuous.

The mid-eighteenth century London gin epidemic is the most notorious example of the potential danger posed by the introduction of spirits. Spirits also seem to have had a particularly adverse impact in Russia. However, their introduction did not cause such problems everywhere. The Dutch, who long had a reputation for excessive drinking and who were one of the principal manufacturers of spirits, seemed not to have been substantially affected by the introduction of spirit drinking. In France, spirit consumption apparently did not engender undue concern, although many did not approve of the new habit. Mercier attacked brandy drinking but no more so than he attacked inebriety in general, and he expressed relief that at least in France the taste for grain spirits had not developed. A moderating factor in France may have been the limiting of on-premise spirit drinking to cafés, which served a wide variety of other drinks, including coffee, as well as food. The introduction of cheap potato spirits into Germany in the late eighteenth century seems to have created more problems than brandy drinking in the sixteenth century. The adverse impact of spirits in London and Russia was due not just to their introduction but to the manner in which, and the conditions under which, they were introduced. Encouraged by government policy, cheap spirits were made easily available in London at a time when there was little stigma attached to drunkenness and when there was a ready market for them among the growing population of lower-class Londoners, many of whom sought refuge in gin from the new-found insecurities and harsh realities of urban life. In Russia, too, vodka was virtually forced upon the people through the state-run *kabak* system.

Government interest in revenues from the alcohol trade and the importance of the trade to national economies in general were also important contributors to high levels of alcohol consumption. As central governments developed, almost immediately they turned to alcohol as a source of revenue, especially during times of war. Revenue generation, not consumption reduction, usually was the purpose for high taxes. Such a reliance often worked against government efforts to reduce consumption. In England, at the same time that King James I was attacking drunkenness as the root of all sin and was passing the first legislation criminalizing it, he tried to establish monopolies and taxes over the trade in order to raise money. The Commonwealth continued the Stuart efforts to regularly tax beer for revenues and even backed away from some restrictions over the alcohol trade when it appeared that revenues would be reduced. Both gin and beer production were viewed as mainstays of the British economy in the eighteenth century. Not only were government fiscal considerations responsible for the expansion of gin production, but efforts to reduce consumption were hampered by pressure from the spirits trade, which emphasized the economic benefits it conferred on the nation. This inconsistency between economic and social concerns was not limited just to the government. Daniel Defoe continued to defend the trade at the same time he attacked the harm gin consumption inflicted. In Russia, the government encouraged the spread of taverns and spirit drinking in order to maximize profits from the alcohol monopoly. Efforts to reform the *kabak* system in 1651 proved futile because of the government's continued reliance on vodka sales, and the *kabaks* were eventually reinstated in 1663 because of revenue losses. In eighteenth-century Poland, the monopoly over alcohol production and sales was one of the prime economic supports of the landed nobility. In France, traditional restrictions over taverns were removed in 1613 because of the need for more revenues from wine sales. Furthermore, one reason for official hostility toward coffee drinking in some areas was the fear of loss of revenues from declining alcohol consumption. This is especially evident in Prussia and in the south of France.

Much of the concern about drinking in the eighteenth century was not directed against drunkenness itself but against drunkenness among the lower classes. In England, upper-class inebriety continued unabated and relatively uncriticized at the same time that the anti-gin campaign was being waged. Drinking in alehouses, taverns, cabarets, and other popular drinking houses was also often more of a concern than inebriety itself. In France, neither Mercier nor Restif de la Brettone considered cabarets respectable: they were haunts of criminals and drunkards, a view shared by the police. Underlying much of this criticism was the abandonment of public drinking houses by the upper classes. With rising wealth and home improvements, the middle and upper classes increasingly partook of their beverages in private, and the public drink house became even more of a lower-class institution. What many upper-class critics really objected to was public drunkenness, and it was among the lower classes that public drinking was most prevalent.

In England and America during the second half of the eighteenth century, there also developed a more general concern about the dangers of inebriety and of spirit drinking. The traditional duality between the positive benefits of drink and the negative effects of drunkenness began to break down. Increasingly it was believed that spirit drinking in itself was harmful and that for many people it was an addictive disease. For the first time in history, the mere drinking of an alcoholic beverage (gin) became stigmatized. Furthermore, many of the traditional beliefs about the benefits and functions of alcohol were being called into question. During the century, several English physicians emphasized the social and physical consequences of high levels of alcohol consumption, particularly of spirits. Worried about inebriety among the upper classes, George Cheyne developed a diet plan popular in the 1740s that called for total abstinence from all alcohol for youths and the aged, and for limits on alcohol consumption and avoidance of spirits by all other people. John Wesley adopted Cheyne's dietary plan for his Methodist followers, and the Methodist Church prohibited not only drunkenness but the selling, buying, and drinking of spirits. At the height of the gin epidemic, between 1735 and 1750, the attacks on spirits were vitriolic. Repeatedly, it was stressed that their very consumption acted as a slow poison that destroyed the mind and body of both the present and future generations. Gin was also held to be a major cause of crime. The anti-gin movement was not characterized by strong temperance motivations; generally, such critics as Thomas Wilson advocated that the "cursed spirit" be placed out of reach of the "lower kind of people" and that good beer be substituted. But the movement had a profound influence on subsequent discussions of the effects of alcohol consumption. In America, the Quaker Anthony Benezet published the first American pamphlet advocating total abstinence from distilled liquors in 1774. At the end of the century, the physician Benjamin Rush formulated the first clear concept of chronic drunkenness as a disease and an addiction over which the drunkard has no control, shifting causal responsibility away from the individual to the addicting quality of the spirits themselves. Together with Benezet, he instigated the process in America by which the Good Creature of God was transformed into Demon Rum and which would give birth to the nineteenth-century temperance and abstinence movements. Also at the end of the century, the British physician, Thomas Trotter, labeled drunkenness a "disease of the mind" that was "produced by remote causes."

In part, these new perceptions reflected the demonstrated dangers of chronic spirit consumption. They also reflected the ideas and influence of the middle sectors of society and the new realities of life. Temperance as a middle-class virtue in early modern Europe has often been exaggerated. In the sixteenth century, most merchants and members of the commercial sector sought to gain status by emulating the eating and drinking extravagances of the nobility. But in eighteenth-century England, inebriety was most prevalent among the extremes of society—the upper and lower classes—and temperance was most characteristic of the middle ranks. As the numbers and influence of the latter expanded, so did the acceptance of the virtue of temperance. The eighteenth-century middle-class world view was characterized by a belief in self-reliance, competition, and the dependence of social control on self-control. The disinhibitory effects of alcohol engendered the concern that chronic alcohol use could deprive drinkers of the ability to control their behavior, threatening both commercial success and social order. In addition, the effects of the industrial revolution began to be felt throughout British society. Industrialists focused attention on the effects of drinking on work efficiency and the need to eradicate the traditional play element in work routines, including drinking. The long-held belief that alcohol aided work performance was being more critically questioned.

Several other developments were beginning to contribute to an overall improvement in English drinking behavior by century's end. Among the upper classes, although there still existed many prominent drunkards, including the future George IV, such behavior was increasingly stigmatized. A new social conservatism with a more refined and "respectable" standard of behavior was developing, eventually to be labeled Victorianism, which disparaged the coarseness and excesses of the Georgian era. The spread of Methodism and Evangelicalism was helping to reform the behavior of the lower classes. Controls over gin instigated after 1750 had reduced its consumption, and restrictions over licensed houses had been tightened. Finally, coffee and tea consumption had substantially increased, due to more readily available supplies at reduced prices, a development that helped to limit alcohol consumption throughout Europe. Coffee, the first hot, nonalcoholic beverage, was introduced in the mid-seventeenth century and promoted as a nonintoxicating alternative to alcohol. Coffee houses provided the first place where people could gather to socialize without drinking alcohol. In England, by the end of the eighteenth century, tea had replaced coffee in popularity and was beginning to supplant beer as the common daily beverage.

Finally, a few observations are in order on the effects of public policies. Throughout the *Chronology* can be seen the difficulties and public hostilities that authorities faced in their efforts to control production, sales, and use. Often, it appeared that the efforts accomplished little in the way of reducing consumption levels or moderating drinking practices. Further, many improvements in drinking situations, such as occurred in Rome after the first century, seem to have occurred because of broad social, economic, cultural, and attitudinal changes rather than because of direct policy efforts. Inebriety among the Jews appears to have been substantially controlled by ritualizing wine drinking. Yet, not all public policies have been ineffective. The decline of the gin epidemic was at least in part due to legislative restrictions enacted after 1750. Licensing in England eventually brought about an improvement in the numbers and quality of alehouses. More attention needs to be paid to why past control efforts failed. In many cases, efforts to enforce laws were undermined because they were directed at or only applied to the lower classes, while the upper classes continued to profit from the alcohol trade and made no effort to reform their own drinking behavior. In Ireland and Scotland, attempts by the English to tax whiskey production fueled the spread of illicit distilling. The history of illicit distilling in these two countries illustrates how conflicts over alcohol controls can merge with, and become symbols for, larger political struggles, particularly when unpopular control efforts are instigated by outsiders. It also shows the difficulties of attempts to reduce domestic alcohol production when it is a significant source of income for many of the population, without supplying alternative economic substitutes.

The ending of the *Chronology* at 1800 disrupts important continuities that exist between the eighteenth and nineteenth centuries. Yet 1800 is, in many respects, a major dividing line in the history of alcohol in Western society. The eighteenth century is far more similar to the seventeenth century than to the modern era. Alcohol still pervaded all aspects of life to a far greater extent than today. For most people, drunkenness was still an accepted part of life, and only a small number had come to reject spirit consumption. Furthermore, the critique of spirits was not yet broadened into a critique of alcohol use in general. Indeed, beer and wine were being promoted as temperance drinks. But by the end of the eighteenth century, trends had already begun that would serve to moderate drinking behavior, heighten the concerns about drinking in the modern era, and call into question many beliefs about the value of alcohol that have existed for over a millennium. In the nineteenth century, attitudes and patterns of use would begin to change rapidly under the influence of industrialization, greater urbanization, and the temperance movement, giving rise to the popularity of the concept of total abstinence.

Antiquity

continued

3000 BC

MIDDLE EAST

BACKGROUND DEVELOPMENTS: BRONZE AGE CULTURE. The transition from Neolithic to Bronze Age culture occurs first in the valleys of the Nile, Tigris/Euphrates, and Indus rivers. The transition is characterized by a loss of self-sufficiency and the development of commercial trade between settled communities. In these river valleys, soil fertility provides the food surpluses needed for the population to expand and to secure by trade the essential raw materials which these valleys lack. States arise embracing various professions and classes which are not involved in the primary task of food production.

BREWING, VITICULTURE: BEGIN. The discovery of fermentation probably occurs when primitive people observe that grapes and fruits naturally ferment when exposed to air. The earliest alcoholic beverages may have been made from berries or honey (mead) (Blum et al. 1969:25; Roueché 1963:168). With the discovery of agriculture, probably in the 8th millenium (Africa 1965:3), Neolithic man learns how to cultivate barley, wheat, and the grape vine, making possible the addition of beer and wine to his diet on a regular basis. The brewing of beer has been traced "almost" to the very beginnings of agriculture and occurs in most areas where agriculture develops (McCarthy and Douglass 1949:3).

In early Quarternary times (at least 500,000 years ago), the wild grape vine (*Vitis vinifera silvestris*) was indigenous to nearly all the present-day great wine-producing countries of Europe and the Near East (Billiard 1913; Jellinek 1976:1721). It is impossible to arrive at any definite date for the origins of viticulture. It has been suggested (Hyams 1965:30-33) that it began around 6000 BC by "a barbarous, Early Neolithic people" in the Transcaucasia (south of the Black Sea, present Armenia). Younger (1966:28) places the beginning of grape wine making sometime in the fourth millenium. The origins of viticulture are believed to be in the Transcaucasia because there the wild grape vine yields an abundance of fruit without any cultivation. That viticulture began in the area around Armenia is further suggested by linguistic evidence, by the Old Testament story of the Noah's planting the first vineyard on Mt. Ararat (Genesis 9:20), and by the fact that the first records of viticulture are found in Sumeria and Egypt, areas south of Transcaucasia in which the grape vine is not indigenous. The Sumerians and Egyptians must have imported the vine and viticulture at an early date. Around 3000 BC the recorded history of alcohol begins with the invention by the Sumerians of writing in the form of cuneiform symbols.

TAVERNS. It is probably in this period that the first alcohol trade and taverns appear. The manufacture of alcoholic beverages becomes a specialized task, with the product sold through outlets. Taverns cater to travelers' needs (Popham 1978:231-232).

ATTITUDES: INEBRIETY. According to Lutz (1922:97), "Intoxication was not yet considered as constituting a moral offence against the drinker's own self and society at large" in ancient Egypt and Mesopotamia. Although warnings for temperance do exist, there is very little antagonism to drunkenness, which is "rather considered in the light of a harmless pleasure."

EGYPT

ATTITUDES: INEBRIETY. Throughout ancient Egyptian history, drunkenness is "a source and a symbol of mirth and forebearance, possibly also of religious ecstasy" (Darby et al. 1977:530). In the Old Kingdom (2686-2181 BC), moderation is urged by "wise men," but is not always observed (Lesko 1977:4). On the walls of one of the tombs of the kings of Memphis is inscribed: "His earthly abode was rent and shattered by wine and beer,/ And the spirit escaped before it was called for" (quoted Crothers 1911:139).

BEER CONSUMPTION. As in Sumeria, barley beer (krithinon) consumption is far more prevalent among both royalty and commoners than is wine (Sigerist 1951:249). Even for offerings to the gods, beer appears to have been used more frequently than wine (Lesko 1977:14). Beer is above all the drink of commoners because of its cheapness due to the abundance of grain and its ease of production in comparison with the extensive technology and care required to make wine (thus the later Greco-Romans equated beer consumption with poverty). It is considered a household and kitchen necessity (Darby et al. 1977:532, 548).

Darby et al. (1977:529) comment: "The invention of beer, even in the earliest Egyptian times, was so deeply lost in the abyss of irretrievable memory, its making and use were so inextricably knit in the fabric of daily life, while its exhilarating powers appeared so mysteriously supernatural that human imagination readily weaved around it the most incredible, albeit significant tales."

BREWING. Brewing is a home industry like bread baking, to which it is closely associated. The method of preparation is similar to present-day bouza or merissa (Darby et al. 1977:531, 534). The earliest Egyptian texts enumerate a number of different beers. Most beer is prepared from barley (hops are unknown to all ancient peoples), although the Egyptians use other plants to improve its taste and keep it for longer periods (Lutz 1922:73-76).

VITICULTURE; RITUAL WINE USE. Considerably before 3000 BC, Egypt has an advanced viticulture in the northern Nile delta. (Sealed wine jars date from the first dynasty, c. 3000 BC [Lesko 1977:11].) As the grape vine is not indigenous to Egypt, viticulture is probably learned from Egyptian colonists in Phoenicia near Mesopotamia. The early vineyards are an expensive luxury for the kings and nobility. They are primarily used for the production of funerary wines and for ritual libations, although possibly also for personal use (Lutz 1922:1-3, 46-61). (On the ritual uses of both wine and beer, see Darby et al. 1977:571-576).

SUMERIA

BEVERAGES CONSUMED. Beer is the primary beverage of the Sumero-Akkadians, and frequent written references are made to it. Large quantities are probably consumed for both medicinal and recreational purposes (Lutz 1922:118; Bubenik 1966). Hyams (1965:40) believes that wine came first to Sumeria and was superseded by beer because of the difficulties of cultivating grapes there. Wine is a luxury drink whose consumption is limited largely to medicinal and sacred purposes.

VITICULTURE. Southern Mesopotamia does not possess wild vines; it acquires the plants and the art of cultivating them from peoples near the native habitat of the wild vines in Transcaucasia to the northeast (Hyams 1965:33). Viticulture is "almost certainly" established before 4000 BC in the northern Tigris-Euphrates area. In the south, the earliest Sumerian reference to a vineyard occurs around 2400 BC (Hyams 1965:37), but viticulture never is successfully established there.

From northern Mesopotamia, viticulture probably travels east to Syria and Palestine and then south into northern Egypt: "There can be no proof, but on the whole it seems...clear that the highly-civilized city-states and later empires of Mesopotamia were the channels through which wine, the vine, and the art of viticulture reached the other civilized centres of the most ancient Near East" (Hyams 1965:52).

MEDICINAL USE. A clay tablet found at the ancient Sumerian city of Nippur, dated c. 2100 BC, contains the cuneiform text of a pharmacopoeia in which beer (possibly wine) is the usual vehicle in the medicaments; this is the oldest physically preserved record of the use of alcohol in medicine (Keller 1958:153; Lucia 1963a:11)

RITUAL USE. Alcohol has a prominent ceremonial and sacred purpose. A portrait of Queen Shu-bad drinking from a vessel next to her throne chair by means of a golden straw has been interpreted as possibly implicating intoxication in the process of ruling and decision-making (Emboden 1977:188). An inscription dated around 2700 BC refers to a Sumerian wine goddess, and wine deities appear in Sumerian myths (Hyams 1965:38-39). In the epic of Gilgamesh, wine is a sacred drink (Emboden 1977:187).

ATTITUDES. When Enkidu in the Gilgamesh epic first drinks seven goblets of beer, described as the customary drink of the land, "his spirit was loosened, he became hilarious. His heart became glad and his face shone" (quoted Lutz 1922:117; see also Contenau 1954:79). Another epic legend tells how in a battle between good and evil gods, when the good gods died, they fell to earth where they spawned the grape vine (Lucia 1963b:152).

1900 BC

EGYPT

VITICULTURE. Throughout the Middle Kingdom (c. 2040-1780), vineyards (which exist only in the north) are still an expensive luxury for the king and great officials rather than a profitable investment. Although many Egyptian gods are tribal or familial, Osiris, the god of wine, is worshipped by all (Lucia 1963b:152).

Because few vineyards exist and the difficulties of transporting wine are great, there are a number of fruit wines. The favorite is date-palm wine, which is widely consumed by the poor, since only barley beer is cheaper.

INEBRIETY PREVALENCE; ATTITUDES; FESTIVE DRINKING. The Egyptians may not be as heavy drinkers as the later Greeks. However, drunkenness from both beer and wine is prevalent, and Egyptian drinking songs "give more warrant for the artistic power of the Bacchic emotion than do their so-called counterparts in

Greece and Rome" (McKinlay 1948a:394-395, who hypothesizes that alcohol consumption was lower in Egypt than in Greece because the water of the Nile needed no flavoring to make it palatable and wine drinking was not as necessary as it was among the Greeks). Most of the evidence of drunkenness derives from the New Kingdom (see 1600 BC/Egypt), but it clearly is commonplace in the Middle Kingdom as well (Darby et al. 1977:583). Generally, intoxication is considered a harmless pleasure. Most drinking occurs at social and religious festivals. Elaborate formulas and rituals for blending and serving wine at feasts are developed. Holidays especially are characterized by great drinking, and the calendar includes a day of drunkenness each month (Lutz 1922:105-107). Banquets frequently end with the guests being sick (Sigerist 1951:250; Younger 1966:54; Petrie 1923:102). An 11th Dynasty tomb picture shows a drunkard being carried by two companions (Darby et al. 1977:583; Lutz 1922:98). Soldiers and students are frequently cited as heavy drinkers; soldiers are given daily rations of beer that are enough to make them drunk. Around 1800 BC, 130 jars of beer are brought to the royal court (Lutz 1922:86, 106).

<u>CONCERNS: INEBRIETY; TAVERNS</u>. Although Egyptians seem to have little antagonism to inebriety (Lutz 1922:97; Sigerist 1951:250), general admonitions warn people to avoid beerhouses (which are often houses of prostitution) and excessive drinking. A Middle Kingdom text, which may be a copy of a 3rd Dynasty work, advises banquet guests to drink only after the drunkard is satisfied, since there would be then less alcohol left (Darby et al. 1977:583). A warning is recorded not to forget oneself in the breweries and become a drunkard, for then one is unable to conduct business (Maspero 1971:31). Controls on beer sales may have been attempted because of concern over its interference with work productivity (Sutherland 1969:123).

It is also "pretty certain that moderation in drinking was recommended to the kings more than any other class" (Lutz 1922:108). Diodorus (1.70.11) relates that the priests, when in contol of the state, made sure the king never drank to the point of drunkenness.

Taverns and tavern keepers are held in low esteem (Lutz 1922:105-106; Maspero 1971:29; Popham 1978:237-238). A Egyptian priest around 2000 BC writes to his pupil: "I, thy superior, forbid thee to go to the taverns. Thou art degraded like the beasts" (Crafts et al. 1900:5; Sutherland 1969:123).

1800 BC

BABYLONIA

<u>BACKGROUND DEVELOPMENTS</u>. The original Sumero-Akkadian empire has been replaced by a Semitic Amorite empire centered at the city of Babylon in central Mesopotamia.

<u>INEBRIETY PREVALENCE; DRINKING FUNCTIONS</u>. Babylonia develops a reputation for heavy drinking (Lutz 1922:115). Barley-based beer and palm wine are rationed to "various recipients" at one gallon per capita per month, and precise descriptions are written of the stages and symptoms of drunkenness (Contenau 1954:72, 79). Both beer and wine are consumed for recreational, ritual, and medicinal purposes, with beer being the principal drink. Although grape wine is a requisite for religious ceremonies, it probably is not commonly consumed by ordinary people since it has to be imported. It has been argued that the

association of wine and religion was so strict that quite possibly ordinary people never tasted it (Hyams 1965:37, 49).

TAVERNS: CHARACTERISTICS, ATTITUDES. Frequenting "wineshops," which are actually beerhouses, is considered disgraceful by respectable persons (Lutz 1922:127; Contenau 1954:73). This suggests that whereas ritual and upper-class drinking is approved, the taverns in which common people drink are not. This may have been more because they are houses of prostitution (as in Egypt) than because alcohol is served (Popham 1978:333). They appear to be run only by women.

TAVERN CONTROLS: HAMMURABI. Around 1750 BC, the famous Code of Hammurabi devotes four paragraphs (nos. 108-111) to "wineshops." Prices and measures to be served are regulated in order to prevent cheating; riotous gatherings are prohibited. The death penalty is prescribed for innkeepers who do not cause the arrest of known outlaws (Lutz 1922:125-129; McKinlay 1948b:398-399; Hyams 1965:50-51; Baird 1944a:540; Popham 1978:323-337). There are no penalties for drunkenness. In fact, drunkenness is not mentioned in it or in any other ancient Babylonian legislation, nor does Babylonian literature contain "anything like moral prescription in which a warning is contained in regard to the excessive use of alcoholic beverages" (Lutz 1922:133-134; see also Popham 1978:236).

Considerable speculation surrounds the references to wine instead of beer in Hammarubi's code, since beer is the national drink while wine production is limited (Lutz 1922:69). Most likely, wine is used in legal texts as a generic term for all alcoholic beverages (Popham 1978:235-236). It has also been suggested that "wine" was the generic term because the drink was introduced into Mesopotamia before beer brewing began (Hyams 1965:40), or that Babylonian beer was so low in alcoholic content that legislation controlling its supply and use was not seen as necessary (Olmstead 1923:558).

1600 BC

EGYPT

BACKGROUND DEVELOPMENTS: NEW KINGDOM. The 18th Dynasty founds (c. 1580 BC) the New Kingdom or Empire; the king of Egypt begins to be called pharaoh and embarks on a militaristic conquest of the Near East. Egypt reaches its height.

FESTIVE DRINKING. One papyrus states that the mouth of a perfectly happy man is filled with wine and beer, and refers to the hilarity caused by wine. There is evidence from tomb-paintings that upper-class men and women often drink to excess (Lutz 1922:98). Based on a picture of an 18th Dynasty feast, Erman (1894:255) observes that at a typical New Empire party, the guests are requested to "celebrate the joyful day" by the pleasures of the moment as they are served their wine. Several 18th Dynasty wall paintings show upper-class women vomiting in drunken stupors at banquets, an act which aparently is regarded as a "trifling incident" (Erman 1894:255; Lutz 1922:109; Younger 1966:54-55). Hieroglyphics record the demand of a femal courtier to be given eighteen bowls of wine because she loves to be drunk. When another woman refuses a drink, she is admonished to "drink unto drunkenness (and) celebrate." Another refuser is told: "Drink! Do not spoil the entertainment.

Let the cup reach me. You know it is due unto the ha to drink" (Lutz 1922:100, 102).

CONCERNS: INEBRIETY. Popham (1978:238) asserts that concerns over intemperance increased under the New Kingdom; he attributes this to a decline in the traditional primitive pattern of drinking centered on banquets, feasts, and ceremonies, and to an increase in nonceremonial drinking such as in taverns. In support, he cites Petrie (1923:102), who observes that until the 19th Dynasty (started by Rameses II in 1320 BC) there was no reprehension of drunkenness. However, Darby et al. (1977:583) emphasize that it "is wrong...[to assert] that the moral necessity of temperance dawned upon Egyptians only under the New Kingdom." Furthermore, most of the drunkenness at this time still seems to center on banquets, and some of the evidence cited regarding taverns is open to other interpretations (see 1200 BC/Egypt).

MEDICINAL USE. In Egyptian medicine, most internal remedies are given as potions with beer and wine (Sigerist 1951:337). Around the mid-16th century BC, the Papyri Ebers and Hearst are written, which record numerous prescriptions that include both beer and wine, many of which may have been copied from earlier medical treatises dating back to 2500 or even 3400 BC (Lucia 1963a:11-12). Out of 250 prescriptions in the Papyrus Hearst, beer is used in 27 and wine in 12, mostly for the relief of pain or to correct urinary problems (Leake 1952:73; Keller 1958:153). (On medicinal uses in Egypt, see also Darby et al. 1977:577-579).

GREECE

VITICULTURE INTRODUCED. Asiatic grape-vine cultivars and the art of growing them probably do not reach Greece until the 2nd millenium, entering through Thrace from Anatolia. Previously, the people of Hellas drank mead and possibly beer (Younger 1966:79). In Euripides' The Bacchae (c. 408 BC), Dionysus says he has conquered all Asia (Anatolia), Arabia, Media, Phrygia, and Lydia before coming to Greece; this may be the order in which viticulture comes to these places. By 1600 BC, the people of Greece have adopted viticulture; by 1200 BC, when the Achaeans conquer Greece, viticulture and wine have long been commonplace there (Hyams 1965:70-72, 84).

HITTITES

DRINKING PRACTICES. The Hittites in northern Syria, whose Indo-European language dominates Asia Minor and adjoining regions, have the oldest cognate for the word "wine" (Seltman 1957:15, 23). The Hittites are also known to have sipped beer through reeds (Lutz 1922:139-142). The Hittite god Ba-al-tars is depicted holding a cluster of grapes and corn to indicate that he is a god of abundance and fertility; he may be one of the forerunnners of Dionysus (Emboden 1977:188).

1200 BC

EGYPT

DRINKING CONTROLS. Under New Kingdom Pharaoh Rameses III (20th Dynasty) in the early 12th century BC, a prohibition of alcohol use by soldiers is repealed (Blum et al. 1969:26). Around this time is written the Wisdom of Ani, which has been called "the first temperance tract" (Roueché 1963:169). The author admonishes, "Take not upon thyself a jug of beer. Thou speakest and an unintelligible utterance issueth forth from thy mouth. If thou fallest down and thy limbs break, there is none to hold out a hand to thee. Thy companions in drink stand up and say: 'Away with this sot.' And thou art like a little child" (as translated in Darby et al. 1977:583; see, however, the alternate translation in Lutz 1922:108).

RITUAL USE; USER CHARACTERISTICS. During the 31-year reign of Rameses III, the royal priests receive 722,763 jars of beer and wine (McKinlay 1948b:396). Wine is still the drink of the upper classes only; the poor consume beer (Lesko 1977:13).

HEBREWS

BEVERAGE CHARACTERISTICS; WINE DILUTION. Around 1200 BC, Moses leads a group of Hebrew tribesmen from Egypt into Palestine (Canaan). The Hebrews are said to regret leaving behind the wines of Egypt (Numbers 20:5), but they find in Palestine a land rich in vineyards. The wines of Palestine and Syria are potent, sweet, and syrupy; when diluted two-thirds with water some are said to be as strong as undiluted wines. The Hebrews often add spices to their wines to make them more palatable (Lutz 1922:26, 30; Hyams 1965:55).

1000 BC

ASSYRIA

BEVERAGES CONSUMED. Viticulture flourishes in the mountains of northern Assyria; already by the first millenium, official interest in viticulture is keen and vineyards abound. Vintages from certain districts become especially prized (Contenau 1954:72; Hyams 1965:37-38, 50). Some imported beer is consumed, but it is not "in much esteem" (Olmstead 1923:558). Beer taverns are probably less common than in southern Mesopotamia (Popham 1978:236.)

900 BC

HEBREWS

ABSTINENCE OF RECHABITES, NAZARITES. Around the mid-9th century BC, the Rechabites, nomads of ancient Hebrew stock, rebel against the pleasures and religious practices found among the urban Hebrews of the new monarchy of kings Saul, David, and Solomon (c. 1025-930 BC). They advocate simplicity and pastoralism as the only way of life for Yahweh's people and they are later praised by the prophet Jeremiah in the 7th century BC for their implicit obedience to ancestral precepts. Their hostility to wine is the closest position to prohibitionism in the ancient world. They say to Jeremiah, "We shall drink no wine," nor build houses, sow seeds, or plant vineyards (Jeremiah 35:1-15; 2 Kings 10:15, 23; Seltman 1957:150). The Rechabites disapprove of cultural refinements and advances, such as viticulture, which they perceive as removing them from God. Thus they attempt to preserve the way of life of the early nomadic days of Israel (Lutz 1922:134; Bainton 1945:47).

Another group known for its abstinence at this time are the Nazarites who, during the period of their vow to pursue holiness, totally abstain from wine and every product of the vine. However, after the vow period, the Nazarites return to normal life and drink wine. For example, Samson's mother, vowing her son to a Nazarite calling, abstains from wine during her pregnancy (Judges 13:18). Although the origins of this rule are obscure, the Nazarites were enjoined from drinking wine by Moses (Numbers 6:1-6). Like the Rechabites, the Nazarites' abstention probably reflects antagonism to the settled life that has developed in Canaan (Raymond 1927:28-29).

800 BC

ASSYRIA

DRINKING PRACTICES, ATTITUDES. The Assyrian empire is established in the upper Tigris valley (northern Mesopotamia above Babylonia) by Tiglath-Pileser III (745-727 BC). The Assyrians conquer almost all the areas of Near Eastern civilization, and dominate the area until the early 7th century BC. They are known for their great drinking (Cherrington 1925, 1:246-247). Assyrian sculptures depict banqueting scenes at which guests drink from huge cups or bowls of wine; food is never evident. The goddess Ishtar is shown biding Assurbanipal (d. 627 BC) to "eat food, drink strong wine, make music [and] exalt my divinity" (Lutz 1922:118-124). In 612 BC, King Sinsharishkun is defeated by a coalition of Medes, Scythians, and Babylonians when they attack his drunken army; the capital Nineveh is sacked and the empire divided. Assyria's defeat is embodied in the Greek legend of the effeminate despot Sardanapalus, who wallows in luxury and, following his defeat, kills himself. On his tomb he is shown snapping his fingers with the inscription, "Eat, drink, and be merry; all else is not worth this [a snap of the fingers]" (Strabo 14.5.9; McKinlay 1948b:400; Africa 1969:47-48).

RITUAL, FESTIVE DRINKING. As in ancient Egypt, most drinking in Assyria appears to be a function of religious and festive occasions, often involving the whole community and lasting for days (Dinitz 1951; Popham 1978:236).

Maspero (1971:368-372) observes: "The Assyrian is sober in ordinary life, but he does not know how to stop if he allows himself any excess....After one or two days [of drinking at the king's festival] no brain is strong enough to resist, and Nineveh presents the extraordinary spectacle of a whole city in different degrees of intoxication."

GREECE

DRINKING REPUTATION. Both contemporaries and later writers have emphasized that the Greeks were among the most temperate of ancient peoples. This reputation rests largely on their custom of diluting wine with water, their rules for moderating drinking, their praise of the virtue of temperance, and their avoidance of indulgences in general, as exemplified in the Delphic counsel "nothing in excess." Gulick (1927:15, 42) emphasizes that Greek life is simple and frugal, and that even at their wealthiest the Athenians never enjoyed a luxury comparable to imperial Rome. Habitual drunkenness was exceptional, although occasional intoxication at banquets and festivals frequently occurred. There is little doubt that, compared to other ancient peoples, the Spartans were moderate drinkers and that the Greek reputation for temperance is largely justified. However, this temperance should not be exaggerated. Drunkenness often becomes a problem in several Greek city-states.

DRINKING FUNCTIONS. Homer and Hesiod describe Greek use of wine for several distinct functions: (1) magico-religious invocations, appeasements of the dead, and offerings (libations) to gods; (2) hospitality rituals; (3) manly displays; (4) formal banquets and informal feasts; (5) healing and pain relief; and (6) nutrition. It would appear that long-standing ideals of temperance exist alongside tendencies toward passionate, drunken outbursts (Blum et al. 1969:30, 40; Billiard 1913:205; Lucia 1963a:20-21; O'Brien 1980b:95-99).

INEBRIETY PREVALENCE (HOMER). Seymour (1914:227-228), among others, emphasizes the temperance of the Homeric Greeks: "The Greeks of Homer's time...were not hard drinkers. Only three instances of drunkenness occur in the action of the Homeric story....Homer knows no drinking-bouts....'Heavy with wine,' addressed in anger by Achilles to Agamemnon, is a grievous reproach and nothing in the poem indicates that it was justified. On Hector's return from the field....he replies [to Queen Hecaba], 'lift not for me the wine, revered mother, lest you weaken my limbs, and I forget my might and valor." Questioning those who cite Homer for evidence of Greek temperance, McKinlay (1953) observes that Homeric heroes are rarely shown getting drunk, but they do drink often and copiously. Whenever the heroes of the Iliad are not fighting they are drinking, and frequently, enough wine is drunk to cause considerable trouble. The ability to drink great quantities of wine is depicted as an heroic virtue. McKinlay maintains that the depiction of copious drinking but little drunkenness is an example of the tendency of epic poets to idealize actions without showing adverse outcomes. Even many later Greeks, such as Plato, criticize the intemperance of Homer's heros (McKinlay 1953:83, 85, 89-90).

DIONYSIAN RITES. In Hesiod appears the first written Greek reference to Dionysus, at this time only a minor peasant god of wine and corn. Hesiod describes him as the "giver of wine" (Emboden 1977:190): "How hath Dionysus given unto men both joy and pain, when one drinketh abundantly; and wine hath come raging upon him and binds feet and hands together, tongue and reason in bonds unforeseen; then soft sleep embraces him" (quoted Athenaeus 10.428c). In the 7th century BC, the cult of Dionysus is introduced.

WINE DILUTION. Hesiod (Works and Days) recommends that one part wine be mixed with four parts water (Sigerist 1951:86). In his treatise on hygiene and diet, he recommends drinking the wine of Byblos diluted with three parts water to combat fatigue in the summer (Lucia 1963a:36). (On the mixing of wine with water, see also Solon's laws, 600 BC/Greece.)

In the Greco-Roman world, wines are customarily mixed with either hot or cold water before consumption. It has been suggested that the primary reason for this was to dilute the thick and syrupy wines; Allen (1961:26) rejects this and emphasizes that the purpose of dilution was to reduce the alcohol content. He observes that the Greeks were aware that unmixed wines were more pleasant-tasting, but "they were generally speaking a temperate frugal people." Drinking unmixed wine was not approved, except as small libations in honor of the gods at the end of dinner. It was considered barbarous and dangerous, leading among other things to madness. The ratio of water added to wine varied considerably depending on the wine, the circumstances, and the users' preferences. The general rule appears to have been that the mixture should contain more water than wine, otherwise it was considered too intoxicating. Hesiod's recommendation of three to one seems to have been the generally accepted formula.

Athenaeus (2.38) also preserves the story of the "upright" Dionysus, who taught a legendary king of Athens how to mix wine and water so that the Athenians could drink and still stand upright, whereas "before they were bent double by drinking their wine neat." At the same time, Dionysus institutes the customs of taking just a sip of unmixed wine after meals "as a proof of the power of the good god" and of repeating over the cup the name of Zeus "as a warning and reminder to drinkers that only when they drank in this fashion would they surely be safe."

HEBREWS

INEBRIETY PREVALENCE; ATTITUDES. In biblical writings, there is considerable hostility to excessive drinking and some opposition to wine itself (Lutz 1922:133-135). Many of these texts appear to date from prior to the Babylonian exile in the 5th century BC, although they were not officially written down until afterwards. Jastrow (1913) argues that these passages reflect an opposition to viticulture and wine drinking as symbols of a higher form of culture and an alien religious faith. They also reveal a serious concern over the prevalence of drunkenness, particularly among the wealthy (Keller 1970; Kelso 1926:192). Drunkenness appears to be more of a personal vice than a social evil (Raymond 1927:26). It is attacked mostly for the moral and spiritual harm it causes. It leads to immorality; is allied with gambling, licentiousness, and indecency; and causes anger, combativeness, and lust (Joel 3:3; Genesis 9:21; Proverbs 20:1, 21:17, 23:21, 29-31). Above all, it produces indifference to religion and destroys the capacity for serious thought (Isaiah 5:12) (Raymond 1927:27). The stories of Noah and Lot contain warnings against viticulture and wine drinking (Genesis 9:21, 19:32). The planting of the vine leads to Noah's drunkenness and shame, and the drunken Lot engages in shameful intercourse with his two daughters. The assumption in these incidents is that he who drinks wine gets drunk and disgraces himself. (Cohen

[1974:9-12] believes that these stories are rooted in the connection between intoxication and sexual potency and in the need to regenerate the earth's population.)

This hostility to drunkenness and (to a lesser extent) to wine is particularly evident in the books of the prophets, who appear beginning in the mid-8th century, starting with Amos and Hosea in Israel and with Isaiah in Judah. These prophets greatly enlarge and deepen the religious concepts of Judaism, emphasizing that moral conduct is required by individuals and the nation and that the external trappings of cult and ritual alone never satisfy God. They protest against the worship of fertility gods, worldliness, and social injustices that have developed under the monarchy. Amos, the first of the prophets (c. 769 BC), attacks the rulers of Israel for drinking wine out of large bowls (like the Assyrians) and women "who press their husbands to join then in a carousal" (Amos 6:6). In Hosea (4:11), Moses is quoted as saying that "wine and strong drink take away the heart." Isaiah (28:3, 7) declares that the "drunkards of Ephraim" shall be trodden under foot and that wine causes one to "err in vision" and "stumble in judgment." He attacks drunkenness as a rampant evil that produces indifference to God: "Woe unto them that rise up early in the morning that they may follow strong drink, that tarry late into the night, til wine inflames them. And the harp, and the viol, the tabret, and pip, and wine, are in their feasts: but they regard not the work of the Lord, neither consider the operation of his hands." Again, "Woe unto them that are mighty to drink wine, and men of strength to mingle strong drink" (Isaish 5.11, 22). That excessive drinking is still considered a major problem in the 7th century is indicated by Jeremiah's (35:13-15) praise of the abstemious Rechabites (see 900 BC/Hebrews).

The Book of Proverbs takes a decidedly unfavorable position to wine, revealing "the strength of the ancient Hebrew temperance movement" (Hyams 1965:59; Jastrow 1913:181). "Wine is a mocker, strong drink is raging: and whosoever is deceived thereby is not wise" (Proverbs 20:1). Love of wine impoverishes (21:17), and causes contentions and madness (23:29-31). A mother warns a royal prince that "It is not for kings to drink wine,/ Nor for rulers to mix strong drink;/ Lest, drinking, they forget the law" (31:4-5). However, at the same time, she recommends: "Give strong drink unto him that is ready to perish, and wine unto those that be of heavy hearts./ Let him drink, and forget his poverty, and remember his misery no more" (31:6-7).

DRINKING CONTROLS. In the late 7th century, King Josiah of Judah (c. 638 BC) initiates a sweeping religious reform based on "the laws of Moses" and the discovery of what is believed to be an early version of the book of Deuteronomy (2 Kings 22), which is probably written by the disciples of the 8th-century prophets in order to correct the conditions which concerned Amos, Hosea, and Isaiah. Deuteronomy (21:18-21) contains "an extraordinary ordinance, which stands without parallel" (Lutz 1922:134). The parents of a disobedient son living in gluttony and drunkenness may accuse him to death before judges. This is the only evidence of punishment for drunkenness in the ancient Near East. (Drunkenness is never mentioned in the Code of Hammurabi [1800 BC] or any other legal regulations of Babylonia and Assyria.) However, the punishment appears not to have been for drunkenness per se but for overt disobedience of parental authority, coupled with gluttony and drunkenness (Baird 1944a:542).

MACEDONIA

WINE CHARACTERISTICS. One Macedonian wine from the town of Maroneia is famous for its strength; Homer (*Odyssey* 9.209) reports that it could be diluted with 20 parts water and still be intoxicating.

PERSIA

WINE: MEDICINAL USE; LEGENDARY ORIGINS. Persian historians say that wine originated in Persia, and one of the earliest of them ascribes its discovery to Djemshid (Jamshid, or Jemsheed), the legendary founder of Persepolis (c. 800). According to legend, Djemshid was very fond of grapes, and, in order to have them in the winter, he ordered a supply packed in jars for his use. After keeping them for a time, he opened a jar to examine the fruit and found that the grapes had changed color and appeared to be spoiled. Thinking that they had become poisonous, he ordered the jars sealed and their contents used for the execution of condemned criminals. A girl at the palace became afflicted with severe pains; and, wishing to die, opened a jar and took some of the grapes. She soon fell asleep and, when she awoke, found herself entirely well. On learning this, Djemshid ordered grapes to be similarly prepared each year and encouraged the use of wine by everyone, giving it to his soldiers with the purpose of making them strong (Cherrington 1925, 1:213).

700 BC

GREECE

VITICULTURE SPREADS. In the seventh century, vineyards spread widely through the Greek islands and mainland. Far more than in previous ancient civilizations, viticulture becomes in Greece a widespread and expert profession involving the growing of wine on a large and commercial scale, and the adoption of a "more or less scientific attitude towards viticulture" (Younger 1966:127).

625 BC

GREECE

DIONYSIAN RITES. Ecstatic, mystical rituals centering on the Thracian (and possibly Phrygian) wine god Baccos Bromios enter Greece some time around the end of the 7th century by way of Macedonia (Winnington-Ingram 1948:1). Baccos Bromios becomes identified with Dionysus, a minor Greek peasant god of wine and corn, who is now worshipped as a savior son of god, despised, imprisoned, and finally killed, who rises again in glory and power (Seltman 1957:50-62). His worship includes wild rites devoted to the production of ecstasy by music and dancing, by bloody sacrifice, and, frequently, but not necessarily, by intoxication from wine drinking. The cult is particularly appealing to Greek women, possibly because they are often left alone by their men during wars and generally hold a secondary status in Greek society. As the cult spreads, many towns are momentarily emptied of women as they dance into the mountains in their revelries. The spread of Dionysian revelries with their destructive potential to social order and property is a subject of grave concern, especially in the developing Greek cities (Seltman 1957:102-103). This concern is reflected in the character of King Pentheus of Thebes in Euripides' The Bacchae (see 425 BC/Greece). Pentheus stands for law and order, common sense, and rationality. Bacchus is alien in origin and spirit to the Achaeans and symbolizes the unleashing of "suppressed immediate apprehensions" (Hyams 1965:82).

SPARTAN TEMPERANCE. Following a slave rebellion in the mid-7th century BC, all citizens of Sparta or Lacedaemon are required to submit to the Lycurgan discipline, the laws purportedly established by the legendary Lycurgus. As part of this reform, Sparta becomes the most abstemious of the Greek city-states and one of the most temperate of all ancient peoples (Plutarch's Lives, Lycurgus 12).

Lycurgus is said to have allowed wine drinking only for relief of thirst, because it was ruinous to body and mind. There are no drinking parties; drunkenness is practically unknown and is in social disfavor (Plato, Laws 637a-b). The wine ration is only a pint or a pint and a half a day (Herodotus 6.57; McKinlay 1951:77-78). To discourage drinking, Spartans periodically force their helots (serfs) to drink to excess as a bad example for Spartan children (Plutarch, Lives, Lycurgus 28).

Spartan temperance is praised by many Greeks in comparison to the excessive drinking in other Greek cities. Spartan dining customs are admired for giving little opportunity for a man to ruin himself on gluttony or wine-bibbing (Xenophon, Const. Lacedaemon. 5.4-6). Plato frequently lauds Spartan temperance over Athenian drinking (Laws 666, 671, 674d). The Athenian Critias praises Spartans for not having the custom of drinking toasts at banquets, pledging loving cups, or setting aside a day to intoxicate the body (Athenaeus 10.433b, 10.432d). This strict regimen lasts down to the time of Philopoemen (c. 200 BC) (McKinlay 1949a:311-312).

600 BC

GAUL

VITICULTURE INTRODUCED. The vine is probably introduced into southern France around 600 BC when Phocaeans from Greece found Marseilles (Hyams 1965:133, 161; Younger 1966:154). Greek wines begin to be imported into Gaul (Allen 1961:87).

GREECE

INEBRIETY PREVALENCE. The Scythian philospher Anacharsis visits Athens and other Greek cities and is amazed at Greek drinking customs and the amount of wine they consume (Diogenes 1.8.103-5; Athenaeus 10.438a).

SYMPOSIUM. The central drinking institution and main form of social activity in the Greek world is the symposium, which means "drinking together" but is usually translated by the word "banquet." As described by Plato and Xenophon (see 400 BC/Greece), the symposium is the occasional gathering of men for an evening of drinking, conversation, and often entertainment, which takes place after a grand dinner. McKinlay (1951:77) emphasizes, "This stress in choice of name on drinking rather than on eating or fellowship is indicative of the part drinking played in Greek consciousness." Contrary to the refined image of the symposium provided by Plato and Xenophon, they often end in drunkenness (see the comments of Critias, 425 BC/Greece). (On the symposium, see also Billiard 1913:222.)

DRINKING CONTROLS. As cities begin to grow and the Dionysian cult spreads, there occurs the first legislation aimed at controlling drinking in the 7th and 6th centuries.

DIONYSIAN RITES REGULATED. Greek statesmen of the early and mid-6th century BC abandon the effort to suppress the Dionysian revelries and attempt to institutionalize and regulate them, by channeling the flood of energies they release. Periander of Corinth (625-585 BC) is the first to introduce controlled festivities centering on drunkenness (Emboden 1977:191). Cleisthenes of Sicyon (c. 580 BC) replaces the cult of the local hero with Dionysian rituals and may have introduced them into Delphi, the center of the worship of Apollo. Later, following Solon's attempts to regulate Athenian drinking (see below), Peisistratus of Athens (governed 560-527 BC) favors the cult, introduces a statue of Dionysus into the Acropolis, and sponsors the six-day festival of the Great Dionysia, from which Attic theater has its origins in masked mummers (goat singers or tragodi) who perform in the god's honor (especially the famous Thespis in 534 BC). Tragedy may have developed out of pageants representing the death of Dionysus; comedy, from rowdy festivals celebrating his resurrection and marriage. At the Great Dionysia, wealthy citizens subsidize dramas and comedies in competition for prizes. Celebrants are expected to get drunk in honor of the god, a tradition even the abstemious Plato supports (see below) (Africa 1965:113, 147; Seltman 1957:102-103; Emboden 1977:191; Younger 1966:117ff).

That the cult was regulated in part by emphasizing the role of wine drinking in it indicates that the concern over the Bacchanals had less to do with inebriety, which often did not occur, than with the orgiastic and ecstatic frenzies the cult released. Goodenough (1956:13-15) emphasizes "that however ribald the fetes of the god might have become...there was something about Dionysus himself which appealed to those who were disgusted at the rites....Dionysus...came for the Greek chiefly to symbolize wine in a religious sense, his character expressed the complex of associations which wine carried in the ancient world." The god reflects the reverence the Greeks feel toward wine as "both a food and a unique means of inspiration." This feeling is also reflected in the ritualistic use of wine. Goodenough (1956:59-60) summarizes: "Greeks and imperial Romans had a vivid sense that wine and love represented their chief hope of immortality, and that these together formed the basis for mystic, even philosophic, notions of the approach to God or the greater life here and now."

DRINKER LIABILITY. Pittacus, the tyrant of the Greek island of Lesbos and one of the Seven Wise Men of Greece, calls wine a mirror of the soul (Athenaeus 10.427e; Rolleston 1927:104). He establishes a law that offenses committed by a man when drunk should have a double penalty and advocates that citizens be instructed to love temperance. His object is "to discourage drunkenness, wine being very abundant in the island" (Diogenes 1.4.76-78; Aristotle, Politics 2.9.9). However, the Lesbians continue to be heavy drinkers (Athenaeus 10.429a, 10.430a-d, 10.442f-443a; McKinlay 1949a:295).

WINE DILUTION. Intemperance is so prevalent in Athens that in 594 BC Solon establishes the death penalty for any drunken magistrate and he requires that all wine sold must be diluted. He also voices his regret that there is nobody to keep the Athenians from excess and to see that they feast properly (Diogenes 1.2.57; McKinlay 1951:63; Rolleston 1927:104). Solon's law would appear to have been little enforced since the first ceremony of the Greek symposium continues to be the determination of how much water to mix into the wine in a large bowl before serving it to guests. Athenaeus (10.431e-f) does quote Solon as saying that because of this practice it was not easy to get drunk at Athenian symposia: "This, you see, is the Greek way of drinking; by using cups in moderation, they can talk and fool with each other pleasantly."

ATTITUDES; INEBRIETY EFFECTS (THEOGNIS). The conservative Greek poet Theognis of Megara critically describes the effects of excessive drinking, warning: "He, whosoever he be, who exceedeth measure in drinking, hath no longer power over his own tongue or his reason....Do thou not, then, knowing this, drink wine to excess, but ere thou begin to be drunk, rise up and depart" (quoted Athenaeus 10.428c; Raymond 1927:59). But in one of his elegies (no. 627), he also writes that "assuredly 'tis a disgrace to be drunken among the sober, but disgraceful is it also to abide sober among the drunken" (quoted Darby et al. 1977:588).

PERSIA

PERSIAN TEMPERANCE; DRINKING PRACTICES. According to several later Greek sources, the Persians before the founding of their empire by Cyrus (see 525 BC/Persia) drink water, not wine, and are known for their austerity (Herodotus 1.71; Strabo 15.3.18). Xenophon (*Cyropaedia* 8.8.4) writes that they drank wine but once a day. Their kings are allowed to get drunk only on days when making religious sacrifices to Mithra (Athenaeus 10.434e). This reputation for sobriety deters King Croesus (c. 550 BC) of Lydia (Asia Minor) from attacking them. Although the claim of abstinence for the early Persians has been rejected by Raymond (1927:65), it is accepted by Modi (1888:5) and McKinlay (1948b:403).

ROME

INEBRIETY PREVALENCE. The city of Rome, whose founding is traditionally dated at 753 BC, is largely a community of shepherds, not farmers. There is general agreement that through to the 3rd century BC the Romans practice great moderation in drinking. References to drunkenness are scant. During this early period, the Roman way of life and an insufficient supply of wine make excessive drinking impossible on a regular basis (Jellinek 1976a:1730). The wine culture develops in Italy very early, as references to Pater Liber, the Roman god of fertility and wine, go back to pre-Greek times (Jellinek 1976a:1722). Viticulture may have been introduced into Rome by the Etruscans, from northern Italy (Hyams 1965:97) or by Greek colonists in south Italy (Allen 1961:125). But Pliny's (14.13.87) statement that the legendary Romulus "made libations with milk and not with wine" reflects the relative importance of these two products to early Romans.

RITUAL WINE USE; AVAILABILITY. The Postumian law of Numa, the purported king after Romulus and founder of the state religion (c. 600 BC), bans the sprinkling of wine at funerals and the use of wine made of unpruned vines in libations. This is interpreted by Pliny (14.14.88) as a response to the scarcity and great cost of wine and to the need to induce farmers to pay more attention to their vineyards. Hyams (1965:96) calls this theory "ridiculous," because "no people ever economize on religious rites"; rather, he argues that Numa was trying to get back to the good old days of primitive virtue, not unlike the Rechabites.

DRINKING CONTROLS; WOMEN, YOUTHS. Numa also purportedly prohibits wine drinking by women, slaves, and men under 30 (Athenaeus 10.440). Pliny (14.14.89-91) provides several examples of the punishment of women for drinking during the early Roman Republic. How strictly this law is enforced is uncertain, but as late as the first century AD, a wife is beaten to death for drinking wine (Sandmaier 1980:42).

525 BC

BABYLONIA

FESTIVE DRINKING. The Neo-Babylonian empire is defeated by the Persians under King Cyrus in 539 BC when, finding the city too fortified to assault, the Persians wait for a festival in which all the Babylonians drink themselves into helplessness. The Persians then secretly enter and capture the city (Old Testament, Daniel 5:1-4; Herodotus 1.191).

JUDAISM

BACKGROUND DEVELOPMENTS: EXILE ENDS. When the Persians overthrow the Neo-Babylonian empire (see above), the Hebrews who have been in captivity there are permitted to return to Palestine. (In 586 BC, the Neo-Babylonian King Nebuchadrezzar or Nebuchadnezzar had conquered Judah, destroyed the city of Jerusalem, and deported the population to Babylon.) It is in postexilic Palestine that Judaism as it is presently known develops and the Hebrews can be said to have become Jews. Their God has become more universal and moral, demanding justice as well as piety, and it is emphasized that piety does not consist of cult practices alone, but applies to every waking moment. In 458 BC, the priest Ezra, called "the Second Moses," leads a mass exodus of Jews back to Jerusalem. Together with Nehemiah, Ezra revives the Mosaic law and begins compiling the Jewish Scriptures out of oral traditions and written materials. In 444 BC, the Torah or Pentateuch, consisting of Genesis, Exodus, Leviticus, Numbers, and Deuteronomy is canonized.

DRINKING PRACTICES, ATTITUDES. During the 200 years following the return to Palestine, it appears that a fundamental change occurs in Jewish drinking practices and attitudes: sobriety increases and the earlier antagonism to wine disappears (Keller 1970). The prophets still protest against excessive drinking, but it is no longer assumed that drinking necessarily leads to drunkenness (Jastrow 1913:182). Among the Jews of Palestine, the vineyard becomes, possibly for the first time, a romantic symbol of poetic mysticism (Younger 1966:69). It is now considered a distinct blessing of God when the vine harvest shows a good crop and the wine-cellars are fully stored with wine. Wine becomes a symbol of joy (Psalm 104; Zacariah 10:7). These changes in drinking patterns and attitudes would appear to be related to three developments in postexilic Judaism: (1) the final banishment of pagan gods; (2) the new emphasis on individual morality; and (3) the integration of existing secular drinking habits into religious ceremonies and their subsequent moderation. Goodenough (1956:130ff) also believes that this change reflects the influence of Hellenism on Jewish thought.

RITUAL CONTROLS. In the Pentateuch, the use of wine in the performance of religious ceremonies is mentioned only in connection with libations, but the rabbis and scribes who follow Ezra set about to integrate wine into everyday Jewish ceremonies, with the apparent intention of establishing ritual controls that will moderate existing drinking customs. One of the acts of the scribes is to fix the rule that the pronouncement of the Sabbath--the kiddish--should be recited over a blessed cup of wine, thus initiating the regular drinking of wine in Jewish ceremonies outside the Temple. It is also stipulated that the wine must be diluted.

In rabbinical Judaism, wine becomes the symbol of the sanctification of the Sabbath and of other Jewish festivals; blessings are made over the "fruit of the wine" and four cups of wine mark the divisions of the Seder on the evening of the Passover festival (Jastrow 1913:190). This is particularly true following the development of Christianity (see 1 AD/Hebrews).

PERSIA

BACKGROUND DEVELOPMENTS: EMPIRE. Following victory over Lydia in 546 BC, the Persians rise to world power under King Cyrus (550-527 BC).

INEBRIETY PREVALENCE; FESTIVE DRINKING. In several battles King Cyrus uses the drunkenness of others to attain victory, and he frequently praises the virtue of temperance to the Persians (Xenophon, *Cyropaedia* 1.3.10-11, 1.5.9, 4.2.38-45, 7.5.75-76). But he also commemorates the victory over the Sacae by establishing a yearly festival at which everybody carouses in unrestrained revelry (Strabo 10.8.5).

Following Cyrus' death, drunkenness increases among the Persians (McKinlay 1948b:404-405; Lutz 1922:118). Xenophon attacks the subsequent physical and moral decline of the Persians, including increased drunkenness. Prior to Cyrus, "it was their custom to eat but once in the day, so that they might devote the whole day to business and hard work"; after Cyrus, they continuously feast and drink from early morning until late at night, and "they drink so much that...they are...carried out when they are no longer able to stand straight enough to walk out" (Xenophon, *Cyropaedia* 8.8.9-11). Although a speech on temperance is made to the Persian king at feasts before drinking begins, they regularly hold parties at which all become drunk, and Cyrus' son Cambyses (d. 522 BC) is widely criticized by his own people for his drinking (Herodotus 3.33-34; Plato, *Laws* 695). Athenaeus (10.434d) preserves the boast of King Darius (d. 484 BC), inscribed on his tomb, "I could drink much wine and carry it well."

RITUAL INTOXICATION. Excessive drinking also appears to be a ritual adjunct to decision-making. Herodotus (1.133) reports that the Persians "deliberate and decide while drunk and reconsider their decision next morning when sober."

ROME

ETRUSCANS, WOMEN/DRINKING. Under King Tarquinius Superbus (534-510 BC), women begin to drink in Rome and drinking in general may have increased (Jellinek 1976a:1723-1730). Tarquinius is of Etruscan origins and, according to the 4th-century historian Theopompus (quoted by Athenaeus 12.517d-e), the Etruscans are "terribly bibulous" and their women drink as freely as men. Overall, however, wine drinking remains moderate.

500 BC

GREECE

TAVERNS: PREVALENCE, ATTITUDES, CONTROLS. Taverns, or more specifically hostelries, appear in Greece due to the expansion of the population and the growth of commerce, both of which have rendered inadequate the traditional custom of hospitality, whereby residents provide food and lodging to strangers. Specialized wine shops exist called kapeleion. They are generally centers of low life and are held in contempt by the upper classes, although some apparently patronize them regularly in secret. Despite this hostility, there is little legislation dealing with tavern operation except for laws regarding false measures or public order (Popham 1978:241-242; Firebaugh 1972:57, 65, 68, 76; Rolleston 1927:113-114; Becker 1906).

PRAISE OF WINE (PANYASIS, BACCHYLIDES). The epic poet Panyasis of Halicarnassus says that in moderation "wine is as great a boon to earthly creation as fire. It is loyal, a defender from evil, a companion to solace every pain." It is "the best gift of gods to men, sparkling wine...It drives all sorrows from men's hearts when drunk in due measure, but when taken immoderately it is a bane." If one "persists to the full measure of the third round and drinks to excess there rises the bitter dome of Violence-Ruin" (Athenaeus 2.37a, 2.37b, 2.36d). Bacchylides refers to "a sweet compelling impulse [that] issues from the cups, warms the heart...sending the thought of men to topmost heights" (Athenaeus 2.39e-f).

INEBRIETY PREVALENCE. Idomeneus (see 300 BC/Greece) refers to the sober days of Themistocles in Athens (Athenaeus 12.532f). However, excessive drinking is not unknown: around 500 BC an archaic red-figured picture on a drinking cup shows a young Athenian vomiting the wine he has consumed at a banquet (Younger 1966:109).

450 BC

EGYPT

FESTIVE DRINKING. Herodotus (2.78) relates that at the end of banquets among the rich an effigy of a dead person is led around and the guests are admonished to "drink and be happy, for when you die, such you will be." (On alcohol use by the Egyptians, see also Herodotus 2.121, 2.173-174.)

WINE AVAILABILITY. Although "their drink is a wine which they obtain from barley," Egyptians import large quantities of grape wine (Herodotus 2.77, 3.6), indicating that home production is not sufficient to meet demand. Artificial wines are made because of the difficulties of transporting grape wines (Lutz 1922:17).

GREECE

DRINKING PRACTICES (HERODOTUS). Herodotus of Halicarnassus (c. 484-425) writes the first major book of history. A widely traveled man, with an insatiable curiosity, he often describes the drinking and eating customs of the lands of the Mediterranean basin. (See some of his observations under 600 BC/Persia, 525 BC/Persia, 450 BC/Egypt, 450 BC/Scythia.)

SCYTHIA

TEMPERANCE. The Scythians, warlike nomads from central Asia who roam the northern Steppes as far west as the Black Sea and Thrace, are said to be abstemious by some. Herodotus (4.79-80) records that the Scythians objected to Bacchic rites on the grounds that it is unreasonable to have a god that leads men to madness, and dethroned one king when he became involved with them. This is consistent with the advocacy of temperance expressed by the Scythian philosopher Anacharsis (see 600 BC/Greece) (Diogenes 1.103). Several other criticisms of wine by Scythians are extant. But much of the evidence regarding the Scythians is contradictory. In the 4th century BC, Plato (*Laws* 637e) and Chamaeleon of Heraclea (cited Athenaeus 10.427a-c) assert that the Scythians took pride in the amount of unmixed wine they could consume.

425

GREECE

ATHENIAN INEBRIETY. There are several indications that drinking increases in Athens during its height in the second half of the 5th century BC. Pericles, who attains political ascendency in Athens in 461 and dies in 429 BC, has been credited with making Athenians "luxurious and wanton" instead of temperate (Plutarch, *Lives*, *Pericles* 9.1; McKinlay 1951:64). Athenians also supposedly become less temperate when Alcibiades (fl. 420-406) teaches them to drink in the morning (McKinlay 1951:67). Beginning in this period, criticisms of Athenian drinking practices and warnings that greater temperance should be exhibited do appear to become more frequent.

TEMPERANCE ADVOCATED (SOCRATES). Socrates (469-399) approves of drinking, but disapproves of drunkenness and is concerned about its effects (O'Brien 1980b:89). He advises Greeks to beware of habit-inducing drinks and those people who tempt one to drink even when one is not thirsty, and not to expect to find good friends in men attached to the wine cup (McKinlay 1951:65). He is quoted in Xenophon's *Symposium* (2.24-26; see 400 BC/Greece) as warning that, although wine tempers the soul and puts our cares to rest, "If we pour ourselves immense draughts, it will be no longer time before both our bodies and our minds reel, and we shall not be able to draw breath, much less to speak sensibly." The guests at Plato's Symposium (176, 213; see 400 BC/Greece) comment on Socrates' ability to drink a great deal without effect or not drink at all.

SYMPOSIUM CRITICIZED (CRITIAS; SOCRATES). The reactionary Athenian aristocrat and sophist Critias (d. 403) describes an Attic symposia at which "upon their eyes a dark mist settles; memory melts away into oblivion; reason is lost completely" (quoted Athenaeus 10.432d-e; McKinlay 1951:64; Africa 1969:167). Plato quotes Socrates as criticizing the symposia of the "vulgar and shallow minded" who are "incapable of entertaining themselves with conversation," whereas the guests at a true symposium entertain themselves with the sound of their own voices, "talking and listening in turn with order and decency, however heavily they drink" (quoted Allen 1961:61).

DIONYSIAN RITES (EURIPIDES). When the frieze of the Athenian Parthenon is completed in 432 BC, among the 12 Olympia gods depicted, Dionysus appears for the first time, in place of Hestia, goddess of the hearth, illustrating the increasing importance of Dionysus to the Greeks (Emboden 1977:190).

Around 408 BC, Euripides leaves Athens and composes The Bacchae in Macedonia, possibly having encountered there the wild religion of Bacchus Bromio in all its occasional violence (Seltman 1957:62). In the play, "Dionysus symbolizes the power of blind, instinctive emotion" which inevitably results in catastrophe (Winnington-Ingram 1948:9). The play reflects the Greek concern over the potential destructiveness of Dionysian revelries; it illustrates how these revelries were gradually controlled 150 years earlier with recognition of the importance of ecstasy and intuition to balance reason and order. The opening chorus calls Dionysus the "spirit of revel and rapture" (Younger 1966:117). While disapproving of Dionysian ecstasy, Euripides shows a profound understanding of it, and at one point refers to "the joys of wine that bring relief from sorrow" and writes that "without wine there can be no love any more, or any other of the things that give men joy" (ibid.).

400 BC

CARTHAGE

DRINKING PRACTICES. The Carthaginians have a reputation as great drinkers but still deny wine absolutely to soldiers in camp and magistrates in office (Plato, Laws 637d, 674a). Among other north Africans, the Numidians and Moors have a reputation for temperance (McKinlay 1948b:393).

GAUL

ESTEEM OF WINE. In 390 BC, the Gauls defeat the Etruscans and capture Rome, ostensibly because once they discovered wine and the other products of Italy they wanted to live in the land that produces them (McKinlay 1948b:389).

GREECE

INEBRIETY PREVALENCE. Xenophon (431-351), who discusses wine and its use in several writings, criticizes the Greeks (the Spartans excepted) for their excessive drinking (Const. Lacedaemon. 5.4-6). He also observes "that many respected men deliberately restrict themselves in the manner of meat and drink" (Hiero 2.1). This suggests that temperance is appreciated but probably unusual (McKinlay 1951:67, 76). Isocrates (fl. 392-338) laments (Areopagiticus 49) that the Athenians have taken to drinking in taverns and advises a pupil to avoid drinking parties altogether, or at least to leave while yet sober. Isocrates and Xenophon reflect "the monarchic and oligarchic strain of fourth-century [Greek] conservative thought" (Africa 1969:184).

ATTITUDES; PHYSICAL EFFECTS (XENOPHON). Xenophon praises wine as a pleasant thirst quencher that brings joy and cheer, provides bravado, heals grief, excites, cultivates friendships, inspires music, singing, and dancing, and is a central part of hospitality, religous rituals, and the nation's economy. He also recognizes that it can drain strength, affect memory, produce madness, fog the mind, evoke violence, bring out bestiality, and lead to ruin (O'Brien 1980b:89, 95-99). He observes that drunkenness hampers diligence and promotes forgetfulness and that sobriety goes with industry (Oeconomicus 1.22, 12.11). In his ideal state, young men are abstemious even to the point of drinking water; tyrants guard against intoxication more carefully than against conspirators; and the prince does not tolerate foolish drunkenness (Hiero 6.3, 9.8; Cypropaedia 1.2.8, 1.3.10-11, 7.5.59).

DRINKING CONTROLS ADVOCATED (PLATO). Plato (429-347 BC) looks upon moderate wine drinking as a beneficial and necessary pleasure, bringing warmth and cheer when practiced in moderation. Its use, promotes health. (See below for his attitudes toward Dionysian rites.) However, he takes a very critical view of drunkenness and proposes that severe restrictions be imposed to promote temperance. This would indicate that inebriety is a considerable problem.

In his Laws, he emphasizes that the state has the right and the duty to protect citizens from the effects of drinking should it put them at risk (Inglis 1975:30). He asserts that hard drinking reveals the goodness of the good man and the rascality of the wicked; experience with excess teaches moderation (Laws 647d-650a). Drinking parties tend inevitably to become uproarious, and "unsober" drinkers commonly quarrel over their wine (671a-d). He proposes that drinking parties be regulated because banquets regularly become drinking bouts; only under tight regulation should drinking be allowed (673e-674d). Boys up to 18 should be totally forbidden wine; moderate use should be allowed up to age 30, but not until age 40 should one be permitted to drink freely at banquets in gratitude to the gods (666a-c). He concludes (674a) that, "unless the state shall make an orderly use of the institution [of drinking bouts], taking it seriously and practicing it with a view of temperance," he would vote against allowing citizens of that state ever to drink at all. In so doing, he recognizes that he would be "going further than the Cretans and the Spartans and even beyond the Carthaginian law [see above] which prescribes water only as the drink for soldiers on the march." (On Plato, see McKinlay 1951:67, 68, 70-72; Rolleston 1927:107; Raymond 1927:66-69; O'Brien 1980b:89-90).

DIONYSIAN RITES (PLATO). Despite Plato's antagonism to drunkenness (see above), he believes that one should get drunk at the Dionysia in honor of the god of wine. He is quoted as having said, "To drink to excess was nowhere becoming...save at feasts of the god who was the giver of wine [Dionysus] (Diogenes 3.41).

SYMPOSIUM (XENOPHON; PLATO). Xenophon provides a portrayal of Greek drinking habits in his Symposium, which probably mirrors more typical gatherings in relaxation than the symposium described by Plato. He shows it to be an informal, gay banquet of entertainment, yet still infused with a sense of grace and restraint. As in Plato's work, alcohol does not play a central role. Although drinking is central to an evening's informal, relaxed entertainment, it is moderate and controlled in accordance with Socrates' admonitions. He even opens with Socrates discussing the virtues of temperance (see 425 BC/Greece).

About 385 BC, Plato writes his Symposium, showing it to be a rather formal gathering principally aimed at lofty, serious, and poetic discussion. Drinking takes second place. However, the text also indicates that such seriousness was not always the case. A discussion is held about how to "drink with least injury to ourselves" as some guests had not yet recovered from the effects of drinking too much wine at a symposium the previous night. Alcibiades arrives at the part already intoxicated. A physician warns that "drinking deep is a bad practice" (sections 176, 213).

MEDICINAL USE (HIPPOCRATES). In the late 5th century, Hippocrates (c. 460-377 BC) introduces the scientific study and treatment of disease and gives the medical profession its ethical ideals. Prior to this, wine and medicine were variously associated with religion and magic (Lucia 1963a:5, 36-39). Hippocrates makes no extravagant claims for the therapeutic value of wine but includes it in the regimen for almost all acute and chronic diseases, especially during convalescence. He describes cases of fever beginning with heavy drinking, observes the power of wine to affect many people adversely, and specifies when wine should or should not be consumed (McKinlay 1950a:234; Cornwall 1939:380). Subsequently, considerable variance exists among Greek physicians in their therapeutic use of wine; not all approve of the practice, but generally the attitude is favorable (Lucia 1963a:8). Theophrastus (see 350 BC/Greece) describes a multitude of medicinal plants decocted in wine (Lucia 1963a:42).

This attitude becomes even more favorable under the Romans. Greco-Roman physicians emphasize the danger of alcohol, but recognize that these dangers depend on the kind, quality, and quantity consumed.

DRINKING EFFECTS; ATTITUDES. The Greek poet Eubulus (c. 370 BC) has Dionysus discuss the effects of increasing amounts of wine, indicating that drinking more than three bowls of wine was disapproved: "Three bowls only do I mix for the temperate--one to health..., the second to love and pleasure, and the third to sleep....The fourth bowl is ours no longer, but belongs to violence; the fifth to uproar, the sixth to drunken revel, the seventh to black eyes. The eight is the policeman's; the ninth belongs to biliousness; and the tenth, to madness" (quoted Athenaeus 2.36b-36c).

PRAISE OF WINES. Xenophon (Anabasis 4.4.9) refers to "old wines with a fine bouquet," which he discovered north of Greece in 400 BC. The storage of wines at this time is made possible by the development of proper cementing or pitching of amphoras and other vessels (Emboden 1977:188).

350 BC

GREECE

CONCERNS: INEBRIETY. Aristotle, Theophrastus of Eresos, Chamaeleon of Heraclea, and Hieronymus of Rhodes all write essays on drunkenness in the 4th century BC (Athenaeus 10.424f, 10.435a). This considerable attention devoted to the subject would indicate that all had seen enough heavy drinking among Athenians to warrant concern (McKinlay 1951:64-65). In a lost work on drunkenness, Aristotle (384-322 BC) is said to have observed that those who get drunk on beer fall on their faces because beer stupefies, whereas those who drink wine fall on their backs because wine makes the head heavy (quoted Athenaeus 1.34b, 10.477a-b; McKinlay 1951:68-69, 74). This passages suggests that drunkenness is common. It also reflects the Greek contempt for beer as a barbarian drink (Rolleston 1927:103). Aristotle recommends for temperance mulled wines (embellished with heat and spices) because they are softer and less intoxicating (Athenaeus 11.464c). He also makes several approving references to Pittacus' law against those who commit crimes while drunk (O'Brien 1980b:93).

Theophrastus (370-285 BC) writes that the mark of a gross man is boasting that he intends to get drunk; habitually drinking pure wine is a sign of a barbarian or a bore. He contrasts the Greeks of his day in their urge to drink with the more temperate Greeks of earlier times (quoted Athenaeus 11.782b, 10.463c, 10.427b; McKinlay 1951:67).

In addition, in the fourth century, the Aristotelian Economists stress the tendency of wine to make even freemen insolent (McKinlay 1951:65).

DRINKING PRACTICES. Chamaeleon of Heraclea asserts that the habit of using large drinking cups was a recent introduction from barbarians.

YOUTHS/DRINKING. Theopompus asserts that Athenians engage in drinking bouts in their youth (McKinlay 1951:64-65).

MEDICINAL USE; ESTEEM OF WINE. A botanist, Theophrastus (see above) describes a multitude of medicinal plants decocted in wine (Lucia 1963a:42). His is the first scientific study of the grape vine (Younger 1966:127).

MACEDONIA

INEBRIETY PREVALENCE. Among the Greeks, the Macedonians are famous for their drunkenness; they view intemperance as a sign of manhood and as a warrior virtue. They are criticized for getting drunk even before eating and for the amount of wine they consume afterwards, including a contest of emptying out six-pint bowls (Athenaeus 3.120e, 4.128a-130e).

INEBRIETY OF ALEXANDER THE GREAT. Alexander the Great, king of Macedon (336-323), develops the reputation as one of the great drunks of history and the "most talked about drinker of ancient times" (McKinlay 1949a:292). As a youth he appears to have been quite temperate in his drinking (Plutarch, Lives, Alexander 4.4, 22.4, 23.1), but his adult life becomes "one unending revel" (McKinlay 1949a:293; see Athenaeus 10.437a, 10.437a-c, 12.537d, 12.539a-b). He may have become a self-destructive alcoholic (O'Brien 1980a, 1980b), but this is controversial (Anastasi 1980). Plutarch insists that Alexander was less addicted to wine than generally believed and attributes his proneness to drink to the heat of his body. Yet he makes frequent references to Alexander's drinking (Plutarch, Lives, Alexander 4.4, 9.5, 13.5, 47.2, 70.1, 75.3-4). Alexander's father, King Philip II, reputedly got drunk daily, was ridiculed by his young son for his drinking, and went drunk into battle (Athenaeus 6.260a-c, 10.435b; Plutarch, Lives, Alexander 9.5; McKinlay 1949a:291-292). Alexander's mother was a fervent devotee of the cult of Dionysus. O'Brien (1980b) views his excessive drinking in the context of the stress Hellenic society places on alcohol's "positive dimensions."

325 BC

GREECE

DRINKING EFFECTS; WINE DILUTION (MNESITHEUS). Mnesitheus (320-290 BC), a Hippocratic physician of Athens, takes a less critical view of wine than his mentor. His treatise on diet and drink is extensively quoted by Athenaeus (2.36a-c, 2.46e, 11.484b). He calls wine the "greatest blessing of mortals" given by the gods, providing nourishment and strength of body and mind. But he is also aware of wine's dangers: "In daily intercourse, to those who mix it [with water] and drink it moderately, it gives good cheer, but if you overstep the bounds, it brings violence. Mix it half and half, and you get madness; unmixed, bodily collapse." He supplies instructions on how to avoid adverse effects from persistent drinking (Athenaeus 11.484b; McKinlay 1950a:246, McKinlay 1951:80; Lucia 1963a:42-43; Rolleston 1927:108; Raymond 1927:56).

ROME

WINE AVAILABILITY; VITICULTURE. According to Pliny (14.14.91), in 320 BC Lucilius Papirius promises Jupiter an offering of a small goblet of wine if he will help him achieve victory in battle. This may indicate that wine is still a scarce commodity in Rome (Jellinek 1976a:1723). Since the 5th century, viticulture has been expanding, possibly as a result of climatic changs which leave Italy warmer and drier (Allen 1961:75-76). But most vineyards until the mid-2nd century BC are very small, cultivated by subsistence farmers who sell a wide variety of crops to small local urban populations. Greek wines remain dominant (Hyams 1965:98-99).

300 BC

EGYPT

MEDICINAL USE. The Alexandrian school of medicine brings into further notice the treatment of diseases by wine (Lucia 1963a:43).

GREECE

INEBRIETY PREVALENCE. Idomeneus asserts that unlike the more sober days of Themistocles (c. 480 BC), Athenians have become devoted to drunken reveling (Athenaeus 12.532f; McKinlay 1951:67).

TEMPERANCE ADVOCATED (ZENO, STOICISM). Zeno (c. 336-264) formulates the philosophy of Stoicism under the influence of Socrates and Crates the Cynic. Stoicism is based on the belief that divine providence controls the universe and that an absolute moral standard exists. A Stoic seeks to reject all emotion which might disturb inner calm. Zeno himself shows no strong feeling regarding heavy drinking, but he lays down the principle that the wise man may drink but will not get drunk. As evidenced in the later writings of Cicero and Seneca, Stoic attitudes toward drunkenness become much more severe. By Cicero's time, Stoic banquets are even held during the day as a simple repast (Raymond 1927:73-74). Zeno's own long life is ascribed to his abstemiousness (Diogenes 7.27).

WINE CHARACTERISTICS. Estimates of the cost of wine indicate that there probably are many Greeks who have to make do with "poor man's drink of *deuterias*, a 'wine' that was pressed out of wine lees or made from grapeskins soaked in water" (Younger 1966:115).

JUDAISM

PRAISE OF WINE. In Psalms (104:14-15), wine is praised as a gift of God "to gladden the heart of man."

200 BC

ROME

BACKGROUND DEVELOPMENTS: RISE OF REPUBLIC. In 509 BC, the last Etruscan king of Rome (Tarquinius' son) is overthrown and the Roman Republic is established. Between 250 BC and 150 BC, republican Rome gains control of the entire Mediterranean basin.

CONCERNS: INEBRIETY. In the early 2nd century, the period of Roman sobriety begins to come to an end. From the 2nd Punic War (201 BC), and increasingly after the mid-century, voices are raised in protest against the decline in the traditional values of austerity, virtue, frugality, simplicity, and temperance, which are being replaced by degeneracy, corruption, avarice, and ambition. The Carthaginian Hannibal (247-182 BC) is lauded in comparison with the Romans for only drinking a half a liter of wine with his meals (Jellinek 1976a:1731).

DIONYSIAN RITES (LIVY). Livy (59 BC-17 AD) relates that in the year 186 BC, Rome first encountered Dionysus (Latin "Bacchus"). A Greek of "mean condition" spread "secret and nocturnal rites" among the men and women of Etruria. "To their religious performances were added the pleasures of wine and feasting, to allure a great number of proselytes" until "debaucheries of every kind began to be practiced." The practice spread from Etruria to Rome, where an inquiry was held and the Bacchanalian rites were banned except on permission from the Senate. Of 7,000 admitted followers, Livy states, some were imprisoned, but most were killed for committing "personal defilements or murders" (Livy, History of Rome 39.17-19; Warmington 4:254-259; Jellinek 1976a:1732).

In Livy's account, frequent mention is made of excessive drinking, but the real cause for his concern is the purported orgies, ritual murders, violations, and other debaucheries occurring. Although Bacchanalian rites often did not involve excessive drinking, and one can suspect the validity of all the charges brought against them, their spread is another indication of a decline in traditional moral behavior.

150 BC

GAUL

FESTIVE DRINKING. Polybius (2.19) writes that the Gauls become so drunk after a victory that they fall to fighting and destroy themselves (McKinlay 1948b:389-390). (On Polybius, see 150 BC/Rome.)

VITICULTURAL CONTROLS. Rome attempts to limit viticulture in Gaul (see below).

ROME

INEBRIETY PREVALENCE. Intoxication is no longer a rarity (Jellinek 1976a:1724, 1731). In the 2nd and 1st centuries BC, most prominent men of affairs are praised for their regular and deliberate temperance in drinking (e.g., Cato the Elder, Scipio, the Gracchi brothers, Marius the Great, Pompey, Julius Caesar). Such praise appears to be a response to growing intemperance in society in general, since prior to this time temperance was not singled out for praise as exemplary behavior (McKinlay 1946/47:149).

INEBRIETY CONDEMNED (POLYBIUS). Around 146 BC, the Greek historian Polybius (203?-120 BC) criticizes the increase in wine drinking and the "general deterioration of morals" among Romans. He asserts that following Rome's conquests in the eastern Mediterranean, a "great many" young Romans began wasting their energies at banquets by drinking wine. This dissoluteness was learned from the Greeks and "as it were burst into flame at this period," owing to the belief that universal dominion had been secured and to the wealth and splendor imported into Rome (Polybius 32.11).

TEMPERANCE OF CATO. Cato the Elder (234-149 BC) advocates and becomes known for temperance. He does not oppose the use of wine, but does frown on those who make a luxury of it and fancy Greek wines (Plutarch, Lives, Cato the Elder 3.2; Ammianus Marcellinus 16.5.2). Cato reflects the reaction of the traditionalists who attack the increasing luxury of Roman life and consider it unworthy of a Roman to admire or profusely drink wines (Ferrero 1909:184).

WINE AVAILABILITY, PRICE. As more land comes to be devoted to viticulture (see below), wine availability increases and prices decline (to an estimated penny for three quarts [Rolleston 1927:110]). While the greater abundance of wine in this period may not in itself bring an increase in drinking, it "does imply acceptance of the product" (Jellinek 1976a:1724, 1733-1734).

LABORERS/DRINKING; WINE CHARACTERISTICS. According to Cato (23.1, 25, 57, 104), wine for the farm hands is made by taking treaded husks used for cattle feed and soaking them in wine. Cato recommends that the "wine" for the farm hands in the winter be made from 10 parts new-wine must (unfermented and probably of bad quality), 2 parts sharp vinegar, 2 parts boiled-down must, 40 parts fresh water, and 1 1/2 parts sea water.

WOMEN/DRINKING. By this time, the law against women drinking no longer exists, according to a lost book by Polybius quoted by Athenaeus (10.444). However, women are allowed to drink only a little passum--a kind of sweet wine made of raisins (Ferrero 1909:183).

VITICULTURE: EXPANSION; ECONOMIC VALUE. The period from about 130 to 30 BC is the golden age of Roman vineyard expansion Italian wines come to play a major role in the economy. Although viticulture is the most expensive kind of farming, by the mid-century the cash return from it is the greatest (Hyams 1965:99). The earliest evidence for the rise of viticulture is provided around 154 BC by Cato the Elder. In evaluating a farm, he assigns first place to the quality and quantity of its vineyard; to grain land he assigns only sixth place (Cato 1.7). The profit per man-hour-acre from vines is at least three times as great as from any other crop (Hyams 1965:136). Cato provides detailed instructions for growing, manufacturing, and marketing wines. Since Cato is the first to speak of the importance of viticulture, it would appear that this is a relatively new development (see Pliny 4.5, 14.5.44-47). The growth of viticulture at this time is related to a decline in the profitability of grain farming following the end of the Second Punic War, which destroyed the small yeoman farmer and left the grain fields desolated and burned.

CONTROLS. In further evidence of the new economic importance of viticulture, in 153 BC Rome enters into a compact with Marseilles designed to head off the planting of vineyards in Gaul and thereby protect the wine trade of Italy. Although this ban is neither honored nor enforced consistently, it is not lifted until the time of Probus (280 AD) (McKinlay 1948b:389). Bans on planting vineyards are frequently imposed on defeated transalpine tribes by the Romans, a practice which Cicero (3.9.16) considers a poor example of Roman justice (Hyams 1965:136).

125 BC

GETAE

DRINKING CONTROLS. King Boeribista of the Getae in Dacia (Rumania) attempts to restore his people after numerous defeats in wars. He creates a theocracy and raises them "to such a height through training, sobriety, and obedience to his commands" that within a few years his empire is viewed as a threat by Rome. He makes his people sober by persuading them to cut down their vines and live without wine (Strabo 7.3.11).

JUDAISM

WINE CONSUMPTION; ATTITUDES. The proper use of wine is discussed in several of the apocrypha and pseudoepigrapha, written for the most part between the 2nd century BC and the 1st century AD, which serve as connecting links between the Jewish and Christian scriptures. The apocryphal book of Sirach, part of the Hebraic wisdom literature, shows a heavy Greek influence and a moderate attitude toward wine. It is considered a distinct blessing of God when the vine-harvest shows a good crop: "I stood in the blessing of the Lord, and like a gatherer I filled my wine-press." In the Testaments of the Twelve Patriarchs, written by a Pharisee at the end of the 2nd century BC. the emphasis is on the evils of drunkenness as a transgression against the Law: "Be not drunk with wine; for wine turneth the mind from the truth, and inspires the passion of lust, and leadeth the eyes into error....he who is drunken reverenceth no man" (quoted Raymond 1927:39-40).

ROME

MEDICINAL USE. In the writings of Nicander of Colophon (190-130 BC) and of King Mithridates VI of Pontus in Asia Minor (132-63 BC) can be traced the origins of theriacs (mainly medicated wines), antidotes, and alexipharmics--"panaceas or cure-alls which influence medicine throughout the western world for more than nineteen centuries." Mithridates is said to have prepared a universal antidote of several dozen ingredients, compounded with honey and taken in wine, which he took daily along with small doses of baneful drugs in order to immunize himself against their effects. Mithridatium comes to be sold in pharmacies throughout ancient Rome, the ingredients varying greatly (Lucia 1963a:48, 50).

TEMPERANCE OF GRACCHI, MARIUS THE GREAT. The Gracchi brothers, Tiberius and Gaius, attempt to reform Roman society by redistributing the wealth of the land and reducing senatorial power. Following their death in 121 BC their efforts are continued by General Marius the Great. All three are praised for their temperance, although Marius is said to have become a nervous wreck and taken to carousing towards the end of his life (Plutarch, *Lives*, *Gaius Gracchus* 1.2, 2.1; *Tiberius Gracchus* 2.3; *Marius* 3.1, 45.3; McKinlay 1948a:146).

100 BC

ROME

INEBRIETY PREVALENCE. By the 1st century, if not before, wine has become the daily drink of all Italians. Through their conquests, the Romans have become familiar with the great vineyards of the Mediterranean world and they bring back the great vintners to Rome. Wine is given free or sold at cost to the large urban mob that has grown greatly following the displacement of the small yeoman farmers. Wine becomes a major industry, article of trade, and contributor to governmental revenues, and it begins to be celebrated by the poets (Hyams 1965:130-131). Widespread, habitual, heavy drinking apart from the festivals occurs in Rome.

SULLA, MILITARY/DRINKING. The prominent general and politician Sulla, dictator between 82 and 79 BC, is one of the first of Rome's public figures to be identified for his excessive drinking, although this purportedly never interferes with his duties (Jellinek 1976a:1734). Sallust (Cataline 2.8, 7.4, 11.5-6; Bellum Jugurthinum 44.5, 66.3, 95) criticizes Sulla's armies for un-Roman drinking behavior and charges that the Romans take more pleasure in harlots and revelries than in war (McKinlay 1949b:27; McKinlay 1950b:35). Plutarch (Lives, Sulla, 1, 2.2) also criticizes Sulla's drinking and the decline in drinking standards under him. However, according to the later account of Ammianus Marcellinus (16.5.1), Sulla restablished many Roman sumptuary laws that had fallen into abeyance (McKinlay 1948c:52).

50 BC

GAUL

INEBRIETY: PREVALENCE, EFFECTS. In 57 BC, Julius Caesar (103?-44 BC) (Gallic Wars 1.1, 12.15) records that in northwest Gaul the Belgae are especially brave because they are uncorrupted by foreign luxuries; the Nervii forbid importing and growing wine on the grounds that it softens and corrupts manhood and spoils military performance (Caesar 1.1, 12.15). Apparently, excessive drinking constitutes a serious problem (see 150 BC/Gaul).

In 36 BC, Diodorus Siculus (5.26.2-3) describes the devastating impact of wine on the Gauls: "Since the temperateness of the climate of Gaul is destroyed by excessive cold, the land produces neither wine nor oil...[But] the Gauls are excessively addicted to the use of wine and fill themselves with the wine which is brought into the country by merchants, drinking it unmixed, and since they partake of this drink without moderation by reason of their craving for it, when they are drunk they fall into a stupor or state of madness. Consequently, many of the Italian traders, induced by the love of money which characterizes them, believe that the Gaul's love of wine is their gift from Hermes...[They] receive for it an incredible price; for in exchange for a jar of wine they get a slave." Similar views on Gaulic drinking are expressed by Ammianus Marcellinus (15.2.4) in the 4th century and by Appian (Gaulic History 4.7) and Polybius (Histories 2.19) in the mid-2nd century (Hyams 1965:133-134; McKinlay 1948b:389-390).

VITICULTURE EXPANDS. Despite Diodorus' (5.26) observations that the climate of Gaul is too cold for viticulture (see above), vines begin to spread rapidly into northern Gaul, through natural adaptation and selection. Julius Caesar's conquest hastens the northward march of vines so that "within an astonishingly short time" some of the wines of Gaul (the Rhone and Garonne/Gironde) are "making a great name for themselves." Within a generation, the direction of the wine trade is reversed (Hyams 1965:133-137).

WINE CHARACTERISTICS. The small, relatively high acid grapes of Gaul produce a finer wine with a lower alcoholic content than the large, ultra sweet grapes of low acidity in southern Europe, which are difficult to keep and mature and have to be preserved with pitch and spices (Hyams 1965:168; McKinlay 1948b:390).

ROME

UPPER-CLASSES/DRINKING. As the republic continues to decay, Pompey, Julius Caesar, and Crassus form the First Triumvirate and take effective control of the state in 60 BC, despite senatorial opposition. Excessive drinking spreads, especially among the increasingly decadent, bored upper class in Rome (as described by Lucretius, Nature of Things 3.1050-1068). Several notable drinkers appear, one of the most notorious being Marc Antony (d. 30 BC), who purportedly writes a book on his own drinking (Pliny 14.28.148; Seneca 83.25). Cimber, one of the conspirators against Julius Caesar, is a "sot" who admits he can not hold his liquor and asks, "What glory is there in being able to hold much?" (Seneca 83.12). This evidence taken as a whole suggests that drinking large amounts of alcohol has become a matter of prestige (Jellinek 1976a:1734-1735; McKinlay 1949b).

In contrast, the temperance of several prominent Romans is praised as a model for others. These include Cicero, Pompey, Cassius, Crassus, Julius Caesar (see Suetonius, Julius Caesar 53), and Augustus (see below).

25 BC

ROME

TEMPERANCE OF AUGUSTUS. In 30 BC , under the guise of restoring the republic, Augustus (63 BC-14 AD) assumes control of the government and armies and establishes the Principate (after his honorary title of princeps of the Senate and first citizen of Rome), which lasts for 200 years. Augustus is personally temperate (Suetonius, Augustus 77), and throughout his reign the major themes are the restoration of traditional values (piety, morality, patriotism, duty) and the establishment of universal peace and order (the pax Romana). In 18 BC, as part of his sumptuary legislation, he forbids spending more than 1200 sesterces on workday banquets and sets limits for their cost on other occasions (Ferrero 1909:8).

FESTIVE DRINKING: SATURNALIA. Augustus limits the celebration of the Saturnalia to three days. Originally a one-day festival (December 17), it had been extended by common usage to as many as seven days. Popularly the festival commemorates the merry and righteous reign of Saturn, god of sowing and husbandry. It always seems to have been especially marked by feasting and revelry, and by the license granted to slaves to get drunk, to rail at their masters, to sit down at the table with them, and even to exchange places entirely with them for the day in role reversal (Frazer 1959:559-562; Fowler 1916:268-273).

INEBRIETY PREVALENCE. Despite Augustus' personal example of temperance and reform efforts, drinking continues to increase in Rome, and excessive inebriety is frequently found within Augustus' family and successors. His daughter Julia is exiled for her reveling and drinking bouts and is forbidden to drink wine during her exile (Suetonius, Augustus 65.3). The populace complains about the scarcity and high price of wine (Suetonius, Augustus 42.1). Contemporary writers such as Livy, Horace, and Virgil continue pessimistic laments that Rome is destined to dissipate itself in incurable social and moral corruption and that the old order is gradually breaking down. Although luxury, avarice, and ambition are the three sins most frequently criticized, excessive drinking is also a component of this new morality (Ferrero 1909:3-15). The writings of Horace (see below) especially indicate drunkenness is common. According to Macrobius (see 375/Rome) citizens are drunk in public assemblies, and even magistrates are intoxicated at the forum (McKinlay 1950b:32; Rolleston 1927:110).

DRINKING PRACTICES (HORACE; OVID). Horace describes Romans passing entire days and nights drinking, brawling over wine, easing one's troubles with wine, drinking strong wine, making plentiful provisions for an evening's drinking, attending elaborate mid-day drinking bouts and lavish banquets in which guests pride themselves on their consumption, even on an empty stomach, and ignore the rules designed to control drinking (McKinlay 1950b:31).

The poet Ovid (43 BC-18 AD) also provides a picture of Roman drinking habits in the late republic and the early empire under Augustus. He cautions against indulging in wine too much at banquets (*Ars amatoria* 1.568), identifies wine and dice as the usual methods of relaxation for Romans (*Heroides* 14.33), and associates wine with the pursuit of love. Frequently he portrays people in an advanced state of intoxication. Drinking among the common people concerns him far less than the increase in indulgency, greed, and drunkenness among the wealthy (McKinlay 1950b:31-32).

The elegiac poet Tibullus (69-19 BC) presents a similar picture. He contrasts the drinking of his day to the sober days of the past and urges Bacchus to abandon his "accursed vats" (McKinlay 1950b:31).

TAVERNS. Horace is the first Roman to distinguish taverns (*diversoria*), which are primarily drinking houses and only secondarily lodges, from inns (*tabernae*), whose primary purpose is lodging. This marks the emergence of the modern tavern as distinct from existing wine and beer shops (Dinitz 1951:53-54).

PRAISE OF WINE (HORACE). The wine-loving nature of the Augustan era is reflected in the glorification of wine throughout the writings of Horace (65-8 BC), particularly in his *Odes*. He counsels men to adjust to an absurd world with frequent recourse to wine (Africa 1965:390). Horace has been accused of being a "mouthpiece" for the Roman wine interests (Ferrero 1909); there is no evidence to substantiate this, although his glorification of wine "may have influenced people in their habits" (Jellinek 1976a:1727). To wine he writes, "You restore hope and spirit to anxious minds, and give horns to the poor man, who after [tasting] you neither dreads the diadems of enraged monarchs, nor the weapons of the soldiers" (*Ode* 3.21). He considers wine to be among those things which, "if denied, would cause pain to human nature" (*Satires* 1.1), He seems to think that drinking and writing poetry go together, that Bacchus is the patron of poets and that wine provides inspiration and eloquence (*Epistle* 1.5.18-19, *Ode* 3.25.9). Yet, as McKinlay (1946/47) emphasizes, Horace also recognizes the dangers of too much wine drinking. He includes *vinosus* in a list of vices (*Epistles* 1.1.38). He writes whole poems on the theme of moderation and contentment, advising that there is a limit beyond which one should not go. He recommends wine be diluted and that a pint is a suitable ration; he is himself a serious and hard worker who draws a comparison between a poet and an athelete. McKinlay suggests that much of the praise of wine in his poetry is simply the following of Greek poetic conventions and that much of it is meant to be satiric and humorous. He concludes that to think of Horace "as a wine poet merely because he wrote odes with a Bacchic background and made himself out to be a devotee of the wine god, is, to say the least, a bit dogmatic" (p. 234).

VITICULTURE: ECONOMIC VALUE. The fortunes of Italy have come to depend on vineyards; this, in turn, greatly increases the desire of Romans for the peace and order offered by Augustus. Whereas destroyed grain fields can be renewed in a year, vineyards represent immense accumulated capital and require years to rebuild. Furthermore, as viticulture increases in economic importance to Italy, it becomes a major factor in Roman policy and concerted efforts are made to expand wine sales. By the second century AD, Roman wines are being exported to India and supplant the national drinks of the Gauls and Getae. Ferrero (1909:189) even maintains that the growth of viticulture is one of the most important foundation of imperial authority and makes possible the Principate; however, Jellinek (1976a:1727) observes that Ferrero's attempt to make viticulture the foundation of imperial authority in Italy is unconvincing.

1 AD

CHRISTIANITY

DRINKING ATTITUDES (GOSPELS; ST. PAUL). References to drinking in the earliest New Testament writings after the death of Jesus (c. 30) are surprisingly meager, and it has been suggested that because drunkenness was an upper-class vice, Christ may not have come into contact with it too frequently (Raymond 1927:79). It has also been suggested that the New Testament shows less animosity to wine than the Old Testament because of the Greek influences on it (Seltman 1957:20). Wine itself is spoken of without disfavor throughout the New Testament. Christ uses wine ("The Son of Man came eating and drinking" [Matthew 11:19; Luke 7:33-35]), sanctions its use by others, and sees nothing intrinsically evil in it as long as it is properly consumed ("Not that which goeth into the mouth defileth a man; but that which cometh out of the mouth" [Matthew 15:11]). Christ's first miracle is to change water into wine at the marriage in Cana (John 2:1-11). In the eucharistic ceremony during the Last Supper, Christ establishes wine as the symbol of "my blood of the new testament, which is used for many for the remission of sins" (Matthew 26:27-29; Mark 14:23-25; Luke 22:20). Hyams (1965:97-98) believes that "perhaps the most extraordinary instance of this special feeling for the vine in the Hellenistic-Roman culture is to be found in the claim made by Jesus (in which he shows the measure of his Hellenism) that he was the True Vine" (John 15:1-2: "I am the true vine, and my Father is the husbandman. Every branch in me that beareth not fruit he taketh away: and every branch that beareth fruit, he purgeth it, that it may bring forth more fruit." The passage also shows an appreciation of pruning in viticulture).

The general attitude of the Gospels toward drinking is one of moderation. John the Baptist is praised for his abstinence (Luke 1:15; Matthew 3:4; Mark 1:6), and Christ's whole life is a model of temperance and of the need to subordinate all activities to divine ends. In the few cases where he discusses drunkenness, he attacks it severely: "And take heed to yourselves lest at any time your hearts be overcharged with surfeiting, and drunkenness, and cares of life, and so that day [of judgment] come upon you unawares" (Luke 21:34; see also Matthew 24:45-51; Luke 12:42-48).

The later writings of St. Paul (d. 64?) deal with drinking and drunkenness in considerable detail and form the major body of writings upon which Christian doctrine on drinking is subsequently based. Paul regards wine as intrinsically good because it is a creation of God (1 Timothy 4:4), and especially recommends its medicinal use as in his famous admonition to Timothy: "Drink no longer water, but use a little wine for thy stomach's sake and thine other infirmities" (1 Timothy 5:23; see Lucia 1963a:68). But he always denounces drunkenness; it produces consequences which bar the gates of heaven and desecrate the body as the temple of the Holy Ghost (1 Corinthians 6:10, 3:16-17; Galatians 5:19-21; Romans 13:3). A Christian should not even associate with a drunkard (1 Corinthians 5:11). Furthermore, he believes that just because there is nothing intrinsically evil about wine drinking does not mean that it is expedient to consume it. He recommends total abstinence as an example to those unable to control themselves. In a passage critical to Christian temperance doctrine, he admonishes all those who seek the kingdom of God, "It is good neither to eat flesh, nor to drink wine, nor anything whereby thy brother stumbleth, or is offended or is made weak" (Romans 14:21; see also 1 Timothy 3:2-3, 8). Nevertheless, in regard to drinking and drunkenness, Paul is far more interested in positive measures than negative prohibitions. He reproaches those who condemn drinking or enjoying food in moderation (Romans 14:2-4), and those who emphasize asceticism as an end in itself. His whole ethical outlook emphasizes a sense of proportion rather than an undue emphasis on suppression (Raymond 1927:87). Bainton (1945:48) suggests that Paul is more explicit in condemning drunkenness because he was confronted with actual Christian drunkenness in the celebration of the Lord's Supper.

JUDAISM

DRINKING ATTITUDES (PHILO). In the first decades of the 1st century AD, the Jewish philospher Philo of Alexandria writes three treatises on the drunkenness of Noah in which he discusses wine drinking: De plantatione (On Husbandry), De sobrietate (On Temperance), and De ebrietate (On Drunkenness). His writings reflect the philosophic blending of Hellenism, mysticism, and asceticism with traditional Judaism and reveal a great interest in wine and the problem of drunkenness. He emphasizes that man has an insatiable desire for wine: "For the passion for wine is extraordinarily strong in mankind; and is unique in this, that it does not produce satiety. [P]racticed topers...drink but do not slake their thirst and, while they begin with smaller cups, as they advance they call for the wine to be poured in larger goblets" (De ebrietate 53.220-221). But wine drinking itself is not evil. Wine is a symbol of many things, of good as well as evil. Sobriety is the cause of good things; drunkenness, of evil things. The life of virtue is superior to the life of pleasure because it produces a higher type of character. Temperance is an instrument by which a virtuous life is maintained and the foundation upon which the soul rests. In De sobrietate (1.1-5), consistent with his emphasis on knowledge, he attacks drunkenness as producing madness and praises soberness as being conducive to clarity of understanding (Raymond 1927:32-35; McKinlay 1948b:403). He often contrasts dangerous physical drunkenness from wine with the spiritual experience, which he calls "sober drunkenness" (sobria ebrietas). Concerning the latter, he writes (De ebrietate 36.147): "For with the God-possessed not only is the soul wont to be stirred and goaded as it were into ecstasy but the body also is flushed and fiery warmed by the overflowing joy within which passes on the sensations to the outer man, and thus many of the foolish are deceived and suppose that the sober are drunk" (see Goodenough 1956:202-203). He denounces the "modern" custom of sottish drinking in contrast with the men of old who drank wine rarely, in small quantities, and as part of a ritual (De plantatione 39.160-164).

RITUAL USE. Following the rise of Christianity, the rabbis seek to preserve traditional Jewish thought and practices of pre-Christian times, a movement that culminates in the detailed regulations concerning the use of wine that are written into the Talmud. The use of wine is integrated into many Jewish ceremonies so that wine drinking ceases to be merely an indulgence of the appetite and becomes a religious rite preceded by a prayer and limited to a single cup (Ginzberg 1923:402; Raymond 1927:45-47).

ROME

UPPER CLASSES/DRINKING. Drunkenness reaches a zenith in Rome during the first half of the first century AD (Jellinek 1976a:1736-1739). Not only excessive drinkers but true alcoholics are frequent among the upper and upper-middle classes, particularly the wealthy freedmen (former slaves) such as Petronius' Trimalchio (see 50/Rome). While drinking undiluted wine is still taken as a sign of excess, meals are followed by many hours during which drinking is the central activity. This custom lasts throughout the empire, but seems to become less common in later centuries. Most of the members of the imperial family indulge in alcoholic excesses.

Pliny (14.28.144) asserts that among the wealthy during the reign of Tiberius (14-37), "the fashion set in of drinking on an empty stomach and preceding meals with a draught of wine--yet another result of foreign methods and of the doctors' policy of perpetually advertising themselves by some novelty."

EMPERORS/DRINKING. Although apparently quite abstemious in his youth, as he ages Emperor Tiberius (14-37) becomes an excessive drinker and a lonely and despondent old man (Suetonius, Tiberius 42). As a young officer, he is nicknamed by the army Biberius Caldius Mero, "the tippler who drinks hot pure wine" (his family name is Tiberius Claudius Nero). Moody and depressive, he may have been an alcoholic "in the modern sense of the word" (Jellinek 1976a:1738).

LOWER CLASSES/DRINKING. Most of the evidence for drunkenness concerns the wealthy, for whom Roman authors write almost exclusively (Jellinek 1976a:1735). Very little is known about drinking among the lower classes and the large population of poor people in Rome. Wine has become the everyday drink of the general population (see, for example, Horace, Satires 1.5.16, 2.7.40-41). Petronius (Satyricon, see 50/Rome) describes drunkenness in the slums of Rome; Dio Cassius (62.15-16) tells of the Roman populace carousing in taverns and scandalous revelries under Nero. Seneca (Epistles 18.4) characterizes the masses as a drunken and vomiting mob on holiday. But the poor do not drink as heavily and frequently as the upper class. They are given only to occasional excesses at official festivals, victory triumphs, and other celebrations, as described by Ovid (Fasti 3.523-542; see 25 BC/Rome), Horace (see 25 BC/Rome), and Martial (12.76; see 75/Rome). During these occasions, wine is frequently supplied to the population. Most of the evidence for drunkenness deals with urban Roman life, but rural drinking, especially on festal days, is noted by Horace (Odes 4.5.31-32, 1.1.22-24) and Tibullus (1.10.51, 1.7.39-40, 2.5.87-90) (McKinlay 1950b:32).

WINE DILUTION, AVAILABILITY. The frequency and level of drinking among the poor is moderated by considerations of price, availability, and the weakness of common wine, which is highly diluted and of low quality (Jellinek 1976a:1732-1733). Called vinum acetum or posca, it is almost vinegar before dilution (Celsus, 1:498). Although the wine of the poor is very diluted, its prices fall throughout this period, and around this time the government begins to accept the wine tax in kind and distributes the surplus wine to the people (Hyams 1965:106).

CONTROL EFFORTS. Faced with demands for controls over gluttony and intemperance, Tiberius writes a letter expressing the need for control and "retrenching excess to the ancient standard," but acknowledging his inability to resolve the problem (Tacitus, Annals 3.52-4; McKinlay 1950b:32).

TAVERNS: CHARACTERISTICS, PATRONS, CONTROLS. Roman tavern life in the early empire differs little from that of the classical Greeks, except that the taverns are more numerous and diversified (Popham 1978:242; see also Firebaugh 1972:119, 130, 193, 121; Carcopino 1940). As elsewhere in the ancient world, gambling and prostitution are common features. Possibly because of his unsavory reputation among the upper classes, the tavern keeper is denied the full rights of citizenship. The patrons of taverns are drawn primarily from the lower strata; however, the discrete, subterranean tavern called the ganea is often secretly frequented by the wealthy (Firebaugh 1972:133). Formal control efforts are few, "although somewhat more common" than in Greece (Popham 1978:245).

Popham sees the concerns over drinking in this period as fitting the pattern which previously occurred in New Kingdom Egypt and fifth-century BC Greece: a shift from traditional ceremonial drinking (largely resticted to the banquets of the wealthy and special occasions) to nonceremonial urban drinking in taverns by the lower classes. He believes that the concerns of the Romans in the early empire were not over drinking per se but over chronic drunkenness, "which in their view was associated with drinking apart from the traditionally approved occasions" (Popham 1978:247).

MEDICINAL USE. Roman medical writers recommend the use of wine and other drugs more extensively than did Greek physicians. The De medicina (On Medicine) of Celsus (25 BC-37 AD), the oldest extant medico-therapeutic document in Roman literature, is "a veritable textbook of the therapeutic uses of wine" (Lucia 1963a:52). He describes the effects and medicinal uses of various wines and recommends wine especially for treatment of fevers (Celsus 1.8, 2.18.11, 2.28, 2.30, 3.13, 4.12).

DISTILLATION. Although Alexandrian chemists discover the general principles of distillation, they do not apply them to the preparation of alcoholic spirits (Forbes 1970:6).

INEBRIETY PREVALENCE. In the mid-century, drunkenness reaches its highest level according to the contemporary writings of Seneca, Petronius, and Pliny (see below). Under Nero's rule (54-68), the Roman population is described as drinking and sporting in scandalous fashion (Dio 62.6.4, 62.15-16; see also McKinlay 1950b:33).

In the Satyricon, Petronius (d. ?66), an intimate of Nero, bewails the greed and drunkenness of his time, particularly among the newly rich freedmen and their followers, as satirized at Trimalchio's famous bibulous banquet. Trying to imitate the aristocracy, Trimalchio serves the most expensive wines available, urges his company to drink until dawn (73), and himself becomes "deep in the most vile drunkenness with his guests following suit" (79). One character laments: "Love of money began this revolution. In former ages virtue was still loved for her own sake, the noble arts flourished....But we are besotted with wine and whores, and cannot rise to understand even the arts that are developed; we slander the past, and learn and teach nothing but vices" (88.1). As noted, Petronius also describes drunkenness among Rome's tenement dwellers (93, 95-96).

VITICULTURE: ECONOMIC VALUE. From the 1st to mid-2nd century, viticulture expands and wine plays a leading role in Italy's economy, contributing appreciably to its great wealth. Columella and Pliny indicate that Italy is now covered with vineyards, even in unsuitable areas. In 65, Columella (3.2.30-31) says that it is impossible to master the long list of names of vines. Unlike Cato the Elder, who emphasized production of wine in quantity as an investment, Columella (3.2.7, 12.19.2) has much to say about the quality and value of fine wine (Allen 1961:130). Pliny the Elder (see below) speaks of the supremacy of Italian wines; of the world's 80 notable wines, two-thirds are Italian (Pliny 14.13.87, 14.5.47, 14.2.8). All told, there are 185 types of wine (Pliny 14.29.150).

However, due to the huge expansion of the wine industry and the growth of viticulture in Gaul and other areas of the empire, overproduction begins to occur and prices decline. Facing foreign imports and declining prices, Italian vintners attempt to maintain profits by increasing their volume of production, which only worsens the economic crisis. Thus, one of the first issues Columella (3.3.2) raises is whether a vineyard can still be profitable, a marked change from Cato's day.

WINE PRICES. In 100 BC, the ratio of the price of wheat to wine was 1 to 2.72; in 65, it is 1 to .72, a 350% decline during a time of continuous inflation. Rolleston (1927:40) also estimates that in the early empire six gallons of wine cost only 15 sesterces, the equivalent of eightpence. In Petronius' Satyricon (76), Trimalchio boasts of the indifference with which he once endured the loss of five shiploads of wine worth its weight in gold, the implication being that wine is now much cheaper (Hyams 1965:105, 138).

EMPERORS/DRINKING. The last three Julio-Claudian emperors are all known for their excessive drinking.

The violent and drunken escapades of Emperor Caligula (37-41) eventually lead to his assassination while drinking and feasting (Dio 59.29.5-7, 59.17.1, 9-10; Suetonius, Caligula 4.5.3, 32, 36, 37, 45.2, 55.2).

In his youth, Claudius (41-54) was frequently inebriated; as emperor rarely leaves the dinner room until drunk and is ready to drink at all times and places (Suetonius, Claudius 5.1, 33.1, 40.1; Dio 60.2.6). (On his control efforts, see below.)

Emperor Nero's (54-68) alcohol-related escapades last all day and night. He haunts the taverns of Rome, carouses in the streets, and stages extravagant revels (Suetonius, Nero 27.1.3, 26.1-2, 31.2, 42.2, 43.2, 47.1; Tacitus, Annals 23.20.1-5, 14.13, 13.15-16, 13.20).

TAVERN CONTROLS. Emperor Claudius tries to close drinking taverns altogether and to forbid the sale of hot wine (Dio 60.6.7; see also Firebaugh 1972:178; Popham 1978:247). This seems to have been less a measure to combat intemperance in general.

FESTIVE DRINKING. The emperors supply Romans "with more festivals than any people, in any country, at any time, has ever seen." So many holidays devoted to rest and leisure have now been introduced that most of the year is given up to festivals. There are 159 holidays of which 93 are devoted to games at public expense (Carcopino 1940:203, 205). Claudius seeks to abolish many holidays in 43 as "the greater part of the year was being given up to them with no small detriment to the public business." However, he also extends to five days the period of the Saturnalia festival (at which all social restraints are suspended) to appease the populous (Dio 60.17.1, 60.25.8). Seneca (18.1-6) criticizes the pervasiveness of holidays and the holiday spirit in Rome. Rome seems to go mad with merrymaking during the Saturnalia, which now seems to last all year. He quotes as correct the remark that "once December was a month; now it is a year."

INEBRIETY CONDEMNED (SENECA; PLINY). Seneca (3 BC-65 AD), the Stoic philosopher and tutor to the young Nero, describes widespread and frequent drunkenness among both men and women in numerous passages in his Epistolae morales (Moral Letters). Although he has some words of praise for wine, he condemns drunkenness severely. In a letter on drunkenness, he praises the abstemiousness of Zeno (epistle 83.9), alludes to the existence of physical tolerance to alcohol ("the drunkard is often free from drunkenness," 83.11), attacks the drunkenness of several prominent Romans (83.12-15), and refers to drunkenness as a kind of voluntary madness (voluntaria insania) that "kindles and discloses every kind of vice" (83.18-19). After describing the many adverse physical effects of drinking, he requests his reader to "think of the calamities caused by drunkenness," and asks, "What glory is there in carrying much liquor?" (83.22, 24). He particularly attacks Marc Antony's drunkenness (83.25). He scolds the people for wasting the blessing of peace on drunkenness and pleasure making, and praises the courage to remain sober when the mob is "drunk and vomiting" (18.1-4). (On the luxurious and drunken life of the rich, see epistle 51.1-6.)

Pliny the Elder (23?-79) discusses drinking practices and wine at length in his Natural History, especially in the 14th and 23rd books. An adherent of Stoicism, he attacks the luxury and corruption of his age, devotes a whole chapter (14.28) to the evils of drunkenness, describes the many varieties of wine (14.6-27), and advocates its therapeutic use in moderation (23.19-26, see below under Medicinal Use). In no aspect of life, he asserts, is more labor spent than on wine: "So much time and labour and outlay is paid as the price of a thing that perverts men's minds and produces madness, having caused the commission of a thousand crimes" (14.28.137-138); "in no part of the world is drunkenness ever out of action" (14.29.149). (See also below for Pliny's and Seneca's descriptions of the physical effects of chronic inebriety.)

WOMEN/DRINKING. Seneca (95.21) criticizes women for imitating men: "They keep just as late hours, and drink just as much liquor: they challenge men to wrestling and carousing, they are no less given to vomiting from distended stomachs and to thus discharging all their wine again."

PHYSICAL EFFECTS (PLINY; SENECA). Most descriptions of inebriety in classical antiquity refer to acute alcoholic intoxication, but the effects of chronic indulgence or alcoholism are also observed, particulary by Pliny and Seneca (Rolleston 1927:111; Hirsh 1953:965-966; McKinlay 1950b:33). Seneca (Moral letters 95.16) writes a detailed description of chronic alcoholism: "The results are pallor, quivering of the muscles soaked in wine, and an emaciation due to indigestion and not to hunger. Hence the uncertain and tottering gait, and constant stumbling as if they were actually drunk; hence the swelling of the skin and distension of the belly, which has taken more than it can hold;

hence the jaundiced and discoloured complexion...and the nerves dulled and without feeling, or, on the other hand, constantly twitching. Why need I speak of the giddiness or the disturbance of vision and hearing and the insidious pains in the head?" A later passage in the same letter refers to a class of fevers accompanied by a feeling of horror and much shaking of the limbs, a description resembling delirium tremens

Pliny (14.28.139-142) describes many features indicative of alcoholism. These include: (1) not being able to wait for a meal in order to drink; (2) never seeing the rising sun; (3) shaky hands; and (4) oblivion of all things. In the most quoted passsage, Pliny (14.28.142) emphasizes the following effects: "Pallor, pendulous cheeks, bloodshot eyes, tremulous hands which spill the full cup, and as an ever-present penalty, sleep disturbed by the furies, restlessness at night, and lastly monstrous passions and even crimes which in their eyes have become the supreme delight. The next day wine infects their breath, and their memory is dead. This is what they call seizing life, whereas each day they lose both that day and the next." The "inevitable result...is that the habit of drinking increases the appetite for it" (14.28.148).

DRINKING INFLUENCES. Pliny (Natural History) describes several customs which clearly fostered heavy drinking (Rolleston 1927:111). These include: (1) wine-drinking contests; (2) drinking as many cups as demanded by a throw of dice; (3) the "outlandish fashion" of drinking on an empty stomach before meals; (4) in toasts, drinking as many measures as the toastee has letters in his name; and (5) the use of a feather to induce vomiting in order to drink or eat more; and (6) the use of raw onions or powdered pumice stones to stimulate the appetite for drink (Rolleston 1927:109). The practice of drinking before meals, he asserts, was introduced during the reign of Emperor Tiberius (Pliny 14.28.143). He also provides numerous remedies for prevention or cure of inebriety (McKinlay 1950a:231).

MEDICINAL USE, WINE. While earlier Greek physicians were divided into those who used wine therapeutically and those who did not, in Rome wine becomes the universal medicine (Lucia 1963a:8). Physicians commonly recommend the use of wine in moderation to their patients; no elaborate system of drugs exists beyond compounded medicines and tonics, and the science of chemistry is elementary (Seltman 1957:154). On the therapeutic value of wine, Pliny (14.7.58) states: "We are justified in saying that there is nothing else that is more useful for strengthening the body, and also nothing more detrimental to our pleasures if moderation be lacking." There are also circumstances in which wine should not be consumed (Pliny 23.19, 15.17, 14.19, 28.14, 23.24).

Dioscorides, a Greek army surgeon in the service of Nero, writes (c. 78) a compendium of medicinal wines in De universa medicina. Considered the founder of materia medica because he is the first to write on medical botany as an applied science, Dioscorides (1934:602-604) advises the use of various different wines for countless ailments. He sees little therapeutic value in beer (Lucia 1963a:61-62; Lutz 1922:75).

WINE DILUTION. Pliny writes that the quantity of water added to wine "depends entirely upon the strength of the wine; it is generally thought, however, that the best proportions are one cyathus of wine and two of water. If, however, there is a derangement of the stomach, and if the food does not pass downward, the wine must be given in a larger proportion." He also advises that water be consumed every now and then when drinking wine and that a long draught of water be taken at the end, "cold water taken internally having the effect of instantaneouly dispelling inebriation" (quoted Lucia 1963a:58).

75

ROME

INEBRIETY PREVALENCE; DRINKING PRACTICES. Heavy drinking continues, particularly under the short reign of Emperor Vitellius (69), who was "at midday...tipsy and gorged with food" and allowed drunkenness to prevail in the army (Tacitus, Histories 1.56, 1.62, 2.2, 2.68, 2.76, 3.76-66, 3.83, 4.36; see also Dio 64.3.1; Suetonius, Vitellius 13.1, 17.2). The writings of Martial (see below) and later Juvenal (see 100/Rome) are replete with references to drunkenness. In his Epigrams, Martial emphasizes the custom of drinking as many cups as letters in a name, and tells of wine tokens being issued to theater goers and of those who spend all night drinking (Martial, 1.11, 1.26, 1.71, 10.47, 11.36; (Jellinek 1976a:1736-37). Drinking still plays an important role in festivals. Emperor Domitian seeks to please the populace of the city of Rome by giving them copious supplies of wine (Dio 61.4.4). However, after the death of Nero in 68, a more temperate era begins; the overall level of drunkenness appears to decline among the upper class and the imperial court.

EMPERORS/DRINKING (FLAVIANS). The Flavian emperors (70-98), and most of the 2nd-century emperors, are generally much more moderate in their drinking than their Julio-Claudian predecessors (McKinlay 1950b:34; Seltman 1957:171). Tacitus (Annals 3.55) gives Vespasian (70-79), the first Flavian, much of the credit for the decline of "spendthrift epicureanism." Emperor Domitian (81-96) is said to be abstemious (Suetonius, Domitian 21).

INEBRIETY CONDEMNED (MARTIAL). The bitter Epigrams of Martial (40-104) criticize widespread drunkenness and its adverse consequences in the late 1st century (see McKinlay 1950b:34 for a detailed analysis). A "mirror of Roman rationality" (Roueché 1960:20), Martial (10.47) includes among the traits of a sensible life "not drunken nights, yet loosed from care." He denies that nights spent over wine make for a happy life. But Martial is not averse to drinking per se; he "delights in taverns and avows it without the least restraint" (Firebaugh 1972:184).

VITICULTURAL CONTROLS. A serious economic crisis occurs in Italy as the price of wine continues to plummet due to the spread of viticulture in the western provinces and the loss of large markets for Italian exports. Concerns are also expressed that too much good land is being unnecessarily devoted to the production of grapes rather than grain (grapes do not require as high a quality of soil as does grain). Around 70 AD, a restoration of Italian viticulture begins as vine acreage increases, improved viticulture techniques lower production costs, the quality of wine improves, and profits rise. The restoration culminates around 92 with an edict of Emperor Domitian (81-96) which attempts to stabilize the industry. The emperor bans new viticulture in Italy and orders that vines be cut down entirely or reduced by half in the provinces. The edict is not repealed until 280 under Emperor Probus (276-282) at which time it is labeled unforceable. The edict is most unpopular, and

Suetonius (Domitian 7) asserts that the emperor "did not persist in the execution of this matter." Hyams argues that Suetonius was "certainly exaggerating or over-simplifying when he said that the Emperor altogether gave up the idea of trying to enforce his wine law." It does seem to have been enforced in some parts of Gaul and Spain. It may be that half the vineyards are destroyed, but only those that are the least important or least profitable, and the edict does restrict the planting of new vines, as viticulture expands noticeably in the provinces following the repeal of the edict by Probus. Furthermore, the edict does achieve the goal of driving up prices for wines and preserving their quality (Hyams 1965:122, 140).

100

GERMANIA

DRINKING PRACTICES. Tacitus (55?-117?) (Germania 22, 23) describes the Germans living simply but drinking heavily, and suggests that a good way to conquer them is first to give them all they want to drink. He observes that the Germans drink a liquor (beer) made from barley or wheat which resembles wine, that they exercise no self-control in drinking, that their drinking bouts often last all day and night, and that quarrels inevitably result and these bouts frequently end in bloodshed. Beer is also consumed during deliberations to encourage frankness.

Jellinek (1945:649) believes that Tacitus exaggerates Germanic drunkenness, but McKinlay (1948b:410-411) finds support for his portrayal in other classical authors. (See below for Tacitus' comments on Roman drinking.)

ROME

BACKGROUND DEVELOPMENTS: "GOLDEN AGE" BEGINS. From the death of Domitian (81-96) through the reign of Marcus Aurelius (161-180), a series of mature, able men of proven ability rule the empire. Dill (1958:116) characterizes this "golden age" as "a period of almost unexampled peace and prosperity, a period of upright and beneficent administration and high public virtue."

INEBRIETY PREVALENCE. The biographers of the emperors of the "golden age" all emphasize their temperance and conscientiousness. The literary sources of the 2nd century are generally silent regarding drinking, indicating that the problem has decreased considerably. Some of this decline may have been related to the religious revival that gathered force throughout the 2nd century, including the spread of Christianity. The sober standards of the court may have done much to moderate the behavior of the upper classes, among whom the previous drinking epidemic seems to have been most pronounced. Nevertheless, early 2nd-century authors such as Juvenal, Tacitcus, and Plutarch still portray drunkenness as a prominent feature of Roman life.

(JUVENAL). Juvenal (55-138) is one of the few 2nd-century writers to indicate that high levels of drinking continued. His bitter Satires attack the pretenses, follies, and vices of Roman society under Domitian and his successors. He describes nightly drinking bouts ending in fights, a drunken Venus, tipplers stopping their friends to tell drunken stories, afternoon drinking by public officials, a woman drinking three gallons of wine while alternately eating and vomiting, other women drinking at sacrifices, and "an officer of the guard [who] frequented all-night taverns and hot liquor shops in a medley of barmaids, drunken guests, and scum of the earth in general" (McKinlay 1950b:34-35, citing Juvenal 1.49,57, 2.87, 5.24-29, 6.300-305, 6.315-320, 427-432, 8.158-162, 8.172-178). Although he attacks the drinking and degeneracy of the Romans (including their preoccupation with free "bread and circuses"), it is generally maintained that, like most satirists, Juvenal was exaggerating his point to stir contemporaries and that his writings cannot be accepted at face value (McKinlay 1950b; Africa 1969:413).

(TACITUS). Tacitus writes bitter histories in which he portrays the 1st-century Principate in dark colors and expresses his belief that cruelty is inherent in the aristocracy (Africa 1969:412). He contrasts the drinking of the early empire through Nero with the sobriety of the republic (Tacitus, Annals 3.53-55, 13.15-16, 13.20, 14.15). He attacks the drunkenness of contemporary Roman soldiers who spend long hours in taverns and carouse all day and night (Tacitus, Histories 1.80, 2.21, 2.68, 2.76, 3.76-77, 3.83, 4.36).

(PLUTARCH). In Sulla (1), Plutarch criticizes the rise of drunkenness in Rome and the abandonment of its earlier sobriety. (On Plutarch's drinking attitudes, see also below.)

TEMPERANCE ADVOCATED (PLUTARCH). In his Parallel Lives, Plutarch (46?-120) writes moral sermons based on the lives of famous men. He praises the temperance and criticizes the intemperance of his famous subjects. Plutarch himself writes that he had never been sorry for having drunk water instead of wine (McKinlay 1951:64). He laments that fathers must see their children drunk and observes that denying men wine has become an accepted cure for chronic drinking (McKinlay 1950b:35).

DRINKER LIABILITY. The Roman senator and legalist Ulpian writes that inebriates are insane and should be considered irresponsible, with the state providing places for their detention and treatment (Crothers 1911:140).

EMPERORS/DRINKING (SUETONIUS). As the secretary to Emperor Hadrian (117-138), Suetonius (67?-140?) writes gossipy tracts about the lives of Roman rulers from Caesar to Domitian, including their drinking habits of which he expresses his disapproval (Tiberius 42, 52; Caligula 4, 32, 36, 37, 45, 55; Claudius 5, 32, 33, 40; Nero 26, 27, 31, 42, 43, 47; Otho 2; Vitellius 13,17; Titus 7; Domitian 21). (His observations of Augustus and the Julio-Claudian emperors are noted in the entries for each.)

175

CHRISTIANITY

ATTITUDES: WINE DRINKING (CLEMENT OF ALEXANDRIA). Beginning in the late 2nd century, several heretical Christian sects place emphasis on abstinence. Gnostic dualists (Encratites, Manicheans) reject all matter as evil and consider wine inherently evil by nature. In Paedagogus (The Instructor), the Christian theologian Clement of Alexanderia (d. 215?) discusses the applications of Christian principles to daily life. Against the Encratites, he emphasizes that it is not the nature of wine that is important but the nature of its use (Raymond 1927:97-98). Moderate wine drinking serves many needs and functions. However, he "inveighs against all excesses and indelicacies in eating and drinking and especially upbraids drunkenness, while recognizing that a moderate use of wine rejoices the heart" (Bainton 1945:48). Wine is "a joy of the soul and heart...created from the beginning when drunk in moderate sufficiency," but drunkenness is shameful and disgusting, and voluntary abstinence is praiseworthy (Clement of Alexandria 2.2.23.3, 2.2.26.1-3, 2.2.28.1-3, quoted Raymond 1927:103).

ROME

MEDICINAL USE (GALEN). The Greek physician Galen of Pergamum (130-193), for centuries the leading medical authority in the West, devotes considerable attention to the benefits and harms derived from wine. His considerable praise of mithridatium, as well as other theriacs and antidotes, lays the foundation for their popularity to the 18th century. He advocates compounding medicaments for internal use with pure, strong, unadulterated wines only (Lucia 1963a:68-72).

ADVERSE EFFECTS (GALEN). Galen is acutely aware of the adverse effects of intemperate drinking. He writes that boys should abstain from wine as long as possible since it moistens and heats the body to excess. He quotes approvingly the poet Theognis (6th-century BC) that "To drink much wine is an evil; but if one drinks in moderation, wine is not an evil, but a blessing" (Rolleston 1927:107). He also discusses the similarities between alcoholic intoxication and other morbid mental states: "The first symptoms of many diseases correspond with those which sometimes characterise drunkenness. When people are comatose they stagger and are heavy in the head, and even when force is applied they are unable to raise their eyelids; and owing to this heaviness they cannot sleep, being restless, turning from side to side and throwing themselves about and changing from one posture to another. We often see these things happen in drunken people when their head is full of drink: the heaviness [in the head] leads to a coma and at the same time prevents sleep. In this way then, when there is an undigested fullness in the head, people are sleepless and comatose at the same time. And when this happens at the beginning [of a disease], sometimes the fullness is well digested, just as if this had happened through wine, and so neither lethargy nor phrenitis results: but sometimes it becomes more acute, and it ends in phrenitis" (quoted Leibowitz 1967:84-85).

200

GREECE

DRINKING PRACTICES, ATTITUDES (ATHENAEUS). Around 230, Athenaeus records an enormous number of quotations on drinking from previous works (many now missing) in his Deipnosophists (Dinner-Table Philosophers). He refers to writings on drinking by Hippocrates, Galen, Mnesitheus, Aristotle, Theophrastus, Chamaeleon of Heraclea, and Menander, to name only a few. The greater part of book 10 (especially chapters 423-447) and all of book 11 are devoted to a discussion of drinking, famous drinks, and drinking cups, of which he describes 100. His work reflects the importance of wine drinking to the age as both a necessity of life and a major social concern.

200 / ROME

TEMPERANCE ADVOCATED. Athenaeus (11.781) advocates temperance as a virtue: "It is the mark of a gentleman to be moderate in the use of wine, not drinking too greedily or drinking large draughts without drawing one's breath, after the fashion of the Thracians, but to mingle conversations with his cup as a sort of wholesome medicine." He mentions several cases of abstinence (2.43-45), but it is clear that there was very little advocacy for total abstinence.

WINE DILUTION. Athenaeus devotes considerable attention to the subject of the mixing of wine with water (2.38c-b, 10.426, 10.431e-f, 10.436c, 11.426c, 11.484b, 11.782).

MEDICINAL USE. Regarding the health-giving properties of wine, Athenaeus (1.22e) holds views similar to the 4th-century BC physician Mnesitheus, whom he quotes. He classifies wine with water, milk, and barley-water as among the nourishing liquids (McKinlay 1951:67; McKinlay 1950a:230).

ROME

EMPERORS/DRINKING. Although the Roman empire enters into a century of political chaos after the death of Marcus Aurelius in 180, the imperial biographies in the *Scriptores Historiae Augustae* (or *Historia Augusta*) regularly refer to the temperance of the Roman emperors (McKinlay 1948a:148). The SHA is a series of biographies of the emperors from Hadrian (117-138) to Carinus (283-285) compiled in the 4th century; many of the biographies are forged or sensationalized, but it is still the major source for Roman history in the 2nd and 3rd centuries.

275

ROME

VITICULTURAL CONTROLS ENDED. In 280, Emperor Probus repeals Domitian's edict of 92 restricting viticulture. This rescission has been attributed to the disregard long shown to the edict. However, there are indications that it still had some force and that the repeal was viewed as of major importance. First, the repeal is mentioned by every biographer and every account of Probus' short reign. Second, "a tremendous planting of vineyards all over the Empire" is said to have occurred following Probus' repeal (Hyams 1965:122, 140).

300

CHRISTIANITY

BACKGROUND DEVELOPMENTS: OFFICIAL RECOGNITION. With the Edict of Milan in 313, Constantine (306-337) grants freedom of worship to all creeds and recognizes Christianity as a legal religion in the territories he governs; the edict is extended to the whole empire when he becomes sole emperor in 324.

DRINKING CONCERNS. The reconciliation of Christianity and the empire, coupled with the enthusiasm of many Church leaders for new converts, leads to a relaxation of Christian standards. Some in the Church countenance drinking in commemoration of Christian martyrs because pagans are hesitating to convert if they have to give up festival drinking. Charges of drunken reveling at Christian celebrations soon cause a scandal (Bainton 1945:49).

DRINKER LIABILITY. The Disticha Catonis, the first work designed to instruct young people in manners and morals, warns: "Do not forgive yourself those sins which were committed while drunk, for it is not the fault of the wine but that of the drinker" (quoted Neuer 1970:28).

ROME

PRICE CONTROLS. The practices of a more or less monopolistic guild of wine merchants leads to profiteering and price fixing, causing Emperor Diocletian (285-305) to fix the maximum price for wine and other commodities in 301. At this time, the best vintages cost only four times as much as the worst, but wine prices overall are considerably more than they are in France today (Hyams 1965:121). A quart of Falerian wine sells for as much as two chickens (Africa 1969:435).

350

GAUL

VITICULTURE. Viticulture is established near Paris for the first time (Hyams 1965:158). The Gaulo-Roman poet and politican Decimus Magnus Ausonius of Bordeaux lyrically praises the bountiful wines of the region (Younger 1966:162, citing Ausonius' fifth epistle).

ROME

BEER CONSUMPTION, ATTITUDES. Beer is used only by the poor people, and sport is made of Emperor Valens (364-378) because he drinks it (Ammianus Marcellinus 26.8.2). Emperor Julian (361-363) is quoted as saying: "I know Bacchus the god of wine, for he smells of nectar; but all I know of the god of beer is that he smells of the billy goat" (McKinlay 1948b:391).

375

CHRISTIANITY

DRINKING ATTITUDES. In response to the challenge of the gnostic heresies, the Church fathers of the late 4th and early 5th centuries emphasize that wine is one of God's gifts to mankind for its use and enjoyment and is therefore intrinsically good; to despise wine is a heresy. At the same time, they vigorously condemn its abuse. They uniformly affirm that the use of wine is permitted, but not always expedient, and it must be used for proper ends (e.g., therapeutic) and in a proper manner (moderation). Excessive use is an evil in itself and leads to the sins of gluttony and drunkenenss; the drunkard lives a life of physical distress and spiritual deadness. Thus, all those Christians who easily succumb to wine should abstain; but if such a course is advocated, it is not demanded of all Christians. In order to avoid drunkenness, God has given us the virtue of temperance, but a Christian may also avoid temptation which leads to sin (Raymond 1927:115-116, 133-124).

(ST. BASIL). Among the Eastern fathers, St. Basil (330?-379) asserts in two sermons on drunkenness that when a person abuses God's gift and disregards his purpose by drinking to excess, he is not to be pitied. The physical and moral consequences of drunkenness are appalling, leading to bodily decay and sin. Therefore, one should avoid all but the necessary use of wine, since the acquisition of virtue, the goal of life, requires voluntary abstinence (Raymond 1927:104-108).

(ST. JOHN CHRYSOSTOM). Similar attitudes are expressed by St. John Chrysostom, bishop of Constantinople (345?-407). While the problem is not in wine but in the drinker, he advocates that Christians follow the example of Timothy and avoid all wine except for reasons of health (Raymond 1927:109-115).

(ST. AMBROSE). Among the Western fathers, St. Ambrose (340?-397) emphasizes St. Paul's position that, although wine drinking is lawful, it is not always expedient because the use of wine in common practice is a poison. Abstinence is an admirable way to remove possible sources of temptation (Raymond 1927:117-122).

GAUL

INEBRIETY PREVALENCE. Ammianus Marcellinus (15.12.4) describes the fondness of Gauls for wine and beer and the drunkenness among them that results. Many of the poor people purportedly go crazy from too much drinking (McKinlay 1948b:389).

ROME

INEBRIETY PREVALENCE (AMMIANUS; MACROBIUS). After the political chaos of the 3rd century, most of the 4th-century emperors are serious, competent men who have a reputation for temperance, as is evident in the writings of Ammianus Marcellinus (*History* 12.16, 15, 16.5.1, 21.16.5). A severe pagan moralist, Ammianus is most critical of the luxury, vanity, indifference, and frivolity of life in contemporary Rome between 353 and 378, but conditions still appear to be less bleak than in the days of Nero. He never specifically attacks the Romans for drunkenness (Ammianus 14.6.26, 18.4.29-32, 4.6.7, 28.4).

Macrobius (fl. 399) claims that at least refined pagan Romans in the late 4th century are free from many of the grosser forms of luxury and dissipation which prevailed in earlier centuries. He emphasizes that, unlike in the days of Augustus, men will no longer be seen drunk in the forum, nor will a judge be found on the bench so drunk that he can not keep his eyes open (Macrobius, *Saturnalia* 3.13, 3.16.14, 3.17.12) (Dill 1958:131, 210; McKinlay 1949b:27; Rolleston 1927:110).

400

CHRISTIANITY

DRINKING ATTITUDES (JEROME; AUGUSTINE). St. Jerome (347-420) is less critical of the proper and moderate use of wine than St. Ambrose, but agrees that it is better to shun what might lead to sin than to expose oneself to possible temptation. He is very critical of drunkenness among pagan Romans (see below).
St. Augustine (354-430) especially emphasizes the goodness of wine in opposition to the Manichaeans. If Christians use things to the proper ends in the proper manner, there is no need for abstinence. But, following St. Paul, he advises voluntary abstinence from wine for those who wish to safeguard the soul and help their weaker brethren (Raymond 1927:126-133).

ROME

BACKGROUND DEVELOPMENTS: BARBARIAN INVASIONS. The invasions of the Mongolian Huns from Asia around 170 sets in motion a wave of Germanic invasions of the empire. Pushed westward by the Huns, the Visigoths defeat and kill emperor Valens at Adrianople and take up residence in the empire in 378. In 410, the Visigoths under Alaric march unopposed through Italy and sack Rome.

INEBRIETY PREVALENCE. St. Jerome and Salvianus (a Christian presbyter of Marseilles at the end of the 5th century) criticize late Roman drunkenness and immorality, but on the basis of very ascetic standards, "which might bear hardly on the most respectable societies of modern Christendom" (Dill 1958:142). Despite the criticism of the Christian ascetics, most of the evidence supports a decline in drinking in the empire over previous generations. Apollinarius Sidonius (see 450/Gaul) and Macrobius (see 375/Rome) congratulate their generation for its temperance. Even among the vices that Jerome and Salvianus catalogue, drunkenness is not given any special emphasis, and Jerome attacks corruption among Christians even more vehemently than among pagans.

450

CHRISTIANITY

DRINKING CONTROLS. The Council of Vannes, held by the Archbishop of Tours in 465, calls for the withholding of communion for 30 days and allows for corporal punishment of clergy found guilty of drunkenness, the mother of all sins (Landon 1909, 2:259).

GAUL

INEBRIETY PREVALENCE (SIDONIUS). In the letters of Apollinarius Sidonius (430-479), bishop of Auvergne, the aristocratic life in Roman Gaul in the years immediately preceding the barbarian conquests is shown to be as self-indulgent and frivolous, but with little indication of loose morals or drunkenness.

"Like the guests in the Saturnalia of Macrobius, Sidonius congratulates his generation on being more decent than their ancestors" (Dill 1958:207-211, quotation at 110).

JUDAISM

RITUAL CONTROLS. Amid the unrest of the late Roman and Sassanid (Persian) worlds, concern grows among rabbis that Jewish learning and way of life are in danger. Detailed rules are established in the Talmud regarding the use of wine: the amount of wine permissable on the Sabbath; the manner in which it is to be drunk; the legal status of wine in any way connected with idolatry; degrees of responsibility for actions committed in a drunken condition; and the definition of strong drink (i.e., wine which retains its strength when three quarters of water is added). Abstinence plays little or no part in the rabbinical discussions. As one rabbi is quoted: "If you become holy by abstaining from wine, why not abstain from everything?" (Raymond 1927:45, 48; Dimont 1962:162-175).

ROME

PHYSICAL EFFECTS: DELIRIUM TREMENS. Cassius Felix (De medicina) describes symptoms of delirium tremens resulting from wine intoxication: "Frenesis is a changing of the mind which persists together with a fever. And it is caused mainly through too much burning...or from intoxication from wine. The sick suffer from acute fever,...unsteady movement of the eyes with tension, wakefulness or insomnia,...mental disturbance and crocydismos, that is to say, plucking small flocks from a garment, for the patients appear to be frequently plucking at the bedclothes. Sometimes...they burst into uncontrolled laughter or fall into great sadness and gloom" (quoted Leibowitz 1967:85).

475

ROME

BACKGROUND DEVELOPMENTS: WESTERN EMPIRE FALLS. In 476, the barbarian Odovacer deposes the last Roman emperor of the West and sends the imperial insignia to Constantinople. This date is usually taken as marking the final end of Roman power in the West. In theory, the empire continues to be administered under the single authority of the eastern emperor at Constantinople. But in the West, classical civilization, government authority, social organization, and the amenities of Roman life and culture gradually disappear.

The Middle Ages

continued

500

EUROPE

BACKGROUND DEVELOPMENTS: EARLY MIDDLE AGES. Encompassing a thousand years from 500 to 1500, the Middle Ages can roughly be divided into three periods: early, 500-1000; central, 1000-1350; and late, 1350-1500. For convenience, in this chronology information about broad developments which can not be specifically dated is placed at the beginning of each of these divisions (500, 1000, 1350).

For the most part, the Germanic migrations come to an end around 565, except for the Lombard invasion of Italy. The Anglo-Saxons have gained control of the eastern part of the British Isles; the Franks control Gaul; the Visigoths, Spain. In the early Middle Ages, urban life virtually disappears in the wake of the invasions, and a thinly populated, underdeveloped, and overwhelmingly rural society emerges. Trade declines, markets are weak, and money is rare. Early medieval communities consciously seek a high level of self-sufficiency.

The basic unit of life becomes the large estate or manor held by a lord (*seigneur*) and worked by peasants or serfs who are fixed to the soil and have to devote a certain amount of time to the cultivation of the lord's own lands (*demesne*) in return for protection. The lord also enjoys certain "banal" rights or monopolies, such as owning the manor's only mill or wine press (see 1000/Europe). The nobility, from whom the manorial lords are recruited, consists of great churchmen and warriors. The fundamental occupation of the lay nobility is warfare.

Cities do not entirely disappear, but they have no economic, political, or cultural importance. Little distinguishes city inhabitants from the rest of society. The centers of life are the castle (generally just a wooden stockard), the cathedral, and the monastery. Cities generally serve only as residences for bishops and as fortresses for the defense of the surrounding peasant populations. They lack the two fundamental attributes of the late medieval city: a middle-class population and a communal organization legally distinct from the rest of society.

The high points of early medieval life are reached during the Carolingian Empire under Charlemagne (see 775/Franks) and the Ottonian German Empire of the 10th century. European civilization during this period largely consists of England, France, western Germany, Ireland, northern and central Italy, and northern Spain. Its vital centers are in northern France and the Rhineland. It is not until after the mid-9th century that divisions between west and east Franks began to give rise to the French and German peoples and that the latter begin expanding further eastward.

BEVERAGES CONSUMED. In the early Middle Ages, the rural population eats simply and frugally. Grain is the basic food. The "barbarian" drinks of mead, "rustic" beers, and "wines" based on wild fruits become popular. A sort of beer and cider made from the fruit of wild apple trees are served in Celtic, German, and Gallo-Roman lands. The Anglo-Saxons and early Scandinavians drink mead (fermented honey and water) and metheglin (similar to mead but with herbs) (Gayre 1948). In the Romance countries, people drink water or a light wine made from the dregs of the vintage (i.e., *piquette*) (Boissonnade 1927:100). Religious institutions are for the most part the only repositories of the ancient techniques of brewing and viticulture. Monks have the time, education, and patience to perfect these skills. They also satisfy the large demand for beer and wine by the brethren and monastery serfs, and by the many guests and travelers who stay at the monasteries, and for the wine used at Mass.

BREWING. With the collapse of the Roman Empire and urban life, brewing at the artisan level disappears along with the existing guilds of brewers (*cervesari*) in Gaul. Rustic beers (*cervoise*, ale) made from barley and oats with the addition of various bitter and aromatic herbs continue to be produced by women using traditional methods. The art of brewing, is essentially the work of monks, who jealously guard its secrets. Up to the 12th century, practically no good beer is made outside the monasteries (Claudian 1970:8, 10; Cherrington 1925, 1:402). All early medieval ales sour quickly and are made for immediate local consumption. It is not until around the 13th century that hops, which are used as a preservative, become a common ingredient in some beers of northern Europe. Ale, the commonest beverage throughout northwest Europe, often has the consistency of a thick soup and can perhaps be considered more of a foodstuff than a drink (Duby 1968:9). It is uncertain how much grain is brewed into alcohol, as the first priority for grain is making bread.

VITICULTURE. In the early Middle Ages, most wine is produced and consumed locally, but the wine trade never entirely stops despite difficulties (Jeanselme 1920:266). There is some controversy as to how the viticulture tradition is preserved and extended. Younger (1966:232-234) believes that too much emphasis has been placed on the role of monasticism in preserving the tradition of Roman viticulture and that the role of private enterprise and the memories of lay *vignerons* have been ignored. He observes that many early churches did not have the manpower for viticulture and that they could have received the needed wine through donations. However, the generally held opinion is that, in the words of Seward (1979:15), the "monks largely saved viticulture": "Throughout the Dark Ages they alone had the security and resources to improve the quality of their vines slowly and patiently. For nearly 1,300 years almost all the biggest and the best vineyards were owned and operated by religious houses." He adds: "To say the least, without their contribution viticulture would have taken far longer to develop." Furthermore, because of the privileges given the monasteries in return for their services, monastic wine is not only the best wine available but also the cheapest (Seward 1979:35; see also Dion 1959:171-187).

Younger (1966:234) also maintains that "the Christian vineyards were not solely--and not even principally--intended to provide wine for the Mass." Their main function is not religious but economic: wine is an important element in the finances of religious institutions and helps contribute to their upkeep. Thus, the monks are in control of the wine trade in early medieval Germany.

TAVERN PREVALENCE. The prevalence of, and drinking in, inns and taverns declines as land transportation and commerce decline (Dinitz 1951:55).

DRINKING FUNCTIONS. Throughout the Middle Ages, the drinking of fermented beverages serves multiple purposes and finds its way into almost every aspect of life. In terms of reasons for use or the functions of drinking, the Middle Ages does not differ from classical antiquity. As in antiquity, drink is so important to life that its moderate consumption is praised even by the most vehement critics of drunkenness. First, beer and wine are the primary thirst quenchers. The water supply is generally unhealthful, scanty, and uncertain; under the unsanitary conditions of life, water is the principal carrier of disease. Coffee, tea, and cocoa are yet unknown. Second, they are important nutritional supplements to the often meager diet of the peasant, providing additional calories and vitamins, particularly in the north where fresh fruit and vegetables are not available during most of the year. Regarded with bread as a necessity of life and a chief source of nourishment for the body, they are consumed with every meal, even if in very weak form. Third, they are widely consumed for medicinal purposes, especially wine. Alcohol is the only anesthesia available in a period in which almost everyone lives with some pain. Fourth, they supply a sensation of well-being and relief from the fatigue of agricultural manual labor. Indeed, most of the weak ale and wine consumed in the course of the day is burned off through hard labor. Fifth, the gathering together of people to socialize over food and drink is one of the few means by which tension can be released, and there are few other forms of entertainment or leisure-time activities available. Even Church holy days are celebrated with much conviviality. McCarthy and Douglass (1949:5) observe: "In a time when life was arduous and existence precarious it is not surprising that alcoholic drinks had a place in the daily routine of people in all walks of life" (see also Hackwood 1909:92; Spiller 1955:92). Thus drinking is an important component of all medieval festivals and community gatherings.

CONTROLS. The importance of beer and wine is signified by the fact that no other products are so regulated. To the end of the Middle Ages, the primary focus of drink legislation is assuring that an ample supply of good quality drink is made available at a fair price. Otherwise, control efforts are directed at reducing the potential adverse consequences of drinking, such as disorderly conduct, rather than against drunkenness itself.

ENGLAND

ALE CONSUMPTION. The principal drinks in England are ale, beer, mead, and metheglin. Until about 1400, "first by tradition and then by law," ale consists only of malt, usually made from barley, water, and yeast. It is unclear what distinguishes ale from beer in early medieval England, but references to beer disappear after the mid-10th century (Monckton 1966:11, 36). Throughout the Middle Ages, the great majority of ale is brewed and consumed at home. All large establishments have breweries. Those of the monasteries become especially famous. Ale is viewed as much as a nutritional supplement and a thirst-quencher and is consumed by everyone. Weak ale is given to children.

ALEHOUSE PREVALENCE. Throughout Saxon times (until the Norman conquest in the mid-11th century), community life is relatively static and travel limited; taverns are not numerous and most drinking occurs on festival occasions during which general drunkenness is customary, as described in Beowulf (see below) (Popham 1978:255). Just how widespread ale selling is in Anglo-Saxon times is obscure. While great quantities of ale are supposedly drunk (at least according to Church chroniclers and later Norman apologists), most of it is consumed "in great households, in gilds, or privately" (Clark 1983:20).

DRINKING PRACTICES (BEOWULF). All the residents of the British isles in the early Middle Ages indulge in periodic binge drinking. Before their conversion to Christianity, the Saxons incorporate beer and mead into their religious ceremonies. Excessive drinking is not looked upon as a vice (King 1947:13). Ale is the drink of the gods and the heroes in Valhalla, who feast daily on copious supplies of food and drink. He who can drink the most is admired. The importance of alcoholic beverages is indicated in the Anglo-Saxon names for feast ("beorscope") and banquet room ("mead hall"). The Saxons are also believed to be among the first peoples to introduce the custom of drinking healths or toasts at public feasts (Hackwood 1909:33, 54, 142). Numerous descriptions of drinking bouts appear in Anglo-Saxon writings such as Beowulf and the Caedmonian poem "Judith." Generally, it is expected that a Saxon hero will be "the last man in the world to be outdone in drinking" (Rolleston 1933:36-37).

The Anglo-Saxon poem Beowulf is probably not composed until the 8th century, but it draws on older Teutonic songs and stories telling of the adventures of Beowulf, nephew of the king of the Geats in southern Sweden. It is the only preserved example of Germanic folk poetry that is close enough to its original version to be usable as a historical source (Cantor 1969:108; Jones 1950:148). It reveals the ideals and mores of the upper strata of Germanic society at the dawn of the Middle Ages. References to drinking wine, ale, and mead appear throughout (Gayre 1948:40-42). A chieftain's honor, in part, depends on largesse--sharing of wealth with warriors, including abundant supplies of food and drink (Jones 1959:23). In the poem, people become drunk (line 531), drink all evening and go to sleep under the influence (119), and engage in noisy carousing (1161). Brave warriors boast over their ale-cups after "quaffing beer" (480-481, 1467). No feast is successful without copious drinking: "It was the best of banquets/ men drank their full of wine" (1232-1233). One passage (1012-1014) describes drinkers behaving and acting courteously at their feasting, suggesting "that the opposite [i.e., raucus, drunken behavior] sometimes was the case" (McKinlay 1953:91). But drinking does not cause any problems or catastrophes, except they run the risk of being slain while drunk (119-124, 1233-1241) (Beowulf 1968:35, 45, 60, 66, 73).

The Britons are also known to indulge heartily in alcohol, which contributes to several defeats by the Saxons. The "ever-recurring subject" of the 6th-century Welsh poem "The Gododin" by Aneurin is the intoxication of the Britons from excessive mead drinking before the battle of Cattraeth (French 1884:14; Cherrington 1926, 3:911).

FRANKS (KINGDOM OF)

BACKGROUND DEVELOPMENTS: RISE AND CONVERSION OF FRANKS. In 496, Clovis, the Merovingian king of the Salian Franks, is converted to Christianity by St. Remi, Archbishop of Rheims. Clovis unites most of Gaul west of the Rhine under his rule (modern France and a large part of the southern half of west Germany) and designates Paris as his capital. By 600, the Gallo-Roman and Frankish societies are thoroughly mixed. Despite their conversion, the Franks remain a brutish people, often prone to drunkenness, as evident in the pages of Gregory of Tours (see 575/Gaul). The Merovingians continue to rule over the Franks until 751 when they are replaced by the Carolingian family.

ESTEEM OF WINE. According to legend, St. Remi blesses and gives to Clovis at his conversion a flask of holy wine and tells him to continue his war against Alaric as long as the flask supplies wine. The jug remains always full (Lausanne 1970:346). It is further said that although the flask of holy wine could never be emptied nor could its contents deteriorate, should some ecclesiastic sell the wine from it in his church, the flask would fail "even as the commonest flask in the vilest public drinking hole" (Firebaugh 1924:27).

Remi himself is a great wine lover and viticulturist. Many of his miracles involve filling empty casks with wine or multiplying meager supplies (Seward 1979:85).

ITALY

VITICULTURE DECLINES. In the early Middle Ages, viticulture in Italy is considerably neglected, although wine production does continue (Jellinek 1976b:17).

525

EUROPE

MONKS/DRINKING. The monastic rule of St. Benedict (d. 547?) which attempts to establish practical guidelines for wine drinking, uphold the value of abstention but urges moderation when wine is consumed and establishes a ration of a half pint a day as sufficient in normal circumstances. Chapter 40 of the rule ("The Measure of Drink") reads: "Each man has his own gift from God....We are therefore hesitant in deciding how much others should eat or drink. Keeping in mind the weakness of the less robust, we consider that half a pint of wine a day is sufficient for everyone. None the less, those to whom God has given the gift of abstaining should know that they will be rewarded. But in a case where the locality or the work or the heat of summer may make a larger allowance necessary, the abbot must decide, taking care that there is no excess or drunkenness. Indeed we read that wine is not a drink for monks, but since monks cannot nowadays be persuaded of this, let us at least agree to drink sparingly not to take our fill, as 'wine maketh even the wise to fall away'" (quoted Seward 1979:23). If a monk is late for grace, wine is taken away from him. A sick monk is allowed a pint and a half of wine per day (Younger 1966:236).

550

ENGLAND

CHIEFTAINS/DRINKING. The monk St. Gildas (c. 570) makes several allusions to the prevalence of drunkenness. He accuses British chieftains of going into battle drunk and thereby leading Britain to ruin (Hackwood 1909:37).

CLERICS/DRINKING. According to St. Gildas, "Not only the laity but Our Lord's own flock and its shepherds...slumbered away their time in drunkenness, as if they had been dipped in wine....If any monk through drinking too freely gets thick of speech so that he cannot join in the psalms, he is to be deprived of his supper" (French 1884:17-18; Longmate 1968:2). In a Church synod (569) held by St. David, four canons are adopted under which priests guilty of drunkenness through ignorance, negligence, or design are required to do various penances of three days or more (Baird 1944a:545; Samuelson 1878:123).

FRANKS (KINGDOM OF)

INEBRIETY CONDEMNED. The Merovingian King Childebert I (d. 558) issues a royal capitulary condemning drunkenness as an offense against God (Jeanselme 1920:2860).

575

ENGLAND

ALEHOUSE CONTROLS. The first Christian king in England, Ethelbert of Kent (560-610), passes laws for the better ordering of alehouses (Spiller 1955:92).

FRANKS (KINGDOM OF)

ESTEEM OF WINE (FORTUNATUS). Fortunatus, a Christian poet and bishop of Poitiers, is a lover of good food and wine, and he frequently describes intoxication: "Mother and sister, ease a fuddled man./ Across a sea of wine the table swims"; "With nectared food and wine, with scholarly discourse/ Choice and abundant, hast thou laden me." Like Ausonius 200 years earlier (see 350/Gaul), he describes the beauty of the hillsides of the Rhine and Moselle rivers covered with vineyards, showing that they have not been harmed by the invasions. On the other hand, he scorns beer (Younger 1966:231, 235; Jeanselme 1920:267; Gottschalk 1948, 1:306). The esteem of wine is also indicated by the many early medieval stories of saints miraculously replenishing empty cellars or stinted supplies (Dill 1926:31, 427). One example is Clovis' conversion cup (see 500/Franks).

INEBRIETY PREVALENCE, ATTITUDES (GREGORY OF TOURS). In The History of the Franks, the only important literary source for these years, Bishop Gregory of Tours (539-594), a wealthy Gallo-Roman, shows that excessive drinking and "alcoholic delirium" are common in the Merovingian times among both clerics and laity (Jeanselme 1920:280). Gregory's History leaves the impression "that coarse sensuality and drunkenness corrupted many lives" (Dill 1926:286). Throughout the book, Gregory violently denounces the drunken excesses of his contemporaries. He praises abstinence and refers to the foul smell of a drunk (book 9.6). However, he does not describe any divine punishment for drunkenness per se, nor does he draw any moral. His attitude is much harsher towards those who break fasts or Lent than towards those who drink too much (Firebaugh 1924:25).

CLERICS/DRINKING. Gregory's predecessor as bishop of Tours drank himself to death: "He became almost half-witted. This weakness so affected his reason that he was unable to recognize guests whom he knew very well, and he would assail them with insults and abuse" (10.31). Two brothers spend the evening in a drinking bout and become "completely intoxicated" (6.13). A village priest invites his neighbors to drink with him (7.47). Abbot Dagulf is a robber, murderer, and adulterer; finally, he is killed by an irate husband after a tipsy Dagulf beds his wife (8.19). He praises one Patroclus for his great abstinence: "He would not drink wine, or cider, or anything else which could intoxicate" (5.10). A Breton priest known for this "extreme abstinence" soon falls into the "very bad" habit of immoderate drinking (8.34). The bishop of

Soissons is "out of his mind...for nearly four years, through drinking to excess" (9.37). A Parisian deacon is so drunk that he falls to his death (10.14).

UPPER-CLASSES/DRINKING. The Merovingian royal family is brutal, and outside the church there is little sense of public morality or human compassion. Nobles are constantly in revolt, ravaging and looting the country. According to Gregory, King Chilperic, "the Nero and Herod of our time," is "extremely gluttonous" and given to every vice and debauchery (6.46). His treasurer, Eberulf, is given to murder and drunkenness, once almost beating a priest to death because the priest refused to give him wine when he was "obviously drunk" (7.22). Sichar of Tours is a drunkard who causes trouble "to all sorts of people" when he drinks, which apparently is always (9.19). A drunken duke orders a girl brought to him in bed (9.27). The drunkenness of the Merovingian army's leader is held responsible for the army's defeat by the Goths (9.31). During a royal banquet (10.27), everyone gets so intoxicated that they all collapse and fall asleep (Gregory of Tours 575/1974:265, 345, 379-380, 403, 428, 468, 486, 502, 513, 524, 566, 600).

GERMANY

DRINKING PRACTICES. Fortunatus comments on the enormous capacity of the Germans for drink (Cherrington 1926, 3:1090).

DRINKING RITUALS. The chronicles of the evangelization of the Allemani in southern Germany (Swabia) relate how the famous Irish missionary St. Columban (see Ireland below) exorcised "diabolical" rites (the worship of Wodan) involving a cauldron of beer, which was offered to the gods and then consumed by the Germans (Claudian 1970:8; Jellinek 1976b:75; Arnold 1911:195).

IRELAND

MONKS/DRINKING. The austere rules of Irish monasticism strictly forbid excessive alcohol drinking, and one Irish monk drinks nothing but water for 30 years. Nevertheless, ale is commonly consumed. St. Columban (543-615) supplies ale to the many European monasteries he establishes, and rules that any brother who spills it must drink water until he makes good the amount wasted (Spiller 1955:92). He also is recorded to have miraculously multiplied beer as Christ had done with wine (Arnold 1911:196).

600

EUROPE

ESTEEM OF WINE. Duby (1974:18) oberves that the Church helped spread the prestige of wine drinking: "The eating of bread and drinking of wine--those two 'kinds' which the chief Christian rites set forth as the very symbols of human subsistence--were regarded...as basic signs of cultural advancement, and were diffused far and wide in the seventh century." In the penal section of the Frankish Salic law, wine already is an object of special protection.

VITICULTURE. Viticulture has spread throughout the suitable parts of France. In Germany, increased ability to navigate the turbulent and shallow Rhine river leads to the planting of vineyards in the Rhine valley to serve the German, Flemish, and English markets. Until the 15th century, this is the sole area of German viticulture (Hyams 1965:162, 192).

CHURCH ATTITUDES. Pope Gregory I (590-604) instructs the missionaries that he sends to Britain (see below) not to destroy pagan altars but to purify them to Christian purposes since "the people will come more familiarly to places to which they are accustomed." In substitution for old sacrifices, the people are allowed to celebrate the dedication of a church or the birthday of a holy martyr with "a festival with religious banquets" (French 1884:25, 40; Arnold 1911:201). Through this doctrine of accommodation, convivial eating and drinking become an integral component of medieval holy days and other church festivities.

ENGLAND

BACKGROUND DEVELOPMENTS: CHRISTIAN CONVERSION. Following the marriage of King Ethelbert of Kent in southeast England to a Christian Frankish princess, the king and his nobles are converted to Christianity by Benedictine monks; the first Latin church in England is built at Canterbury ("Kent town").

FRANKS (KINGDOM OF)

INEBRIETY PREVALENCE. The sermons of St. Eligius (Eloi), bishop of Noyon, whose diocese comprises all the semi-heathen lands of the Low Countries, principally inhabited by Frisians, refer to drunkenness as a common vice (Dion 1959:475).

650

ENGLAND

DRINKING CONTROLS: PENANCES. In the *Liber poenitentialis*, Archbishop Theodore of Canterbury (668-693) decrees that if a Christian layman should drink to excess, he must do penance for 15 days, thus indicating that ecclesiastical authority regards drunkenness per se as an offense against morality (Baird 1944a:545; French 1884:17, 18).

INEBRIETY DEFINED. Archbishop Theodore may also have declared that a man should be considered drunk "when his mind is quite changed, his tongue stutters, his eyes are disturbed, he has vertigo in his head with distension of the stomach, followed by pain" (Cherrington 1926, 3:913). (French [1884:28] identifies this description as one of the Exceptions of Ecbright [see 750/England].)

675

ENGLAND

DRINKING CONTROLS: DISORDERLY CONDUCT. The law codes of the English kings of this period uniformly have something to say on the subject of drink and drinking, generally stipulating penalties for offenses committed while drinking in order to promote peace and order (Monckton 1966:29-30).

ALEHOUSE CONTROLS. Regulations are established over "ale-booths," which appear to be small but numerous (Cherrington 1926, 3:912).

VALUE OF ALE. One law of the king of Wessex stipulates that payments in ale and other goods should be made to the king in exchange for holding land. The act presupposes that all families who have land brew their own ale, illustrating its importance (Monckton 1966:30).

700

EUROPE

CONCERNS: CLERICS/DRINKING. The west Saxon missionary Winfried (680-755), better known as St. Boniface, the "Apostle of the Germans," on several occasions expresses his concerns over the prevalence of drunkenness, particularly as a component of the general debasement of the clergy. In 742, he complains to Pope Zachary (741-752) that many Gallic clerics are drunkards (Arnold 1911:199). (On Boniface's concerns see also 700/England.)

ENGLAND

CLERICS/DRINKING. King Wihtred of Kent decrees in 695: "If a priest...is too drunk to discharge his duty, he shall abstain from his ministrations, pending a decision from the bishop" (Monckton 1966:30). Archbishop Egbert of York (c. 735) writes to the Venerable Bede deploring the revelry, drunkenness, and fornication prevalent in monasteries and recommending episcopal inspection (Rolleston 1933:39).

CONCERNS: INEBRIETY PREVALENCE (ST. BONIFACE). The Church in England begins to adopt a much stronger line in its attitude toward the abuse of drink (Monckton 1968:31). In a letter to the archbishop of Canterbury, St. Boniface observes that drunkenness is especially peculiar to the Anglo-Saxons: "It is reported in your dioceses the vice of drunkenness is too frequent. This is an evil peculiar to pagans and to our race. Neither the Franks, nor the Gauls, nor the Lombards, nor the Romans, nor the Greeks commit it" (quoted Monckton 1966:31; see also Rolleston 1933:39 and French 1884:27). (Similar complaints are made by John of Salisbury and William of Malmesbury in the 12th century.)

750

ENGLAND

ALEHOUSE REPUTATION. Because alehouses and taverns have a bad reputation, Ecbright, archbishop of York, commands bishops and priests to have a *hospitium* or guest house not far from each church for the care and entertainment of strangers. This type of hospitality spreads throughout the country. He also enjoins any priest from entering or drinking at a tavern (Monckton 1966:32; French 1884:29).

775

EUROPE

DRINKING ATTITUDES (ALCUIN). Alcuin (735-804), the most distinguished scholar of the late 8th century, detests intemperance and often warns ecclesiastics against excesses of the table (Jeanselme 1920:289-290). But he loves to drink in moderation, and he reminds his monastic brothers that sin lay not in the use of alcohol, but in its abuse (Monckton 1966:32). While giving pious instructions, he himself would drink beer and wine so as "to be able to teach and sing the better." Many of his letters indicate a fair amount of jovial drinking. Aware of his responsibility for the drinking of one student, Alcuin laments: "Woe to me, if Bacchus should drown [the student] in his floods" (quoted Fichtenau 1964:92). After Alcuin is given the task by Charlemagne of reforming the Frankish schools and of educating the nobility, he develops a great fondness for French wines. In one letter, written during a return from Gaul to York, where he is head of the cathedral school, he laments his inability to find any Frankish wine to drink (Allen 1961:164).

FRANKS (KINGDOM OF)

BACKGROUND DEVELOPMENTS: CAROLINGIAN EMPIRE. In 768, Charlemagne (d. 814) becomes the Carolingian king of the Franks and the greatest monarch of the early Middle Ages; on Christmas day 800 he is crowned Roman Emperor. (With papal approval, the Carolingian family officially replaced the Merovingians in 751, after effectively ruling behind the scenes for almost a century.) Charlemagne extends the original Gallic kingdom of the Franks to north and central Italy (Lombardy, Ravena, and Rome); to Gascony, Septimania, and the Spanish March to the southwest; and to Saxony, Bavaria and Slavic lands to the north and east. The heartland of the empire lies between the Loire and Rhine rivers. As Charlemagne completes the political goals of the Carolingians, he turns to improving the cultural and intellectual level of Frankish society, working closely with the church. In this "Carolingian Renaissance," he attracts scholars from all over Europe to his court, most of whom are churchmen, and sets up schools to educate the Frankish lords.

TEMPERANCE OF CHARLEMAGNE. In his biography of Charlemagne modeled on Suetonius' biographies of the Roman emperors, the Frankish historian Einhard emphasizes (in chapter 24) that Charlemagne "was moderate in his eating and drinking, and especially so in drinking; for he hated to see drunkenness in any man, and even more so in himself and his friends....He was so sparing in his use of wine and every other beverage that he rarely drank more than three times in the course of his dinner" (Einhard 1969:78).

DRINKING CONTROLS. Both by his personal example of temperance and his edicts, Charlemagne tries to reform the drinking habits of his subjects. He forbids anyone to appear in court intoxicated and forbids priests to offer drinks to penitents. Soldiers found drunk in camp are placed on water as a beverage until they publicly ask forgiveness for their offense (Samuelson 1878:105; Jeanselme 1920:286). In his *Admonitio generalis* of 789, he tells all citizens that drunkenness (*ebrietas*) is among the grave sins. In several capitularies he bans monks and clerics from entering taverns and warns them to avoid drunkenness and laxity (Dozer 1980:333-338). Though he exhorts all people to be temperate, his laws against drunkenness do not always carry a penalty for drunkenness, probably because matters of immorality are left to Church discipline (Baird 1944a:544).

ALCOHOL PRODUCTION. Although Charlemagne emphasizes temperance, he makes sure that an adequate supply of good drinks are available. The capitulary *De villis* (Concerning Estates, c. 800), a set of regulations concerning the management of royal manors probably issued by Charlemagne himself, the calls for careful attention to viticulture and wine making. In chapter 8 he orders: "That our stewards take care of our vineyards which are in their territory, and make sure that they are worked well. Let them place the wine in good containers and let them diligently see to it that nothing is lost in shipping it. Let them acquire through purchase special kinds of wine, in order to send it to the royal estates....They should have vine slips from our vineyards sent for our use. The rents from our estates which are paid in wine are to be stored in our cellars." In chapter 14 and 41 he stipulates that grapes must not be pressed by feet, that "everything should be clean and orderly," and that wine presses be carefully constructed. The capitulary includes a discussion of new and old grape wine, boiled wine, mulberry wine, beer, perry, cider, and mead.

Less attention is directed toward brewing, but among the "good artisans" that every steward should have in his territory are "brewers (that is, those who know how to make beer or cider, perry or other liquid fit to drink)" (ch. 45). He also instructs (ch. 61) "masters know how to make good beer" to come to the palace (De villis, quoted Herlihy 1968:42-52, who translates the entire capitulary; see also Dion 1959:190; Munz 1969:15-16; Arnold 1911:204-205).

DRINKING PRACTICES. Drinking occurs among clerics and by the laity within churches. The custom of communal religious drinking (pre-Christian in origin) continues. Hymns for holy days conclude with invitations to drink to the health of the saints. Paul the Deacon (one of Charlemagne's scholars) composes a drinking song "the verses of which are very reminiscent of ancient oaths." Many a parish priest is said to have drunk with the laity well into the night and to have failed to hold divine services the next day while sleeping off the effect (Fichtenau 1964:92, 163).

800

FRANKS (KINGDOM OF)

NOBILITY/WINE DRINKING. Munz (1969:16) states that the main drink of wealthy Carolingians is *cervisia* made from sprouting barley and that only the poor drink water. But wine is replacing beer and mead as the drink of the nobility, who come to view beer drinking with disdain (Firebaugh 1924:45; Jellinek 1945:650). Among the nobles, it becomes a point of honor to produce, drink, and offer their guests plenty of the best wine (Dion 1959:190-191). The choicest gift is to bestow wine on friends. Great men boast of the quality of the wine from their own vineyard and bask in the honor they derive from it (Duby 1968:8-9, 42). The prestige of wine is evident in the attention that is directed toward it in the capitulary *De villis* (see 775/Franks).

VITICULTURE. The spread of wine drinking among the nobility leads to an expansion of viticulture in the 9th century and considerable competition for good vineland. Vineyards are established around the cities, which are the abodes of the nobility and higher clergy. Still, neither vines nor meadows cover more than a very limited part of the cultivated area, since cereal crops are so important and require all the available agriculture activity. Not all Carolingian villas are closed, self-sufficient communities. Many lords own several demesnes which specialize in different crops because of different natural conditions. Those villas that specialize in wine bring in other foodstuffs from elsewhere (Duby 1968:8-9, 42-43).

MONASTIC WINE PRODUCTION. The monastery of Saint-Germain-des-Prés produces over 11,000 gallons of wine a year from its various vineyards for use in communion, and by monks, serfs, pupils, and guests (Seward 1979:33).

ATTITUDES: TEMPERANCE. The *Specula principis* (Mirrors of Princes), discussions on proper princely behavior) composed during the Carolingian Renaissance all place the greatest emphasis on Christian goodness and portray Charlemagne as the perfect prince. Prudence, kindness, justice, personal virtue, and moderation are all especially admired qualities. However, temperance or abstinence from alcohol is not praised especially, indicating that it probably does not constitute a matter of great concern. These works are largely composed by Franco-German churchmen such as Alcuin, Smaragdus of St. Mihiel, Jonas of Orléans, Sedulius Scotus, Hincmar of Rheims (Born 1933; Bell 1962).

CLERICS/DRINKING. Legend tells of an Abbot of Angers in the 9th century who becomes so saturated with local wine that it preserves his body after his death (Waddell 1949:1999-200).

DRINKING CONTROLS: PENANCES. A number of Church councils and Carolingian capitularies denounce excessive drinking; most take place in the early 9th century in conjunction with the Carolingian Renaissance (Jeanselme 1920:283-284). In 813, the Council of Mayence (Mainz) bans priests and monks on penalty of excommunication from engaging in gluttony and inebriety, the latter of which is called the mother of all vices. The council is held by order of Charlemagne to restore Church discipline. That same year, the Council of Tours classifies drunkenness as among the works of the devil, along with murder, fornication, and adultery, and warns that it is physically harmful. In

827, a Carolingian capitulary orders communion suspended for 40 days for ecclesiastics guilty of drunkenness, following the canons of the Council of Vannes in 465 (Dozer 1980:304-306, 320-321, 333). In 847 (Council of Mayence?, called by Hrabanus Maurus to deal with scandal involving drunken monks at inns), it is reaffirmed that a 40-day penance be imposed on any priest who gets sick after excessively indulging at the table (Hopkins 1899:105; Landon 1909, 1:363-364, 2:175-176).

IRELAND

DRINKING SONGS; MONKS/DRINKING. Most of the drinking songs of the 9th-century come from the Irish, who as monks and scholars wander across Europe (Waddell 1949:64-65, 179-180).

SWITZERLAND

MONASTIC BREWING; BEER RATIONS. The monastery of St. Gall builds the first full-fledged brewing plant in Switzerland. The monastery, incorporating three buildings, becomes famous for its great breweries, which produce two, possibly three, different types (strengths) of beer. Each monk is allowed five quarts of beer daily, besides occasional drafts of wine (Arnold 1911:207-215; Cherrington 1925, 1:402, 1926, 3:1090; Claudian 1970:10; Jellinek 1976b:76).

850

EUROPE

BACKGROUND DEVELOPMENTS: EMPIRE DIVIDED. After years of bitter civil war, the Treaty of Verdun (843) divides the Carolingian empire among the three sons of Emperor Louis the Pious (d. 840), Charlemagne's son. This gives rise to the two major divisions of east and west Franks, the origins of modern Germany and France. The treaty also shows the early differentiation of the German and French languages. The Carolingian line rules over the east Franks until 911 and in the west until 987, but these Carolingian rulers are nonentities, no longer exercising any effective control.

The Frankish kingdom disintegrates; social and political power becomes localized. This disintegration is accelerated by the pressures from invading Scandinavians, Muslims, Slavs, and Hungarians. Between the first Norse excursion into Ireland in the 820s and the establishment of the rule of the Normans in Normandy (c. 912) the Vikings continually raid northern Europe. Fortresses are built everywhere. In the mid-10th century, however, the Holy Roman Empire is revived under the Ottonians in Germany.

FRANCE

DRINKING CONTROLS: PENANCES. In the 9th century, the Church authorities of Angiers (Angers?) declare: "If any bishop, or any man serving under his orders is an habitual drunkard, if any priest or any deacon shall vomit after excesses at table, he shall be sentenced to undergo a penance of 40 days duration....Any layman convicted of drunkenness shall submit to 30 days of penance, and in addition shall be deprived of bacon, beer and wine. If he has led his neighbor astray, he shall fast 10 days additional for the offense" (quoted Firebaugh 1924:24).

CLERICS/DRINKING. In 852, Bishop Hincman of Rheims issues a pastoral letter forbidding priests to get drunk at anniversary and memorial days of the dead (Arnold 1911:200).

GERMANY

INEBRIETY CONDEMNED (MAURUS). In the mid-9th century, the German theologian Hrabanus Maurus complains of drunkenness among the Germans. A favorite pupil of Alcuin, and later archbishop of Mainz, Maurus writes: "Amongst these vices, feasting and drunkenness especially reign, since not only the rude and vulgar people, but the noble and powerful of the land, are given up to them. Both sexes and all ages have made intemperance into a custom;...and so greatly has the plague spread, that it has infected some of our own order in the priesthood, so that not only do they not correct the drunkards, but become drunkards themselves." He further asserts that if anyone who is ordained has the habit of drunkenness, he must either resign or be deposed. Thirty to 40 days penance are required if one drinks until he vomits, unless it is through inadvertence or carelessness, in which case the penance ranges from 7 to 15 days (quoted Samuelson 1878:114).

900

EUROPE

ATTITUDES. An anonymous 10th-century poet sings the praises of Bacchus in classical terms: "Hither your goblets brimming deep with wine,/ That Love aflame may his long vigil keep./ The fire of Love kindles from burning Bacchus,/ For Love and Bacchus are like-minded gods" (quoted Waddell 1949:88).

ENGLAND

ALEHOUSE PREVALENCE. By the end of the 9th century, a considerable number of alehouses exist not only in sizeable towns, but also in villages and by the side of old Roman roads, serving refreshment to travelers. These "alehouses" are often no more than wooden huts probably adjacent to the brewer's house (Monckton 1966:33; see also King 1947:20).

BEER CONSUMPTION. Reflecting the importance attached to ale and beer, sources from the 10th century mention both drinks at the top of lists of products to be given to lords for rent. About this time (c. 950), the use of the word "beer" disappears from the English language for 500 years (Monckton 1966:36). However, it has also been suggested that beer disappeared because it was an upper-class drink, stronger and more costly than ale. It also lost patronage with the arrival of the wine-drinking Normans in the mid-11th century (Gayre 1948:83-84; Simon 1948:146).

GERMANY

BACKGROUND DEVELOPMENTS: HOLY ROMAN EMPIRE ESTABLISHED. In 918, Duke Henry of Saxony is elected king of Germany, to be succeeded by his son Otto I (936-973), who becomes the greatest military leader since Charlemagne, extending his rule to the north and east. Otto is crowned emperor by Pope John XII (955-964) in 962, establishing the German or Holy Roman Empire.

LOWER CLASSES/DRINKING. From the 10th through the 12th centuries, there is no evidence of conspicuous consumption among the lower classes (Jellinek 1945:650-651).

950

ENGLAND

CONTROLS: ALEHOUSES, PEG DRINKING. Archbishop Dunstan (925?-988) issues a number of canons for checking intemperance, and convinces King Edgar the Peaceful (959-975) to take similar actions. Edgar closes many alehouses, decreeing that there should be only one to each village. He also orders that pegs should be fitted at regular intervals to the insides of drinking horns, with the instruction that the drinker not go beyond a fresh mark at each draught (about a modern half pint). This innovation is not a great success; peg (or pin) drinking soon becomes regarded as a challenge because no limit is imposed on the number of times the cup may be filled. The pegged tankards are handed around, with each person drinking the ale down to the next peg. This practice of "sharing a pot" goes on to the end of the 19th century (French 1884:37; Longmate 1968:3; Rolleston 1933:39; Baird 1944a:549; Monckton 1966:33-35).

PLEDGING. It has been suggested (King 1947:14-15) that Dunstan's reforms are a response to increased drunkenness caused by the influence of the Danes who are in possession of a large part of east England (the Daneland). The Danish invasions are also said to have fostered the habit of pledging to provide a measure of mutual protection while drinking.

RUSSIA

DRINKING ATTITUDES. Legend has it that St. Volodimir (978-1015), grand prince of Kiev, under whom Rus´ is converted to Christianity, rejects conversion to Islam in part because of its ban on wine drinking; he declares that "drinking is the joy of the Russe. We cannot exist without that pleasure" (Jellinek 1943:663; Clarkson 1969:33). (Kievan Rus´ is the predecessor of Russia, Ukraine, and Byelorussia.)

1000

EUROPE

BACKGROUND DEVELOPMENTS: CENTRAL MIDDLE AGES. Europe as we know it in the Middle Ages has taken shape, with distinct groupings of peoples permanently bound together by a common allegiance to Christianity. Profound changes begin to occur in almost every aspect of life, inaugurating the vigorous and creative central Middle Ages (c. 1000-1350). Although these changes begin in the 11th century, it is in the 12th century that the evidence of change is most evident. Population expands, and with it commerce, the area of land under cultivation, and urban life. German colonists penetrate beyond the Elbe until the area of German settlement is triple that of Carolingian times. Trade in kind begins to be replaced by a money economy. Throughout the central Middle Ages a gradual erosion of serfdom. As the labor supply expands it is no longer necessary or possible to bind men to the soil; labor mobility is also encouraged by extensive external and internal colonization. Landlords who wish to receive higher returns from their properties also are forced to free their serfs. The isolation and self-sufficiency of the manor declines and, as transportation improves, for the first time a vigorous court life develops as nobles gather together at courts away from their homes.

The first centers of commerce are Venice and Flanders; it is in Italy and the Netherlands that cities first revive as merchants establish permanent residences there. With the growing prosperity of the 12th century, merchants build cities into fortified enclosures, and trade and population growth stimulate industry to concentrate there. Men from all over are drawn to the cities to take advantage of the new jobs they offer. Totally outside the manorial and feudal systems, the bourgeois or burghers (*burgenses*) of the cities gradually gain rights to independence and establish their own laws (Pirenne 1956:72-129).

OVERVIEW: DRINKING IN THE MIDDLE AGES. As the national groupings of Europe begin to emerge, distinct drinking reputations also appear. Throughout the Middle Ages, the Germans and English have reputations as heavy drinkers; the Spanish, French, and, especially, the Italians, as light drinkers. South Europeans do not seem to indulge in the convivial parties and drinking matches which characterize northern Europe. Nevertheless, the similarities in drinking levels, patterns, and attitudes between peoples are as important, if not more so, than the differences, particularly since medieval Europeans share a common adherence to Christian tenets and a predominately poor, agricultural existence. How extensive drunkenness is in different periods and areas of medieval Europe is impossible to determine precisely. Rolleston (1944) argues that drunkenness was more common in medieval England than in antiquity, but he fails to make a convincing case; certainly concerns about it are more frequently expressed due to the influence of the Church. What is apparent is that during the central and late Middle ages excessive drinking is common, but it is largely an occasional problem associated with periodic festivities and with specific populations who have the greater wealth, availability, or free time to indulge more regularly, such as nobles, students, and clerics. Inebriety is not a widespread concern except as it affects individual salvation. To some extent the problem of drunkenness is moderated by the influence of Christianity, the simplicity of life, the weakness of the beverages consumed, and the local and small-scale nature of production. Because of the importance of beer and wine to the diet, most drink controls focus on protecting the drinker from unscrupulous sellers, maintaining a good supply and a fair price, and preventing adverse consequences from too much drinking, such as public disorder. As taverns grow in numbers, they are continually criticized, but often less because drunkenness occurs there than because on Sundays and holy days people gather at taverns instead of at the church, because they are the source of dancing dancing and a wide variety of disapproved activities, and because they are viewed as centers of criminality and idleness. The Church is far more concerned over attendance at taverns instead of divine services than with drunkenness per se. Medieval authorities and laws are more concerned with reducing the number of outlets, hours of sale, closing times, drink availability and quality, and maintenance of order than with promoting temperance.

NOBILITY/DRINKING INFLUENCES. Eating and drinking in the Middle Ages occupies a far more central position than they do in today's social life, and many treatises are written on table manners. But social controls are relatively mild and manners lax. Since the primary occupation of the upper classes is warfare, they are constantly ready to fight, and without a central authority strong enough to compel the populace to restraint, people give way to drives and emotions "incomparably more easily, quickly, spontaneously, and openly" than today (Elias 1978:60, 62, 106, 191, 201, 214). Bloch (1964, 2:411) emphasizes the emotional instability of the age, its insecurity and irrationality: "Violence was an element in manners. Medieval men had little control over their immediate impulses; they were emotionally insensitive to the spectacles of pain, and they had small regard for human life, which they saw only as a transitory state before Eternity; moreover, they were very prone to make it a point of honour to display their physical strength in an almost animal way."

The valiant knight of the 11th century is above all praised for prowess as a fighter and for his athletic build. Since strength must be sustained, he is also known for his mighty appetite and thirst. In the old French "Chanson de Guillaume," a knight is praised as follows: "By God! fair sire! he's of your line indeed,/ Who thus devours a mighty haunch of boar/ And drinks of wine a gallon at two gulps;/ Pity the man on whom he wages war!" (Bloch 1964, 2:294-295). Furthermore, drinking provides one of the few diversions in life for many nobles who have little else to do. Of the virtues of chivalry, temperance in drinking hardly is mentioned, except in the writings of churchmen who seek to create a religious chivalry or priestly order of knights, most notably John of Salisbury (see 1150/England).

Contributing to the importance of feasting and drinking in upper-class medieval culture is the ideal of *largesse* by which the knight and nobility gain prestige and legitimacy by expending the wealth they acquire on their followers. Often this takes the form of costly banquets. "Prestige through extravagance" is one of the fundamental notions of honor that separate social groups; by squandering his fortune, the noble affirms "his superiority over classes less confident in the future or more careful in providing for it" (Bloch 1964, 2:311). This practice is an inheritance of the Germanic war band in which the chieftain gained the loyalty of his warriors through the giving of his bounty through his generosity and protection. Throughout the Middle Ages, fortunes are often ruined because of excessive expenditures on feasts. Anyone who hopes to retain authority and influence has to show himself to be the source of all good things for his dependents and in this to be equal, if not superior, to his allies and enemies. If banquets are costly, they are nevertheless required; frugality brings contempt: "To get into debt was bad, but any attempt to cut down expenses by cutting out dinner parties was far worse, an open invitation to the brutal mockery of one's peers and the knowing titters of subordinates" (Henisch 1976:12).

At the same time, however, moralists criticize feasts as a waste of money; as a result, medieval men "stumbled through a maze of contradictory assmumptions about food and its role in life" (Henisch 1976:14-15).

PEASANTS/DRINKING. Bishop Burchard of Worms lists drunkenness among the tenants on a papal estate as one of three causes for the murders committed everyday (Bloch 1964, 2:411).

FESTIVE DRINKING. Virtually all work is seasonal in character and punctuated by the feasts of the Church. Most of the major festivals in the central and late Middle Ages occur in slack periods of the agrarian year when one kind of work has finished and another has not yet begun. Periodic festivals provide needed release in a rigid hierarchical society, making more tolerable the miserable existence of most peasants. Family and community festivities are special occasions when people gather together to eat and drink and to wear their best clothes. Eating and drinking are essential components of all festivals, particularly the Carnival, and are central to the utopian vision of the land of Cockaigne ("kitchen-land"), where life is one long banquet. Amid the daily struggle to live, the major festivals give people something to look forward to; by them people reckon time and events. They also contribute to a sense of communal spirit since they celebrate the community itself and its abilities to put on a good show. The excitement and drunkenness that occur often lead to riots and expressions of hostility; although this causes concern to authorities, it is generally recognized that festivals help give vent to spirits and anxieties in a somewhat controlled manner (Burke 1978:178-179, 200-203; Burke's discussion deals with popular festivities in early modern Europe, but the origins of these festivities lie in the Middle Ages).

CARNIVAL. The Carnival preceding Lent is the festival par excellence in southern Europe. Lasting usually three days (in February or March) and ending on Shrove Tuesday or *Mardi gras*, it is a time of plenty and indulgence, as opposed to the abstinence of Lent. The figure of Carnival himself is usually depicted as a fat man with food hung around his neck, sitting on a wine or beer barrel (e.g., Pieter Bruegel's depiction in 1559). Carnival is *the* time of ecstasy and liberation, "a time of institutionalised disorder," when the everyday world is turned upside down. In northern Europe, the Carnival tradition is weak because of bad weather, but other festivals fulfill its function (Burke 1978:182-192). Frazer (1959:569), among others, suggests that the Carnival developed out of the Roman tradition of the Saturnalia, which the Church allowed to "linger unmolested in the countryside" in a disguised form following the fall of Rome. With the revival of cities, the Carnival tradition emerges, establishing itself fully in the late Middle Ages (see also Bakhtin 1968:5-13; Rudwin [1920:1-2] observes that Carnival festivities were popular within the church because they offered recompense for the 40 days of Lent that followed.) The origins and historical development of the Carnival are obscure. We know little about many of its traditions, such as Carnival comedies, only from the attacks of reforming ecclesiastics beginning in the 13th century (Rudwin 1920:vii). Most of the descriptions of Carnival activities occur in the late medieval and early modern periods when concerns mount about Carnival revelries.

WAKES. At the parish wake (French *veille* or *fête patronale*, Italian *veglia*, German *Kirchenweihtag*) celebrating the eve of their patron saint, parishioners often spend the night in the church eating and drinking, singing and dancing. The wake illustrates the important role of the church as a facility for social gatherings and the intimate and familiar sense of the sacred, the blending of the sacred and the profane, that often occurs in the Middle Ages (Burke 1978:109). Wakes often last for days or even weeks (Thomas 1964:54; for England, see Homans 1941:373, 392).

CHRISTMAS; FEAST OF FOOLS. The period from Christmas through Epiphany is the greatest holiday of the year in England and northern France. Christmas festivities are "the very epitome of exuberent self-indulgence" (Henisch 1976:54). They often become orgies which end in riots. It is a time when everyday rules no longer apply, when a Lord of Misrule or a Bishop of Fools for the Feast of Fools is elected. There is dancing, gambling, and drinking in the streets and in churches, and a mock mass is held at which ecclesiastics dress as mimes or buffoons (Burke 1978:192-193). The altar itself is turned into a tavern (Frazer 1959:566-567). In England, Christmas is a time of license in which all partake heavily in food and drink "without another thought for the days before harvest when food might be hard to come by" (Homans 1941:357-359, 279). Around 990, Theophylact, patriarch of Constantinople, is said to have sanctioned the Feast of Fools in Greek churches as part of an effort to wean the people from pagan ceremonies; it flourishes in France as early as the 11th century; it does not appear to have been quite so popular in England. In the festivities of the Feast of Fools may be traced some of the origins of medieval drama (Hastings 1902:96-98).

CHURCH ATTITUDES: FEAST DAYS. Although the church continences drinking and feasting on holy days, such material pleasures are intended to be merely incidental to the primary purpose of reverent celebration and church attendance.

AVAILABILITY. Although "the dietary history of the medieval world remains wholly unexplored," it appears that the variety of foods and the availability of ale and wine increase in the 11th century as the area of arable land undergoes a considerable extension (Duby 1968:65). Barley cultivation expands as it has a higher yield than oats. In the North Sea countries, the growth of brewing also brings an increase in the price of barley. The most actively marketed commodities are wine and animal products (cattle, wool, leather, etc.).

VITICULTURAL INFLUENCES. Throughout the Middle Ages, since bread constitutes half to three-quarters of caloric intake, an increase or decrease in bread prices brings about a corresponding decrease or increase in vineyard acreage (Hyams 1965:195).

PRODUCTION, SALES PRIVILEGES: BANVIN. In viticultural areas, the lord almost invariably has a monopoly (*ban*) of the wine press. The lord also usually has a monopoly during certain seasons of the year (usually about three times) on wine sales in his village for a fortnight or so (*droit de banvin*). For example, the king of France has a monopoly of wine sales in Paris when his wines are ready (*ban le roi*). Often ban wine is sold for more money than wine normally available at the taverns. Nor can peasants avoid drinking it, since many manors have requirements of how much ban wine each tenant has to drink or else they will incur punishments.

In ale/beer-consuming areas, lords generally have a monopoly on brewing or exercise rights to a certain percentage of each brew, though at times arrangements get very complicated (Coulton 1925:59-60; Bloch 1966:80).

ENGLAND

ALEHOUSE PREVALENCE. Although the number of alehouses increases in the 12th century, throughout the Central Middle Ages the role of ale sellers remains narrow and restricted. Ale sales are generally irregular and largely consist of the sale by part-time female brewers of small quantities for off-premise consumption to supplement the family income. They generally only brew when supplies are available and/or necessity arises. It is "an amateur, small-scale enterprise without a real institutional base" (Clark 1983:23). This situation is due to the following factors. First, most of the ale sold is still a poor commercial product: rough, thick, flat, and lacking in palatability. Second, because it deteriorates within a few days, it cannot be stored and transported. Third, good quality ale is relatively expensive; poor villagers and townspeople can probably purchase it only occasionally. Fourth, alcoholic alternatives exist in cider and mead. Fifth, household servants receive ale as part of their "wages," as often also do day laborers and tenants. Sixth, festivals and ales provide people with many of the opportunities to drink without going to the alehouse (see below). Seventh, hospitality to travelers is offered free, especially at monasteries, as one of the corner-stones of Christian charity. Eighth, some lords seek to restrict alehouses to preserve their monopoly over brewing, forcing tenants to purchase ale from them. Finally, homes are too small and rudimentary to allow on-premise consumption (Clark 1983:23-26). (See also the comments about alehouses in the early Norman period under 1050/England.)

DRINKING CONTROLS: DISORDERLY CONDUCT. King Aethelred II (978-1016) issues a code of laws which includes punishments for breach of the peace in an alehouse (*eala-hus*) (Monckton 1966:36; Lennard [1959:405] suggests that the enactment may only relate to towns and not villages).

FESTIVE DRINKING: ALES. The importance of drink in the social life of medieval England is signified by the name of communal gatherings as "ales." Ales are the most popular form of medieval festivities and take a variety of forms; the most prominent are bride-ales, scot-ales, and church-ales. These ales are a kind of bazaar at which all have to attend and all have to buy drinks. Their main function is to raise money for some purpose, and they are not always welcomed by those invited (Bennett 1969:266). At bride-ales (from whence the word "bridal") the bride benefits from the sale of ale on her wedding day. Bid-ales are general "benefit feasts" at which all neighbors come to aid the object of the "bidding." At funerals, the parish invites all the friends and relatives of the deceased to make tribute by the sale of a wake-ale (King 1947:43-45). Great meals and drinking, accompanied by boisterous behavior, are an inseparable part of funeral ceremonies throughout the Middle Ages. Ales are officially frowned on by the Church, but since it cannot suppress them, it recognizes church-ales in an effort to exercise some control over their extravagances (Bennett 1969:265-266). Church-ales serve as a means to facilitate the collection of parish dues and to pay church expenses. Thus, the gallery of a Norfolk church carries the inscription "God speed the plough/

And give us good ale enow..../ Be merry and glade/ With good ales was this church made" (quoted King 1947:44).

The most notorious and criticized ales are the scot-ales, originally communal festivals to raise money for some cause, at which all participants share, or bring money to cover, the drinking expenses (the scot). By the 13th century, this "friendly and beneficent feast" degenerates into a burdensome system used by sheriffs, under whose authority they are held, to collect a form of compulsory and illegitimate tax (Coulton 1925:28).

Although ales are frequently held and frequently condemned, Bennett (1969:269) warns that "we must not exaggerate the number of such ales and drinking bouts." Furthermore, it is evident that they are often condemned not because of drinking but because dancing occurs, because they keep people away from the church, and because they are provocative of sin.

FRANCE

VITICULTURE. The best vineyards and vintages remain in the hands of rich men--high Church dignitaries and nobles--who take great pride and personal involvement in wine production. In regions where viticulture does not flourish, only the lord's residence includes any enclosures planted with vineyards. In areas such as Laon, Paris, and the slopes above the Moselle, the light white wine that is the most appreciated at aristocractic tables is produced in quantity for consumers in neighboring provinces where the climate is unfavorable for viticulture. Originally cultivated for personal consumption, vines come to be cultivated for wine export. Vineyards spread along the waterways, which alone offer a means of transporting casks cheaply and without excessive jolting (Duby 1968:65, 97, 129, 134, 138, 140, 216).

CIDER CONSUMPTION. Cider originates in Biscay, home of the cider apple tree; it appears in Cotentin, the Caen region, and the Pays d'Auge during the 11th or 12th century. These regions are north of the commercial boundary of the vine. Cider competes with beer and meets with some success, since making beer sometimes means going without bread (Braudel 1981:240-241). French (1982:10) argues that the cultivation of cider apples and the technology of making cider spread from Moorish Spain to southern France and then Normandy and, from there, to England.

GERMANY

NOBILITY/DRINKING. Following the establishment of the German (Holy Roman) Empire in the second half of the 10th century, court life begins to be more refined, but how drinking behavior among the upper classes is affected is uncertain. According to Cherrington (1926, 3:1090) and Jellinek (1945:651), the knightly life which characterizes the German court between the 10th and 12th centuries moderates wine drinking among the nobility. While the German nobility still indulge in much wine, they are now told that drunkenness is disgraceful. "The amenities of the court," Jellinek concludes (but without evidence) "were not compatible with coarse and drunken behavior." (See also the discussion under 1100/Germany)

MEAD CONSUMPTION. Mead is still widely drunk in northern Europe. Ulm, Riga, and Danzig are all known as mead-making towns. In 1015, a fire in Meissen, Saxony, is extinguished with mead for lack of water (Gayre 1948:87).

POLAND

TAVERN FUNCTIONS. Beginning in the 11th century, among the western Slavs in Bohemia and Poland, there is always a tavern (*korchma*) in the market place. Some cities have more than one. The Old-Slavic korchma was "the place to which people came to eat and drink, to talk and to celebrate with songs and with music....Among the western Slavs, officials proclaimed orders of the government in the korchma, judges held court, affairs with travelers were threshed out, and for a long time the taverns served as council chambers and guest houses" (Pryzhov 1868, quoted in Efron 1955:491).

RUSSIA

BEVERAGES CONSUMED. Drinks are made of barley or honey: beer, small beer, mead, ale, and *kvas* (a purely Slavic beverage with a very low alcohol content, which is usually made from bread fermented in water). Each family brews its own beer and mead as needed for everyday life. The wealthy have cellars for the storage of mead, beer, and imported wines (Efron 1955:490).

DRINKING PRACTICES. Every discussion, every community affair or public occasion begins with (often copious) drinking or a festivity (Efron 1955:490).

INEBRIETY PREVALENCE. Drunkenness is widespread in medieval Russia. It is one of the few vices to which entire sermons are exclusively dedicated, while other vices are usually dealt with collectively. Russian priests dwell at length on the virtues of moderate drinking, far more so than their contemporary European counterparts. The idea of total abstinence, however, is rejected as heretical (Jellinek 1943:663).

1050

EUROPE

MEDICINAL USE, WINE. Translations are made of Greco-Roman and Arabic medical manuscripts. At the medical school of Salerno, Italy, an organized art of healing begins to be taught that eventually spreads throughout Europe; wine is one of the therapeutic agents most frequently mentioned in the school's code of health (the *Regimen sanitatis Salernitatum*). It is prescribed as a nutrient, internal antiseptic, and restorative, but the code warns against drinking between meals or beginning a meal with drinking, a warning that seems to reflect common Italian practices of the time (Jellinek 1976b:3). Wine, the code asserts, can do much good or harm depending on whether it is used or abused (Lucia 1963a; Claudian 1970:17). (On the Salerno school and the origin of alcohol distillation, see 1100/Europe.)

ENGLAND

BACKGROUND DEVELOPMENTS: NORMAN CONQUEST. William the Conqueror (1066-1087), duke of Normandy, asserting himself as the rightful heir to the English crown, invades and conquers England in 1066.

INEBRIETY PREVALENCE; DRINKING PRACTICES. Norman temperance is often praised by Anglo-Norman historians (e.g, William of Malmesbury, see 1150/England). While temperance may have characterized William the Conqueror, his son William Rufus (1087-1100) is known for his intemperance, and the Normans are soon said to be drinking like Saxons; on the whole there is no evidence of an appreciable change in drinking levels. However, at the close of the Saxon period, and especially after the Norman Conquest, traditional drinking patterns grow beyond festival occasions, taverns spread and wine-drinking expands (see below).

WINE CONSUMPTION. Prior to the Norman conquest, many English monasteries had vineyards (the Domesday Book of 1086 mentions 38), but not all of them necessarily produced wine, total production remained small, and quality was poor. Wine was probably only drunk by the monks themselves and a few nobles. The Normans bring their wine-drinking habits with them, marking the beginning of wine's lay history as an upper-class drink in England (Sutherland 1969:82-83; King 1947:15). In the centuries after the conquest, wine imports from France rise (see below); French wine does not seem to be unduly expensive until price increases in the 14th century (Younger 1966:262). A more intensive planting of vines also occurs, although domestic wine production never fulfills the island's needs. Most vineyard ownership is aristocratic, not monastic (Barty-King 1977:17-23).

WINE TRADE. The merchants of Rouen, Normandy, are among the principal wine importers to England in late Saxon, early Norman days. They probably bring with them mostly the wines of the Seine basin and of Burgundy (Simon 1906, 1:30; Carus-Wilson 1967:266).

ITALY

WINE MERCHANTS. The first reference to a wine seller (*vinadro*) appears in Florentine records of 1070, indicating that the business of specialized wine merchant has come into existence. Prior to this time, wine was not sold separately from other commodities (Staley 1967:360).

1100

EUROPE

BACKGROUND DEVELOPMENTS: 12TH-CENTURY RENAISSANCE. Greater stability and peace materialize, as commercial and urban life are revitalized. An unprecedented proliferation of culture and thought occurs beginning around 1050 and culminating in what is known as the 12th-century Renaissance. In almost every area of life, there is an attempt to investigate and analyze the problems of society (Cantor 1969:335-338).

STUDENTS/DRINKING; TAVERN LIFE (GOLIARDS). As an outgrowth of the 12th-century humanistic Renaissance, monastic and cathedral schools flourish and give rise to Europe's first universities. Students travel from school to school throughout Europe, taking or teaching classes. Many gain reputations as drinkers and tavern haunters. Drunkenness is rarely treated as an offense at the universities, where the violence of life is "almost equalled by its bibulosity" (Rashdall 1895:613, 686). A typical Parisian tavern frequented by students opens to the street and is filled with benches and stools, with an occasional table board laid on two trestles. Gambling is as customary as drinking, and taverners also serve as pawnbrokers. Little food is sold but it can be brought in. Because students are clerics, and because it is against the law to strike or arrest them unless on the bishop's authority, it is difficult to keep order. The amount of wine students drink, however, is often limited by their finances (Holmes 1952:81, 113-114).

The wandering scholars or *goliardi* (from Golias, the name of their fictional leader) compose songs praising the delights of wine and expressing their love of life and physical enjoyment (Firebaugh 1924:134-138). The two main sources for these Goliardic songs are the *Carmina Burana*, a 13th-century manuscript found at the monastery of Benedictbeuern, and a collection attributed to Walter Map in England. Both sources reveal a common stock of songs widely diffused throughout Europe. These audacious Latin drinking songs often rise to lyric rapture in praise of the wine-god, tavern life, and women. The songs celebrate wine as "the source of pleasure in social life, provocative of love, parent of poetry" (Symonds 1899:41-42). A favorite refrain is "let us drink deeply, then drink once more." The emphasis is on the quantity consumed, rather than the quality (Allen 1961:163). One poem, "The Debate between Wine and Beer," asserts of beer that "its reign is universal and everybody drinks it--kings, hermits, bishops, popes, matrons and maids, old and young, sick and well." But wine is superior (Hanford 1913:327). In "The Contest between Wine and Water," water is attacked as impure and foul and is ultimately defeated by wine (Symonds 1899:151-153). The most famous of the Goliards, the Archpoet, observes that "Never yet could I endure/ Soberness and sadness," and that the quality of his verses varied with the quality of wine he drank. (On the Archpoet, see Haskins 1957:181; Waddell 1949:37.) One poet even declares that he loves the tavern better than the church. Another advocates that "when your heart is set on drinking/ Drink on without stay or thinking."

The "Confessions of Golias," the most famous of the drinking songs, is probably written in Pavia by a German. It proclaims: "In the third place, I will speak/ of the taverns pleasure;/ For I never found nor find/ There the least displeasure..../ In the public-house to die/ Is my resolution;/ Let wine to my lips be nigh/ At life's dissolution:/ That will make the angels' cry,/ With glad elocution,/ Grant this toper, God on high,/ Grace and absolution!" (quoted Symonds 1899:64).

These wandering scholars quickly develop an unsavory reputation, and in the late century a reaction sets in against the liberal, purely humanistic studies of the early century. Giraldus Cambrensis calls the fictional Golias "a certain parasite...notorious alike for his intemperance and wantonness." The word "goliard" soon degenerates into "one of the vilest in the language" (Waddell 1949:204).

In assessing the significance of these songs, Haskins (1957:178, 183) observes of one that it is so contrary to the conventional view of the Middle Ages that it appears almost "unmedieval," and that "the pleasures of love and youth and spring, the joys of the open road and a wandering, carefree life, an intense delight in mere living--this spirit runs throughout Goliardic poetry. Its view of life is frankly pagan, full of enjoyment of the world rather than ascetic anticipation of the next." They are a new, more secular note in medieval life, but wine, women, and song are only peripheral pursuits and did not preclude deep religious devotion. Cantor (1969:377) emphasizes that "in assessing the social significance of the Goliardic and similar student poetry of the twelfth century, it must be emphasized that these same writers who declared that it was their resolution to 'drop down dead in the tavern' also listened in rapt attention to the lectures of Abelard and the sermons of St. Bernard" (see also Waddell 1949:120).

<u>CHURCH ATTITUDES: DRINKING (BERNARD OF CLAIRVAUX)</u>. St. Bernard of Clairvaux (1090?-1153), called "the self-appointed conscience of the mid-twelfth-century church" (Cantor 1969:369), believes in wine as a source of wealth for the Church and as a restorative medicine, but rejects drinking for pleasure. One of the criticisms he directs at the wealthy and powerful monastery of Cluny, in the heart of the rich vineland of Burgundy, is that the monks drink too much and obviously enjoy it (Lichine 1951:65). Bernard is the spokesman for the new Cistercian order and those who seek to reemphasize ascetic tendencies in Western monasticism. He calls for the moral reformation of Europe and a strict reordering of life according to Christian teachings. Bernard reemphasizes the concept of <u>sobria ebrietas</u>, found in the early Church fathers (Bainton 1945:49).

<u>ALCOHOL PRODUCTION</u>. As towns and artisan industries develop, drink production expands from its previous concentration in noble and clerical estates. Urban centers slowly begin to acquire the right to make, sell, and trade wine and beer. Alcohol revenues are an important source of support for town governments. Brewing becomes an urban trade in northern Germany and for the first time good beers are made outside the monasteries.

<u>WINE CONSUMPTION</u>. As wealth increases, so does the demand for wine of high quality. Urban artisans specialize in making the two principal adornments of noble life: wine and precious fabrics. The rise in seigneurial incomes and the "popularization of princely habits" regularizes wine drinking among the whole aristocracy. Merchant fraternities also acquire a taste for wine "in their periodic drinking-bouts" (Duby 1974:236).

<u>VITICULTURE</u>. In the areas of viticulture, especially those where quality wine can be easily transported in bulk, efforts are made by many lords to expand the areas under vines (Duby 1968:216; Duby 1974:238). A great revival of vineyards occurs in the 12th and 13th centuries in the Rhine and Moselle lands, French Switzerland, and Burgundy (Jellinek 1976b:72). Duby (1974:238) comments: "The expansion of wine-production represents a vital aspect of

twelfth-century rural growth. One of the most outstanding investments that lords allowed themselves was to create and improve vineyards. They were thinking primarily of the reputation of their tables, but also of the profit promised by the sale of surplus output."

DISTILLATION DISCOVERED. Several texts indicate that alcohol distillation (from wine) is first discovered in Europe around 1100, probably in Italy and most likely at the medical school of Salerno, an important center for the transfer of medical and chemical theories and methods from the East to the West (Forbes 1970:32). The Arab technique of distillation, especially for the preparation of pharmaceutical and medical compounds, is inherited by the Salernitian school between the 10th and 12th centuries, but there is no evidence of the earlier discovery of alcohol distillation by Muslims, although it is not impossible. Salernus (d. 1167), a physician at Salerno writes a summary of pathology and therapeutics which contains one of the earliest recipes for distilled alcohol; he is the first to actually use the word aqua ardens (Forbes 1970:58, 87-88). Two other, possibly earlier, recipes in the form of cryptograms are made by Marcus Graecus and the author of the Mappae clavicula (Little Key to Printing). Alcohol becomes praised as the best medium for the preparation of pharmaceutical distillates as well as a panacea in itself. However, it is not until the mid-13th century that interest in the process really begins.

ENGLAND

BREWER CHARACTERISTICS. The largest breweries are still located in monasteries, which remain the principal source of ale. As the monasteries grow in number and size after the Norman conquest, brewing also expands (Spiller 1955:92). In the Domesday Book (compiled in 1086), William I's account of his British land and population, references are made to cerevisiarii, who appear to be tenants whose dues were paid by a rent of ale. In the 12th century, references appear regarding the "brewster," or the alewife. For the next few centuries, women do most of the brewing outside monasteries (Monckton 1966:39, 228n1).

ALE CHARACTERISTICS. Alewives produce at least two qualities of ale; monks produce three, with the strength of the brew indicated by single, double or triple X's. Country alewives probably brew beverages that are "thin and often far from having any intoxicating qualities" (King 1947:3).

ALEHOUSE PREVALENCE, CHARACTERISTICS. It is unknown how many alehouses exist in the rural villages of early Norman England: "The historian cannot well say whether it was usual, or rare, or wholly unknown for the rustic inhabitants of Norman England to have a village ale-house to which they could resort for gossip and refreshment" (Lennard 1959:387). However, in the 12th century positive evidence of ale selling at the local level appears for the first time and increases steadily thereafter, causing ecclesiastical authorities to voice greater disapproval of them (Popham 1978:250, 256; Clark 1983:20). Underlying this growth is a population expansion which increases demand and promotes urban renewal and trade. As trade and travel expand, the community and the parish priest can no longer adequately provide their traditional hospitality to travellers, who turn to the alewife for refreshment. Ale selling is especially active in market centers. Special guest houses are built at manors, convents, and monasteries, which later become commercial establishments and form the forerunners of the English public house. Alcoholic beverages are commonly available in monastic and parochial inns (Popham 1978:250-251; Monckton 1966:43). Nevertheless, although the evidence is patchy, it appears that until the 16th century, there are relatively few alehouses in England. Most victuallers continue to sell drinks out of their homes for off-premises consumption. Only in the larger towns such as London or Bristol do alehouses exist in any number and then often under constant threats from the magistrates to close them down. Parish churches and courtyards, not alehouses, are the principal centers for communal life (Clark 1978:50, 61).

CHURCH CONCERNS: INEBRIETY, TAVERNS. Throughout the extant homilies of 12th-century England, there is "ample denunciation" of the vice of drunkenness. The medieval English pulpit is "clear and emphatic" that drunkenness is the mother of all vices (Owst 1966:426, 431). Taverns are especially condemned: "The tavern and the ale-house, apart from the acknowledged fact that they are the occasion of much gluttony and drunkenness,...stand for a very definite menace to the common weal. They have established themselves as deadly rivals to the ordinance of the Church, to the keeping of holydays and fast days, above all to the attendance at divine service" (Owst 1966:435). (For example, see the comment of William Langland, 1375/England).

CLERICS/DRINKING CONTROLS. Archbishop Anselm decrees (1102): "Let no priest go to drinking bouts, nor drink to pegs." Between this and a similar decree by Archbishop Edmund Rich (1236), there are six ecclesiastical regulations passed in a concerted effort to reform and control excessive drinking by members of the clergy (Baird 1944b:128; French 1884:61; Monckton 1966:39, 44-45).

FINLAND

BEER PRODUCTION, CONSUMPTION. Poem no. 20 in the collection known as the *Kalevala* (meaning poems from the district of Kaleva), praises beer and describes in detail the techniques for its brewing, including the addition of hops. The greatest literary monument of the Finns, much of the *Kalevala* dates from the 12th century and before. The poem describes the preparation of the food and drink to be served at a wedding feast and "drinking bout," with a decided emphasis on the importance of the drink and drinking. One passage mentions making beer from barley, hops, and water which can be stored away. This early reference to hops could indicate that the making of hopped beer originates in Finland. (On the significance of hops, see 1200/Europe.) The beer of the Kaleva district is called "the elixir of man" and "a good sort of beer, a drink good for the righteous./ It set women to laughing, put men in a good humor,/ the righeous making beer, fools to joking" (*Kalevala* 1963:131, 137-138; see also Arnold 1911:254-259).

FRANCE

TOWNS/DRINKING. The new town dwellers follow the ecclesiastical and lay aristocracies in the honor they give to the vine and wine, which become status symbols of bourgeois wealth. When vineyard laborers migrate to the cities, they continue to drink wine, even though they are poor and must now purchase it themselves (Dion 1959:475-476).

VITICULTURE; WINE TRADE. At the beginning of the 12th century, a few regions begin to specialize in viticulture (see Salimbene's comments, 1250/France). Vines, along with cereals and livestock, are among the great sources of wealth (Boissonnade 1927:235). Wine has a unique value as a commodity; it is easy to transport and is certain of a market wherever no wine or poor wine is produced. This departure from a subsistence economy is only possible in areas of favorable climate within reach of great waterways or highways. Viticultural specialization spreads slowly, even into the 16th and 17th centuries (see 1600/France). Since commerce and communication remain uncertain, wines are produced for local consumption, even in areas of poor climate (Bloch 1966:22). For example, in areas of Poitou where grapevines are today nonexistent, laymen and clergy set aside bits of land for vines in order to produce their own wines, probably of dreadful quality, but preferable to no wine at all. Probably because of the very damp climate, viticulture is considered a less financially rewarding business than agriculture (Beech 1964:38-39, 104).

TAVERN REPUTATION. As in contemporary England, the taverner often participates in a variety of disreputable callings, and "on his head the cumulated infamy of the whole district" descends in cases of trouble. The poet Daudouville refers to the taverner as the "perverse imp of evil life" and the tavern as "but a lurking hole for things,/ Thou universal fence, where Evil brings/ The spoils of crime to decorate thy table" (quoted Firebaugh 1924:64).

GERMANY

INEBRIETY PREVALENCE, CONCERNS. Hauffen (1889:482) and Jellinek (1946a:540) argue that there is no evidence that drunkenness is a problem in Germany. There is apparently no drink literature: neither songs or stories in praise of drinking nor any sermons directed against inebriety. Neither are there administrative or legislative documents dealing with alcoholic excesses. Folk and courtly epics mention wine only incidentally and without emphasis in their descriptions of banquets (see the *Nibelungenlied*, 1200/Germany). However, this argument ignores the prominence of the praise of wine and drunkenness in the Goliardic poetry which is common throughout 12th-century Europe (see 1100/Europe).

NOBILITY/DRINKING. According to Neuer (1970:9-10), during the Central Middle Ages, the German knightly-courtly class strives for a more elegant form of life, but its members are still rooted in the coarse behavior of pre-courtly customs in which the "sensual life of pure materialism" is the sole ideal. Germans feel inadequate in regard to their manners and decorum compared to the French. Germans embrace the external decorum of chivalry without grasping the inner content of etiquette. There emerges a popular literary genre of texts dealing with table manners (*Tischzuchtliterature*) which is part of the more extensive literature of etiquette and ethics (*Anstandsliteratur*). These texts concentrate on proper adult behavior. In the 12th-century they show that "the festiveness of a social gathering centered on the liquid refreshments rather than on the intake of food. Moreover, while gluttony is frowned upon, it is still more acceptable than excessive drinking. Moderation in drinking is stressed again and again in all texts of etiquette while overindulgence in food is still tolerated to some extent" (ibid., 35-36).

BREWING PRIVILEGES. As towns are established, they are granted the privilege of brewing and selling beer by their imperial or territorial lords. The privilege usually covers the city and a radius around it of one German mile (a biermeile, about 4 3/5 miles). Within the city, brewing usually is a privilege granted to those who own homes. Gradually, professional brewers are appointed to direct the numerous small municipal brewers, and special breweries are erected by the towns, usually outside town limits (Arnold 1911:268-272). A flourishing artisan brewing industry grows up in many towns (e.g., Bremen, Hamburg, Danzig), around which considerable civic pride develops. Princes and nobility own most of the breweries. Because of the danger of fire, the trade is quickly placed under severe restrictions (Cherrington 1925, 1:402, 407).

ITALY

VITICULTURE EXPANDS. After centuries of decline, wine production and exports expand, especially after the Norman conquest of Apulia and Sicily in the 11th century reactivates the wine trade to France. In the 12th century, Italy recovers its old fame for wine and an enormous extension of vineyards occurs (Jellinek 1976b:1).

INEBRIETY PREVALENCE. While many books on viticulture are written in Italy during the Middle Ages, there are few writings on drinking, indicating that the latter is not an important issue. Drinking apart from meals appears not to occur commonly (Jellinek 1976b:1, 3, 4).

NETHERLANDS

BACKGROUND DEVELOPMENTS. Medieval Netherlands (or the Low Countries) comprises 17 provinces. The north is dominated by Holland and Zeeland; the south by Flanders and Brabant (modern Belgium) and by Luxembourg. The wool-manufacturing towns of Flanders make it the greatest textile and trade center and one of the wealthiest areas of Europe.

WINE CONSUMPTION, IMPORTS (FLANDERS). Flanders becomes the largest wine market for northern (septentrionale) French wines in the 12th century. Already at the end of the 11th century, it was the drink preferred by urban merchants. Wine imports have become so substantial by 1125 that during a famine Charles the Good fixes the maximum tariff for retail wine sales in the hope that merchants will switch from importing wine for the rich to importing food for the poor (Craeybeckx 1958:2-3).

RUSSIA

ILLICIT TAVERNS. In the course of the century, the korchmas come to be owned by the government, by princes, by the clergy or monasteries, or by those who run them. The people begin to organize secret korchmas, known as taberna occulta (Efron 1955:492).

1150

EUROPE

CHURCH ATTITUDES. Gratian's Decretum, which systematizes and harmonizes the body of canon law and which becomes the main canonical text for the rest of the Middle Ages, provides canonical authority to the effect that drunkenness is sinful and that frequent and continued drunkenness is a mortal sin.

ENGLAND

MONASTIC BREWING. In the mid-century, the most numerous and best brewers are monks (Cherrington 1925, 1:407). Many parish priests are also brewers and publicans. The monks show the best methods of brewing to their lay assistants who later begin manufacturing beer in their own homes (King 1947:20).

INEBRIETY PREVALENCE; DRINKING PRACTICES. William of Malmesbury (Chronicles of the Kings of England), in the first part of the 12th century, writes a grim description of the drinking habits of the Anglo-Saxons in explaining and justifying the Norman conquest: "Drinking in particular was a universal practice, in which occupation they passed entire nights as well as days. They consumed their whole substance in mean and despicable houses; unlike the Normans and French, who in noble and splendid mansions lived with frugality....They were accustomed to eat till they became surfeited, and to drink till they were sick. These qualities they imparted to their conquerors [the Normans]; as to the rest, they adopted their manners" quoted in Monckton 1966:40; Samuelson 1878:128). (William, it should be noted, is a Norman historian with an anti-Saxon bias.)

Similarly to St. Boniface (750/England), John of Salisbury (d. 1180) contends in a letter that "the constant habits of drinking made the English famous among all foreign nations," and that "both nature and national customs make you drunkards" (quoted French 1884:68-69). (Similar charges are leveled against English crusaders at the end of the century, see 1175/England.) In De nugis curialium, he criticizes Normans for allowing themselves to be overcome by Saxon luxury and for not promulgating laws of temperance (French 1884:57; Jellinek 1944:468).

DRINKING ATTITUDES (JOHN OF SALISBURY). In books 8.6 and 8.8 of the Polycraticus, a treatise on the art of rulership, John of Salisbury provides rules for conduct at banquets, mainly drawn from classical and biblical sources. He "distinguishes between vulgar [plebian] feasts, when the mightiest tippler is considered the best man; and polite [civil] feasts, where sobriety becomes joyous, and plenty does not lead to excess" (French 1884:69). In book 8.6, he writes: "The theory of civil banquets is one of moderation that opulence may enliven sobriety and yet, though satisfying, avoid intoxication; such banquets supply abundance of food and drink, and dispensing as from a horn of plenty, they are sparing in all things in such a way that they may be lavish in all things and are lavish in such a way as to be sparing" (John of Salisbury 1159/1972:318). John's main guideline is that all things must be done in moderation, but he is not opposed to drunkenness per se and writes: "I myself am a drinker of both [beer and wine], nor do I abhor anything that can make me drunk" (quoted Waddell 1949:65n.1).

<u>MONKS/DRINKING</u>. Welsh churchman Giraldus Cambrensis criticizes the luxury of the refectory table of the monks at Canterbury, where there was "wine, mead, mulberry juice, and other strong drink...in such abundance...that beer...found no place there" (quoted Younger 1966:235). He accuses the otherwise praiseworthy Irish clergy of excessive wine drinking at night (King 1947:16; Samuelson 1878:307; Monckton 1966:40).

<u>WINE TRADE (GASCON)</u>. The rich vinelands of Gascony (Bordeaux) are joined together with England after the marriage in 1152 of Eleanor of Aquitaine to Henry Plantagenet, duke of Anjou and Normandy, who becomes King Henry II of England in 1154. Gascony is ruled by the kings of England for three centuries, and Gascon wines capture a monopoly of the English market. (Prior to this time, the wines of Bordeaux served only the local population.) They are given special privileges and the profits of this wine trade help finance the Crown (Pirenne 1933:232; van Werveke 1933:1098; Simon 1920:8; Barty-King 1977:35, 44). As a result, there is a general decline of viticulture in England, "in marked contrast to the continuing progress shown by most branches of agriculture" (Carus-Wilson 1967:266-267).

1175

ENGLAND

<u>FESTIVE DRINKING</u>. In <u>Architrenius</u> (c. 1184), a long satiric poem on the vice and miseries of the age, John de Hauteville relates how the members of an English drinking party emulated each other in their drinking with as much zeal as Ajax and Ulysses contending for the arms of Achilles (Rolleston 1933:41).

<u>INEBRIETY PREVALENCE</u>. William Fitzstephen (<u>Description of the City of London</u>, c. 1180) calls "drinking and fires the plague of London" (quoted Glatt 1958:52).

<u>TAXATION</u>. The first national levy on ale in England occurs in 1188 when King Henry II (1154-1189) introduces a tax on "movables" to support the Crusades: a grant of one-tenth of the value of the stock-in-trade owned by people engaged in trade, including brewers. Previously, ale has not been the subject of regular taxation; any charges which were made upon the brewing or selling of ale were purely local or exceptional in character (Monckton 1966:40-44).

<u>DRINKING REPUTATION; CROSSCULTURAL COMPARISONS</u>. Richard of Devizes, a chronicler of the crusade of King Richard I (the Lionhearted, 1189-1199), criticizes the intemperance of the English troops and its effects on their French allies. He indicates (section 86) that the French are not used to such daily and excessive feasting and drinking and learned from the English "the well-known custom" of draining their goblets to the dregs on a signal. The Palestinian merchants who supply the armies are astounded at such consumption (Richard of Devizes 1192/1848:56). King Richard's own intemperance may have played a role in inflaming the wound that led to his death (French 1884:76-77).

Continentals consider the English drunkards because of their toasting practices (see below) (Holmes 1952:52).

TOASTING. Alexander of Neckham (De nominibus utensilium, c. 1180) comments on the "silly" English custom of "Wassail! Drink-hail!" in which the first drinker picks up the mazer or bowl filled with wine, salutes his companion with a kiss and a cry of "Wassail!" The companion bestows a kiss in return and cries "Hail!" as both get drunk in turn (quoted Holmes 1952:52).

TAVERN PREVALENCE. According to one translation of Walter Map's De nugis curialium, every parish (diocesis) may have at least one alehouse; however, this translation is questionable. The extent to which each village has an alehouse remains unclear (Lennard 1959:405).

FRANCE

TEMPERANCE CITED. French crusaders are said to have been more temperate than their English colleagues (see 1175/England).

WINE RATIONS. The daily pay of carpenters at the abbey of Saint-Germain-de-Prés includes four loaves of bread, four pints of wine, and usually beans, with meat on special occasions (Holmes 1952:256).

IRELAND

WHISKEY ORIGINS. Tradition holds that Henry II of England, when he invades Ireland in 1172, finds distillation of spirits from grain occurring there (Daiches 1969:2). It has been claimed that some of the soldiers anglicized the Gaelic word for aqua vitae uisca beatha into whiskey. But Murphy (1979:142) finds this "extremely dubious on both historical and etymological grounds." The first definite reference to Irish whiskey does not occur until 1406.

1200

EUROPE

<u>CHURCH CONCERNS: INEBRIETY</u>. In the early 13th century, the Franciscan and Dominican orders of mendicant preachers are given papal approval to preach. Oriented towards the towns, from its beginnings mendicant oratory emphasizes "the follies of the inebriate both ludicrous and tragical...in all their grim reality" (Owst 1966:426).

The English <u>Speculum laicorum</u>, setting forth <u>exempla</u> of proper lay conduct, contains a whole section "On Inebriation and its Evil," including 10 warning anecdotes. Inebriety must be shunned because it "befools, enfeebles, and impoverishes man and hastens his death" (ibid.)

<u>DRINKING CONTROLS</u>. At various regional councils and synods, monks and nuns are banned from attending the Feast of Fools (Paris, 1212), clerics are forbidden to enter taverns (Paris, 1197; London, 1200; Salzburg, 1274; Tours, 1282), goliards are ordered to have their tonsures shaved off (Rouen, 1231), and penances are called for priests, deacons, and clerics found drunk (Angers, 1219-1220, reiterating that priests must do 18 days penance if drunk by negligence, 40 days if <u>per contemptum</u>. At the Fourth Lateran (Ecumenical) Council in 1215, the clergy is ordered to abstain from drunkenness and gluttony and to discourage others from the same, "since drunkenness doth banish wit and provoke lust." Further, it is decreed "that abuse shall be utterly abolished whereby, in divers quarters, drinkers bind one another to drink healths or equal cups, and he is most applauded who quaffs off most carouzes." The clergy is again banned from frequenting taverns (quoted French 1884:193; see also Dozer 1980; Le Grand d'Aussy 1782, 3:274). The Synod of Rouen in 1235 specifies three days penance without meat, wine, or <u>cervisia</u> for a lay person guilty of excessive drunkenness that results in vomiting.

<u>SALES PRIVILEGES</u>. Evidence exists that some peasants are resisting the lord's privilege (<u>banvin</u>) on wine sales (Coulton 1925:59).

<u>HOPPED BEER INTRODUCED</u>. In the 12th century, an artisan brewing industry develops in the flourishing regions of northern Europe. In the 13th century, towns such as Hamburg, Danzig, and Munich begin to specialize in making good beers which are exported. These beers are increasingly being brewed with hops, particularly in northern Germany. Hops give beer a clearer appearance and a less sweet (more bitter), aromatic taste; they also act as a preservative, lengthening its life and enabling stronger beers to be brewed. The beer produced is actually "a new drink altogether, a product of the technique of precise fermentation using only barley, and in which addition of hops ensured an agreeable taste and the possibility of better conservation" (Claudian 1970:10). Together with improvements in transportation, the growth of cities, and capital accumulation, the introduction of hops makes possible the development of a commercial brewing industry as large quantities of beer are produced, stored, and transported to meet rising urban demand.

Precisely when and where brewing with hops began is uncertain. Apparently the plant comes from east Europe or Finland, and it has been suggested that the Finns may always have brewed hopped beer (see 1100/Finland). Evidence of the cultivation of hops exists before evidence that hopped beer is brewed, although Monckton (1969:29) cannot imagine any other use to which they would be put than brewing. The first reference to hops in Bavaria appears around the mid-8th century. King Pepin (d. 768), the father of Charlemagne, was given a gift of hop yards (Cherrington 1925, 1:135, 402). But no Carolingian capitulary mentions them. In the 9th century hop, cultivation is reported around the convents of the Rhine, Brabant, and the Ile-de-France (Claudian 1970:10). Several 11th-century texts mention hops without mentioning beer, but it was probably being made in the monasteries of Roman Gaul. Hopped beer appears in northern Germany around 1070 (Spiller 1955:146; see also Jacob 1935:83; Cherrington 1925, 1:402). The Physica Sacra of St. Hildegard of Bingem (d. 1179) discusses their preservative qualities (book 1) and refers to the brewing of beer from oats and hops (book 3.27) (Arnold 1911:229-231). The use of hops in beer is retarded and provokes centuries of controversy in some areas because it threatens the monopolies over "gruit," a bitter, more costly mixture of vegetable substances used to flavor beer, where compounding is a closely guarded secret (e.g., see the archbishop of Cologne's antihop decree, 1375/Germany). Such interdictions last until the 16th century when the gruit monopolies are abolished (Arnold 1911:241). Many popular German beers, such as that of Bremen, are still made with gruit in the 13th century, but it is probably the use of hops that accounts for the popularity of Hamburg beer in the second half of the century and the quick rise in its export. Hopped beer does not reach England until the beginning of the 15th century.

ENGLAND

LOWER CLASSES/DRINKING. Thirteenth-century peasant life is a "rather miserable existence." For daily consumption, the ale is thin and not very heady. "Moist and corny" ale is consumed only on special occasions, as when peasants are entertained by the lord. The drinking of this weak ale accompanies all meals: at breakfast it accompanies a lump of bread; the midday meal, a lump of bread or cheese, possibly with an onion for flavor; and the main meal at the end of the day. It is considered a presumption for landless laborers to drink anything more than the weakest penny ale (Bennett 1969:5, 13, 236).

TAVERN, SUNDAY DRINKING. Many villagers brew their own ale but there is room for some village alehouses where "bride ales" or convivial meetings are held. These alehouses are outside the manorial rule, except for the payment of a tax on the ale brewed and the general supervision of its quality. Unless disorders arise there, the lord cares little that his men spend their time "lounging on the tavern-bench, and drinking and merrymaking" til the stars appear. On Sundays, some peasants might leave mass early (as soon as communion finished) to go to work; others, "sons of Belial," go to the alehouse, which the peasants are told by the friars is the "devil's chapel." After mass, villagers often dance until dusk, at which time journeys to the alehouse become more common (Bennett 1969:17, 266-268).

DRINK AND WORK. Often the manor lord holds "wet boons," days when special labor on his lands are required by the peasants at which the lord furnishes ale, generally a weak secunda cerevisia (not the melior cerevisia made for the priory). Custom specifies that the peasants are allowed to "drink at discretion" (Bennett 1969:17, 110-111).

CONTROLS: PRICE, QUALITY. The production of ale increases during the century, and a need for more careful control of, and safeguards over, a product considered a national necessity arises. The 13th century marks the start of a much more vigorous attitude on the part of the authorities towards controlling the price and quality of ale. Between 1225 and 1697, some 83 statutes are passed which have to do with measures, standards of price and quality, import and export regulations, licensing, and taxation of alcohol (Baird 1944b:149). Most early licensing legislation is designed to protect the drinker, rather than the public at large, and to ensure the quality of all ale sold. Control at this early stage is essentially local (Longmate 1968:3-4; Monckton 1966:47-48).

WINE TRADE, CONSUMPTION, PRICE CONTROLS. Eager to encourage wine trade and consumption as a source of revenues, King John passes the earliest statute dealing with the foreign wine trade (Assize of 1199). He grants protection to French wine merchants and establishes a low maximum price for which the wines of Anjou, Poitou, and France (meaning the region around Paris) may be sold in order to promote their use. (No mention is made of Gascon wines, indicating they have yet to become important imports.) John himself seeks to profit by buying and selling wine (Simon 1907, 2:73-79; Longmate 1968:3-4; James 1971:172-173). Violators will be arrested and imprisoned, and their goods seized until their release. To enforce the assize, two inspectors are to be selected in every borough and city. According to the contemporary Annals of Roger de Hoveden (1200/1968, 2:446-467), the price controls were abandoned because of the harm it inflicted on merchants, but "by this means, the land was filled with drink and drinkers" (quoted Pirenne 1933:234). (On the antagonism toward the crown's price-fixing policies on wine in medieval England, see also Simon 1906, 1:308.)

INEBRIETY OF KING. King John (1190-1216) is said to drink copiously and to have died in drunkenness (French 1884:78-79).

BREWERS GUILD. The ale brewers of London probably first join together in a guild in the 12th century; by the mid-13th century they are a wealthy group (King 1947:37-38).

DRINKING CONTROLS: SCOT-ALES BANNED. Concerns grow over the prevalent practice of scot-ales. At least in some 13th-century estates, the lord has the right and duty to hold three a year; sometimes they last three days. Tenant customs specify how much they are allowed to drink and how much they pay, and often the "scot" fails to cover the expenses of the lord. People living near alehouses are often compelled to come in and join the scot-ales. As complaints against the system mount, regents of King John (during his absence in 1213) forbid scot-ales on the grounds that they are occasions for extortion. The legislation appears to be directed at the officers of the forest, which are outside common law, who keep alehouses and draw customers by intimidation. The regents' decree is unsuccessful, and throughout the Middle Ages the Church continues to issue many injunctions against these ales (French 1884:83-84; Monckton 1966:45-46; Coulton 1925:28-29; Coulton 1957:287).

FRANCE

ESTEEM OF WINE. Philip II Augustus (1180-1223) orders the provinces to send specimens of their wines to Paris, where a national exhibition is held in 1214 (Butler 1926:146). The fame of the wines of Auxerre, Anjou, St. Pourcain, Aunis, and Saintonge spreads (Duby 1968:138).

PRICE CONTROLS: WINE CRIERS. King Philip issues an edict that any member of the corporation of criers may enter an inn, select some wine, fix the price at which the wine should be sold, and announce it to the citizens. The taverners are soon continually at odds with the criers. Taverners are obliged to contribute well for this service, which becomes a racket (Hopkins 1899:122-123; Emerson 1908, 2:162; Holmes 1952:80). (On the criers, see also Boilleau 1268/1879:21.)

WINE TRADE. The wine trade flourishes between producers in southern France and consumers in northern Europe, particularly England, Flanders, and other Low Countries. By 1213, a corporation of *marins de Bayonne* is transporting wine from La Rochelle to Flanders and probably England (Pirenne 1933:233).

SALES PRIVILEGES. In 1192, the king grants to the citizens of Paris the sole right to import wines into their city via the Seine and to sell it directly from their boats. Any one else who wishes to bring wine into the city has to first associate himself with a Parisian (DiCorcia 1978:215).

TAVERN LIFE. In three 13th-century vernacular dramas--the *Jeu de Saint Nicolas* by Jehan Bodel, *Courtois d'Arras*, and *Jeu de la feuillée* by Adam de la Halle--the tavern plays a central role. They show that the tavern is a convivial meeting place of a heterogeneous urban society and that it is viewed either as a *castellum diaboli* or, conversely, as the utopian center of urban life. Drunken excesses and games of chance are common, as are thievery and prostitution. In *Courtois d'Arras*, the main character's entire inheritance is stolen by two prostitutes who convince him to give up his money so he will not be tempted to drink; at the same time, the tavern resembles the legendary *pays de cocagne* (Dozer 1980:4-6, 211, 249; Firebaugh 1924:139-151).

STUDENTS/DRINKING; CARNIVAL FESTIVITIES. The University of Paris is banished for a while because of a Carnival riot in which students beat up an innkeeper and poured his wine stocks into the street (Henisch 1976:39).

GERMANY

DRINKING PRACTICES, ATTITUDES (THE NIBELUNGENLIED). The *Nibelungenlied* is composed in Germany between 1190-1205 and immediately gains great popularity. Although the work is based on Germanic folk poetry, it is "heavily overlaid by anachronistic chivalric sentiment" (Cantor 1969:108). Thus, it probably more accurately reflects the drinking practices and attitudes of the late 12th century than actual old Germanic drinking practices. Feasts are held throughout the epic, some lasting a week or two, during which considerable drinking occurs. Surprise is expressed at the absence of wine at a hunting feast, and explanations for this absence are offered. This would "warrant the assumption that wine was a significant part of the meal in such [aristocratic] circles of the time." Yet drinking and drinking prowess are not stressed, and "the poet gives little inkling of anyone's having had too much" (McKinlay 1953:90-91).

In the etiquette literature created by and for the courtly circle, the topic of drunkenness is largely ignored in the early 13th century (Neuer 1970:55-56).

BEER TRADE. Bremen is exporting gruit beer to Bruges and other cities in the Low Countries. For a long time, this is the most famous commercial beer (Arnold 1911:290-291; Dollinger 1970:41, 75)

ITALY

TAVERN CHARACTERISTICS. In 1189, the first Florentine reference occurs to a taverner (*tabernarius*), who, unlike the operator of a wine shop, sells both food and drink. In 1211 appears the first reference to a host (*oste*) or innkeeper, probably "a superior and prosperous sort of *tabernarius*." In 1236, wine-shop operators, taverners, and innkeepers are all merged into the guild of wine merchants (Arte de' Vinattieri), from whose members will come some of the great Florentine families (Staley 1967:361, 363, 368; Jellinek 1976b:1).

NETHERLANDS

WINE CONSUMPTION. In the 13th century, the Flemish increasingly consume the wines of France. By the mid-century, the bourgeois of Ghent are differentiated from peasants and the artisans (*gens de métier*) as people who drink wine at their table. Frenchmen marvel that Flemish nobles have so much wine that they do not mix it with water; Guillaume de Breton (*Philippidis*, 1214) comments on the large amount of wine imported from La Rochelle and the profits the merchants derive from it (Pirenne 1933:234-236; Pirenne 1937:153-154; Craeybeckx 1958:2-3). By mid-century, Bruges emerges as the central market of the Western world.

1225

ENGLAND

CHURCH CONCERNS: SCOT-ALES, PEG DRINKING. Between 1230 and 1250, prominent churchmen such as Robert Grosseteste, the bishop of London, seek an end to scot-ales, peg drinking, tavern drinking by priests, and other abuses of drink (Monckton 1966:44-45; Samuelson 1878:140).

GERMANY

TAXATION. The first tax on beer in Germany is levied in the city of Ulm. It pertains mostly to imported beers, since the local water cannot be used for brewing. Revenues from the tax go to the municipal authorities, who use a portion of it to set up *Rathskellers*, or drinking cellars in or near the city hall (Dorchester 1884:52).

1250

EUROPE

LOWER-CLASSES/DRINKING. Throughout Europe, it is evident that the peasantry and lower-classes rarely consume much wine. The story of Meier Helmbrecht tells of the lack of wine at the festivities thrown by a rich Bavarian peasant when his son returns home from court. For the yearly feast of one estate, it is declared that fowls and wine should be on the table, but "the serfs shall be fed without fowls and wine" (Coulton 1925:314, 316; Haufen 1889:482).

SPIRITS: DISTILLATION SPREADS; MEDICINAL USE. In the second half of the century, knowledge about the process of distillation begins to spread slowly throughout medieval Europe. Knowledge is retarded because the world of alchemy is jealously guarded (it is believed that any substance prepared according to hermetic tradition loses all virtues if the mode of preparation is revealed to the uninitiated). Early manuscripts are rare and couched in mysterious terms. Doctors faithful to the Galenic tradition also regard spirits with suspicion (Claudian 1970:11, 18). At first, alcoholic spirits of wine (containing enough alcohol to burn) are produced only on a small scale as costly medicines, or as a menstruum for medicine, mainly by monks, physicians, and apothecaries. Distilled alcohol is extravagantly praised as the elixir or water of life, aqua vitae. The "spirit" is seen as inducing a sense of youth, vigor, and buoyant health (Leake and Silverman 1966:10). In its original form, aqua vitae is made only from wine, what becomes known in English as "brandy" by the middle 17th century (from the Dutch brandewijn or burnt wine) (Lucia 1963a:99; Jellinek 1976b:77; Roueché 1960:21-25). It is also called aqua ardens; a stronger spirit produced by repeated distilling becomes known as quintessence.

ALBERTUS MAGNUS. The Dominican theologian and scholar Albertus Magnus (1193-1280) is possibly the first person in western Europe to describe in detail the process of distillation. A text ascribed to him (De secretis mulierum) considers distilling a most important method in alchemy and provides two recipes for aqua ardens. However, these may have been written by one of his pupils around 1300 from Arab recipes (Forbes 1970:58).

ENGLAND

ALE PRICE CONTROLS. In 1267, King Henry III (1216-1272) introduces the Assize of Bread and Ale (Assiza pani et cervisiae), which becomes the means by which the prices of these products are controlled for the next 300 years. The assize consists of a periodic local announcement fixing the price of these two staples according to the prevailing price of corn and malt. In addition to insuring that a fair price is charged, it also seeks to prevent adulteration. There are heavy penalties for brewers who fail to comply with the terms of the current assize. Enforcement is the responsibility of the manorial court. In the following years, other assizes are decreed for London which adjust prices and amend brewers' obligations according to the needs of the time (Monckton 1966:48, 50; Monckton 1969:20-21).

To determine if the royal assize is being met in terms of quality and price, the job of official aletaster (known as an ale conner in London) develops. The aletaster assesses the ale before it is sold. The alehouse keeper is required to be summoned whenever a new brew is prepared. If the assize is violated, the manor lord takes a fine from the offending brewer, although not all lords are given the privilege of holding the assize from the Crown. The assize is so regularly broken that the fines amount to as regular an income for the landlords who collect them as any excise system could have produced. The aletasters are also given the authority to reduce the price of the ale if the quality is too low. In London, the first reference to the duty of the civic post of ale conner does not appear until around 1377 (Coulton 1925:85; Homans 1941:238, 312-313; Monckton 1966:52-53; Monckton 1969:24).

DRINKING PRACTICES (SALIMBENE). On English drinking practices, Fra Salimbene (1287/1961:234) writes: "The English...delight in such things [wine drinking] and like to drain full cups. For when an Englishman takes a large glass of wine, he will drain it and say...'You must drink as much as I.'" He believes he is saying and doing a great act of courtesy and he will take offense if some one does otherwise." He finds the pleasure with which they drink wine somewhat excusable since there is so little wine in England. (Salimbene is discussed in more detail under 1250/France.)

CIDER CONSUMPTION. The making and drinking of cider spreads in England as new varieties of apples are introduced (French 1982:11).

FRANCE

VITICULTURE; WINE TRADE. The demand for wine has not only created a market economy for it, but the wine trade has helped stimulate the entire rural economy by creating a demand for cereal supplies in viticultural regions and by animating currents of exchange between wine and grain areas (Duby 1968:139-141). Fra Salimbene of Parma, who lived in France from 1247 to 1249, observes in his Chronicle (written 1282-1287) that the three great wine producing areas of France are Beaune and Auxerre in Burgundy and La Rochelle in Charente. Passing through Auxerre in 1245, Salimbene is amazed at the degree of viticultural specialization that exists there: "The people of the region do not sow, or reap, or gather into barns, but they send their wine to Paris, because next to the village, there is a river that flows to Paris. They sell it at a good price and from this they get all their food and clothing" (Salimbene 1287/1961:230 [author's italics]).

The wines of the Midi and southeast are too far from the large centers of consumption to be well known, and their trade is blocked by the wine interests of Burgundy and Languedoc. The dukes of Burgundy promote the popularity of Beaune wines, which had earlier also not been well known because the area is not situated near waterways. They were also considered too heavy and strong (Duby 1968:138). Bordeaux, which is still an English possession, doubles in size due to the expansion of the wine trade with England. So profitable is the trade, and so certain is the market, that no restrictions are placed on the growing of vines in favor of cereals in Gascony, in contrast to other areas in France (Carus-Wilson 1967:270). The vineyards of the Bordeaux country penetrate far up the course of the rivers (Duby 1968:138).

ESTEEM OF WINE (SALIMBENE). In describing his stay in Auxerre, Fra Salimbene (see above) shows a great appreciation of the local wine and the value of wine drinking in general: "Note...that the wines of Auxerre are white, sometimes golden; they give off a delicate aroma; they are very comforting and very delicious; they give to all who drink them peacefulness and cheerfulness, and they transform them entirely; thus the passage of the Book of Proverbs (31, 6-7) can be applied to the wines of Auxerre: 'Give strong drink to one who is perishing, and wine to the sorely depressed; when they drink they will forget their misery, and think no more of their burdens.' And note that the wines of Auxerre are so strong that, when they have been for a time in jugs, tears gather on the outside....The French...enjoy drinking good wine; and this is not astonishing, for good wine 'cheereth God and men,' as is said in the Book of Judges (9, 13). For good wine rejoices the heart of man and gives peace and joy to his spirit" (Salimbene 1287/1961:231, who goes on to praise the virtues of wine based on numerous loose quotations from various Scriptural passages).

TAVERN CONTROLS. In 1256, King Louis IX (St. Louis, 1226-1270) bans taverners from serving drinks for consumption on premise to anyone except travelers passing though the town. Residents may only buy drink *à pot*, to be consumed off-premises. This is one of many tavern rules, most of which deal with clients and their activities (gambling, prostitution), not drinking per se. The purpose of the ban is "to restrain the criminal proclivities of the tavern-haunting criminal classes," to keep down the population frequenting the inns, and to render them less dangerous to the peace. A later law requires innkeepers to maintain a book of guests by name (Dion 1959:487; Firebaugh 1924:124; Hopkins 1899:109-110). Enforcement may not have always been strict. In 1320 poet Watriquet de Couvin tells of "Trois Dames de Paris" who go to a famous tavern and get drunk (Dion 1959:485, 487).

SALES PRIVILEGES. A law of 1268 declares that when the king sends his wine to the market, only it can be sold, and criers must announce its availability morning and evening at the crossways of Paris (Hopkins 1899:123).

INEBRIETY PREVALENCE. At least in part because of the tavern controls of 1256 and their repeated renewal, there is "no trace of the profound and general unrest that goes by the name of alcoholism" in France through to the end of the 16th century (Dion 1959:487-488). Subsequent tavern controls are justified on the basis of health, security, and religion, but not on the need to prevent public drunkenness.

According to Salimbene (see above), "The French and English are persistent in their draining full cups. As a result, the French have purulent eyes, because they drink so much wine. Their eyes are red, bleary, and rimmed around. Early in the morning, after they have slept off their wine, they go, with eyes like that, to the priest who has celebrated Mass and ask him to put into their eyes the water he has washed his hands in....[The priest would respond] "Go! and God give you an evil lot! Put the water into your wine when you drink it, and not into your eyes....One must forgive the English if they drink good wine with such greater pleasure whenever they can, for they have but little wine in their country. The French are less excusable, for they have much more wine; but of course some will want to plead that it is hard to give up a habit" (Salimbene 1287/1961:233-234). (For his comments on English drinking practices, see 1250/England).

BREWING. In 1268, Louis IX issues statutes for the guild of *brasseurs* or *cervoisiers*, indicating the existence of "small breweries run by specialized artisans beyond the ordinary domestic trade form" (Jellinek 1976b:76). Ingredients are specified for strong and weak beers, and mixing the two qualities is forbidden (Boilleau 1268/1879:21; Gottschalk 1948, 1:306; Arnold 1911:364-366).

CHURCH CONCERNS: WAKES. A Church council in Cognac declares: "Since the wakes which are held in churches frequently lead to many foul acts and end very often in bloodshed...we decree and firmly command that no wakes be held henceforth in the aforesaid churches or cemeteries" (quoted Coulton 1925:272-273).

GERMANY

INEBRIETY PREVALENCE. In the mid-13th century, excessive drinking appears to increase and become more of a problem, possibly in conjunction with the deterioration of the lower nobility and knightly class. Ulrich von Lichtenstein (d. 1275/6), one of the last minnesingers, in commenting on the deterioration of courtly love, complains that wine is taking the place of women. The preacher Berthold von Regensburg (d. 1272) warns against drunkenness to large crowds throughout Germany. The first specimens of German drink literature also appear. In these stories, disapproval of individual excessive drinkers is expressed, but none of them allege widespread intemperance, and their tone is light and approving of the pleasures of drinking (Jellinek 1945:651-652; Jellinek 1946a:541; Hauffen 1889:483-484).

In the Wiener Mervart (c. 1254-1283), written by "der Freudenleere" (the Joyless One), it is suggested that drinking is a means of recovering that simple, joyful bliss that a commercial age has deprived man. The story deals with a tavern drinking party of some wealthy Viennese merchants whose lives are focused on the acquisition of money. The author's indignation is directed not against the drunken behavior that occurs, which is portrayed as a comical situation, but against the conditions in the world which make drunkenness necessary for relief, whereas in "olden days...people could enjoy themselves in virtuous ways." While the pleasures of drunkenness are praised, it is also recognized that drinking can cause harm: "Now hear what wine will do: It brings good cheer to men and women when they're sad and if they use it with good sense. But those who overdrink get hurt in honor, body and in their possessions." The participants themselves are not habitual drunkards but responsible citizens who go too far on this one occasion (Jellinek 1945:652; Jellinek 1946a; Hauffen 1889:482).

According to Jellinek (1945:651-652), this increase in drinking occurs because the stable political and social conditions which moderated drinking in the 10th through the 12th centuries begin to crumble during the 13th century with the gradual break up of the Holy Roman Empire into independent princely states. Knights, deprived of their former standing and prestige, turn to force and drunkenness to reassert their manliness; peasants and city dwellers seek in drink and the tavern some consolation from the resulting disorder, uncertainty, and violence of the time.

BEER TRADE. The city of Hamburg develops a flourishing beer trade as a result of the city's use of hops in the brewing process instead of gruit which is still being used in the lower Rhine and the Netherlands. Hamburg also has a flourishing hop market. In 1241, Hamburg and Lübeck form the first Hansa (compact) to obtain control of the Baltic trade. Eventually some 85 towns between the Netherlands and Livonia are members of the Hanseatic League. Beer is one of the most important products exported from Hamburg, along with Rhine wines and linen. City charters and remains of old houses in Hamburg show that brewing is an integral part of the town's industries in the mid-century. All the burghers engage in commercial brewing, which soon furnishes the largest proportion of the city's revenues and becomes one of the main sources of the city's wealth. Beer has become a specifically civic trade subject to town regulations. By 1270, output has increased beyond domestic requirements and beer is being shipped to Norway and Sweden. In 1276, the Brewer's Ordinance is established. Soon Hamburg beer is being shipped throughout the Hanseatic world, especially Holland (Arnold 1911:274, 277-285; Cherrington 1925, 1:406). Hamburg's main competitor later becomes Lübeck.

The first reference to a private brewery in Freiburg occurs in 1262, marking the beginning of commercial brewing separate from the monasteries (Klemm 1855:332).

1275

EUROPE

SPIRITS PRAISED; MEDICINAL USE. In the late 13th and early 14th centuries, the knowledge of distillation spreads in monastic, medical, and alchemical circles. Three individuals play especially prominent roles in this spread: Raymond Lull, Thaddeus of Florence, and Arnald of Villanova.

RAYMOND LULL. The Franciscan monk Raymond Lull (1235-1315)--a visionary, alchemist, teacher of Arabic, philosopher, and missionary to the Muslim world--describes the preparation of alcohol, which he considers a "marvellous medicament" to be used in serious illness (Claudian 1970:18). He advises its consumption before battles to stir and encourage soldiers' minds (Simon 1948:131). It is "an emanation of the divinity, an element newly revealed to man but hid from antiquity, because the human race was then too young to need this beverage destined to revive the energies of modern decrepitude" (quoted Lichine 1967:6). He recommends three rectifications as sufficient but seven if one wants to make a "quintessence." This is perhaps the first reference to absolute alcohol (Forbes 1970:59-60, 89). (While no doubt Lull distilled alcohol, many of the texts ascribed to him are of a later date. It has been suggested that Lull introduced spirits into the English court when he visited there in 1300).

THADDEUS OF FLORENCE. The founder of the Medical School of Bologna, Thaddeus of Florence (1223-1303), praises the therapeutic value of aqua vitae in *De virtute aquae vitae, quae etiam dicitur aqua ardens* (On the Virtues of the Water of Life, Which is also Called Fiery Water) (Lucia 1963a:99). He is a pioneer in the method of cooling the distillate after it leaves the still-head, therefore paving the way for the modern method of cooling vapors outside, instead of inside, the still-head and collecting the condensate in the alembic itself. He describes a "worm cooler," or cooling coil, which runs through a tub of water (instead of relying on the cooling of a long delivery tube by air). His method "is the only efficient way of producing low boiling distillates like alcohol" (Forbes 1970:60-61, 83-84).

ARNALD OF VILLANOVA. Arnald of Villanova (alchemist and professor of medicine at the University of Montpellier) (d. 1315) allegedly may have been among the first to praise aqua vitae as a medicine and possible key to everlasting life, asserting that "it is well called water of life...because it strengthens the body and prolongs life" (quoted Lucia 1963a:107; Forbes 1970:62). Arnald's authorship of this passage has been questioned, although his books on wine (see below) and the preservation of youth show he was familiar with aqua vitae and considered it as having miraculous therapeutic properties (Lucia 1963a:108; Arnald of Villanova 1310/1943; Braudel 1981:242-242).

MEDICINAL USE, WINE (VILLANOVA). Arnald of Villanova recommends the medicinal use of 49 different wines in his Liber de vinis (1310), which becomes one of the most popular medical works of the 15th and 16th centuries. Arnald helps establish the use of wine as a recognized system of therapy during the late Middle Ages, asserting: "If wine is taken in right measure it suits every age, every time and every region. It is becoming to the old because it opposes their dryness. To the young it is a food, because the nature of wine is the same as that of young people....No physician blames the use of wine by healthy people unless he blames the quantity or the admixture of water....Hence it comes that men experienced in the art of healing have chosen the wine and have written many chapters about it and have declared it to be a useful embodiment or combination of all things for common usage. It truly is most friendly to human nature" (Arnald of Villanova 1310/1943:24-25; also quoted Lucia 1969a:104).

MEDICINAL INTOXICATION. Arnald of Villanova also discusses the medical value of periodic intoxication, a belief that remains prominent through the 18th century: "There is undoubtedly something to be said for intoxication, inasmuch as the results which usually follow do certainly purge the body of noxious humours. But one should not do so too often; for a normal man, no more than twice a month" (quoted Emerson 1908, 2:140-141).

ENGLAND

CONTROLS: TAVERNS, CURFEW. An act of King Edward I (1272-1307) in 1285 forbids London taverns to operate after curfew because unsavory and criminal elements meet at night and "hold their evil talks in taverns more than elsewhere, and there do seek for shelter, lying in wait and watching their time to do mischief." The act also regulates the price of wines sold in taverns, stipulating that "if the taverners exceed, their doors shall be shut" (Monckton 1966:48-51). This is another example of how medieval tavern controls are not simply or even largely concerned with temperance (Popham 1978:256).

PRICES. It appears that by the end of the century, the cost of a gallon of even cheap ale is equal to a third of the daily earnings of a craftsman and two-thirds of a laborer (Clark 1983:24).

ATTITUDES: LAND OF COCKAIGNE. According to a popular late 13th-century English poem, "In Cockaigne is meat and drink/ Without care, trouble, and toil/ The meat is choice, the drink is clear,/ At dinner, draught, and supper" (quoted French 1884:90).

GERMANY

NOBLE YOUTHS/DRINKING. Konrad von Haslau's Der Jügling attacks the inceasingly coarse table manners of courtly youths, including their spilling wine all over their clothing. This is seen as reflecting the replacement of the former ideal of temperance and "austerity of morals and manners" by "profligacy" and a taste for epicurean living as the knightly order begins to decline (Neuer 1970:64).

IRELAND

WHISKEY CONSUMPTION. In one of the earliest possible references to whiskey, Sir Robert Savage, landlord of Bushmills, fortifies his troops before battle in 1276 with a "mighty draught of aquavita, or water of life" (Tuohy 1981:7). However, the reference may also be to imported brandy.

ITALY

TAVERN PREVALENCE. In large part because of the numbers and affluence of visitors to Florence, the number of innkeepers and retail wine merchants in the city and the *contrado* has risen to 86 (Staley 1967:371).

1300

EUROPE

BACKGROUND DEVELOPMENTS: OVERPOPULATION, DEPRESSION. Around the mid-14th century, medieval civilization enters its late declining phase. As early as 1300, characteristics of this late period are already becoming apparent, particularly the beginnings of an economic depression and population decline which lasts until 1450. This crisis is apparently triggered by an imbalance in the relationship between food production and population. Due to overpopulation, by 1300 almost every child born in western Europe faces the probability of extreme hunger at least once or twice during the expected 30-35 years of life. Food crises become common, beginning with the famine of 1315-17. Mortality rates rise, and the population begins to decline, a situation aggravated by the mid-century Black Death and other outbreaks of epidemic diseases, whose devastating effects may have in part been due to the already-weakened resistance of the population. The decline in population has several ramifications for viticulture and drinking patterns (as discussed under 1350/Europe).

SUNDAY, FESTIVE DRINKING. Church leaders voice concerns over the drunkenness and irreverence that occurs on Sundays and other holy days. At the General Council of Lyons (1274), Cardinal Humbert de Romans recommends that no new holy days be added to the Church calendar "both because on feast-days there is a greater increase of sin in taverns, dances and brothels, by reason of evil idleness, and also because the working-days are scarce enough for the poor to get their livelihood." Bishop Guillaume Durand of Mende, speaking before the pope at the General Council of Vienne (1311), asserts that "more evils are sometimes committed on Sundays and festivals than in all the rest of the week." Durand is scandalized by church-ales "at which both clergy and layfolk swill (se ingurgitant)." On the same occasion, Guillaume le Maire of Angers concurs: "On those holy-days, on which God ought more especially to be worshipped, the Devil is worshipped; the churches are left empty, the lawcourts and taverns and workshops resound with quarrels, tumults, blasphemies, and perjuries, and almost all sorts of crimes are there committed" (quoted Coulton 1925:28, 272).

DISTILLATION. Knowledge of distillation spreads at different times in different places, with Italy, Germany, and north Europe in the vanguard. The outbreak of the Black Death at the mid-century accelerates the spread of the medicinal use of spirits. Monasteries still play the principal role in the dissemination of distilling techniques, followed by apothecaries, who later become the first large-scale producers. Whereas apothecary shops were previously closely connected with town governments or local universities, they are now more often privately owned. There may also be some professional alcohol distillers among vintners and innkeepers, but this is uncertain (Forbes 1970:90-91, 95; Claudian 1970:12).

BREWING. Commercial beer brewing expands, in part as the result of a concentration of trades which were formerly performed in the house or monastery (Forbes 1970:57). Hanseatic merchants from Hamburg and Danzig export and help spread the taste for hopped beer throughout northern Europe.

ENGLAND

CONTROLS: DRINKING, ALEHOUSE. By the 14th century, there is no clear evidence of secular laws against drunkenness by laymen, but several kinds of laws have been issued to control the traffic in and use of liquor: (1) penalties for breach of peace; (2) tax and revenue measures; (3) penal provisions against clerics thought derelict in their duties; (4) fixed standards of price and quality; (5) protection of established settlement and reconciliation procedures; (6) supporting provisions to fixed standards (i.e., limiting the number of alehouses); and (7) drinking-to-pegs regulation. National regulations are "merely restrictive in an indirect manner," being primarily concerned with ale purity and prices. Local regulations over alehouses are more restrictive (Iles 1903:252). The primary action against lay drunkenness is still the penances required by Archbishop Theodore in the 7th century (Baird 1944a:556). French (1884:106) characterizes the 14th-century legislation as "of an in-and-out character. It enacted and repealed, repealed and enacted." As the ale industry grows in importance, authorities increase their attempts to exert control over it. Alehouse keepers are required to pay some sort of fee for being in business. The Liber Albus (compiled in the reign of Henry V [1413-1422]) sets out various assizes with which alehouse keepers and taverners must comply. Brewers are punished for defrauding the public by using measures which are illegal or which fail to hold their reputed content (Monckton 1966:52, 56, 60). The book also includes in full the oath of the ale-conners, the officials appointed to look after the quality of ale (French 1884:114).

ALEHOUSE, TAVERN PREVALENCE. Ale selling has become a common economic activity both in towns and countryside. In one village around 1300, an estimated 60% of all families in some way appear connected with brewing and probably selling ale. In 1309, London (population 30,000-40,000) has 354 taverns and 1,334 alehouses, possibly one drink shop for every 12 inhabitants (Mead 1931:243n.7; Clark 1983:21). The urban ale trade is acquiring a more organized and regular character, offering drinks for consumption on-premises and various forms of social services, including serving as a social center. Underlying this expansion is the continued growth of population, trade, and towns. At the same time, manor lords are abandoning their efforts to preserve their monopolies over brewing and are less resistant to private ale selling, partly because this supplies a source of income to poor villagers and because the fines from the assize provides useful revenues (Clark 1983:21-23, 39).

The expansion in the number of taverns, which are mainly engaged in vending wine, is promoted in the early century by an increase in wine imports, the result of an alteration in the manner by which duty is collected from ships arriving with wine at English ports. The complaint is heard in 1330 that there are "more taverners in the kingdom than there were wont to be" (Dowell 1888, 4:110).

ALE PRICES, CHARACTERISTICS. The price of ale remains fairly steady in the 14th century. Three kinds of ale are available: the cheapest costs about a penny a gallon; the medium strength about a penny and a half; and the best, about two pence. These prices refer to towns; they are somewhat lower in the country (Monckton 1966:62).

UPPER CLASSES/DRINKING, DIET. A large proportion of food consumed is "heavy," much of it being meat. Banquets are lengthy; with three courses of 11 to 12 different dishes a course. Many different wines are served, some--called pigments--are mixed with honey and spices. One of the most popular is hypocras, served with dessert and wafers. Breakfast is generally only bread and wine or ale (Abram 1913:143-144).

WINE TRADE. The growing importance of the wine trade in the 14th century attracts the constant attention of the king and is the subject of considerable legislation (Simon 1906, 1:194).

FRANCE

CABARET CHARACTERISTICS. The term cabaret, as the name for a commercial wine sales outlet distinct from the taverne, first appears in the 14th century. The tavern is generally for purchase of wines to be drunk off-premise; the cabaret serves food and drink for consumption on-premise. The distinction is often blurred at first, but this development marks the beginning of the social decline of taverns (Dion 1959:484).

LOWER CLASSES/DRINKING. A great part of the wine consumed by servants is produced by vines cultivated especially for this purpose; grown on trellises in arbors in the lord's garden, they produce abundant grapes in a small area but make a poor wine. The making of this wine, called jus de la treille or garden wine, continues through to the end of the 16th century (Dion 1959:479-480).

WINE TRADE (GASCONY). In the early decades of the century, the Gascon wine trade flourishes, largely on the strength of the English market. Between 1305 and 1309, 90,0000-100,000 tons of wine are exported annually, about a quarter going to England. The outbreak of the Hundred Years War between France and England in 1337 seriously, if intermittently, disrupts the Gascon wine trade, bringing about a substantial increase in the cost of wine. The region, which is the last of England's continental possessions, becomes more and more dependent on the English market, which virtually monopolizes the trade by the end of the century (James 1971:20ff).

BREWER PREVALENCE. There appears to be about 33 cervoise brewers in Paris throughout the 14th century and until 1428 (Gottschalk 1948, 1:306).

ITALY

LIQUEUR DISTILLING. Distilling begins to be "more or less an industry" in Italy, the site of most of the original experimentation on the process. In 1320, a burgher of Modena produces large quantities of alcohol for sale. Venice is a center of spirit production. Italy specializes in producing sweetened alcoholic beverages ("liqueurs") consisting of alcohol, sugar or syrup, and some flavoring (Forbes 1970:95). It is not until the 16th century, however, that the liqueur industry really develops (see 1500/Italy).

NETHERLANDS

BEER CONSUMPTION, BREWING. The Hanseatic merchants of Danzig and Hamburg increasingly export beer to the Low Countries, one of the most important markets for Hansa beer. Soon thereafter domestic commercial brewing expands (Spiller 1955:146). At first the nobles who hold the gruit monopoly attempt to prevent the use of hops, but after 1325 hopped beer begins to be brewed in many parts of Holland. For a while both beers are brewed. Holland leads the field after 1350, with Haarlem its most important brewing center. Hopped beer is brewed in Bruges in 1351 and Louvain in 1378 (Houtte 1977:89; Dollinger 1970:193).

WINE CONSUMPTION. A significant decline in wine imports and wine drinking occurs in the south at the beginning of the 14th century, which is the result of both the introduction of hopped beer and the economic decline at this time. A comparison of prices and wages up to the end of the 16th century indicates that the wealthier artisans can afford two to five times less ordinary wine than 20th-century workers. A liter of wine per day costs a family a third or more of its daily wage. Thus the majority of the population drinks wine only occasionally. Consumption is higher in prosperous cities such as Bruges than in declining cities such as Gand (Craeybeckx 1958:36-43).

SWITZERLAND

WINE CONSUMPTION. In the 14th century, wine becomes for the first time "a commodity of importance and wide use." (In the 13th century, there was no daily use by local populations with the possible exception of the viticultural regions.) There is no legislation against excessive drinking, though it does appear that "interest groups carried on their deliberations accompanied by copious drinking in some favourite drinking place, perhaps one created especially for a given group."

In Bern, in the first half of the 14th century, vineyards expand under special protection, the price of wine begins to decline, wine becomes a popular beverage among artisans, and it is one of the main presents given by the city to honored guests. By the beginning of the 15th century, wine has become important enough to be included in the system of tithes; by the 17th, vineyards have spread all over Bernese territories, especially the Vaud (Jellinek 1976b:72-74, 80-81).

1325

ENGLAND

ALEHOUSE CONTROLS. The Statute of Frankpledge (1325) is directed against those who "continually haunt Taverns and no Man knoweth whereon they do live." A London proclamation of 1329 prohibits a taverner or brewer from remaining open after curfew lest the establishment become a haunt for misdoers at night. Related to these actions, the Statute of Laborers (1349) is the first in a series of statutes against vagabonds, vagrants, and beggars, motivated by the need to control a labor force greatly depleted by the plague and the breakdown of the feudal system (Monckton 1969:23; Iles 1903:252; Baird 1944b:142).

PRICE CONTROLS. The Statute of Frankpledge of 1325 requires all jurors to indict violators of the Assize of Bread and Ale. The Assize of Wine of 1330 first establishes the maximum price for which a taverner can sell wine (Baird 1944b:150-151; Iles 1903:252).

FRANCE

LIQUEUR DISTILLING. The secret of making sweetened liqueurs flavored with various herbs and spices is introduced by Italian distillers to Paris in 1332 (Forbes 1970:95-96). From France, the making and drinking of liqueurs spreads. However, liqueur drinking is very limited until the fashion is set at court by Catherine de' Medici in the mid-16th century (see 1530/France).

1350

EUROPE

BACKGROUND DEVELOPMENTS: LATE MIDDLE AGES. The 150 years of the late Middle Ages (1350-1500) are a "long death agony of chaos and malaise" as the medieval intellectual, moral, social, and religious order disintegrates (Cantor 1969:504). This is a time of violence, extremism, and a rejection of fundamental values. Traditional and established ways of thinking, living, and associating give way to new and different concepts, institutions, and relationships. Medieval institutions and beliefs no longer seem able to cope with the mounting problems of the age. The Church itself is wracked by controversy and schism, which began in 1309 when the papal court moved to Avignon. Its prestige is at a low point. The Holy Roman Emperor has little authority, and the kings of England and France are engaged in the Hundred Years War (beginning 1337). Late medieval piety is touched with a loss of proportion, an intense preoccupation with death, and a profound mystical bent stressing the importance of the inner life. Men and women all over Europe search for safeguards against changes, catastrophes, and unknown dangers that seem to portend the end of the world. Many seek solace in a variety of forms of popular piety including prophetic and apocalyptic movements that look forward to a future millenium. Others give themselves over to excessive living and luxury.

PLAGUE; AGRICULTURAL DEPRESSION. The Black Death and subsequent plagues reduce the population in some areas by as much as two-thirds and plunge Europe further into the economic and demographic crisis begun in the early century. Agricultural prices are in decline until the mid-15th century. Land is diverted from cereals to products with greater demand elasticity. Low cereal prices and agricultural rents make it advantageous for laborers to give up farming and take up some industrial occupation, leading to a wholesale flight from country to town or to the development of rural industries. Rural masses descending on population centers cause a further disruption in urban supplies. In England, whole villages disappear and lands are enclosed for grazing (Slicher van Bath 1963:136-144; Miskimin 1969:26, 32, 71).

At the same time, from 1350 to 1550, Europe probably experiences "a favourable period as far as individual living standards [are] concerned" (Braudel 1981:193). Living conditions for workers who survive rise as manpower has become scarce. Real salaries may have never been as high as in this period, especially among artisans and in the cities. Greater wealth is accumulated due to inheritances and scavenging the goods of the dead. At every social level, per capita wealth rises. This is a period of large meat consumption even among the poor, since animal husbandry expands at the expense of agriculture. High-value luxury goods are in great demand in those cities still benefiting from commercial and industrial prosperity (see below) (Braudel 1981:190-193; Miskimin 1969:86ff).

SUMPTUARY LAWS. Sumptuary laws are passed to regulate people's material expenditures (such as for food, drink, clothing, and furnishings) and their conduct and activities. While much of the sumptuary legislation concerns proper dress and expenditures on banquets, it also deals with drunkenness, consumer protection, occupational monopolies, poverty, and idle vagabonds. These laws are especially prominent in Italy and in German territories and imperial cities (Vincent 1898; Vincent 1935; Greenfield 1918). These laws are a response to the unique problems created by the socioeconomic conditions of the times, which increase the desire for superfluous display while at the same time reduce society's ability to satisfy this demand due to high mortality (Miskimin 1969:136-149). These problems include excessive luxury, extravagance, and social uppishness. An outgrowth of "medieval paternalism," in which the prince or city council took personal concern for the right and proper conduct of the individual, sumptuary legislation is motivated by a combination of moral, social, and economic concerns: to protect individual welfare, morality, and class structure--all of which are challenged by the new wealth (Dorwart 1971:11, 25-26).

BEVERAGES CONSUMED. The nobles of the late Middle Ages wash down their meals with "priceless wines"; burghers and peasants drink wine and beer at feasts only, "though very light beer and wine mixed with water [are] drunk everyday" (Boissonnade 1927:222, 235). Spirits consumption is limited and exclusively medicinal.

FESTIVE DRINKING. Throughout France and northern Europe, free distribution of wine often occurs in towns as part of pageants, jousts, contests, royal entries and marriages, and other great public festivals. These festivals are vital and obligatory collective demonstrations of local culture and town pride (Heers 1973:201, 205; Heers 1971). At the pageant celebrating the accession of King Henry IV (1399-1413) of England, as described in Jean Froissart's Chronicles, fountains in the streets of London run with red and white wines (French 1884:111).

BREWING, VITICULTURE EXPAND. The agricultural depression encourages the expansion of viticulture and brewing, since the consumption of wine and beer does not decline as much in proportion to falling population. Hops, barley, and vine production rise significantly, although in some areas of France, where viticulture cannot stand the pressure of higher labor costs, vineyards contract. In Germany, vineyards are planted in areas which otherwise would have been regarded as unprofitable (Slicher van Bath 1963:143-144; Duby 1968:344; Miskimin 1969:32, 52-58).

DRINKING INFLUENCES: BLACK DEATH. According to Boccaccio (see 1350/Italy), the plague encourages some people to increase drinking; others to become more temperate. In both cases, drinking behavior is part of a wider world view: "Some thought that moderate living and the avoidance of all superfluity would preserve them from the epidemic....They shut themselves up in houses where there were no sick, eating the finest food and drinking the best wine very temperately, avoiding all excess....Others thought just the opposite. They thought the sure cure for the plague was to drink and be merry, to go about singing and amusing themselves, satisfying every appetite they could, laughing and jesting at what happened. They put their words into practice, spent day and night going from tavern to tavern, drinking immoderately, or went into other people's houses, doing only those things which pleased them. This they could easily do because everyone felt doomed and had abandoned his property....Many others adopted a course of life midway between the two just described. They did not restrict their victuals so much as the former, nor allow themselves to be drunken and dissolute like the latter, but satisfied their appetites moderately" (Prologue, Boccaccio c. 1350/1949:3).

DRINKING PRACTICES; STUDENTS/DRINKING. Medieval students drink heavily; college statutes frequently recognize the mug of wine as forfeiture for minor offenses and accept it as a symbol of initiation into the fraternity of students, trying at the same time to regulate abuses. Lecturers frequently offer students a friendly drink. Initiation rites frequently involve heavy drinking bouts. Such practices occur in other medieval societies and groups; one of the essential features of many initiation rites is a *potacio* or drinking bout as well as violent ragging and even sexual orgies. The purpose of these rites is to break a man down and bind him irrevocably to the community that masters him and at the same time make him a brother through shared feasting. This fraternity is renewed by periodic feasts and drinking bouts. Among students, such fraternal groups provide an informal system through which students are controlled. However, late medieval university and ecclesiastical authorities increasingly condemn the prevalence of student drinking bouts and initiation rites; the growth of this opposition leads to the establishment of an authoritarian hierarchy in the schools (Ariès 1962:244-246, 252, 321, who comments [p. 246]: "People could not imagine a society that was not cemented by public recognition of friendship--maintained by the common meal and the *potacio*, and sometimes sealed with intoxication."

MEDICINAL USE, SPIRITS. The spread of spirit drinking seems to follow very closely in the wake of the Black Death of 1348-1349. Because aqua vitae gives a temporary feeling of warmth and well-being, it is prescribed by doctors in cases of plague (Forbes 1970:95; Wilson 1973:381; Claudian 1970:18-19).

DISTILLATION. Franciscan chemist and prophet John of Rupescissa (Roquetallaide) writes *De consideratione quintae essentiae*, in which he argues that the supreme remedy against corruption is the fifth essence, that is, pure alcohol (*aqua vitae rectificata*). It is identifiable by its marvelous odor, very different from ordinary aqua ardens. Influenced by Raymond Lull (see 1275/Europe), John emphasizes the medicinal and preservative properties of spirits and is perhaps the originator of the doctrine of the fifth essence in each thing. His rectified aqua ardens is a cordial made by repeated distillations (Forbes 1970:64-65). Around 1360, Ortholanus' *Practica* includes a section on "the spirit of the fifth essence" and a recipe for the preparation of alcohol (ibid., p. 67).

ENGLAND

SUNDAY DRINKING. Archbishop Islep complains in 1359 that on Sundays men neglect their churches and hold unlawful meetings, commit "various tumults and other occasions of evil," and practice revels, drunkenness, "and many other dishonest doings" (quoted French 1884:122).

DRINKING CONTROLS: SCOT-ALES BANNED. In 1366, Archbishop of Canterbury Simon Langham makes another attempt to suppress scot-ales, which he observes are banned because of concerns over the common health of bodies and souls. He charges priests to frequently exhort parishioners to obey the prohibition. He defines drinking bouts as a gathering of 10 or more men in the same house "for drinking sake." Those who organize scot-ales are to be excommunicated until they have "merited the benefit of absolution" (Monckton 1966:64).

INEBRIETY, ALEHOUSES CONDEMNED. In the second half of the century, ecclesiastical criticism of inebriety and popular drinking establishments increases (Clark 1983:28). In his guide to preachers (Summa predicantium), the 14th-century Dominican John of Bromyard vigorously condemns gluttony, excessive drinking, and the drink trade. He attacks drunkenness within the clergy, taverners, the "race of swindlers" who inhabit the taverns, feasters who boast of how much they have drunk, and workmen and craftsmen who, at least once a week and on holidays, get drunk, "whereby their goods are diminished." (The last is a common complaint among preachers all over England and continental Europe.) He places particular emphasis on the larger implications of drunkenness: it disturbs the realm and impoverishes the nation both materially and spiritually. He categorizes one by one the idle excesses of the intemperate and the ills that drunkenness produces: impoverishment, "paralysis, blindness, headaches and many inconveniences," a bad taste in the mouth, trembling of the body, derangement of the mind, facial transformations, and quivering of the tongue. He explicitly attributes the extravagances of dress and diet then in vogue to the "ever-growing desire of the glutton to indulge his appetite, even beyond the bounds of natural capacity" (Owst 1966:427-434, 445-448).

The same charges and warnings appear in the preachings of Master Robert Rypon. He complains that during the Lenten fast "very few people abstain from excessive drinking: on the contrary, they go to the taverns, and some imbibe and get drunk more than they do out of Lent." They justify heavy drinking on the thirst created by the fish they must eat. Similarly, on feast days they go to the taverns "and more often than not seek food such as salt beef or a salted herring to excite a thirst for drink. At length they get so intoxicated that they fall to ribaldries, obscenities and idle talk, and sometimes to brawls" (quoted Owst 1966:435, 444, and Henisch 1976:41).

John Wycliff (1320-1384) makes several references to drunkenness in general and to its prevalence among priests. In "The Seven Deadly Sins," he says that the drunken man surpasses the beast in folly and shame. In the "Order of Priesthood," he alludes to priests who, among other faults, haunt taverns (Rolleston 1933:39-40).

WINE PRICES. Due to the Hundred Years War against France, the price of Gascon wine is now 50% higher than at the start of the century. Although the price stabilizes and England comes to virtually monopolize the Gascon trade, wine is never again as cheap (James 1971:30-31).

ALEHOUSE PREVALENCE; BREWING. Alehouses appear to be fairly numerous in late medieval towns, and there is a growing tendency for alehouse keepers to purchase supplies from larger producers. The new importance of the drink house is indicated by the practice of using ale-stakes (usually long poles with a bush at the end) to identify them. Several towns gain a reputation for making a better quality of ale, including London and Burton. This reflects the growth of larger-scale wholesale brewers with more sophisticated equipment (Clark 1983:29).

FRANCE

WINE TRADE. A million cases of wine are exported from the port of Bordeaux, although as late as the 16th century Bordeaux wines are considered by the French to be of very poor quality (Lichine 1951:4, 19). Wines from the Charente area are shipped from La Rochelle to northern countries, especially England, Scotland, and Scandinavia (Butler 1926:147).

TAVERN CONTROLS. A gambling law passed in 1350 prohibits tavern keepers from receiving or lodging "any dicers or other persons of vicious calling" or from receiving drinkers after curfew (Firebaugh 1924:87, 91). No tavern keeper can refuse the wine he sells to those who demand it (Dion 1959:484).

GERMANY

INEBRIETY CONDEMNED. Emperor Charles IV (1346-1378) states that the vice of drunkenness is greatly on the increase, that it leads to blasphemy, murder, and manslaughter, and that such vices and crimes have rendered the Germans "despised and condemned of all foreign nations" (Samuelson 1878:105-106). Henceforth, concerns about inebriety and charges that it is increasing seem to escalate with each generation, culminating in the 16th century. Hauffen (1889:482-483) sees this as rooted in the deterioration of the moral code of the lower aristocracy in the Late Middle Ages.

BREWING. The prosperity of Hanseatic towns is due to a considerable extent to beer exports. Around this time, brewing in Lübeck especially expands. Brewing is the main local industry of north Germany, utilizing much of the large supplies of grain produced there. Brewers have combined into leagues and guilds, stronger beers are becoming common, and governments are becoming patrons of brewing. Hamburg still dominates the market. With about 450 brewers, nearly half of all master artisans registered in the city, Hamburg is known as the beer city. Many brewers produce for specific foreign markets. In 1369 beer represents a third of its total seaborn exports; half of this goes to Amsterdam (Cherrington 1926, 3:1090; Arnold 1911:280-282, 287; Dollinger 1970:118, 193, 221, 225, 251).

BEER CONSUMPTION. In Bavaria, beer drinking is promoted with the slogan, "A Beer Equals Half a Meal." In the Late Middle Ages, beer consumption probably peaks at 300 liters per capita a year, compared to 150 liters today (Abel 1981:24).

WINE AVAILABILITY, PRICE, CONSUMPTION. Vineyards now cover all of Germany, although the better wines are produced in the south and Alsace. In the Late

Middle Ages, wine is almost as cheap as beer and people (including servants) consume up to 2 liters of wine a day (Abel 1981:20-21).

BRANDY CONTROLS. The first official references to brandy occur. The city law books of Frankfurt for 1361 contain a statute forbidding the making of wine mixed with brandy or any other "unnatural" ingredients. Similarly, Würzburg in 1372 outlaws the making of "false" wine, and specifically refers to brandy (Rau 1914:8).

TAXATION. In 14th-century Austria, "certain classes" are given permission to levy taxes on beer instead of the privilege of printing money (Cherrington 1925, 1:230-231).

ITALY

INEBRIETY PREVALENCE. Judging from Boccaccio's (1313-1375) Decameron, eating and drinking well is one of the delights of living to medieval Italians (book 10.10). A person who drinks only water is considered a miser or an ascetic (1.8). Habitual drunkards do exist (9.4); Boccaccio refers to the clerics of Rome as gluttons and tipplers (1.2); and wine drinking frequently occurs during amorous seductions (2.7, 3.4, 9.6). But medieval Italy appears to be more sober than other European nations. The ancient Roman drinking bout after the meal is rare; there is no instance in the Decameron of drunkenness at a banquet and only a few references to drinking outside of meals during which some intoxication occurs (see book 9.6) (Boccaccio 1351/1949; Jellinek 1976b:3). (For Boccaccio's comments on the influence of the plague on drinking, see 1350/Europe.)

ATTITUDES (BOCCACCIO). Boccaccio (1351/1949:560) compares the value of his tales in the Decameron to the effects of wine: "But, such as they are, they [these tales] may be amusing or harmful, like everything else, according to the persons who listen to them. Who does not know that wine is a most excellent thing..., while it is harmful to a man with a fever? Are we to say wine is wicked because it is bad for those who are feverish?."

WINE CONSUMPTION, CHARACTERISTICS. Around 1350, the average annual wine consumption in Florence is about 10 barrels per person. Wine is cheap, but only the well-to-do drink the first press (primo vino). Most people drink mezzo vino, made from the grape mark by the addition of water and refermentation (Jellinek 1976b:3). Staley (1967:366-367) describes Florentine drinking habits as follows: "The Florentines of old were a pleasure-loving race despite the many serious traits in their character. Nothing pleased them more than to sit in the wine-shops after their meals, and there to sing and dance, to wager and to drink, to their hearts' content; but, like sensible men, they knew when they had had enough."

SALES CONTROLS. Florentine statutes of 1357 prevent innkeepers from selling wine or potables of any kind to the poorer people. They can sell them only to guests and persons in their houses. In no way are they to do a rival trade to that of the wine merchants (Staley 1967:371).

MEDICINAL USE, SPIRITS. A Florentine document of the 14th century, discussing the powers of spirits when used both externally and internally, asserts that they are a remedy against graying hair and lice, and that they cure body aches, sterility, gout, and deafness (Patrick 1952:30).

NETHERLANDS

BREWING. Hop cultivation and the brewing of hopped beer expands. Dutch beer begins to oust imported German beers from the domestic market and even begins to be exported to England. But 35,000 barrels of Hamburg beer are still imported in 18 months into Amsterdam (Houtte 1977:95-96; Dollinger 1970:193).

VITICULTURE. Although the taste of beer has improved with hops, local viticulture reaches a peak in the later Middle Ages both at Louvain and Huy (Houtte 1977:68).

WINE TRADE. Sailors from the city of Kampen in northern Netherlands travel to the south French Atlantic coast and bring the salt and wines of Poitou and Gascony to the Baltic (Houtte 1977:93).

WINE CONSUMPTION. In Hainaut, wine drinking has become common among the peasantry as well as the wealthy and the bourgeois. As the rural textile industry expands, the wine commerce extends even to the small villages. The average annual consumption per village family is probably about 37 litres; this increases to 63 by the end of the century. In the city of Valenciennes, an increase in the tax collected on wine and beer provokes a popular uprising in 1345. Rural wine consumption leads to the establishment of regulations in villages following the example of urban magistrates. For example, women are banned from going to cabarets and taverns. In Maroilles, the only people who can drink in a tavern at night are guests of the taverner, lodgers, or foreign merchants, and then on the condition that no more than one lot of wine is purchased. If this quantity is exceeded, a fine must be paid (Sivery 1969:40-45, 178). Van Werveke (1933) estimates that in the Flemish city of Gand in 1360 an average of 37-38 liters of wine are consumed annually per capita, about four times more than in 1926. Craeybeckx (1958:6) believes that the rate is more like two or three times as much as in the 1950s. However, wine consumption is limited to the wealthy, and throughout the 14th and 15th centuries it varies with the level of taxes imposed and the degree of prosperity. Wine is a highly-taxed luxury drink which furnishes as much as three-fifths of Gand's revenues. In 1360, the wine tax furnishes three-fifths of Gand's revenue (van Werveke 1933:1100). In 1333, Bruges receives 61.5% of its communal resources from the wine excise and 26% from beer taxes, a total of 87.5%. Between 1350 and 1421, the wine tax supplies about 45% of the city's revenues. In Anvers, beer drinking throughout the Middle Ages is greater than wine drinking. Why wine consumption is consistently less there than in Bruges or Gand is uncertain (Craeybeckx 1958:5-10).

1375

ENGLAND

PRICE CONTROLS. Gradually, the price of corn begins to rise, which results in a corresponding rise in the price of ale. Fears grow that both ale and bread might become too expensive for the very poor. The mayor of London decrees in 1381 "that in order to assist the poor, bakers shall make bread at a farthing the piece and brewers shall sell ale by a farthing measure" (Monckton 1966:61).

ALEHOUSES. By the late century, the alehouse is emerging as a significant institution (Clark 1983:29).

DRINKING/PRACTICES; ATTITUDES (CHAUCER). In the Canterbury Tales (composed 1366-1400), Chaucer (1340?-1400) provides a panorama of the drinking habits and attitudes of the late-medieval English. Drinking, often excessive, occurs throughout. Chaucer himself is from a family of vintners and wine merchants, and on the whole he views drinking and occasional drunkenness as a natural aspect of life, but habitual drunkenness is condemned (Firebaugh 1924:219-247). He demonstrates that one function of wine drinking is simply to pass the time (Henisch 1978:26).

The very pretext for the book is a contest suggested in the "Prologue" by the host of the Tabard Inn, who serves the pilgrims strong wine that they were glad to drink and offers a free meal to whomever has the best tale. All seek to win the contest and "drink good wine and ale" (Chaucer 1400/1963:37, 39, 279). Among the heavy drinkers is the wealthy, good-natured Franklin (a nonnoble landowner) who exhibits his wealth through the food and drink he serves and consumes, who loves a morning sop of cake in wine, and who indulges in the finest beer and wine (p. 26). The biggest drunk is the Miller, who has a collection of filthy tavern stories, who travels along "very drunke and rather pale" on this horse, and who tells his tale while drunk (p. 32). The Summoner loves "drinking strong wine til all was hazy" (p. 34). Chaucer also describes the care with which the refined Prioress wipes her lips when she drinks from the wine goblet, and how the Skipper embezzled Bordeaux wine (pp. 21, 28).

The Wife of Bath, who has outlived five husbands, personifies female wiles and a life of sensual joy in which drinking is a frequent and natural accompaniment. Of the ancient character Metellius, who put his wife to death for drinking wine, she asserts: "If I had had a draught of sweetened wine./ Metellius, that filthy lout--the swine/ Who snatched a staff and took his woman's life/ For drinking wine--If I had been his wife/ He never would have daunted me from drink" ("The Wife of Bath's Prologue," pp. 286-287).

At the same time, he writes in the Summoner's tale that "drunkenness is filthy to record," and the Parson warns of the dangerous effects of drunkeness, "the horrible sepulture of man's reason," and recommends abstinence, temperance, and sobriety as the remedy ("The Summoner's Tale," p. 329; "The Parson's Tale," p. 504). The strongest denunciation of drinking is made by the Pardoner, who tells of the foul deeds men commit under the influence: "Wine is a lecherous thing and drunkenness/ A squalor of contention and distress./ O drunkard, how disfigured is thy face,/ How foul thy breath, how filthy thy embrace!" However, the Pardoner is hypocritical and his knowledge of drinkings' harm appears gained from much personal experience ("The Pardoner's Tale," pp. 260, 263).

Summarizing Chaucer's attitudes, Carter (1976:6) describes him as "the early voice of most of the poetic attitudes that have been taken toward wine: wine as the symbol of poetry, of the creative spirit itself; wine as the servant of love and sensuality; wine as the symbol of social manners and grace; yes, and...wine, at least intemperate use of wine, as the destroyer of reason and its controls over the emotions."

LOWER-CLASSES/DRINKING: TAVERN LIFE (LANGLAND). The English villagers described by William Langland in Piers the Plowman are badly fed and live on milk and gruel; eating bread and drinking beer is a luxury (Duby 1968:337). Hunger and poverty are characterized by the lack of even weak penny ale, night-old worts, and a piece of bacon. Langland complains that beggars will not drink cheap beer at any price and that day laborers who demand and get high wages are not satisfied with "draught-ale" (book 6.296-314). Although the poor man longs for good ale, he goes "to his chill bedding" without it, nor does he eat rich food or drink much wine (14.229, 14.250).

In an allegorical vision, Gluttony is waylaid by an alewife on his way to Confession (book 5). In this passage, Langland provides a look into the alehouse and the lower-class drinking habits of the period. Given a "good welcome" by all in the alehouse, the Glutton is started off with a pint of the best: "Then there were scowls and roars of laughter and cries of 'Pass round the cup!' And so they sat shouting and singing till time for vespers. By that time, Glutton had put down more than a gallon of ale, and his guts were beginning to rumble like a couple of greedy sows. Then, before you had time to say the Our Father, he had pissed a couple of quarts....He could neither walk nor stand without his stick...and when he drew near to the door, his eyes grew glazed, and he stumbled on the threshold and fell flat on the ground. Then Clement the cobbler seized him....But Glutton was a big fellow, and he took some lifting; and to make matters worse, he was sick in Clement's lap." The next morning, his wife berates him for sinning in word and deed, after which he confesses his guilt: "And I have let myself go at supper, and sometimes dinner too, so badly that I have thrown it all up again before I have gone a mile....On fast-days I have eaten the tastiest food I could get, drunk the best wines....And to get more drink and hear some gossip, I've had my dinner at the pub on fast-days, too" (Langland 1375/1966:71-72). Rolleston (1933:43) calls the full version of this story "a realistic picture of acute alcoholic intoxication."

Langland has nothing commendable to say about the producers and dispensers of ale (11.413-25). In chapter five, the wife of Avarice is a brewer who cheats the poor. The brewster is among the people for whom the pillory is recommended (Spiller 1955:143). But he also shows the alehouse as functioning as a common meeting place of the village (Firebaugh 1924:151).

MONKS/DRINKING. William Langland (see above) describes the ale drunk by monks as feeble (Younger 1966:270).

SPIRITS CONSUMPTION. Though Chaucer (see above) shows considerable knowledge of the techniques of distillation and alchemy in the "Canon's Yeoman's Tale," no references to brandy or distilled alcohol appear in the Canterbury Tales, indicating that even its medicinal use is limited.

WINE AVAILABILITY, CONSUMPTION. The capture of 19,000 tons of French wine in 1387 leads to a great abundance of wine in England and a temporary rise in inebriety (Simon 1906, 1:251-252).

FRANCE

VITICULTURE. In 1395, Philip the Bold, duke of Burgundy, orders the destruction of all vineyards planted with the Gamay vine because the "disloyal plant makes a wine in great abundance but horrible in harshness" (Lichine 1951:67). Such wines are intended for popular, mass consumption.

WOMEN/DRINKING. The Goodman of Paris (subtitled "A Treatise on Moral and Domestic Economy"), written by an elderly high-bourgeois gentleman for his new, young wife around 1393, suggests that female drunkenness is not uncommon (Jellinek 1976b:83 says it includes "a description of a true alcoholic woman"). In the section on the deadly sins, the author provides a vivid picture of the female glutton. As in Langland's Piers Plowman, the tavern is described as the "Devil's Church" and gluttony interferes with being a good Christian. The Goodman writes: "God commands us to go to church and rise early and the glutton saith: 'I must sleep. I was drunk yesterday....' When she has with some difficulty risen, know you what be her hours? Her matins are: 'Ha! what shall we drink? Is there nought left over from last night?' Then says she her lauds, thus: 'Ha! we drank good wine yestreen.' Afterwards she says her orisons, thus: 'My head aches; I shall not be at ease until I have had a drink.'" (Here is shown "a good knowledge of withdrawal symptoms" [Jellinek 1976:83].) Such gluttony, the Goodman asserts, puts a woman to shame, makes her ribald, wanton, and a thief. When people join the devil's disciples and visit a tavern they enter "upright and well spoken" and leave unable to hold themselves upright nor speak--"all fools and madmen" (Goodman of Paris 1393/1928:83-84).

At the same time, this household book shows the moderate use of wine is well integrated into everyday life. It provides details on the preservation, serving, and use of wine as a medicine and beverage, and as an ingredient in cooking. Special attention is devoted to how to prevent wines from going bad and remedying them if they do (pp. 216-217). In serving wine at the mid-day meal, he recommends "one drink nourishing but not intoxicating," adding: "Forbid them to get drunk, and never allow a drunken person to serve you nor approach you, for it is perilous" (p. 218). Reviewing one dinner menu, he comments that the specified portions of a half pint of wine for every two persons for the main and dessert courses was too great; a half pint for three persons was sufficient (p. 237).

GERMANY

HOPS BANNED. Archbishop Frederick of Cologne issues a decree in 1381 on behalf of the gruit monopoly which the bishopric holds, forbidding anyone to import "hopped beer" (hopfenbier) from Westphalia under the severest penalties of the Church, and requiring all those who wish to brew--whether brewers, clergymen, or housewives--to buy gruit from the episcopal gruit-houses. It is not until 1495 that Cologne stops using gruit in brewing (Arnold 1911:237-238; Cherrington 1925, 1:403).

SPIRITS CONSUMPTION. Around this time, spirit drinking comes "into vogue" among German miners (Forbes 1970:96, who supplies no further information about this).

1400

EUROPE

BACKGROUND DEVELOPMENTS. Fifteenth-century Europe, particularly in the north, is characterized by widespread fears, insecurities, violence, and morbidity as the medieval order disintegrates and society is racked by plagues (Huizinga 1954:28-30). The heightening of religious sensibility continues, and a growing desire for greater separation of the sacred and the profane develops. At the same time, the Renaissance flourishes in Italy.

DRINK AND WORK. In reward for the extra-heavy work that the peasants of the monastery of St. Peter in the Black Forest must perform at vintage time, the monks decree that the peasants be given "meat and drink in abundance...and, if so be that the peasants wax drunken and smite the cellarman or the cook, they shall pay no fine for this deed; and they shall drink so that two of them cannot bear the third back to the waggon" (quoted Coulton 1925:27, who observes that medieval lords often allowed drunkenness as a gift for periods of hard work).

DISTILLATION. By the end of the 14th century, distillation methods have been gradually perfected by alchemists. Alcohol is one of the main substances they produce (Forbes 1970:69). Several medical tracts praise it; one asserts that distilled alcohol will preserve the virtues of herbs all year so that apothecaries need no longer sell stale and outworn drugs (Forbes 1970:68).

SPIRITS CONSUMPTION. Spirits are drunk all over Europe, but their popularity is largely confined to Germany, and they are consumed as medicinal preparations and only on a limited basis (Forbes 1970:96; Younger 1966:324; Braudel 1973:17). The high cost of medicinal spirits probably limits their general use, but Forbes (1970:91) doubts this: "There is no doubt that brandy was not an expensive drink used by the higher classes only before 1500 as some authors have claimed. It was consumed by all classes." He cites as evidence the brandy legislation that begins to appear in Germany in 1450.

WINE TRADE (BALTIC). Better wines begin to arrive in the Baltic countries from southern France as the increasing purchasing power of northern peoples stimulates demand. However, the Baltic market remains insignificant compared to that of the Low Countries (Bizière 1972:112; Craeybeckx 1958:128). Dutch traders carry most of the wine.

ENGLAND

UPPER CLASSES/DRINKING. For the nobility, a quart of beer and a quart of wine are given to each couple for breakfast and a gallon of beer and a quart of wine for the liverie, a repast taken in their bedrooms before bed. At a banquet for the new archbishop of York in 1464, 300 tuns of ale and 100 tuns of wine are consumed (French 1884:111, 121).

ALE CONSUMPTION, PRICE. Ale demand rises along with living standards in the aftermath of the plague. A building craftsman can probably afford nearly three times the amount of ale in 1400 than a century earlier (see 1275/England) (Clark 1983:32).

ALEHOUSE CHARACTERISTICS. The alehouse "is starting to come into focus for the first time" (Clark 1983:30). Alehouses become more formalized and operate on a more regular basis. The trend continues for concentration of production in fewer but larger hands, developments which are added by the introduction of hops. The alewife becomes a stock figure in popular literature, but the predominance of women in the trade declines as it becomes more established. There occurs a "winnowing-out of some poor, part-time ale-wives" (ibid., pp. 31-32). These developments are aided by the introduction of hops (see below), the rise in demand (see above), a possible decline in other drink alternatives as spring and river waters become more polluted due to town and industry growth, and the decline in the practice of furnishing drinks as part of labor renumeration as landowners give up direct arable farming and lease out lands (ibid., p. 33). Also by the beginning of the century, victuallers increasingly refuse to sell drinks for off-premises consumption as had been common throughout the Middle Ages (ibid., p. 94).

HOPPED BEER INTRODUCED. As trade with overseas markets expands, hopped beer is introduced in England from the Low Countries, probably at the end of the 13th century. (However, hopped beer may have been imported in the mid-14th century for Philippa of Hainaut, the wife of King Edward III [King 1947:43].) Initially, it is brought by Dutch ships, English soldiers and camp followers returning from foreign service, and Dutch brewers (Kerling 1954:112, 114-115; Mathias, 1959:37. Imported hopped beer is sold in London in 1418 (Wilson 1973:375). Opposition to the new beverage is strong because the public does not appreciate its bitter taste and views it as a foreign threat to England's ale interests, especially as production is controlled by foreigners. Opposition diminishes over the century as hop brewing develops in England and as it is discovered that the process lengthens the life of beer (Monckton 1966:66-67; King 1947:42-43, 53; Spiller 1955:146; Mathias 1959:3). This promotes the change in brewing from a domestic trade to an organized industry; with this change, the male brewer replaces the female brewster. (According to King [1947:42], hops are at first mainly considered to be a flavoring herb; only later are their preservative functions recognized. On the importance of hops in the development of commercial brewing, see 1200/Europe.)

WINE AVAILABILITY, CONSUMPTION. The supply of wine, chiefly imported from France, becomes so abundant that the price per ton declines and wine consumption increases considerably (French 1884:109; Rolleston 1933:34). It is estimated that average per capita wine consumption in the 15th century may be three times as great as in the 20th century (Carus-Wilson 1967:271n.1).

MEAD CONSUMPTION, CHARACTERISTICS. The English shift from drinking the dry wines of France (claret and burgundy) to sweet wines. This brings about the degeneration of the character of mead. As sweeter mead is produced, the art of making the original dry mead is lost. Honey becomes less available and increases in cost, due to increased population. In the 16th century, a taste for spiced wine also develops (Gayre 1948:92-98, 124).

DISTILLING. Spirits distilling is still largely confined to monasteries, which are centers of medical science. There is some home distilling, but in 1404, King Henry IV (1399-1413) inhibits the growth of professional distilling by prohibiting alchemists or "multipliers" (Forbes 1970:101).

FRANCE

LOWER CLASSES/DRINKING. Christine de Pisan (d. 1431) describes the fare of the laborer's wife as "black bread, milk and water" (Coulton 1925:315). In the late century, a lament on the hard life of the common people by poet Jean Meschinot ends with the observation that they have only water to drink. Meschinot is a prime example of 15th-century despair and pessimism (Huizinga 1954:33-34, 63, 291-292).

TAVERN PREVALENCE. During his occupation of Paris, Henry V (1413-1422) of England suppresses 26 taverns as centers of espionage. After the liberation of Paris, their numbers increase considerably as bishops and kings in need of revenues constantly authorize new taverns (Gottschalk 1948, 1:320).

DRINK AND WORK. In addition to providing free wine on feast days and celebrations (see 1400/Europe), French cities often offer free wine to the workers on public projects. Nothing is said to have a better moral effect on workers (Dion 1959:476).

WINE TRADE. During the early 15th century, in Rheims, wine has become the chief article of trade and the first references to wine brokers (*courtiers de vin*) occur (Forbes 1967:93).

GERMANY

BACKGROUND DEVELOPMENTS: PROSPERITY. The 15th century inaugurates Germany's "most luxurious and self-indulgent epoch." The commercial prosperity of Nuremberg and other south German cities expands tremendously. Great personal fortunes are accumulated by merchants. For the first time, artisans are also able to put aside surplus capital. A great number of small, independent dealers appear. Men of all classes become better off and are able to satisfy appetites beyond the demands of necessity. As the century progresses, complaints increase about the abandonment of the simplicity and frugality of the past and about the rise of luxury and extravagance (Greenfield 1918:48-49).

FESTIVE DRINKING. The village wake remains the one bright point in the oppressive lives of the peasants, at which they spend all their savings. At the bridal-feasts of the well-to-do, all eat and drink so freely at the cost of the young couple that it is often long before the couple can recover the expense (Coulton 1925:26, who observes that there is evidence that the lay and ecclesiastical lords willingly encouraged such extravagances because they kept the serf in good humor.)

A scene in Heinrich Wittenweiler's *Der Ring*, a comical epic parody of knighthood and a precursor of later satirical *Tischzuchtliteratur*, depicts immense greediness, gluttony, and drunkenness at a feast, including drinking directly from a serving bowl (Neuer 1970:91-92).

CONTROLS: TAVERNS, TREATING. In the course of the century, more numerous and stricter sumptuary laws are passed in Nuremburg to regulate drinking. Penalties are raised for those who serve drinks after curfew. On holy days, drinking places are closed and it is forbidden to serve drinks to others at one's home. At the close of the century, the first laws against treating are passed (Greenfield 1918:88-89).

IRELAND

<u>WHISKEY</u> <u>CONSUMPTION</u>. Reference to whiskey occurs in <u>The Annals of the Four Masters</u> in 1406 (Murphy 1979:142) cites this as the first real reference to the drink, but see 1275/Ireland). There is no evidence that whiskey distilling is a common practice before the 15th century; during the century it appears to be purely domestic and very limited. A statute of 1450 regulating drink measures refers only to "Irish wine, alcohol or other liquor" (McGuire 1973:91-92).

ITALY

<u>BACKGROUND</u> <u>DEVELOPMENTS:</u> <u>RENAISSANCE</u>. The Renaissance originates in Italy in the 14th century and reaches its height there in the 15th and 16th centuries, spreading into northern Europe where it lasts into the mid-17th century. Characteristic of the period is a rise in more secularized values and ideals, a new importance given to individual expression, self-consciousness, and worldly experience. It is also a period of emerging nation-states, geographic exploration, commercial expansion, and wealth. The focus of Renaissance society and culture is the glittering new, urban court of the prince.

<u>CONTROLS:</u> <u>SALES,</u> <u>TAVERNS</u>. The Florentine government shows "a tendency towards controlling consumption through closing-hours, a law against games of chance in inns, taverns, and hostels and a ban on the sale of wine to the poor and in the vicinity of public buildings" (Jellinek 1976b:3). In the revision of the Florentine guild statutes in 1415, the wine merchants are strictly forbidden to sell wine and other such beverages within a distance of 50 yards from the porch of San Giovanni Battista, a favorite lounging place for the poor. Wine shops are forbidden to take in travellers and to sell beverages to be drunk on the premises. Neither may they stand opposite the Palace of Priors, the House of the Captain of the People, nor closer than 200 arms' lengths, nor in the neighborhood of the monastery of San Giovanni Evangelista. No one can sell wine to citizens after the final stroke of the Compline bell, with a penalty of 100 lire. The sale of any provisions within 50 arms' lengths of a wine shop is banned (Staley 1967:362-363).

<u>FESTIVE</u> <u>DRINKING</u>. According to Burckhardt (1958, 2:402, 408, 418-425), popular festivals such as Carnival do not reach their full development in Florence and the rest of Italy until "after the decisive victory of the modern spirit" in the 15th century (quote at p. 402).

NETHERLANDS

<u>BACKGROUND</u> <u>DEVELOPMENTS:</u> <u>UNION</u>. During the 15th century, all 17 provinces of the Low Countries (including Holland, Zeeland, Flanders, Brabant, Luxembourg) are inherited, purchased, or conquered by the dukes of Burgundy.

<u>BREWING</u>. Commercial beer brewing becomes a specialty of Holland and one of its most important industries for the next two centuries (Forbes 1970:94-95). This is largely due to the area's proximity to hop-growing areas, good water, and cheap fuel in the form of peat (Slicher van Bath 1963:274). Haarlem, the main brewing center, has a hundred master brewers (Houtte 1977:89). To protect the brewers of Haarlem, beer imports from northern Germany are prohibited (Jacob

1935:86). However, the 15th-century wars in France cut imports of corn and fuel, ruining many of Netherland's brewers, who then immigrate to England and establish beer breweries there (Spiller 1955:146).

SCOTLAND

WHISKY CONSUMPTION. The practice of distilling barley and other cereals with malt is believed to have originated in Ireland and then to have been transported to Scotland. But nothing definite is known about the origin of Scotch whisky (Lockhart 1951:5). The process of distilling was probably "well-developed long before men got around to writing about it" (Murphy 1979:9). The death on Christmas Day, 1405 of Richard Magnarell, chieftain of Moyntyreolas, in one of the first possible references to whisky, is attributed in the *Annals of Clonmacnoise* to drinking too much *aquae forte*, perhaps a synonym for whisky (Ross 1970:2; Murphy 1979:9). The earliest definite reference to aqua vitae as whisky occurs in 1494 (see 1475/Scotland). (The traditional difference in spelling between the Irish "whiskey" and Scottish "whisky" is preserved in this chronology; the lack of the "e" in the Scottish spelling has been interpreted as indicating that the word originated in Ireland and then was simplified when it was introduced into Scotland.)

WHISKY CHARACTERISTICS. The original "pure" whisky of Scotland is made from only barley malt, kiln dried over peat, and distilled twice in a pot still. It differs in several respects from Irish whiskey, which is distilled three times, is generally dried over coal, is made from unmalted barley, has always been a blend, and has a less smokey flavor. The home of malt whisky is the remote Highlands, where there exists an abundance of clear water, barley, and peat bogs, and where the damp, cold climate makes such a warming drink especially appealing. Yet until the mid-19th century, Scotch whisky is considered a coarse drink suited only to the harsh northern climate. Even within Scotland, its production and consumption remain localized in the northern Highlands until the end of the 18th century.

1425

EUROPE

CHURCH CONCERNS; FESTIVE DRINKING. Archdeacon Nicholas de Clémanges (*De novis festivitatibus non instituendis*) protests against the introduction of new holy days because the existing ones are defiled by drunkenness and irreligion. Nicholas speaks of peasants as "the most innocent of men" and asks how the Church can create new holy days when even now, if a peasant works on all the existing ones, he cannot satisfy all the demands placed on him by greedy lords. He characterizes peasant life as miserable and sober on workdays and licentious on feast days. Money and leisure is ill spent by men who have little chance to learn the best use of either. Similar sentiments are made in Zürich by Felix Hemmerlin (*Tractatus*) (Coulton 1925:113, 275).

ENGLAND

SUNDAY DRINKING CONTROLS. During the reign of Henry VI (1422-1461), a number of towns attempt to stop drunkenness on Sundays and even to bring about Sunday closing "here and there" (French 1884:118). In 1448, Henry curtails fairs and markets on Sundays because they lead people to "drunkenness and strifes" and to miss divine services (Baird 1944b:142).

BEER BREWING PROMOTED. On the south coast and in London, beer is accepted by the authorities fairly early in the century as a wholesome beverage. In 1436, King Henry VI issues a writ proclaiming that all brewers of beer should "continue to exercise their art as hitherto, notwithstanding the malevolent attempts that were being made to prevent natives of Holland and Seland and others who occupied themselves in brewing the drink called biere from continuing their trade, on the ground that such drink was poisonous and not fit to drink and caused drunkenness, whereas it was a wholesome drink, especially in summertime. Such attacks had already caused many brewers to cease brewing, and would cause greater mischief unless stopped" (quoted Monckton 1966:67).

The first beer brew-houses are established in London in 1440. There are no controls over brewing until 1441, when Henry VI appoints two ministers for life to survey and correct all the beer brewers in England (Mathias 1959:3).

FRANCE

BEER INTRODUCED. As a result of the expansion of the north German beer trade, beer (*bière*) appears in France; it soon shares popularity with *cervoise*. Claudian (1970:10) dates the appearance of the term at about 1435, but a Parisian's journal in 1428 says that wine has become so dear (probably due to the Hundred Years War) that people are resorting to drinking beer, cervoise, or cider. The number of brewers increase in Paris for the first time since 1300 (Gottschalk 1948, 1:306).

GERMANY

BREWING CONTROLS. In 1420, Munich declares that all beer must be well fermented by top fermentation and not sold for seven days after brewing (Cherrington 1925, 1:406).

SCOTLAND

INN FUNCTIONS. In the reign of King James I (1406-1437), two statutes (1424, 1425) ordain the establishment of hostels for the hospitality of travellers and the provision of food and drink to them. These are the first measures to be concerned with alcoholic beverages in Scotland (*Scottish Law Review* 1943:85).

TAVERN CONTROLS. The Scottish Parliament passes a measure in 1436 closing taverns at 9:00 p.m. and prescribing imprisonment for violators. The law only refers to burghs (towns), indicating that there are few taverns in the countryside (shires) or that taverns are not considered a problem there, perhaps because rural life naturally requires retiring earlier (Cherrington 1929, 5:2383; *Scottish Law Review* 1943:85-86).

BEVERAGES CONSUMED. The 1436 law governing tavern closing hours mentions the drinking of beer, ale, and wine, but not whisky or aqua vitae, indicating that spirit drinking does not occur at taverns. The reference to beer implies that it was introduced into Scotland, as in England, by the early century.

1450

EUROPE

BACKGROUND DEVELOPMENT: RECOVERY. The economy begins to recover. The next hundred years constitute a period of transition during which the population grows, silver mining resumes, commerce and trade expand, more money becomes available, cereal prices increase, and real wages remain the same (Slicher van Bath 1969:144). By 1500, Europeans are enjoying a "remarkable prosperity" (Rice 1970:38-39).

ENGLAND

CHURCH CONCERNS: TAVERNS. The anonymous 15th-century moral treatise *Jacob's Well* (c. 1440), a huge work designed for use on the pulpit, is a prime example of the low esteem of the brewer and the publican. (Rolleston [1933:46] calls it "the most severe denunciation of the mediaeval tavern which I have yet encountered.") The tavern is considered the "well of gluttony," the devil's school house and chapel, where he instructs his disciples in gluttony, lechery, foreswearing, slander, backbiting, fighting, robbery, heresy, and other sins. Unnecessary drinking is attacked as a venial sin, and immoderate drinking by oneself and making others drunk are deadly sins (Owst 1966:426).

Another preacher attacks "leches, physicyons, taverners, and tollers" as four examples of those who by false subtleties take men's goods falsely (Owst 1966:351).

FRANCE

WINE TRADE (GASCON): ECONOMIC VALUE. At the close of the Hundred Years War (1337-1453), King Louis XI of France (1461-1483) allows the Gascony-England wine trade to resume without punishment, even though Gascony joined with England against France in the years of warfare. The king's counselor, M. Regnault Girard, argues that to impoverish one of France's richest cities (Bordeaux) would gravely menace the peace and prosperity of his realm, that the whole basis of Bordeaux wealth was the wine market in England, that no other country could take England's place as the purchaser of its wines and supplier of its needs, and that the English should therefore be allowed to trade as freely as they would like (Carus-Wilson 1967:278).

STUDENTS, POOR/DRINKING (VILLON). Student, vagabond, criminal, and poet Francois Villon (1431-1463) takes his Licentiate of Arts from the University of Paris and goes on to chronicle in his poems the life of the urban poor in the years immediately following the end of the Hundred Years War (Firebaugh 1924:101-115). (Fox [1976:xvi-xvii] describes Villon as "a flippant character, fond of gay company, thumbing his nose at life and everything in it, with all the couldn't-care-less, pleasure-at-all cost of a post-war generation.") Throughout many of his poems Villon alludes to the drunken escapades of his friends and neighbors. Life among the poor at the University of Paris and on the Left Bank is shown to be riotous, disorderly, and often violent. In the "Ballad of Good Doctrine" he tells how to procure wine by trickery when one is without money and how he and his friends spend all their money in taverns and on women. His "Ballade and Prayer" describes Master Jean Cotart, "who drank only the best and the dearest/Without a red cent to his name./ Yes, this man was the finest of guzzlers; no one could make him let go of his mug/-/ He never was lazy in this business of drinking." This ballad has been described as "reeling up to us out of the Middle Ages full of drink and life" (Stacpoole 1916:86, who also comments [p. 33] that "it is very difficult to estimate the influence of wine on medieval man, for at times he did most exceedingly drunken things when he was sober"). Villon also characterizes the life of the poor as life with only water to drink ("Ballade Franc Gantier Refuted") (Villon 1462/1960:87-89, 101, 111-113).

TAVERN PREVALENCE, CHARACTERISTICS. It is estimated that there are 4,000 taverns in Paris during the days of Villon. Nearly everyone in Paris sells wine, from the highest to the lowest persons; some wholesale, others retail. Some taverns are respectable and frequented by the bourgeoisie; others are owned by men who have no fear of the law and are centers of gambling, which is forbidden in the taverns. The most celebrated Left Bank tavern is the Pomme de Pin, frequented by Villon (Stacpoole 1916:45). The atmosphere of the taverns frequented by Villon appears to be similar to that in *Piers the Plowman*, ribald and somewhat unsavory.

GERMANY

NOBILITY/DRINKING. According to a papal official to the court of Emperor Frederick III (1440-1493), "Living here is nought but drunkenness" (Samuelson 1878:112).

WINE SALES, INEBRIETY PREVALENCE. Describing life in Vienna, Pope Pius II (1458-1464) observes: "To sell wine in the home is in no way damaging to the reputation. Almost all citizens operate taverns for wine-drinking. They...fetch in the drunkards and harlots, and give them free some of the food...so that they will drink more, but then they're given shorter measures in wine. The common people worship their bellies and are gluttonous" (quoted Ross and McLaughlin 1953:212).

BRANDY CONTROLS: SALES, CONSUMPTION. In 1450, the Nuremberg city council issues an ordinance against brandy consumption. This is the first legislation concerning the drinking of brandy to appear in Germany (Baader 1868:315). A Brandenburg code reads, "Nobody shall serve aquavit in his house or give it to his guests" (Forbes 1970:97).

DISTILLING/DOMESTIC. By this time, a common sight is the wise women, the *Wasserbrennerinnen* (water-burners), of whom Michael Puff writes (see 1475/Germany). They combine the professions of distiller, confectioner, fortune-teller, and procuress, making the product in simple stills in kitchens. Apothecaries remain the specialists for all kinds of alcoholic medicines; monasteries become less important centers of medicine (Forbes 1970:102, 108).

MEAD CONSUMPTION. At Eger in Bohemia, there are still no fewer than 13 mead-houses in 1460, producing as many as 384 barrels a year. After the Thirty Years War (1618-1648), there is only one (Gayre 1948:87).

ITALY

LIQUEUR DISTILLING (SAVONAROLA). The influential Italian doctor Michael Savonarola (1384-1464) writes several books on the art of distillation. In his principal work, De arte confectionis vitae simplicem et compositam (The Art of Making Waters), first published in 1532, he discusses in much detail the making and pharmaceutical application of many aqua ardens compositae, alcohol distilled in the presence of aromatic herbs. This latter invention gives rise "to a small liqueur industry" in 17th-century Italy (Jellinek 1976b:4). In a treatise published in 1484, he accepts the conception of alcohol as the fifth essence. Whereas most of the early stills are made from metal or pottery, Savanorola's writings indicate that in the 15th century distilling apparatus are being made of glass (Forbes 1970:65-66; Crombe 1959:137).

1475

EUROPE

SPIRITS CONSUMPTION. Beginning in the last years of the 15th century, the drinking of brandy for recreational purposes becomes apparent, most particularly in Germany (Braudel 1973:171). Also by this time, professional distillers and sellers appear (water-burners, or Wasserbrenner) (Forbes 1970:91).

ENGLAND

COURT/DRINKING. During the reign of Henry VII (1485-1509), there is little to indicate that excessive alcohol drinking poses a problem at court (French 1884:127).

WINE AVAILABILITY. The Navigation Act of 1490 stimulates the wine trade with Bordeaux, but total English imports are still half that of the prosperous years in the early 14th century and the price is now at least double. "To what extent these two facts are directly related it is not possible to say, for we have no precise knowledge as to the size of the wine-drinking population...; nor do we know whether the gradual diminution of supplies had in any way affected the habit of wine drinking" (James 1971:53).

BREWING CONTROLS; BEER CONTROVERSY. Norwich bans the use of hops in brewing in 1471. The Ale Brewers' Company (founded by royal charter in 1437), concerned that hops and other herbs are being added to ale, petitions the mayor of London in 1488 to prevent the practice. A fine is imposed on every barrel of ale not properly brewed. The ale brewers do not object to beer, but to putting hops into ale. In 1493, the brewers of beer are officially recognized as a guild (Monckton 1966:69-70).

SALES CONTROLS (LOCAL). In the latter part of the century, in the absence of parliamentary action to regulate the retail liquor trade, authorities in various towns take steps to bring order to the sale of liquor: establishing character standards for publicans, enforcing price and quality controls, restricting hours of sale on Sundays (Monckton 1969:31-32).

ALEHOUSE CONTROLS. The ordinary public-house trade is put under statutory control for the first time in 1495 as part of Henry VII's act "against vagabonds and beggars." This marks the beginning of legislative control of the trade. The act includes a provision empowering any two justices of the peace to suppress useless alehouses and "to reject and put away common ale selling in towns and places where they should think convenient, and to take sureties of keepers of alehouses on their good behaving." Acts of Parliament to control the liquor trade become progressively more frequent (Monckton 1966:82).

FRANCE

BREWING CONTROLS. A royal ordinance of 1495 recommends that brewers "make good and loyal *servaises* and beers with nothing but grain, water and hops." Soon the term "cervoise" (*servaise*) is dropped completely from everyday usage (Claudian 1970:11).

PEASANTS/DRINKING. In the last quarter of the century, the farm worker in Languedoc insists on money in his pocket and white bread on his table, and is "a big drinker of good white wine." The annual ration of wine for an adult is one *muid* of red wine plus .4 *muids* of piquette (1 *muid* = c. 650 liters). The average farm worker consumes about 1.74 liters of pure wine a day. Due to this high level of wine consumption, vineyards now cover 10% of the productive land area of Languedoc (Le Roy Ladurie 1974:43).

GERMANY

BRANDY CONSUMPTION, MEDICINAL USE (PUFF). Literary and legislative evidence indicates that brandy is becoming popular as a recreational drink or at least that a considerable expansion of medicinal use is occurring. Michael Puff (von Schrick), an Austrian, publishes the first significant printed book on distilled spirits, *Hienach volget ein nüczliche Materi von manigerley ausgepranten Wasser* (Afterwards Follows a Useful Material of Distilled Water, written in 1455 but not published until 1478, after which it undergoes many editions, 44 by 1500). In it he praises spirit drinking for medicinal purposes but complains that this medicine is now being abused. He states that one of the main reasons for writing the book is that "the abuse of medicaments is so great that no one would wonder if stupid physicians, wise women, vagrants, and apothecaries were to poison the whole world with them." The only way to combat this danger is to treat illness with distilled waters properly and to show their useful effects. He includes recipes (many from other books) for making brandy (*geprannten weyn*) and states that anyone who drinks half a spoon of brandy every morning will never be ill. If one is dying and a little brandy is poured into his mouth, he will speak before he dies (Simon 1948:130; Forbes 1970:108-109).

CONCERNS (FOLZ). A poet (probably Hanz Folz, a Nuremberg surgeon/poet) publishes a pamphlet at Bamberg in 1493 advising people to use spirits moderately: "In view of the fact that everyone at present has got into the habit of drinking _aqua vitae_, it is necessary to remember the quantity that one can permit oneself to drink and learn to drink it according to one's capacities, if one wishes to behave like a gentleman" (quoted Braudel 1981:243; Forbes 1970:97; Janssen 1910, 15:417).

SALES CONTROLS. In addition to literary evidence, increased consumption is indicated by sales controls. Because "many persons in this town have appreciably abused drinking brandy," a Nuremberg ordinance in 1496 forbids the selling of brandy on Sundays and feast days and stipulates that brandy bought during weekdays can be drunk only at home and then not more than a pennyworth a day (Baader 1868:316). Between 1498 and 1500, the city of Leipzig issues in succeeding years ever stronger laws pertaining to the making, selling, and drinking of brandy (Rau 1914:13) but these and similar ordinances in other places do little to curb brandy drinking (Forbes 1970:97). Frankfurt town authorities request the clergy and the medical faculty to warn people against spirits consumption (Arnold 1911:322).

QUALITY CONTROLS. In a treatise on brandy and pharmacy for the common man (_Von dem geprannten Weyn und Apothek für den gemeinen Mann_), Michael Schrieb (1484) warns the public against very serious adulterations of spirits of wine, but praises the pure product and gives details on proper pharmaceutical applications (Forbes 1970:103). In 1487, Frankfurt requires brandy to be checked for quality not only in the market place as a finished product but also in the distilleries while in the making (Rau 1914:11).

INEBRIETY CONDEMNED (BRANT). In 1494, Sebastian Brant's popular _Das Narrenschiff_ (Ship of Fools) vividly satirizes the drinking customs of his day, giving rise to a whole genre of navicular satirical literature (Hauffen 1889:486, 487-488). He attacks the striving for pleasure and luxury in all classes so that no one is any longer content with their position and all try to outdo each other in dress and drink. In section 16, "Of Gluttony and Feasting," he declares that "demon wine" is a "very harmful thing" that kills reason, makes the head and hands shake, speeds death, and leads to neglect of friends, fornication, and "grave offense." He concludes: "I censure those who tipple beer,/ A keg of it per man, I hear/ Becoming so inebriate/ That with them one could ope a gate./ A fool shows no consideration,/ A wise man drinks with moderation,/ Feels better, illness too defies,/ Than one imbibing bucketwise" (Brant 1494/1944:96-99; see also Jellinek 1945:653).

In chapter 72, the gluttonous Grobianus, the patron saint of all ill-mannered, indecent, drunken and gluttonous people, makes his first appearance in literature, quickly passing into popular mythology, giving rise to a whole genre of grobianish ("boorish") literature in the 16th century (Coupe 1972:112). (The word "grobianus" first appears in 1482 as a translation of _rusticus_ or peasant [Neuer 1970:118].) Brant (1494/1944) asserts: "A new St. Ruffian [Grobian] now holds sway,/ Men celebrate him much today/ And honor him in every place..../ Now grossness everywhere has come/ And seems to live in every home."

In chapter 95, "Of Being Misled on Holidays," he describes the forsaking of the church for the tavern and the feasting that occurs on holy days in which they spend more money on wine than they earn by working all the week. Brant attacks the carnival tradition as displeasing to God as a waste of time and money. He objects to the eating drinking, dancing, and gaming that occur at church feasts (Burke 1978:213, 217).

A spokesman for the "enlightened but conservative" middle-class burghers (Pascal 1968:35), Brant seeks to raise moral standards by "indiscriminately condemning" all deviations from his own "puritanical, middle-class norm" as "foolishness," drawing no distinction between unwise behavior and actual vices (Coupe 1972:93). He castigates uncouth behavior far more radically than any predecessor, with the exception of Henrich Wittenweiler (Neuer 1970:120).

DRINKING CONTROLS: INEBRIETY. At the Diet of Worms in 1495, Emperor Maximilian I (1493-1519) orders "all electors, princes, prelates, counts, knights, and gentlemen to discountenance and severely punish drunkenness" (Samuelson 1878:105). This is the first of several imperial diets over the next 50 years that specifically mention the problem of alcohol use (Cherrington 1926, 3:1092; Stolleis 1981:103).

Also at the diet, the reforming Archbishop Bertold of Mainz proposes to place under an imperial regency the internal matters of civil administration (called Pollucy or "police matters") that establish good order, welfare, peace, and security. These include excessive drinking, adulteration of wine, blasphemy, cost of clothing, itinerant musicians and entertainers, and usury. As part of the great effort to reform the constitution of the Holy Roman Empire between 1495 and 1512, the term Polizei comes into use to designate the broad areas of individual and community activity which require legislative regulation in order to preserve good order in society and promote general welfare and common good. The greatest concerns are centered on individual conduct, with religious and moral overtones. The idea that moral and spiritual welfare is the proper concern and duty of the prince originates in the religious and paternalistic nature of medieval conceptions of government (Dorwart 1971:3-7, 14; Vincent 1935:1-2).

BREWING. Beer begins to be brewed municipally in Nuremberg; monasteries and private individuals still brew their own beers (Strauss 1979:201).

QUALITY CONTROLS. Quality control concerns and regulations over alcohol beverages are common in Germany throughout this period (see above for brandy). In 1487, Munich passes an ordinance forbidding the use of anything but barley, hops, and water in the brewing process (Cherrington 1925, 1:406).

SUMPTUARY LAWS. In 1465, Brandenburg restricts the number of guests who may attend betrothals and wedding celebrations because of concern over excessive food and drink consumption and extravagances that threaten to impoverish burghers (Dorwart 1971:34-35).

In 1485, Nuremberg issues a new wedding manual detailing the limits on wedding expenditures. Reflecting the rising prosperity, it permits more guests and more numerous and costlier goods than previously. It also establishes for the first time regulations for lower-class weddings, indicating that the increasing standard of living was filtering down the social structure (Greenfield 1918:46).

SCOTLAND

QUALITY CONTROLS. In 1492, Parliament forbids the importation of mixed wine and prohibits every sort of adulteration of wine or beer. The penalty for the offense is death (Cherrington 1929, 5:2383).

WHISKY CITED. The oldest certain written reference to aqua vitae that clearly refers to the making of a malt liquor, or whisky, occurs in 1494 in the

Scottish Exchequer Rolls. It states that "eight bolls of malt [be given] to Friar John Cor wherewith to make aqua vitae." This is the first of 19 references to aqua vitae between 1494 and 1512 in Exchequer rolls and Treasury accounts that clearly refers to malt whisky (Daiches 1978:3; Ross 1970:2; Wilson 1973:32).

SWITZERLAND

COMMERCIAL BREWING. Commercial breweries run by tradesmen appear in Switzerland in the 15th century. Previously, beer was produced on a commercial scale only at monasteries, convents, and large noble estates. Widespread wine drinking had also previously discouraged brewing, but the destruction of vineyards in the late 1400s by storms creates favorable opportunities for breweries. In 1488, the first commercial brewing plant is established in Basel, with a second following in 1491. Wine, however, remains the favorite drink of the Swiss, beer drinking being considered "unseemly, foreign, and even revolutionary" (Jellinek 1976b:76-77).

PLEDGING BANNED. In 1492, the state of Bern forbids the custom of pledging drinks to someone's health, but to little effect. This is one of several pre-Reformation proscriptions designed to reduce excessive drinking (Jellinek 1976b:81).

The Sixteenth Century

1500

EUROPE

BACKGROUND DEVELOPMENTS. The economic recovery begun about 1450 continues. The population increases, probably due to a decline in recurrent epidemics and more effective government order. At the same time, an inflationary spiral begins, which escalates rapidly in the second half of the century and lasts until the mid-17th century, undercutting the recovery. In the early century, complaints are already raised that parts of Europe have become overcrowded and that goods have become more costly. The increase in urban population encourages building. Everywhere towns grow and production rises; new trade routes are opened up and merchants begin exchanging a larger volume of goods over greater distances. A new standard of luxury is established; commerce, banking, and industry become sources of great wealth, and the enriched middle-class seek to buy into and live like the nobility. In the expanding urban centers, the majority of citizens have very few rights and are very poor, living in horrible slums. These developments create considerable tension within society since religion glorifies poverty and social status is still based on birth. Furthermore, conflicts are sharpened between landlords and tenants and between employers and workers. In both town and country, class relations begin a process of increasing fragmentation. Complaints are frequently heard of the newly enriched bourgeoise not staying in its place, but seeking to emulate and even surpass the nobility in luxury and extravagance. At the same time, as commercial and business life quickens, a new ethic emerges which centers on thrift, self-restraint, and sobriety. These developments become most pronounced in the second half of the century (Rice 1970:48-62; Koenigsberger and Mosse 1968:55-57; Medick 1983:85; Kamen 1972:48-50, 57).

INEBRIETY PREVALENCE. The 16th century, particularly the second half, is an age of ostentatious display, hugh appetites, gluttony, and copious drinking. For example, on a visit to Nuremberg, Philip Melanchthon (1497-1560) feasts on an eight-course meal, all courses consisting of meat, fowl, or fish. At such meals, each guest consumes a third of a liter of wine with each course (Koenigsberger and Mosse 1968:313-314). In Valladolid, Spain, beverage alcohol consumption reaches 100 liters per person per year in the middle of the century; in Venice, new and severe actions against public drunkenness are taken in 1598; in France and England, complaints about drunkenness increase in the late 16th century; in Poland, peasants consume up to three liters of beer a day; in Germany, this period has been called the century of drunkenness (Braudel 1981:236, 238). In 16th-century Europe, high levels of drinking are mostly characteristic of towns (Braudel and Spooner 1967:407-408).

DRINKING ATTITUDES. Attitudes toward drinking from this period through to at least the beginning of the 18th century are characterized by a continued assertion of the positive nature of drink alongside a greater emphasis on the negative effects of drunkenness not only on individual salvation but on society in general. A greater effort is made to reduce drunkenness through external force and imposed discipline. On the whole, drunkenness is viewed as a component of a more general propensity toward self-indulgence.

Legnaro (1981) speculates that in this period Western attitudes toward drinking change because of the rise of the individual spirit of early capitalism and the rationalization of reality. The communal spirit of the Middle Ages, with its more impulsive, emotional reality, is replaced with an ideology which places new importance on time, efficiency, and order, and which views drinking as disruptive. In the Middle Ages, intoxication "was not looked down on extensively" because "intoxicated loss of individual control did not provoke either fear, feelings of guilt or any significant social reaction." At the beginning of the modern era, however, loss of psychic and physical control through intoxication begin to be viewed negatively because it is no longer compatible with the "rational mastery of reality." This brings a greater ambivalence in the attitude toward intoxication.

PROTESTANT ATTITUDES, DRINKING PRACTICES. From the beginning of Christianity there have been occasional advocates of total abstinence, but not until after the Protestant Reformation in the 16th century does abstinence become a general movement (Hirsh 1953:964). Nevertheless, there is not a sharp break between Protestantism and Catholicism in regard to attitudes toward alcohol consumption. Like the early Church fathers, Protestant leaders such as Luther (see 1520/Germany) and Calvin (see 1540/Switzerland) sanction moderate drinking of wine and beer, viewing alcohol as one of God's creations intended for man's benefit and enjoyment. They attack drunkenness and the abuse of alcohol, but no more so than did the Church fathers or many contemporary Catholics. In fact, Harrison (1971:87-88) suggests that in the Reformation period abstinence was more a characteristic of Catholics than Protestants, who sometimes "made a point of distinguishing themselves from Catholics by advocating moderation rather than abstinence." Nevertheless, the Protestant Reformation contributes to a less tolerant attitude toward drunkenness and popular drinking practices by reemphasizing fundamental Christian doctrine, by seeking to end the close association between the secular and the sacred within the Church, and by relying more on civil authority to institute and enforce moral behavior (especially among the Calvinists). Similarly, while "strict inquiry into the habits and customs of people in their private capacities" (e.g., sumptuary laws) has long been looked upon as the duty of government authorities, the Reformation brings about "a marked change in the moral sentiments and the social customs of people wherever it [is] introduced." Daily habits and practices become affected by moral law "in a way hitherto unknown"; old ordinances are "renewed, amplified, extended"; penalties are made more severe; and many communities become more staid and even sober (Vincent 1898:363). (See also Max Weber's [1958:95-154] discussion of "worldly ascetism" among some Protestants.)

The most significant shift in attitudes toward alcohol occurs among the radical Protestant groups such as the Anabaptists and Hutterites (see 1520/Germany), who insist that the true Church must be a community of saints and must exclude all those who fail to fulfill its austere moral code. Bainton (1945:53) argues that the origins of modern prohibitionism are rooted in a combination of the Anabaptist moral code and the Calvinist reliance on civil authority. The Anabaptist doctrine heavily influences later German Pietism, which in turn influences English Methodism and the 17th-century Quakers, the real pioneers in the modern temperance crusade, and, following their lead, the Calvinist churches of the late 18th century.

CONCERNS: POPULAR CULTURE, FESTIVE DRINKING. At the beginning of the century, the upper classes still participate in popular culture, particularly the communal festivals and other times of "ritualised collective rejoicing" (Burke 1978:25). But in the course of the century, the clergy and nobility, as well as merchants and professional men, begin to abandon the popular culture of the lower classes, and seek to reform it, inaugurating a process in which the word "people" will come to mean "common people" instead of "everyone." The clergy begins to withdraw with the Protestant and Catholic reformations. Whereas in 1500 most parish clergy are of a similar social and cultural level as their parishioners, afterwards they are better educated, of higher social status, and more remote. The nobility withdraws as new aristocratic modes of behavior and manners develop (ibid., pp. 270-271).

Although complaints about popular culture increase in the 15th century, as evident in the writings of Sebastian Brant (see 1475/Germany), among others, most of the bans on traditional customs begin in the mid-century, as early as 1530 in Protestant Bern and Zürich. The movement is most prominent among Protestants and in England and France, but to some extent it involves all European nations and Catholics as well, although Catholics tend to be less extreme. Particularly after the Council of Trent (see 1560/Europe), it can be found in Catholic Strasbourg, Munich, and Milan. Reformers insist on a much sharper separation of the sacred and the profane than existed in the Middle Ages, and particularly object to profane elements in religious festivals, in the belief that the traditional familiarity with the sacred breeds irreverence. The movement is linked directly to the shift in religious mentality and sensibility that begins in the late Middle Ages and to the Protestant and Catholic Reformations, since the reform of the Church as it exists means reform of popular culture, for the church is still the major cultural center used for secular as well as religious purposes. In addition to parish wakes and other church festivals, complaints are raised about pilgrims becoming drunk at the fairs held in conjunction with pilgrimages to holy places.

Reformers also object to many secular aspects of popular culture, especially the tavern. Whereas popular culture stresses the values of generosity, spontaneity, and greater tolerance of disorder, the ethic of the reformers is one of "decency, diligence, gravity, modesty, orderliness, prudence, reason, self-control, sobriety, and thrift." Although this ethos later becomes most associated with middle-class shopkeepers, in the 16th and early 17th centuries it is championed largely by the clergy and justified on purely theological grounds (Burke 1978:207-218; Bercé 1976:133-134).

Another reason authorities increase their vigilence over popular festivities is concerns about the violence that frequently occurs during them, fueled by copious supplies of wine or beer. For example, brides are often taken away and held prisoner in a cabaret by mischievous friends, only to be released when drinks are purchased for the mischief makers. The close connection between popular feasts, violence, and drink is also demonstrated by an outbreak that occurs when a rumor of a massacre of Swiss troops sweeps through the Oberland villages of Bern in 1513 during church dedication celebrations. The intoxicated crowds fall in large numbers on the innkeepers who as recruiters of these troops were responsible for the massacre. Only by furnishing drinks do the innkeepers calm the crowd (Bercé 1976:14, 20, 73-74). Le Roy Ladurie (1980:308) demonstrates that the Carnival, with its role reversal lubricated by plentiful drinking, was a means by which social change and protest could be practiced.

Authorities are also concerned that the feasts, which often cost several months' earnings, are ruinous to the poor (Bercé 1976:164).

<u>CONCERNS: IDLENESS, POVERTY</u>. During the 16th century and the early 17th century, especially in England and France, concerns grow over idleness, which is viewed as the mother of all vices. Although such concerns are not new, they become more prominent as attention is focused on putting people to work so that labor might increase national productivity and wealth (Cole 1939, 1:14). It is

only in the 16th century that the fundamental economic importance of labor as a factor of production seems to become "explicitly recognized," and with this recognition a new insistence upon the duty of every man to work follows. Although an element in Christianity from its origins, the positive merit of work is most clearly asserted in the religious teachings of the post-Reformation period, among Catholics as well as Protestants (Thomas 1964:58-59). This new concern for idleness contributes to the movement to reform popular culture as there are over a hundred holy days of nonwork.

This is a period of chronic underemployment for laborers. Most unemployment is involuntary; but some of it is voluntary--a continuation of the medieval tradition of a preference for leisure instead of higher earnings (Coleman 1956:290). Since there are few affordable consumer goods, laborers have little incentive to earn more than subsistence wages or to save money. Thus, throughout the early modern period, the complaint is heard that when real wages are high, so is idleness which produces excessive drinking (Hill 1964:125).

CONSUMPTION PATTERNS. Among all classes, the long-term needs of the household have a relative low priority in the monetary sphere, whereas the demand for public consumption--including expenditures on feasting, drinking, and other public rituals--is "extraordinarily high." Any money income beyond that required for the short-term needs of subsistence are "to some degree" viewed as surplus. Consumption functions as a vehicle of plebian self-consciousness, a way people are joined together (Medick 1983:91-94).

REFINEMENT OF MANNERS. In the course of the century, among elites the most exact observance of differences of rank in behavior becomes the essence of courtesy, the basic requirement of the concept of civility (*civilitas* or *civilité*), which society adopts to designate "good behavior." Whereas the same elementary rules of behavior were repeated throughout the Middle Ages without producing firmly established habits, in the 16th century all problems concerned with behavior take on new importance; books providing guidelines and rules for good behavior multiply (see Castiglione 1520/Italy; Erasmus 1530/Europe; della Casa 1550/Italy). These books emphasize how the virtue and perfection of the gentleman are no longer embodied in chivalry and military prowess but in ways of speaking, eating, and gesturing. As the medieval knightly-warrior nobility declines, a new aristocracy emerges at the semiurban courts of kings and princes, with a new social space, a new function, and a different emotional structure. The nobleman is no longer as free a man as he once was, the master of his own castle; he now lives at court, only one of a heterogeneous group of people serving the prince. Forced to live with one another in a new way, people become more sensitive to the impulses of others, the need for self-discipline, and the avoidance of giving offense to others (Elias 1978:70, 79-83, 201, 203, 216-217; Ariès 1964:376).

CONTROLS: STUDENTS/DRINKING. Beginning in the 16th century, student drinking practices are increasingly suppressed. Drinking is officially prohibited in French and German schools. Students drink in secret in rooms or do all their drinking outside the school at taverns. They are viewed as libertines in the mold of Villon and ordinances are passed forbidding tavern and inn keepers from putting them up. Students maintain more of their freedoms in England (Ariès 1964:321, 323).

ATTITUDES (ERASMUS). In Erasmus' *In Praise of Folly* (1509/1964:104, 112), Folly says she was nourished by two charming nymphs--drunkenness and ignorance--and claims credit for having invented banquets, drinking good healths, and sending the cups around.

DRINKING INFLUENCES: DIET. One of the factors probably accounting for high levels of alcohol consumption is that greater quantities of salts and spices which are added to foods to preserve them and make them more palatable. This is due to the declining availability of fresh meat and to the increasing availability and fashionability of spices. The result is "an oceanic thirst in many parts of Europe" (Minchinton 1976:124; see also Slicher Van Bath 1963:84; Thomas 1971:18; Strauss 1979:20).

SPIRITS CONSUMPTION, PRODUCTION. The "great innovation, the revolution in Europe," is the spread of distilled spirits made from wine (brandy) and, to a much lesser extent, grain. The extent of brandy drinking in the 16th century is somewhat uncertain, as are the reasons for its use. There is general agreement that spirit drinking spreads and that distillation becomes fairly well established, especially in Germany and northern Europe (Keller 1976:13). As Braudel summarizes, alcohol is created in the 16th century; the 17th consolidates it; and the 18th popularizes it (Braudel 1981:241). The printing press enables descriptions of the technique and the virtue of spirits to be more widely disseminated. During the first half of the century, a whole class of texts (called in German _Arznei-_, _Kräuter-_, and _Distillierbücher_) appear showing the increased interest in distilled spirits and their medicinal value. These tracts are written by physicians, apothecaries, and botanists such as Hieronymus Braunschweig (see 1500/Germany), Valerius Cordus, Walter Ryff, and Conrad Gesner (see 1550/Germany). The number of apothecaries increases under the influence of Paracelsus (see 1520/Europe) and his school of iatrochemists, especially in Germany, Sweden, and Russia. Spirit production begins to expand beyond monasteries, apothecaries, and alchemists into the hands of artisans and merchants. In every country, the preparation of brandy and other alcoholic beverages begins to become the special job of associations of distillers (Forbes 1970:97-98, 101).

Jellinek (1976b:77-78) believes that it was only at the very end of the century that brandy use begins to pose a serious problem: "Only at the end of the 16th century did some use of liqueurs spread from Italy and not before the beginning of the 17th century did distilled spirits have any importance as beverages [as is evident in Rabelais' and Montaigne's failure to mention them]....The impression that aqua vitae may have been prominent in earlier centuries comes from some versified encomia of distilled spirits for their medicinal properties, and from some Nuremberg police regulations against the 'adulteration' of wines with 'burnt waters.'" The evidence suggests that while his conclusion is valid for France and England, he underestimates the extent of spirit drinking in 16th-century Germany, although even there it is uncertain how much is consumed for medicinal purposes and how much purely for intoxication. The latter, however, clearly expands as the century progresses.

BEVERAGE CHARACTERISTICS: WINE. In the first half of the century, commercial capitalism strongly influences the wholesale wine trade and its techniques (Craeybeckx 1958:207). Wine is still largely a luxury drink; consumption by peasants is limited even in viticultural regions. But to supply rising urban populations, coarse types of vines with high yields are becoming general in commercial vineyards. The wine consumed is still new wine; it easily turns sour and cannot be kept from year to year. For example, Emperor Charles V (1519-1558) is advised in 1539 not to buy large quantities of wine for the navy because it most likely will turn to vinegar. Clarifying, bottling, and the regular use of corks are unknown. Wine is transported in wooden barrels, a procedure invented in Roman Gaul, in which it does not always keep well (Braudel 1981:244). However, the wines of the Renaissance and Reformation periods are probably better than those of the Middle Ages due to production improvements (Younger 1966:310-311).

BEER. Hopped beer is firmly established in the northern lands of England, the Netherlands, Germany, Bohemia, Poland, and Muscovy; the varieties and strengths of beers increase; and numerous books appear praising their virtues (for example, Heinrich Knaust, see 1570/Germany). Beer is no longer the drink of poverty. The Netherlands has a luxury beer for the rich, as well as a popular type. In Germany, Bohemia, and Poland, a large growth in urban brewing of strong beers pushes the light beer made by lord and peasant, often without hops, into second place. At the end of the century in England, too, there are frequent complaints about the increasing strength of beer (Braudel 1981:238-239).

SPIRITS. Until the 18th century, it is not always possible to determine exactly what type of spirits are being consumed on any occasion, as all are labeled "waters" (Younger 1966:324). But almost all spirits consumed in the 16th century are wine based. Many experiments are conducted to create new kinds of spirits with herbs and fruits, but the only new kind to take a lasting hold are those made from cereals. The technique of making grain spirits was first developed in the 15th century (Forbes 1970:91). However, grain-alcohol distilling encounters considerable public resistance until well into the 17th century because of (1) concerns that it increases the cost of corn for the poor and uses up limited supplies; (2) a preference for the taste of wine spirits, especially French brandy; and (3) the rectification defects of older apparatus, which make the taste of grain spirit objectionable (Forbes 1970:103; Cherrington 1926, 3:1091).

ENGLAND

BACKGROUND DEVELOPMENTS. The Tudors succeed in establishing a royal monopoly over political and military power and make the monarchy the overriding focus of the country's allegiance, resulting in an enormous explosion in the prestige and size of the court and central administration. The Tudors, especially Henry VIII and Elizabeth I, realize the importance of an alluring court as a stabilizing factor that fosters the dependency of the nobility. Court attendance and service to the state become "an ideal, a social convention, a pleasure and a necessity" (Stone 1967:97, 183, 190, 217).

During the 16th and 17th centuries, the population of London increases tenfold. Perhaps one-sixth of the total population of England spends some time living in London, "many of them returning to their rural communities with newly acquired urban habits of living" (Thomas 1971:3-4).

INEBRIETY PREVALENCE; CROSSCULTURAL COMPARISON. The English may drink and eat less than their continental counterparts. In the words of Younger (1966:314): "For the first time since the Dark Ages the English lost their Crown of Inebriation. By almost universal consent it was awarded to the Flemings, the Dutch and the Germans." But the difference is only a matter of degree. Drink is still built into the fabric of social life (Thomas 1971:17), and concerns about drinking rise markedly in England, especially in the course of the second half of the century. According to Bretherton (1931:199), it is during the

Tudor and Stuart periods that "the habit of drunkenness [gains] its first real hold upon the nation."

It is among the rising commercial sectors that a change in drinking levels is most evident. Simon (1907, 2:149) comments: "Although the consumption of wine in the royal household and the houses of the nobility was still considerable, it does not increase during the fifteenth and sixteenth centuries in anything like the same proportion as amongst the mercantile classes, whose wealth and political importance was far greater than in preceding ages."

ATTITUDES. In 1509, Alexander Barclay publishes Ship of Fools in imitation of the German original Das Narrenschiff (1494) by Sebastian Brant, satirizing and attacking the widespread drunkenness of the period (Jellinek 1944:469).

BEVERAGE CHARACTERISTICS. One reason for increased drunkenness under the Tudors and early Stuarts is the "changing character of popular drinks," specifically in the availability and strength of beer and wine (Bretherton 1931:168).

WINE; SACK. In the Middle Ages wines were rare, weak, and diluted; as commerce and naval power expands, stronger imported wines now become more plentiful, particularly strong sack and malmsey. References to sack first appear in the early century as the name for wines from Spain. The origin of the word "sack" is unclear, but most likely it is derived from the Spanish word for export (Allen 1961:171, but see Younger 1966:289). They were probably exported because they were too heavy and sweet for Spanish tastes. Allen (1961:176) and Younger (1966:295) believe that they were fortified, although the later hesitates "to express an opinion on how much, how often, and how well." Certainly they were stronger than contemporary northern wines but probably less strong than the fortified sherry of today. The popularity of sack is well established by the mid-century, and it goes on to become one of the most popular Elizabethan wines, much praised by Shakespeare's Falstaff (Henry IV).

BEER. As elsewhere, the rise in the brewing of hopped beer creates a stronger, better tasting, and longer-lasting beverage. Small beer under 14-days-old is still available, but even this is stronger than medieval ale. Strength and age become the most important criteria of quality (Bretherton 1931:168-169; Monckton 1966:83). In the course of the century, ale also is brewed with hops so that it becomes difficult to draw a clear distinction between the two brews, although ale is perhaps sweeter and is made with less hops, while beer is more bitter (Monckton 1966:71). Although the strength of beer rises, the poor generally still drink ale of low strength and poor quality. (On the rising strength of beer, see also 1570/England.)

DRINK RATIONS. The average naval and military allowance seems to be one gallon of beer per day per person (Thomas 1971:18).

DRINK HOUSES: PREVALENCE, CHARACTERISTICS. Three types of drinking establishments exist in Tudor England: (1) an inn or hotel with sleeping accommodations, food, and all alcoholic beverages; (2) a tavern, open to all of the public, selling wine as well as ale and beer; and (3) an alehouse or tippling house, where ale and beer are brewed and retailed on the premises.

ALEHOUSES. There are about nine alehouses to every inn (Monckton 1966:96). Most are still squalid premises located in ordinary houses, in a backroom, cellar, or back alley. For the most part, they are run by the poor (laborers, petty craftsmen) for the poor, playing an important role in domestic economy of the poor and providing them with a myriad of services. (See Skelton's description of alehouse life given below). Although justices were instructed in 1495 to reduce useless alehouses and insure the good behavior of ale sellers, alehouses are still largely uncontrolled. Town councils continue to complain that their numbers are increasing (Clark 1978:49-50, 53; Clark 1984; Wrightson 1981:2). Often alehouse operation is referred to as "tippling" (Clark 1983:39), a word also used in the sense of drinking outside the home and mealtime.

INNS, WINE TAVERNS. The "Golden Age" of the English tavern begins. "Scarcely a writer of importance [fails]...to extol the virtues of the inn and the wine tavern" (Popham 1976:257; see also Everitt 1973). Taverns and inns become cultural centers and the semi-official center of towns, their importance growing as travel expands. The keeping of an inn or wine tavern is regarded as a respectable, if not honorable, calling. Most tavern keepers are men.

LOWER-CLASSES/DRINKING, ALEHOUSE LIFE (SKELTON). John Skelton (1460-1529), tutor to the future King Henry VIII and one of the ablest satirists of his age, describes the kind of folk who gather at lower-class alehouses in his popular poem "The Tunnyng [Brewing] of Elynour Rummyng." He shows that the alehouse has changed little from the days of Langland and Chaucer (see 1375/England). He shows it to be an informal, dirty, unhygienic place where "lewd [i.e., lowly] folk" come to gossip as well as drink, but often they would "drynke tyll they stare." Some drink so habitually that, as in the case of one woman, "The dropsy was in her legges" (quoted Jellinek 1945:102-110; see also Firebaugh 1924:214-218). In his "Bouge of Court," Skelton also satirizes the corruptions of court life (French 1884:135).

CONTROLS. In Tudor and early Stuart England, the production and consumption of drink is "perhaps more fully regulated and controlled than anything else." A variety of factors contribute to increasing regulation: (1) all wine is imported and is subject to high duties; (2) the barley used for beer is important for the food supply; (3) alehouses and taverns are suspect because artisans and apprentices are believed to waste time and money there and because they are natural centers of conspiracies against the Crown, or, at the minimum, sites of disorder and immorality; (4) drinking and gambling compete with the parish church on Sundays; and (5) after the dissolution of the monasteries (see 1530/England), alehouses and inns became the natural resort of travellers and wayfarers, people still regarded in Tudor England with distrust by authorities and as fair game by thieves, highwaymen, and roguish innkeepers (Bretherton 1931:147).

DRINKER LIABILITY. During the century, drunkenness is ruled to be no excuse for crime but rather an aggravation of the offense. Subsequent mental impairment does not erase the legal import of the initial choice. This rule, which prevails in the Anglo-American world until the 1960s, asserts that the "blameworthiness of the offender lay in the voluntariness of his first drink." In the 17th century, however, consideration is given if habitual drunkenness has led to insanity (Bonnie 1975:26-27).

FRANCE

VITICULTURE. From the end of the 15th through the 17th centuries, French rural life is still characterized by great difficulty and hardship in which all energies are devoted to survival. The prevailing economic system remains that of the late Middle Ages. The typical peasant seeks only self-sufficiency. Even in the north, peasants grow vines among their crops to make their own wine. The vignerons (vine-growers) of Aquitaine and Burgundy are already producing well-known vines and making wines which they sell to the towns. Though wedded to the land, their commercial activity makes them actually town people (Duby and Mandrou 1964: 207-217).

WINE CONSUMPTION. Wine consumption even among vignerons is still limited, "both by custom and because they could not afford to do so," although the citizens of Bordeaux and other towns drink it. Peasants drink pin-pin (piquette), made from the weakest of the strainings (Francis 1972:247). To southerners, northern Frenchmen are drunkards (Le Roy Ladurie 1974:64). One contemporary observes of Vezelay: "The common people...rarely drink wine, eat meat three times a year....[they are] weak and unhealthy people" (quoted Duby and Mandrou 1964:213).

URBAN POOR/DRINKING, DIET. Sixteenth-century France witnesses an astonishing urban expansion; towns grow in wealth and population, accelerating large-scale commerce. Yet towns still have a rural aspect, characterized by moderation at the table, outdoor living, pastimes, beliefs, and customs, and the rugged quality of life. Except for the wealthy, diet consists of breads of mixed flours and warm gruel. Meat is rare; wine of inferior quality is more common. Moderation at the table seems ordinary: "contemporaries present it without excessive praise (or, for that matter, blame)" (Duby and Mandrou 1964:228-229, 233-234).

SALES PRIVILEGES. All over France, bourgeois owners of vineyards seek (successfully) to gain the same permission given to the great lords to sell their surplus or wastes in retail outlets (Dion 1959:482). Parisians gain this right in 1633.

DISTILLING. Distilling in France is more closely bound to the monasteries and apothecaries than in Germany, and the popular consumption of brandy spreads less rapidly. It is not until 1514 that Louis XII (1498-1515) grants the privilege of distilling eaux-de-vie (aqua vitae) to the guild of vinegar-makers, but 20 years later the distillers form a separate corporation. Vignerons do not yet distill themselves. Little is known about the geography and chronology of the first brandy industry. An early distillery is probably operating at Gaillac in the Bordeaux region in the 16th century; brandy may have been sent to Antwerp as early as 1521 (Braudel 1981:241-243; Delamain 1935:16; Le Grand d'Aussy 1782, 3:67).

From 1506, Colmar on the Rhine, in present-day northeastern France (Alsace), places controls over wine distillers and brandy merchants, whose product begins to figure in the city's fiscal and customs returns. Conflicts during the next century and a half over who has the right to distill indicate the profitability of the expanding industry in Colmar (Braudel 1981:242; Delamain 1935:16).

MEDICINAL INTOXICATION. Like many during the century, the popular physician Jacques Dubois (d. 1555) regards wine as an admirable remedy, even against fever, and does not hesitate to affirm that it is good to awaken the forces of the stomach once a month by an excess of wine (Franklin 1889, 6:126).

GERMANY

BACKGROUND DEVELOPMENTS. By the standards of other countries of the 16th century, Germany is a prosperous country, with towns rich from trade, respected artisans, and silver-producing mines. Many German merchants, bankers, and mine-owners are among the richest individuals in Europe. However, politically and socially Germany is the most unstable area. It is composed of hundreds of large and small political units with little central authority. For a generation, the efforts of princes to extend their territories and authority and to raise taxes has caused chronic unrest. In 1517 this instability is wracked by the Reformation.

INEBRIETY PREVALENCE, CONCERNS. The rise in drinking and drunkenness that has been developing since the 14th century reaches a peak during the 16th century, which has been called "the century of drunkenness in German-speaking Europe" (Blanke 1953:175 see also Loffer 1909:5; Hauffen 1889:480; Krucke 1909:13). The flourishing drink literature of the time suggests "a fantastic excess hardly ever approached at any other time or by any other nation" (Jellinek 1945:648). Although Germans had a reputation for drunkenness throughout the Middle Ages, the literature of the early 16th century clearly shows that contemporaries felt the level of drunkenness in society had recently increased greatly. The onslaught of criticism directed toward drunkenness is unprecedented. For the first time, entire books are devoted to the subject of drinking; some praising it, most condemning it. Of particular concern is the custom of pledging health, attacked by Johann von Schwarzenberg in one of the first of these books (see below). Moreover, additional laws appear attempting to control the extent of drunkenness, whereas before the laws were largely limited to price, quality, and good measure (Jellinek 1945:648, 654, 658).

Braudel (1981:233) comments that in this century southern Europeans "looked jeeringly upon those [northern] drinkers who, in their opinion, did not know how to drink and emptied their glasses in one gulp." Contemporaries generally emphasize that excessive drunkenness is a recent occurrence and that it is a vice peculiar to the Germans and differentiates them from other Europeans (Flandrin 1983:67-68). Aventinus (d. 1534) asserts: "All other nations speak evil of us, scolding us as...senseless drinking Germans, always intoxicated, never sober" (quoted Reinhardt 1962:183). Similar statements are made by, among others, Luther, Dedekind, Cellini, Platter, Montaigne, Albertinus, and Sommer (see 1520, 1550, 1570, 1590, 1610/Germany).

DRINKING LITERATURE. The 16th century is characterized by a preponderance of satiric literature, much of it directed against the morality of the time, especially drunkenness. According to Hauffen (1889:480) and Jellinek (1945, 1946a), a new and distinct genre of drink literature develops because the situation had deteriorated so that it now required more attention. Drinking is condemned as damaging to one's soul, honor, possessions, and body (Hauffen 1889:503). There also develops a related genre of "boorish" (*grobianische*) literature which satirizes ill-mannered popular behavior. Although the roots of the genre can be found in the late Middle Ages, it is not until the end of the 15th century that the theme of extreme indulgence in food and wine emerges, with proportionately larger sections being devoted to it as the century progresses. The sudden growth in the drinking literature is at least partly a result of the invention of the printing press. The 16th century provides a flood of pamphlets and tracts on a much wider and diverse range of topics than ever before. The Reformation also draws all writers into the religious controversy and makes them propagandists of morality (Coupe 1972:91). The preponderance of satire is in part an inheritance from the Middle Ages, in part the influence of Sebastian Brant and Humanism. It also reflects the growth in importance of cities and city dwellers. Sixteenth-century literature strongly reflects the interests and values of the burghers. This literature is both directed toward and reflects the personality of German burghers, it reflects both their coarse behavior as well as their desire to transcend it. Indeed, through the new printed media burgher culture is consolidated and communicated to the public (Pascal 1968:9-10). Intended to both instruct and entertain, the satiric literature on manners often praises inebriety in order to ridicule it. Instead of giving serious advice on behavior, these works limit themselves to describing atrocious manners to discourage bourgeois readers, without providing positive examples of how they should behave (Neuer 1970:12-14).

HEALTH CONCERNS. Excessive drinking becomes a medical concern in German-speaking countries. Both doctors and ministers warn of the dangers to health from alcohol and point to specific diseases associated with brandy and beer (Diethelm 1965:53-54).

DRINKING INFLUENCES. Inebriety may have increased in the 16th century due to the following factors. First, for some people alcohol may serve as a form of anesthesia in response to widespread spiritual, social, and political insecurities, as Jellinek (1945) suggests. However, the lower classes and peasantry, who suffer the most in this period, do not appear to have engaged in as much drunkenness as their superiors. Second, as elsewhere in Europe, the decline in the availability of meat and the increased trade with the East results in most meat and other foods being heavily salted and spiced, giving rise to greater thirsts. Third, it is likely that the overall availability of alcohol increases due to the commercialization of the alcohol trade, improved brewing and distilling techniques, better transportation, the growth of towns, and the development of capitalist industrial development--all combining to stimulate large-scale production of alcohol and promote consumption. The alcohol trade is in the forefront of the commercial and monetary changes occurring at this time. Town governments and private citizens turn to it as a source of income as inflation increases costs and reduces the return from long-term rents. Throughout the century, there are numerous attacks directed against the nobility and town magistrates for erecting distilleries, breweries, and taverns. The commercial brewing industry is far more developed in Germany than in other nations, with the exception of the Netherlands. Fourth, the number of taverns or drinking outlets increases in response to population growth, urbanism, and financial conditions which make the monetary profit from taverns attractive. Fifth, brandy drinking spreads. Production is still small scale, but it is in Germany that evidence of the rise of recreational spirit drinking is most apparent in the 16th century. Sixth, there is continued growth of conspicuous consumption. Blanke (1953) emphasizes that Germany in this period is characterized by a general concentration on excessive living, of which drunkenness is only one manifestation. There is greater wealth to make such consumption possible, with the newly enriched merchant and commercial families seeking to imitate the extravagance of the nobility. This in turn generates social insecurities and competition among the nobility to become even more extravagant to maintain its superior position. Among the primary manifestations of this conspicuous consumption are great banquets and excessive eating and drinking in general. That drunkenness is related to this consumption pattern is evident in the frequency with which it is attributed to the nobility and the nouveaux riches. Neuer (1970:85-86, 139) also suggests that drunkenness and boorish behavior in general increase due to the rise in the standard-of-living of lower-class burghers, the population to which the grobianish literature is directed.

CONCERNS. Several other factors appear to influence the concerns over and attacks on drunkenness. The Reformation causes the issue of inebriety to become more charged and causes competition between Protestant and Catholic leaders to instill higher moral standards in their respective congregations and to attack the immoral practices of each other. The population growth causes a rise in vagrancy, class conflicts, and social disorder which heighten fears of drink-provoked outbreaks. Furthermore, as the population growth puts a strain on land and resources and as prices escalate, the economic system becomes increasingly more precarious and concerns rise that such extravagant consumption of food and drink, which often places grave financial burdens on families and depletes grain supplies.

NOBLES/DRINKING. Many contemporary commentators criticize the nobility for their drunkenness, their involvement in the drink trade, and their influence on the rise in drunkenness throughout the population. Such attacks become especially strong at the end of the century. This occurs in the context of the increasing brilliance and magnificance of court life and of the tendency of the enriched bourgeoisie to imitate the lifestyle of the nobility. Most of the tracts against drunkenness center on wine, which leads Jellinek (1945:658-659) to suggest that drunkenness in this period was primarily a problem among wine drinkers, although "some inebriety may have occurred among those who could not afford anything else but beer." Since the price of wine triples between the years 1540 and 1610, keeping it a wealthy man's drink (Pfaff 1892:167; Strauss 1979:201), the attacks on wine probably reflect the prevalence of drunkenness among the nobility and wealthy, as well as dissatisfaction with their moral conduct and in the excessive expenditures on luxury goods.

PEASANTS/DRINKING. Less frequently, the drunkenness of peasants is also attacked, but most often this is in the context of an attack on drinking within all classes, or it is charged that in their inebriety peasants are only following the example of their superiors. Peasant inebriety appears to be only periodic, and often tends to be excessive because it is rare.

MEDICINAL USE, SPIRITS (BRAUNSCHWEIG). An Alsatian army doctor, Hieronymus Braunschweig (or Brunschwig) (d. 1533), publishes the first major work to deal at length with the process of distillation. It appears in two versions, known as the small and the big (or great) books. The Liber de Arte distillandi de composites (Big Book on Distillation, 1512) includes illustrations of distilling apparatus and an extensive discussion of the medicinal uses of distilled spirits. Braunschweig is writing for a popular audience as well as for physicians and apothecaries; he writes in German in the hope that the book will reach the general public through the vernacular. The book is so successful that it remains in print in Germany until after 1554, and it is quickly translated into all European languages (Anderson 1977:117; Forbes 1970:110-111).

Braunschweig popularizes the medicinal use of spirits, especially _aqua vita composita_, a mixture of strong Gascony wine, brandy, and herbs (Lucia 1963a:116-117). He writes: "Aqua vitae is commonly called the mistress of all medicines. It eases the diseases coming of cold. It comforts the heart. It heals all old and new sores on the head. It causes a good color in a person. It heals baldness and causes the hair well to grow, and kills lice and fleas. It cures lethargy....It eases the pain in the teeth, and causes sweet breath. It heals the canker in the mouth, in the teeth, in the lips, and in the tongue....It heals the short breath. It causes good digestion and appetite for to eat, and takes away all belching. It draws the wind out of the body. It eases the yellow jaundice, the dropsy, the gout, the pain in the breasts when they be swollen, and heals all diseases in the bladder, and breaks the stone. It withdraws venom that has been taken in meat or in drink, when a little treacle is put thereto. It heals all shrunken sinews, and causes them to become soft and right....It heals the bites of a mad dog, and all stinking wounds, when they be washed therewith. It gives also young courage in a person, and causes him to have a good memory. It purifies the five wits of melancholy and of all uncleanness." He warns, however, that "it is to be drunk by reason and measure...five or six drops in the morning, fasting, with a spoonful of wine" (quoted Roueché 1963:173 from the first English translation Braunschweig 1527).

BRANDY CONSUMPTION, DISTILLING; GRAIN SPIRITS. In the course of the century, distilleries begin to spring up throughout Germany, playing an important role in the economy (Janssen 1910, 15:417-420). The trade is less closely bound to monasteries than in France, and more bound to towns. Many distillers in towns develop their own brands in simple home stills (Forbes 1970:103). As a result, the consumption of brandy gains ground in towns and villages. Drinking in apothecary shops becomes notorious, prompting attempts to regulate it. Grain spirits (made largely from rye but still going under the name of brandy) also become common in Germany, since wine for making brandy has to be imported from Italy and France (Forbes 1970:103). This practice, which originated in the 15th century, makes the manufacturing of spirits less expensive, but at the same time it consumes large amounts of wheat and rye that could be used for bread. Faced with tenuous grain supplies, the resistance to grain distillation remains strong, and several cities of Germany forbid such distillation throughout the 16th century (Rau 1914:8-9). Examples are the ordinances of Regensburg in 1530 (see 1534/Germany) and Bavaria in 1553 (see 1550/Germany).

BRANDY SALES CONTROLS. In 1506, Munich bans brandy sales on Sundays and feast days. Anyone caught selling or drinking brandy will be fined 75 crowns (Rau 1914:13).

BEER CHARACTERISTICS. By the 16th century, the use of gruit in Germany to flavor beer is almost completely replaced by hops (Cherrington 1925, 1:402).

BREWING. Hamburg loses its reputation as the leading beer town and both domestic consumption and exports decline. This is largely attributed to greater competition from other cities, especially Lübeck with whom the city engages in a series of "beer wars," to heavy beer taxes instigated to pay for the wars, and to the spreading fashion of wine drinking (Arnold 1911:287).

PLEDGING CONDEMNED. The custom of pledging healths, which exists throughout medieval Europe but is particularly prominent in Germanic countries, is vigorously condemned. The custom requires one to empty the cup each time someone's health is toasted. The person whose health the group has drunk to has to respond. These rounds of toasts frequently lead to drinking competitions in which the winner is the last person to fall under the table. Sometimes the drunkenness that results leads to acts of violence (Blanke 1953:174; Samuelson 1878:104). In 1516, Johann von Schwarzenberg's _Das Büchlein von Zutrinken_ attacks this custom and inaugurates a flourishing period of literature on drinking. Nearly all writers on the subject (whether they approve of drinking or not) oppose the practice (Jellinek 1945:648, 658; Hauffen 1889:492).

IRELAND

WHISKEY CHARACTERISTICS. The term "usquebaugh" probably refers to both grain and wine-based spirits, with distillers making both depending on availability. Both are compounded spirits, as plain spirits are still harsh and unpalatable (McGuire 1973:93). The same is true of Scotch whisky.

ITALY

INEBRIETY PREVALENCE. Castiglione's _The Courtier_ (see 1520/Italy), Cellini's autobiography (see 1550/Italy), and other works of the period indicate that Renaissance and early modern Italians do not engage in "such unbridled drinking as is reflected in contempory English, German and French literature of those times" (Jellinek 1976b:4).

CARNIVAL CONDEMNED (ERASMUS). Reflecting the desire for a sharper separation of the sacred and the profane, Erasmus attacks Siena's Carnival in 1509 as un-Christian because it contains "traces of ancient paganism" and because people indulge in license during it (quoted Burke 1978:209).

NETHERLANDS

BREWING: FLANDERS, BRABANT. As imports of good quality wine increase, poor quality local wines can find no satisfactory market and viticulture virtually disappears from the Belgian countryside. Breton shippers prosper from the carriage of Poitevin and Gascon wines for consumption in Flanders and Brabant. Better-tasting beer has replaced wine in popular consumption and commercial brewing has expanded appreciably. Louvain has begun to emerge as "the great brewing town." The number of Louvain brewers falls from 72 in 1477 to 27 in 1520, but output rises "by more than universe proportion." With the rise of Antwerp, one person builds 24 breweries after 1552 in the new industrial quarter of the old town (Houtte 1977:147, 170, 172, 179).

HOLLAND. Beer production increases markedly in Haarlem at the same time as industrial centralization leads to a decline in the number of brewers (Houtte 1977:170).

DRINKING REPUTATION. The expression "the Dutchman for a drunkard" becomes common (Simon 1948:133).

<u>WINE PRICES</u>. In the early century, wine is so expensive in the Netherlands that one wine-drinking Frenchman living there sometimes spends 20 crowns in one day when he holds an open house (Braudel 1981:234).

<u>DISTILLING; SPIRITS TRADE</u>. Although the Low Countries produce neither wine nor cereal in sufficient quantities to provide cheap base materials, the spirits trade grows and becomes a "national asset" by the 17th century. In 1497, impost or excise duty is placed on brandy in Amsterdam. The early 16th century finds ordinances regulating spirits distilling, sales, and retail trade in Amsterdam, Dordrecht, Leiden, Haarlem, Delft, and other towns. The rise of distilling in the course of the century is due to several factors: (1) Emperor Charles V favors the rise of towns in the Low Countries and brewers profit by his policy; (2) the well-established brewing industry facilitates the development of distilleries, which at first are dependent on the breweries for their yeast; (3) German soldiers in the Dutch army introduce spirit drinking during the Dutch wars against Spain later in the century; and (4) German migrants bring the distilling art to the Low Countries. However, the growth of domestic distilling is limited by its reliance on breweries for yeast and a strong dislike for wasting grains in making spirits. The early Dutch distilling trade is first dependent on imported French wine brandy and cheap German spirits; French wines long remain serious competitors of corn as a base material (Forbes 1970:104-105). It is not until the mid-17th century that a domestic commercial distilling industry really expands.

POLAND

<u>NOBLES/DRINKING</u>. The 16th-century Polish nobleman fears a loss of status if he is "satisfied" with the home-brewed beer of the peasants (Braudel 1981:234).

RUSSIA

<u>SPIRITS INTRODUCED</u>. In the early 16th century, distilled spirits appear in Muscovy for the first time, possibly introduced by Italian merchants. In the course of the century, they become known as <u>vodka</u> ("little water") (Pryzhov 1868; Efron 1955:493; Wasson 1968:332).

SCOTLAND

<u>BEVERAGES CONSUMED</u>. Ale is the main beverage in Scotland, as elsewhere in the British Isles. Breweries are attached to monasteries; in addition, every barony has one. The baronial courts fix the prices of grain and ale. The wealthy drink French wine. Whisky is still largely taken for medicinal purposes only (see below).

<u>MEDICINAL USE, SPIRITS</u>. Spirits are still associated with alchemy and viewed as a medicine (<u>Scottish Law Review</u> 1943:85; Cherrington 1929, 5:2384). In the latter half of the reign of King James IV (1488-1513), references to aqua vitae often appear in the work of John Damien, abbot of Tungland and an alchemist patronized by the king (Ross 1970:2). In 1498, treasury accounts record that nine shillings were given to the "barber that brought aqua vitae to the king in Dundee." At this time, barbers are also surgeons. Sources for social habits of

the Highlands are scarce, but King James IV calls for whisky when he is in the Highlands. In 1505, the king grants to the surgeon-barbers of Edinburgh a monopoly to dissect and the associated monopoly to make and sell aqua vitae within the city. The charter specifies that "no persons, man nor woman, within this burgh, make nor sell aquavitae except the said Masters, Brethren or Freemen" (Wilson 1973:32).

SPAIN

VITICULTURE EXPANDS. As Spanish wealth rises, wine making in the south assumes vast proportions and much of it is exported. During the century, the Cortes complains that viticulture is undermining the cultivation of grains because it is more profitable. An attempt is made to prevent vineyards from encroaching on grain acreage (Sombart 1967:138, 139).

1510

GERMANY

DRINKING CONTROLS. In 1512, Emperor Maximilian I reissues a strongly worded imperial edict at Cologne demanding that all local rulers, lay and clerical, enforce the penalties against drunkenness and set a good temperate example for their subjects. All those refusing to abide by this standard are to be brought before the imperial court for punishment. This edict has a modest effect as some of the German princes begin to make an effort to curb the obvious abuses (Petersen 1856:82, 83).

The city of Bamberg establishes in 1516 a fine or a three-day diet of bread and water for the inebriate and the person who serves him. For a second offense the fine is doubled (Stollies 1981:102).

SUMPTUARY LAW. A Brandenburg police ordinance of 1515 for the regulation of wedding festivities stipulates the maximum number of guests allowable for different social classes and limits the amount of food and drink that can be consumed in any one day to no more than the amount a household consumes in one year (Janssen 1910, 15:411-412).

TEMPERANCE SOCIETIES. In 1517, the St. Christopher's Society is founded in Austria by 79 nobles and knights in order to combat heresy and drunkenness. All members pledge to moderate their drinking habits or pay a fine. The elite nature of the group prevents it from having any impact on society as a whole (Neuburger 1917:118; Krücke 1909:22)).

In 1524, a "temperance" brotherhood of princes, bishops, and nobles is organized in Heidelberg; its members pledge to abstain from "full swilling" and health-drinking, except on visits to Lower Saxony, Pomerania, and Mecklenburg, where moderation cannot be carried out (Krücke 1909:23; Cherrington 1926, 3:1092).

RUSSIA

DRINKING PRIVILEGES. Tsar Vasily III (1505-1532) grants to his courtiers permission to drink as much alcohol as they wish. These nali ("drinkers") must live in a special suburb of Moscow, where their habits will not corrupt the lower classes (Johnson 1915:135).

1520

EUROPE

MEDICINAL USE (PARACELSUS). The noted Swiss physician and surgeon Paracelsus (Aureolus Theophrastus Bombastus von Hohenheim, 1493-1541) becomes a pioneer of chemical pharmacology, popularizing the medicinal use of tinctures and alcoholic extracts. (His school is often called iatrochemistry.) A large part of his pharmacopoeia is based on the distillation of wine, plants, and mineral substances. Each malady, he believes, has a specific quintessence capable of curing it. Revering Hippocrates, he denounces and publicly burns the works of Galen and teaches that nature, the "natural balm," heals wounds (Claudian 1970:22; Allbut 1905:28-31; Lucia 1963a:117). Paracelsus also espouses the pharmaceutic uses of iron and antimony. The usual method of preparing the former is to immerse iron filings or iron wire in wine, allowing the mixture to stand for a period of weeks while the metal rusts. As a remedy for anemia, this *vinum ferri* is popular until the late 19th century. Wine of antimony, admitted in 1637 to the published list of remedies in France, continues in use at present as a diaphoretic, expectorant, and emetic (Peters 1906:129; Lucia 1963a:143).

The word "alcohol" is used in the sense of spirits by Paracelsus and the German iatrochemist Andreas Libavius early in the 16th century, but the meaning changes frequently. Libavius tries all sorts of herbs and fruits in order to find new base materials for the distilling trade (Forbes 1970:103, 107).

DENMARK

BREWING CONTROLS. In 1522, minimum requirements are established in Denmark for commercial breweries in order to increase their capacity, since larger and fewer breweries limit the fire danger in towns and simplify the control of excise payment. It is also stipulated that brewers should be chosen from among the most prosperous citizens (Glamann 1962:133).

ENGLAND

KING/DRINKING. According to French (1884:133), King Henry VIII (1509-1547) is "constantly intoxicated." He takes great pride in the quantity and quality of his wines (Simon 1907, 2:135).

BEER CONSUMPTION. The average amount of beer or ale consumed in Coventry is about 17 pints per head per week. This compares with an estimate of three pints for 1965. Coventry is the fifth largest town in England, with a population of 6,600 and 60 brewers (Monckton 1966:95, 230n.12).

MEDICINAL USE, SPIRITS. By 1521, distilled spirits begin to acquire more of a reputation as medicine. The earliest English translation of Braunschweig's small book on distillation is made by L. Andrew (*The Vertuose Boke of Distyllacyon of the Waters and All Maner of Herbes*) in 1527. This is the first important publication in England "to render the art of distillation more generally understood" (Simon 1948:131). The preface reads: "Beholde how moche it excedeth to use medicyne of eficacye naturall by god ordeyned then wycked wordes or charmes of eficacie unnaturall by the devil envented" (Braunschweig

1527, quoted Anderson 1977:119; this text is also quoted in the discussion of Braunschweig under 1500/Germany). Under Henry VIII, Irish settlers are said to introduce distilling in Pembrokeshire, England (Williams and Brake 1980:1).

FRANCE

NOBLES/DRINKING. The reign of King Francis I (1515-1547) inaugurates a great era of sumptuousness among the wealthy. Francis personally eats and drinks copiously, holds great banquets, and dresses elaborately. Through him and his Italian daughter-in-law, Catherine de' Medici (see 1530/France), the luxurious and refined lifestyle of Italy enters France, although the ideal French gentleman of the period is less temperate and restrained than Castiglione's courtier. In terms of food and drink, refinements occur without any appreciable diminishing of quantities consumed. The French are probably the biggest eaters in Europe at this time, eating four to five times a day. The wealthy hold huge suppers which begin around 4:00 or 5:00 and last until all guests have lapsed into a "state of greasy stupor." Wine gives "lubrication and scintillation" to the "dullness and stupor of gastronomic excesses" (Wiley 1954:67, 70, 76-78, 90-96).

GERMANY

BACKGROUND DEVELOPMENTS. Charles V becomes the Holy Roman Emperor (1519-1556) and ruler of Austria, Netherlands, and Spain. Martin Luther ignites the Protestant Reformation in Germany with his "95 Theses" in 1517.

INEBRIETY CONDEMNED (LUTHER). Attacks on the particular propensity of Germans towards inebriety are frequent in Protestant tracts. In "An Open Letter to the Christian Nobility" (1520), Martin Luther (1483-1546)) refers to "the abuse of eating and drinking which gives us Germans a bad reputation in foreign lands, as though it were our special vice. Preaching cannot stop it; it has become too common...The temporal sword can do something to prevent it" (Luther 1520/1960:108). In a "A Sermon on Soberness and Moderation against Gluttony and Drunkenness" (1539), he observes: "The Italians call us gluttonous, drunken Germans and pigs because they live decently and do not drink until they are drunk. Like the Spaniards, they have escaped this vice." Germans "are the laughing stock of all other countries, who look upon us as filthy pigs." The problem, he says, first began with the peasants, then spread to the citizens and finally to the nobility, who once considered drunkenness a great shame but are now the worst drinkers of all (Luther 1539/1954, 51:291-294). (Similar comments are made in Luther's Table Talk, nos. 3468 and 4917, in Luther 1967, 54:205, 371.)

Philip Melanchthon, Luther's colleague and author of the Augsburg Confessions (1530) which establishes the fundamental doctrines of Lutheranism, similarly asserts that "we Germans drink ourselves poor, drink ourselves sick, drink ourselves to death, and drink ourselves to hell." Ulrich von Hutten, one of Luther's main supporters, contrasts the drunkenness of Germans to their bravery and trustworthiness (Cherrington 1926, 3:1092).

LUTHER/DRINKING. Although Luther attacks drunkenness as an obstacle to salvation in his "Sermon on Soberness" (1539), he does not forbid moderate drinking: "It is possible to tolerate a little elevation, when a man takes a drink or two too much after working hard and when he is feeling low. This must be called a frolic. But to sit day and night, pouring it in and pouring it out again, is piggish. This is not a human way of living, not to say Christian, but rather a pig's life....This is a great sin, and everybody should know that this is such a great iniquity, that it makes you guilty and excludes you from eternal life" (Luther 1539/1954, 51:293). Later he says: "God does not forbid you to drink, as do the Turks....But do not make a pig of yourself; remain a human being. If you are a human being, then keep your human self-control" (p. 296).

Luther is sympathetic towards popular traditions and he himself enjoys good food and drink. In 1517, he writes: "Who loves not women, wine and song,/ Remains a fool his whole life long" (quoted Seward 1979:168). In his Table Talk (no. 139), he defends his drinking practices: "If our Lord God can pardon me for having crucified and martyred him for about twenty years [by saying mass], he can also approve of my occasionally taking a drink in his honor. God grant it, no matter how the world may wish to interpret it" (Luther 1955, 49:20). Consequently, Luther is accused by his adversaries of too much beer drinking. For example, Zwingli, prior to his breakup with Luther in 1529, refutes Luther's arguments regarding drinking. He deprecates Luther's boastfulness and intemperance (Coupe 1972:56-63).

DRINKING PRACTICES (ERASMUS). In his Colloquy on Inns (1523), Erasmus complains of the discomfort of German inns and observes that the Germans "admire heavy drinkers, although the one who downs the most wine pays no more than the one who drinks the least." Some Germans spend twice as much on wine as on dinner, and "the uproar and tumult after they've all begun to grow heated from drink is astonishing" (Erasmus 1523/1957:19).

BRANDY CONTROLS, CONCERNS. In the state of Hesse, a general prohibition is issued in 1524 against the selling and retailing of brandy. In 1527, the council of Nuremberg complains that many people in the village of Altdorf "had no shame in disgracing themselves with brandy drinking and in other ways in the public streets and also in inns and taverns on Sundays and holy days whilst preaching was going on" (quoted Baader 1868:317; Janssen 1910, 15:409, 418). This indicated that brandy drinking, no longer occurs only in towns.

PEASANTS/DRINKING. Hans Beham (Joannes Boemus), a chaplain to the Teutonic Knights, describes the life of a German peasant as bleak and sober except on special occasions (Mores leges et ritus omnium gentium, 1520): "Their condition is very wretched and hard....they drink water or whey....On holy-days they come altogether in the morning to the church....[In the afternoon] the younger folk dance to the sound of the pipe, while the elders go to the tavern and drink wine" (quoted Coulton 1925:22-23, who calls Beham's work the "most explicit and trustworthy account of the German peasantry" at this time.)

Taverns appear to play an important role in the peasant revolt of 1525, and charges of drunkenness are frequently leveled against the rebels. The first "Evangelical Brotherhood" of peasants is created in 1524 by Hans Müller under the cover of a local church-ale, knowledge of the uprising spreads through the village inns, and several of its leaders are inn keepers. Peasants frequently appear to indulge themselves in the wine captured from the cellars of nobles, priests, monks, and burgermeisters (Bax 1968:37-38, 55, 113, 116). There are records of popular disturbances being quelled by gifts of wine. Opponents particularly emphasize the drunken propensity of the rebels. While these hostile accounts are biased, there is good evidence that peasants frequently are overcome with a carnival spirit and, seeking to turn the world upside down, indulge themselves in the food and drink denied them. For example, a contemporary writes that "the peasants [at Frauenberg] were always full [drunken]; showed much ill-behaviour in word and deed"; at Bamberg, "no one was certain of his life and goods, after the multitude had bedrunk themselves in the wine-cellars of the churchmen, as continually came to pass"; at Lautenberg, the women from the villages entered the captured castle and "did drink themselves so full of wine that they might no more walk" (quoted Bax 1968:166, 175, 181, 185, 227, see also pp. 111, 147-149; Burke 1978:110, 189). Lorenz Fries, private secretary to Konrad von Thungen, bishop of Würzburg and duke of Franconia, writes: "Where they came, or where they lay, they fell upon the monasteries, the priests' houses, the chests and the cellars of the authority, consuming in gluttony and in drunkenness that which they found, and it did exceedingly please this new brotherhood that they might consume by devouring and drinking their fill, and had not to pay withal. More drunken, more full-bellied, more helpless folk, one had hardly seen together than during this time of rebellion" (quoted Bax 1968:165, who observes that this is a hostile source. It must still be admitted "that gluttony and wine-bibbing contributed as potently as any other influence to the politically unproductive character of the peasant successes and to that lack of cohesion and discipline which lead the way to the final catastrophe and soaked the German soil of the blood of its tillers").

CONCERNS: POPULAR CULTURE. As social discontent grows among the peasantry, efforts are made to curtail the popular festivities which bring the inhabitants of different parishes together, such as weddings, pilgrimages, church wakes, and guild feasts (Bax 1968:36).

The 37th of the 100 Grievances of the German Nation presented to the emperor at the Diet of Nuremberg complains about the "great abundance of church feasts and holy days" which oppress the people, prevent them from working, and lead to "innumerable transgressions" (Coulton 1925:274).

IRELAND

DISTILLING CONTROLS. During the time of Henry VIII, it is decreed that "there be but one maker of *aqua vitae* in every borough-town upon pain of 6s 8d" and that no wheaten malt should go to any "Irishman's country." This last restriction is doubtless intended to limit the use of wheat in the manufacture of ale and spirits (Morewood 1838:618).

WHISKEY CHARACTERISTICS. The quality of aqua vitae in this period is estimated in proportion to the spices and other aromatics they contain (Morewood 1838:618).

ITALY

TEMPERANCE ADVOCATED (CASTIGLIONE). Baldessare Castiglione (1478-1529) publishes *Il Cortegiano* (The Courtier, 1528), the most successful of several Italian Renaissance books on proper etiquette for a gentleman in which a new code of refinement is delineated. Translated and circulated throughout Europe, the book gives wide publicity to a new social ideal--the educated, refined, worldly, and urban Renaissance man. Castiglione hardly mentions wine drinking and makes only a few references to drunkenness, all of which are derogatory. A man of bad drinking habits "not only may not hope to become a courtier, but can be set to no more fitting business than feeding sheeps" (book 2.31, quoted Jellinek 1976b:4). In book four, Castiglione emphasizes that the courtier should be a man of good birth who has been trained to mix agreeably in the company of others, and who is characterized by the neatness of his dress, graceful movements, good conversation, wit, proficiency in sports and arms, literacy, and temperance in drinking. "I affirm of those Things which we term good, some are simply, and in their own Nature always good: such are Temperance, Fortitude, Health and those Virtues which conduce to the Tranquility of the mind" (Castiglione 1528/1724:358; also quoted Wiley 1954:90).

In book 4, Castiglione also argues that because some people get drunk is no reason to prohibit drinking by law (Wiley 1954:90). When told that temperance may perhaps be suitable for a monk or hermit but not a prince, "who is of a noble Soul, bountiful and courageous in War," Castiglione reponds: "Wherefore to free the Mind from disturbance, it is not necessary totally to root up the affections; this would be as if a Man to prevent Drunkenness should procure a Decree, that none should drink Wine, or to forbid that any one should run, because a man sometimes falls by running." However, he emphasizes that "Temperance gives birth to many other virtues" (Castiglione 1528/1724:376-377). In general, his ideals imply a rejection of popular culture and reflect the increasing separation between the amusements of the rich and poor in 16th century Italy (Burke 1978:275).

1530

EUROPE

TEMPERANCE ADVOCATED (ERASMUS). Erasmus advocates temperance in the fifth chapter of his *De civilitate morum puerilium* (On Civility in Children, 1530), a short treatise for the instruction of youths on good behavior, including manners in food and drink. He emphasizes that nobility arises by virtue of the mind and behavior, rather than rank and wealth. He condemns those that drink out of habit rather than thirst or who drink before meals, and he recommends water for youths and the aged (Neuer 1970:216-217). He writes: "To begin a meal with wine is the part of a drunkard and a youngster should drink water or very weak wine or beer. A passion for wine...blackens the teeth, makes the cheeks droop, and dulls the intelligence." One should also be sure to wipe the lips before drinking from a communal cup (quoted Wiley 1954:90). The work is significant as the starting point for the formulation of the concept of *civilitas* or *civilité* to designate "good behavior" in early modern Europe. It is the first prose work devoted solely to good behavior, providing rules of conduct that had previously been uttered chiefly in mnemonic verses or scattered in treatises on other subjects. It acts as a "cursor of new standards of shame and repugnance" which is beginning to slowly form among the secular upper class. Immensely popular, it is reprinted more than thirty times in the first six years after publication and is translated into every European language. This success can be seen as indicative of the growing importance of the question of what defines good behavior in society (Elias 1978:53-54, 79, 135). The book is social in orientation; the stress is on conduct within the community. Although addressed to a prince, it draws no class distinctions. Erasmus emphasizes that education should be given to all. It has been suggested that the book reflects the ethos of Rotterdam's bourgeosie (Neuer 1970:181, 203-210).

ENGLAND

BACKGROUND DEVELOPMENTS: REFORMATION. In a series of statutes (1531-1533), Henry VIII abrogates the ties between Rome and establishes the English (Anglican) Church, with himself as its "Supreme Head." Monasticism is ended by the suppression of some 376 religious houses.

BREWING. Monks still have the reputation for making the best brews, and many ecclesiastics may have been more involved in the drink trade than in their vocation. An act passed in 1529 stipulates that they can no longer be brewers. The dissolution of the monasteries leaves a considerable number of monks available to apply their talents to brewing and distilling in the commercial world. The exact impact of the dissolution on the brewing trade is unknown, but in the course of the century there is a general improvement in the quality of ale and beer (Monckton 1966:96; Forbes 1970:101).

HOPS CULTIVATION. Hops are first grown on an appreciable scale in Kent by Flemish immigrants in the 1520s. Only in a few areas of Britain will hops grow successfully (Monckton 1966:193).

DRINKING REPUTATION. Rabelais (see 1530/France) describes a man overcome with drink as "drunk as an Englishman" (Legouis 1926:4-5).

LABORERS/DRINKING. In 1538, it is recorded that a London workman is suppose to work a 12-hour day beginning at 6:00, with a quarter-hour break for breakfast, an hour for midday dinner, and a quarter-hour afternoon break "to go to his drinking" (Younger 1966:309).

MEDICINAL USE. Thomas Elyot writes in book 2.19 of The Castel of Helth (London, 1530; augmented 1541) that wine moderately consumed nourishes and comforts the body as well as the spirit, and was therefore ordained by God for mankind. Taken "out of order or measure," however, it transforms a man or woman into a beast. He cites Plato, Galen, and the Bible as evidence. In book 2.21, he discusses the medicinal properties of ale, beer, and cider, but regards all of them as inferior to wine.

FRANCE

LIQUEURS INTRODUCED. Drinking aromatic spirits or liqueurs, the making of which is an Italian art, becomes fashionable through the influence of the Italian Catherine de' Medici (1519-1589), who in 1533 marries the future King Henry II. Among the most popular are rossolio or rossolis--originally from Turin and made with rose petals, orange, and jasmine, among other ingredients--and the weaker populo (Le Grand d'Aussy 1782, 3:76-77; Gottschalk 1948, 2:47-48, 134). From France, sweet liqueurs spread to England and Germany (Delamain 1935:16).

DRINKING PRACTICES. Catherine de' Medici is also said to have helped introduced sugar and (around 1550) the use of individual glasses for wine drinking, replacing the single glass or goblet from which each person sitting at a table drank, emptying the contents at one draught so no residue was left when it was refilled for the next person (Franklin 1889, 6:104, 146; Emerson 1908, 2:140).

BENEDICTINE INTRODUCED. Francis I (1515-1547) visits the abbey of Fecamp, Normandy, and is delighted by a cordial served to him which he calls Benedictine, first made by the monks around 1510 (Seward 1979:152).

DRINKING CONTROLS. Francis I, in reaction to disorders in Brittany caused by drunkards, issues a general edict in 1536 stipulating that drunkards (ivrognes) will be jailed with only bread and water (time period unspecified). A second offense is punishable by whipping; a third, by whipping in public; and subsequent offenses, by banishment and amputation of the ears (Franklin 1889, 6:124; Le Grand d'Aussy 1782, 3:274). The act is passed to prevent the "idleness, blasphemy, homicides, and other damage and harm which comes from drunkenness" (quoted Brennan 1981:221-222).

The ordinance is enforced only fitfully; police inspecting taverns at night make little effort to arrest those found drunk. Furthermore, its provisions are not incorporated into other ordinances dealing with cabarets and morals generally. There is no evidence that it is reissued in the 17th or 18th centuries (Brennan 1981:222).

DRINKER LIABILITY. The edict of 1536 (see above) further states that "if by drunkenness or the heat of wine drunkards commit any bad action, they are not to be pardoned, but punished for the crime and in addition for the drunkenness, at the judge's discretion." Drunkenness is seen as an aggravating circumstance not connected to other crimes.

DISTILLING PRIVILEGES. In 1537, Francis I divides the privilege of distilling between vinegar-makers and victuallers and creates a separate guild devoted solely to making and selling eau-de-vie. Numerous quarrels result, indicating that the profits from distillation are already considerable (Delamain 1935:16; Braudel 1981:243).

PEASANTS/DRINKING (LANGUEDOC). Wine is considered a basic food in the wine-growing area in southern France, not a luxury as in the north. Around 1530-1540, the wages of farm laborers in Languedoc begin to decline and the wine ration is reduced. Since 1480 it has been 1 muid (625 liters) of wine and 0.4 muids (250 liters) of prenses (piquette) per person per year or about 2 liters of wine a day. The pure wine ration is now reduced to 0.8 muids per person (521 liters) and the piquette ration increased to 0.65 muids (416 liters). These rations remain stable until 1585. Whereas laborers had white bread and red wine in the late 15th century, they have black bread and weak piquette in the late 16th (Le Roy Ladurie 1974:43, 102).

ATTITUDES: WINE, INEBRIETY (RABELAIS). In Rabelais' Gargantua and Pantagruel (1533), wine is so praised as the promoter of good fellowship, wit, and wisdom that Rabelais has been called "the great priest of Bacchus" and his work "the Bible of drinkers" (Legouis 1926:4-5). This "tapestry of contemporary life" shows the coarseness of the period and its huge appetites. A former friar and monk who left the orders and studied medicine, Rabelais emphasizes the goodness of man and his instincts, and advocates accepting life wholeheartedly and living it to the fullest. He is a passionate advocate of learning and of extending man's potential in every direction, not only in learning but also in sex and food (Koenigsberger and Mosse 1968:313-314). His characters are continually thirsty. The 5th chapter of Gargantua (1.5) includes "one of the greatest paeans in praise of drinking that has ever been written" (Seward 1979:163), and Gargantua himself is born telling the world to "drink, drink, drink" (1.7).

A central episode, the hesitations of Panurge over marriage, is resolved (in book 5, ch. 44) after an appeal to the "divine bottle" for illumination. On this bottle is printed the words of the grape-harvest song, which includes the lines: "Thy twinkling liquor holds the whole/ Of truth, which India's conqueror/ Lord Bacchus gave thee to control,/ and which all honest men adore" (Rabelais 1533/1936:831). A priest further explains: "We folk here hold drinking, not laughter, to be the essence of mankind. Mind you, I do not mean drinking in the simple and absolute sense in which any beast may be said to drink. No--I mean the drinking of cool, delicious wines. Remember, friends, that by the vine we grow divine" (5.46, ibid., p. 834).

Rabelais follows the classical tradition of viewing wine and intoxication as a source of creativity and inspiration. He writes in the prologue to the third book: "As I drink I here deliberate, discourse, resolve, and conclude. After the epilogue I laugh, write, compose and drink again" (quoted Weinberg 1972:37; see also Rabelais 1533/1936:296). According to Weinburg (1972:53), "The effect of the Rabelaisian drinking bout is not, as it appears on the surface, vulgar drunkenness, but a type of inspiration, or ecstasy, modeled closely upon the classical ideas of furor...[in which] the normal faculties of the intellect and senses are in a state of suspension, while the god speaks through the person possessed."

Weinberg (1972) views Rabelais' message as a syncretic combination of Christian and humanistic Neoplatonism and Renaissance Epicurean ideas, in which wine intoxication symbolizes that truth is revealed to the faithful through intuition, not reason. Bakhtin (1968:279) emphasizes the pervasiveness of food and drink in the book: he views Rabelais as the embodiment of the "material bodily principle" of medieval folk humor and the carnival spirit in which the body and bodily functions such as eating and drinking have a cosmic as well as popular character. Through these functions, humanity encounters, tastes, and triumphs over the world in the sense that it partakes of the world instead of being devoured by it (ibid.,p. 285). Through this "prandial libertinism," mankind demonstrates that it is not afraid of the world. However, Bakhtin emphasizes that this should not be confused with the glorification of gluttony of itself: "Banquet images in the popular-festive tradition (and in Rabelais) differ sharply from the images of private eating or private gluttony and drunkenness in early bourgeois literature. The latter expresses the contentment and satiety of the selfish individual, his personal enjoyment and not the triumph of the people as a whole" (ibid.,p. 301).

GERMANY

PROTESTANT ATTITUDES; TEMPERANCE OF ANABAPTISTS, HUTTERITES. Although the Anabaptists allow moderate drinking, they are ethical rigorists who criticize the Lutheran Reformation for its failure to produce a manifest change in moral demeanor and emphasize that one of the marks of a true Christian is sobriety. Anabaptists are more rigorous because they insist that the church must be a community of saints, no matter how small, and that all who are unworthy must be excluded. A Lutheran minister testifies in 1531 that the best way for any person identified as an Anabaptist to disprove the allegation is to indulge in frequent drinking bouts. In 1545, the rule of the Hutterian Brethren forbids any member of the society to be a public innkeeper or to sell wine and beer (Bainton 1945:52).

INEBRIETY CONDEMNED (FRANCK). In *Von dem greulichen Laster der Trunkenheit* (On the Horrible Vice of Drunkenness, 1531), Sebastian Franck emphasizes the widespread prevalence of inebriety in Germany, which he says is a product of the habit of wine drinking introduced by the French. He castigates drunkenness for producing blunted emotions, economic irresponsibility, untruthfulness, brutality, and "loss of interest in the finer aspects of life." It is a source of depravation, poverty, and crime (quoted Jellinek 1941:394). Like others, Franck believes that excessive eating and drinking are responsible for shortening the life span of the Germans: "It is complained that no one nowadays grows old. For this we have to thank the fact that we spoil more wine than our forefathers drank, and that we eat like hogs; how can nature stand it? I firmly believe that every tenth person dies no natural death. The women overeat, the men overdrink themselves" (quoted Janssen 1910, 15:420).

He also alludes to the failure of legislative controls to reduce the problem and criticizes the nobility for setting a poor example: "Much has been tried against drinking among Germans but nothing has been achieved. The legislators have failed, although they have made promises....It [drinking] is too deeply rooted and sin has become a habit." The only remedy is "personal self-restraint and temperance, and zealous study of the Scriptures, which condemn so sternly such wickedness. The clergy must be sober, apt to warn, to teach; but courage is lacking to many. Princes and rulers must teach temperance by example, and sternly punish drunkenness. Now, they are 'full' day and night, examples of shame." Franck believes that a prince's example is 10 times more forceful than a decree in influencing people's behavior (quoted Jellinek 1941:393, 394).

Although morally opposed to drunkenness, Franck believes that alcohol is a creation of God and therefore is not in itself an evil, nor can drinking it be evil so long as moderation is followed. A little drinking is healthy. As an analogy, he observes that flowers are not poison because of the poison that spiders make out of them (Jellinek 1941:392, 395; Jellinek 1945:655-656). Franck is considered one of the leading evangelical reformers and spiritualists of his age. To him, the essence of Christianity is the spirit of love and selflessness, and he encourages an inward mystical piety opposed to all forms of rigid dogmatism. He is highly sympathetic to the poor and oppressed and critical of the intolerant and mighty (Coupe 1972:84).

ATTITUDES: INEBRIETY (OBSOPOEUS). In 1536 appears Vincent Obsopoeus' *Ars bibendi* (The Art of Drinking), a Latin manual on how to be a virile drinker without becoming a habitual inebriate. He writes: "There is no task which skill does not conquer. And we must cultivate Bacchus with skill in order not to drink the pleasant wine crudely at feasts. Unless Bacchus is worshipped with particular skill, as is proper, the worshipper will see the god angered....My spirit doesn't burn at all to write of a crowd of gluttons....I sing of the lawful banquet and permissible drinking. There'll be no bibulous glutton in my song. I don't unseemly dine in frequent drunkenness....Drunkenness is a furious evil....Flee it as though it were a plague on body and mind" (Obsopoeus 1536/1945:663; see also Hauffen 1889:495-496).

Inspired by him, Leonard Schertlin (*Künstlich Trinken* [Artful Drinking], 1538) praises moderate drinking, even though he himself has become a servant of Bacchus. In his *Die vol Brüderschaft* (The Drunken Brotherhood, 1543), Bacchus declares that all men recognize him as their god (Jellinek 1945:654, 656; Janssen 1907, 12:216n.2; Hauffen 1889:497).

INEBRIETY PREVALENCE; NOBLES/DRINKING. Although "every land far and wide is full of bibulous men," Obsopoeus (1536/1945:678) describes drunkenness as particularly characteristic of nobles and the wealthy: "Don't let the example of any prominent man induce you, under his influence, to follow his leadership to any ugly ruin. No other vice now so possesses great halls, though the court at this period abounds in all vices, as Drunkenness, which has seized the highest strongholds. Courtly life is merely continual intoxication....The leading men drip with constant drunkenness."

In the preface to the German translation (1537) of the work, the Colmar actuary Gregory Wickram complains of the recent increase in the abuse of drinking among young and old, high and low "in these anxious and grievous times" (Janssen 1907, 12:212).

DRINKING PRACTICES (SACHS). Hans Sachs (1494-1576), the most prolific German dramatist, moralist, and poet of the mid-century, immortalized by Richard Wagner in *Die Meistersinger*, often writes about German (Nuremberg) drinking customs and practices. A staunch Lutheran and spokesman for the burghers, he never overlooks pointing out a lesson in morality. By profession a shoemaker, Sachs embodies the bourgeois virtues of thrift, piety, sobriety, and forthrightness. But he is also good, natural, humorous, and tolerant. Excelling in the portrayal of the common man, he is very popular with the lower middle class. He views drunkenness with disfavor; nothing but harm can come from a life of drinking and dissolution. In one of his dialogues on drunkenness (1536), Sachs describes a typical German drinking contest in which 12 "beer heroes" drink a tun of beer in six hours (Samuelson 1878:108; Jellinek 1945:656; Hauffen 1889:498). In order to raise a big thirst, he asserts, people eat salted bread or highly peppered and spiced little cakes (Strauss 1979:201).

In his realistic and natural Fastnachtspiel, or short comic dramas prepared for the Carnival, most of which are written in the 1550s, Sachs shows that tavern keepers are generally poor and occupy a low social position, not just because of their poverty but because of their association with beggars, rascals, and thieves and their own common trickery in diluting drinks and overcharging. He makes frequent references to drunkenness within all classes and groups. Although he is sympathetic to the laboring classes, he shows that the dissolute habits of artisans, including their love of drink and gambling, in which they often indulge on Mondays as well as Sundays, prevent them from being able to save money. Peasants are also criticized for drunkenness and gluttony (French 1925:9, 26-27, 29, 34, 36, 56).

WINE PRICES. According to one proverb: "In fifteen hundred and thirty-nine/ The casks were valued at more than the wine" (Samuelson 1878:113). This might indicate that that wine is relatively cheap, as Hans Sachs asserts. According to him, because of the low price of wine and beer it was considered a disgrace to drink water, although milk was consumed in rural areas (French 1925:57). However, other evidence indicates wine is still expensive. Abel (1981:53) estimates that between 1551-1600, a liter of wine in Hamburg costs 9.1 times more than a liter of beer. The proverb probably is just a comment on the expense of the casks. About this time, enormous casks (one 24 feet long and 16 feet high at Tübingen) are erected.

DRINKING CONTROLS. The ruler of Bavaria issues a territorial mandate against inebriety in 1526. Other condemnations of drunkenness and efforts at reform are made by the elector of Saxony (1530), the towns of Constance (1531) and Augsburg (1548), and the imperial Diet of 1548, which decrees that drinking standards must become more moderate throughout Germany (Stolleis 1981:102; Krücke 1909:15-16). A 1540 Brandenburg ordinance bans the serving of alcohol on Sundays and holy days (Dorwart 1971:63).

QUALITY CONTROLS. In addition to regular wine, many special wines are prepared from herbs, spices, and other ingredients. In 1539, the Leipzig council passes an ordinance regulating such wines because, "owing to the adulteration of wine, illness increased in the towns from day to day, and the doctors complained that they could not get a drop of good pure wine for their patients." In 1562, Cologne prohibits "new-fangled wines prepared with bacon, which were never heard of formerly, and which are highly injurious to health" (Janssen 1910, 15:415-416).

DISTILLING CONTROLS: GRAIN SPIRITS. A Regensburg police ordinance (1530) is issued banning the making of grain brandy (Rau 1914:8-9).

BREWING: CONTROLS; ECONOMIC VALUE. In 1536, Brandenburg forbids illicit brewing in order to protect the local municipal economy which relies on beer revenues. The 1540 ordinance (see above) bans brewing on Sundays and holy days. In 1549, Brandenburg imposes a stiff beer tax and more restrictions on illicit production (Dorwart 1971:28, 63).

SWITZERLAND

TAVERN CONTROLS. Under the influence of Ulrich Zwingli (1484-1531) the Zürich evangelical synod petitions for a reduction in the number of taverns; the town council passes the resolution in 1530. Secondary taverns and cabarets are closed, and only one tavern per city is permitted. Although the law remains in place, excessive eating and drinking continue and become the focus of the concerns of Heinrich Bullinger in the 1560s. Zwingli moves the Reformation in a more radical direction when he attempts to establish a "holy community" in Zürich, in which "Christian discipline is systematically imposed on the community and is tightly and jointly supervised by church and state." The theocracy composed of magistrates and pastors is his invention, which is passed on to John Calvin (Blanke 1953:181-182).

1540

ENGLAND

ATTITUDES: DRINKING, BEER (BOORDE). Andrew Boorde, an ex-Carthusian and wandering physician, discusses the therapeutic benefits and harms of different beverages in his Dyetary of Health (1542). He attacks the rise in the drinking of hopped beer, "the naturall drynke for a Dutche man" but a detriment to the English people. It is unhealthy and makes one fat and inflated, as is evident from the appearance of Dutchmen. He emphasizes that "ale for an Englysshe man is a natural drynke" (quoted Mendelsohn 1963:30; Sutherland 1969:15). Among other therapeutic benefits, moderate wine consumption actuates and quickens a man's wit and comforts the heart. Furthermore, the better the wine, "the better humours it doth engender" (quoted Seward 1979:91). As for water, it "is not wholesome, sole by itself" (quoted Wilson 1973:383-384).

FRANCE

BACKGROUNDS DEVELOPMENTS: CONCERNS OVER IDLENESS. Between 1545 and 1547, laws are enacted in Paris in an effort to reduce the number of beggars, to "take away all opportunities for idleness from the healthy" and to put them to work (Cole 1939, 1:13). Any man able but unwilling to work is to be sent to the galleys; any woman, whipped and driven out of Paris.

TAXATION. An entry tax of five sous per hogshead of wine entering Paris is established. This tax is designed to help pay for public works and to increase the price of inferior suburban wines destined for popular consumption. The tax is repeatedly raised (Dion 1959:500).

GERMANY

DRINKING INFLUENCES: AVAILABILITY. Protestant reformer Martin Bucer asserts in 1549 that "the vice of drunkenness, which has invaded the country, is seen at its worst in Marburg, for there the town councillors are wine sellers" (quoted Janssen 1910, 15:410).

INEBRIETY PREVALENCE; DRINKING REPUTATION (DEDEKIND). In 1549, Frederick Dedekind writes Grobianus, von sroben sitten und unhöflichen Gebärden (Grobian, Of Coarse Morals and Impolite Manners), translated from Latin into German by Casper Scheid. Widely circulated, the book gives a vivid but satiric, moralistic, and exaggerated picture of the "filth and licentiousness of the time." It provides a detailed description of loutish table manners, dress, appearance, and behavior, exemplifying the whole German genre of middle-class grobianisch (boorish) literature in which mockery and scorn is used to express a serious need for a "softening of manners" (Elias 1978:74-75). Grobian defends his coarse behavior with great gaiety. In the dedication of his translation, Scheid writes: "From other nations we have received, with regard to drinking, very fine, subtle, and elegant names, such as...German drunken sows, and coarse drunken Germans." He laments: "Drunkenness of the very worst and most heinous description has grown to such a pitch that life has come to be a long drinking bout with us" (quoted Janssen 1907, 12:213-214).

HEALTH CONCERNS (PICTORIUS). In a 1549 tract on the preservation of health, Georg Pictorius criticizes the German drinking customs and stresses the serious dangers to health from alcohol abuse (Diethelm 1964:53-54).

ESTEEM OF BEER. In 1549, Johann Brettschneider (Placotomus) issues *De natura et viribus cerevisiarum et mulsarum opusculum* (On the Nature and Usages of Beer), first tract of its kind. In it he praises the nourishment values of the different beers present in Germany at the time and declares that some people live more on this drink than on their meals. He further alleges that, although excessive beer consumption can be injurious, it is not as harmful as wine drinking and declares that some people live more on this drink than on their meals. He further alleges that, although excessive beer use can be damaging, it is not as harmful as wine drinking (Löffler 1909:6-8).

NETHERLANDS

CROSSCULTURAL COMPARISON. According to Andrew Boorde (see 1540/England), Flemings get drunk as a rat, Rhinelanders get drunk often, and the Dutch get very drunk or "cupshoten" (Younger 1966:314).

SWITZERLAND

ATTITUDES (CALVIN). In 1536, John Calvin (1509-1564) publishes the first edition of his *Institutes of Christian Religion* in Basel, Switzerland; it is continually revised until the final edition of 1559. Like Zwingli, Calvin is a second generation reformer "anxious to create and to operate ecclesiastical, social, and even political mechanisms which would at least attack corrupting influences and help to clear the channels for the operation of divine grace" (Dickens 1966:159). Calvin writes that God has not forbidden the drinking of wine, since wine is a part of God's creation, appointed for man's use and enjoyment. But "to wallow in delights, to gorge oneself, to intoxicate mind and heart with present pleasures, and be always panting after new ones--such are very far removed from a lawful use of God's gifts" (Calvin 1559/1961:140). To a greater extent than Luther, Calvin stresses an upright, austere, moral life as a test of election, including the avoidance of obscenity, gambling, and drunkenness. Calvinists also differ from Luther and the Church fathers in the extent to which they are ready to make use of civil authority to impose this code of conduct on the community in general. Such ecclesiastical discipline is meant to encourage holiness (Bainton 1952:115-116). (On the implementation of these principles in Geneva, see below.)

TAVERN CONTROLS (CALVIN'S GENEVA). After an initial failure to reform Geneva in the mid-1530s, John Calvin is asked to return to the city in 1540, and a year later the city adopts his Ecclesiastical Ordinances, which serve as a model wherever Calvinism takes root for establishing a holy community to honor God. The church and the morality of all citizens is controlled by a consistory composed of ministers and 12 elder church members who meet every Thursday. The consistory is formed "to admonish in love" anyone in error and if necessary to report them to the ministers for "brotherly correction." Dissidents are driven out of the city and Calvin's followers flock to Geneva, turning it into a "city of saints." In 1546, Calvin persuades the town council to replace the taverns with evangelical refreshment houses or hostels at which no drinks can be served on Sundays, during church hours on weekdays, or after 9:00 at night. Drinks can only be served to those who look like they can say prayer before and after eating. However, the taverns are restored in three months because of public protest and the failure of the new refreshment houses to make a profit (Blanke 1953:181; Dickens 1966:163; Bainton 1945:53).

Bainton (1952:119) comments with regard to the many regulations imposed in Geneva, that "one must bear in mind that they were simply a continuation of the sumptuary legislation of the late Middle Ages which the Calvinists, with their new found energy, proceeded to enforce."

1550

EUROPE

BACKGROUND DEVELOPMENTS. The Commercial Revolution is in full swing, and the market economy is beginning to erode traditional agrarian habits of local self-sufficiency. Europe also is startled by a "price revolution" which lasts for at least a century. The exact causes for this price inflation are still debated, but it is at least partly rooted in the increase in population since 1450, which has created more competition for limited land and food. During the next hundred years, prices all over western Europe double or triple, straining government budgets. Agrarian prices increase the most, while industrial prices, real wages, and employment decline.

One result of these developments is that everywhere the distance between propertied and propertyless widens, while relatively low productivity helps keep the social structure sharply stratified. The growth of capitalism does allow many to better themselves; price inflation particularly benefits the commercial classes, who mount a vigorous challenge to the status of the aristocracy. The wealthy continue to spend their money in conspicuous consumption, with the increased wealth making possible an extravagant lifestyle undreamed of by medieval kings. Luxury and waste become necessary badges of high social status. But most of the rewards go to a few, and the economic disparities of upper, middle, and lower classes grow sharply. For most people throughout Europe, living standards and diet begin to decline. The propertyless bottom half (or more) of the population can expect nothing beyond bare survival. In general, the laboring classes are worse off at the end of the 17th century than during the mid-15th, and are viewed as hapless and burdensome to society. Cereal prices, and thus bread prices, rise so high that little money is available for anything else. Until 1850, meat consumption declines throughout Europe, with the possible exception of England.

Overpopulation also has left Europe encumbered with a vast increase of vagrants, beggars, and the unemployed. Many towns must protect themselves from regular invasions of poor peasants, who often arrive from far away in veritable armies. An alarmed bourgeoisie begins to devise ways to prevent the poor from doing harm (Dunn 1970:89-112; Koenigsberger and Mosse 1968:27-34; Minchinton 1976:93-99; Braudel 1981:54, 76).

Price inflation, coupled with the emphasis on conspicuous consumption among the wealthy, severely strains those in the nobility who exist on fixed incomes. As Hill (1961:1516) observes for England, because inflation favors those producing in order to sell, the safest road to prosperity lies in careful attention to the market, in checking and shortening leases, in avoiding unnecessary extravagance, and in developing more bourgeois virtues. Failing to meet this challenge, many a noble ends up in ruin because of his extravagance. The tension between the upper and middle classes over the ability of the latter to challenge the former in expenditures on luxury adds to an increase in sumptuary legislation designed to keep each class in place. These laws also are designed to keep down overall expenditures on increasingly limited food supplies and on luxury items, which often are purchased from other countries and undermine national self-sufficiency (Cole 1939, 1:5-6, 10-12).

VITICULTURE. With food and especially cereal prices rising, the area of agrarian cultivation expands. Lands are reclaimed, and even in areas of viticulture, such as Maine in France, vineyards are replaced by cornfields (Slicher van Bath 1963:197-98, 204).

ENGLAND

BACKGROUND DEVELOPMENTS. As a result of land enclosures, growing instability of employment, and a rise in prices and population, poverty increases and people are forced off the land and out of traditional occupations, creating a class of unemployed, vagabonds, and "masterless men." They are later joined by disbanded soldiers who have fought against the Spanish and Irish. These vagabonds are repeatedly accused of drunkenness, idleness, and lawlessness, and are of grave concern to the government (Hudson 1933:4-5). Other major social problems of the day include, first, the rapid urbanization of London, which triples in population during the reigns of Elizabeth and James I. The existing civic institutions cannot absorb this population increase, and the dangers of plague, fire, and social disorder are intensified. Second, a religious vacuum is left by the slow decay and then abrupt collapse of the medieval church. Third, society begins to face the problem of developing a new social organization to replace the medieval manorial and parochial systems. At the same time, the nobility and newly enriched indulge in unprecedented luxury and ostentation. Thus, the age produces an abundance of "tiresome moralizing treatises, sermons and plays itemizing contemporary sins," virtually all of which mourn the increased mobility of the lower orders, the ostentatiousness of the nouveau riche, and the effeminate and mannered elegance of the court (Walzer 1971:200-205).

Commenting on this situation, Walzer (1971:199) observes that in the Tudor period there was an "intense awareness" of mutability, danger, and the "ever-present threat of social disorder." One of the sharpest manifestations of the many changes England is undergoing is the appearance of the "masterless man," who was "alien from the feudal world, vagabond and criminal, hero of the new picaresque....In the eyes of their sober, prosperous and fearful fellows, these uprooted peasants, disbanded soldiers, and discharged retainers were the most hated villains of the age, carriers of the social diseases of violence and crime."

POPULAR CULTURE: DECLINE OF PARISH. The once vigorous parish social life, with its feasting, dancing, and church-ales, begins a slow eclipse due in part to the Reformation, but also to the Tudor transformation of the parish into an administrative unit (Walzer 1971:202, who suggests that one reason for the appearance of Puritanism was that it supplied an alternative set of social and spiritual activities).

PRICES; PRICE CONTROLS. The costs of foodstuffs rise steadily until about mid-century, by which time they have increased some 25%. After 1550, the rise is much more rapid, with prices doubling in 20 years. Thereafter, prices become more stable and rise slowly. The prices of ale and beer generally follow the same pattern, although Henry VIII does much to keep their prices under strict control for as long as possible (Monckton 1966:105).

ALEHOUSES CONDEMNED. Beginning in the 1540s the first major attacks appear on the number of alehouses and their adverse effects on society. These attacks increase in frequency and intensity over the next hundred years. Combined with concerns over vagabonds and idlers and with shifting attitudes toward the poor, the expansion in the number and social role of alehouses seems to pose a threat to general order and the conventions of established society. With little to do the "masterless men" spend much of their time in public houses. Of particular concern was the amount of time the "masterless men" and the unemployed spent in alehouses. This prompts the first licensing controls (see below). A broad consensus begins to emerge among the middle and upper classes that too much drunkenness and too many alehouses are a social evil, and that alehouses are "a new and increasingly dangerous force in popular society." Alehouses are viewed as centers of lower-class crime, disorder, drunkenness, promiscuity, and popular opposition to established religious and political order (Clark 1978:48). Critics elaborate on the view of the medieval Church that taverns are the source of all disorder (Wrightson 1981:12, 16).

On the whole, authorities in the second half of the century are still more concerned about the alehouses and breaches of the peace than about drunkenness per se. Encouraging the local drunkards seems to be a less flagrant offense than operating an unlicensed house or "entertaining" wandering rogues and "other undesirable characters." Magistrates appear far more concerned with the economic aspects of running alehouses than with the number of drunkards. Thus, more alehouse keepers get into trouble than customers, unless the latter are vagrants (Emmison 1970:211-212).

ALEHOUSE PREVALENCE, FUNCTIONS. Throughout the second half of the century the number of alehouses expands, their size becomes larger, and the alehouse gains a new importance as a "vital and widespread institution in the popular landscape." Under the Tudors and early Stuarts, the number of alehouses grows at a faster rate than the population in some areas, especially towns. As early as 1544, Coventry magistrates declare that brewers and alehouse keepers have so increased that they now constitute "a great part" of the city's inhabitants (Clark 1983:39, 59, 72). A variety of factors promote expansion in the numbers of occasional alehouses "beyond the economically necessary minimum" (Wrightson 1981:3). Many of these factors are associated with the urbanization of English society. First, artisans and laborers resort to ale selling to supplement their income, which has been undercut by inflation. Alehouses are a particularly appealing source of income because they require a minimal capital investment and licensing is often lax, even after new controls are established in 1552. Most unlicensed sellers are only briefly involved in the trade in order to acquire money. Thus alehouse keeping serves as a "system of circulating aid in which economic activity, neighbourly assistance and festivity [are] subtly blended" (Wrightson 1981:5). Second, the socioeconomic changes of the period create powerful demands for the services provided by alehouses: because of migration and urbanization more people are drinking outside of home and are no longer brewing for themselves; the demand for beer and ale rises because the population is growing and because these beverages supply the poor with a fairly cheap source of nutrition as grain prices inflate; alehouses supply rudimentary accommodation for growing numbers of vagabonds and travellers at the same time that the traditional means of hospitality are declining, especially after the dissolution of the monasteries; finally, as the Reformation erodes the traditional sense of community focused on the local church, the alehouse assumes many more functions as a center of popular culture. Thus alehouses begin to offer a broader range of services to meet the needs of their essentially lower-class customers. They also provide a new sense of commonality and fellowship to their clientele through drinking rituals (Clark 1978:52-53; Clark 1983:137-139, 156; Wrightson 1981:9-10; Hudson 1933:3-5).

ALEHOUSE LICENSING (1552). Alehouses are put under a formal national licensing system for the first time in 1552. The statute is passed primarily to strengthen the 1495 statute (which only gave the justices of the peace the power to withdraw permission to sell), to suppress useless alehouses, and to prevent "the intolerable Hurts and Troubles" which "doth daily grow and increase through such Abuses and Disorders as are had and used in common ale-houses and other Houses called Tippling Houses." Justices of the peace are authorized to license competent persons in each county or borough to keep alehouses. Alehouse owners must post sureties to enter into a pecuniary recognizance or bond binding them to maintain good order, prevent drunkenness, refuse to serve drinks during divine service, and refuse sanctuary to rogues and vagabonds, as well as to provide wholesome drink and to use true measures. Annual licenses can be granted only on conditions laid down by Parliament for regulating the conduct of the trade. Parliament makes it clear that the number of licenses are to be kept down just enough to supply the legitimate needs of each neighborhood, thus creating a sort of monopoly (Baird 1944b:155; Longmate 1968:5; Webb and Webb 1903:5-6; Emmison 1970:202-204).

TAVERN LICENSING (1553). In 1553, an act is passed to regulate the "many taverns of late newly set up in very great numbers in back lanes, corners, and suspicious places within the City of London" and elsewhere, which are the "common resort of misruled persons." The act sets down price regulations for wine, stipulates that no one can operate a tavern without a license, and names those towns which can have more than two taverns and specifies the number (Wilson 1940:95).

BREWING. In 1556, beer and ale brewers unite; beer exports "slowly but surely" increase (Monckton 1966:96-97).

FRANCE

WINE CONSUMPTION; DIET. Complaints over declining standards of living appear throughout the second half of the century. As early as 1548 an old Breton peasant contrasts current scarcity with past abundance. Similarly, in 1560 a Norman asserts: "In my father's time, we ate meat every day, dishes were abundant, we gulped down wine as if it were water." Another witness at the end of the century contrasts the abundance of foods and goods before the Religious Wars (1562-1598) with the poverty afterwards (Braudel 1981:195).

ATTITUDES: WINE, TEMPERANCE (RONSARD). Pierre de Ronsard, the favorite poet of King Henry II (1547-1559), constantly praises the wines of Vendome and wine drinking in general. He "would have his page pour out just a little bit more [wine] in order that all ennuis might be effaced" (quoted Wiley 1954:96). (However, he makes it clear that he sips, never gulps, his wine.) In 1554, he writes an epitaph for Rabelais portraying him as a big drinker. In summing up the qualities of an ideal gentleman in a discussion of moral and intellectual virtues given in the presence of King Henry III, he recommends moderation--the midpoint between too much and too little. Temperance he defines as being between "debauchery and stupid asceticism" (quoted Wiley 1954:39). Like Rabelais, he adopts the ancient idea that Bacchus is the civilizing god from whom all good accrues to the earth and man. Bacchus inspires man's mind, refines his morals, and raises his imagination to heaven (Legouis 1926:5).

TOASTING. Toasting exists in France as elsewhere in Europe. Ronsard tells how he would toast his beloved Cassandra nine times, once for each letter of her name. Such a refrain appears also in Colletet's bacchic poetry. Toasts often are drunk to one's own name, which leads another poet to say that if the wine is not good, his name his Jean, but if it is good, his name would be Marc Antony (Franklin 1889, 6:121-122). Besides having more letters in the name, Marc Antony was a famous drinker (see 50 BC/Rome).

DISTILLING: BORDEAUX, CHARENTE. Distilling has become so active in Bordeaux that in 1559 the city bans it for fear of fire (Delamain 1935:16; Younger 1966:326). Most of this brandy is exported to England or used to fortify exported wines. Within a century, Bordeaux becomes the main supplier of brandy to England (see 1660/France). Brandy distilling also appears to have begun in the area around the port of La Rochelle in Aunis, Charente: a merchant in La Rochelle is recorded (in 1549) buying four casks of good eau-de-vie. In 1559, twelve casks of brandy are sent from La Rochelle to England (Dion 1959:443; however, Delamain 1935:15 does not believe that distilling was important in Charente until the early 17th century [see 1630/France]).

INEBRIETY: CROSSCULTURAL COMPARISON (PLATTER). While studying at Montpellier's famous school of medicine, the Swiss-born Felix Platter records in his journal on May 8, 1553, that he never saw anyone drunk in the city other than Germans, although the city lies within the viticultural region of Languedoc (Platter 1557/1961:58). Those instances of drunkenness he does describe involve Germans, and the one example of a drunkard among the city's natives is a man who once lived in Germany and was tainted there by inebriety (pp. 32-33, 56, 78). Le Roy Ladurie (1974:64) comments: "Platter, who could not find words strong enough to condemn the drinking habits of his compatriots at Montpellier compared to the sobriety of the local populace, describes Germans, red as lobsters, dousing their heads with wine--wetting their breeches, dead drunk, while their playful comrades cut off their beards."

WINE DILUTION. Although wine is plentiful in the city, Platter (1557/1961:53) emphasizes that the citizens of Montpellier dilute their wine before drinking it. He records on February 12, 1553: "There is wine for those who want it; it is deep red in colour, and is drunk mixed with water. The servant first pours as much water as your wish, and then adds the wine. If you do not drink all of it, what is left is thrown away. This wine does not keep more than a year, and turns sour." His description of peasant life indicates that they rarely can afford to drink wine. (On Platter, see also 1570/Switzerland.)

GERMANY

INEBRIETY PREVALENCE. Between 1550 and 1600, the recorded number of drunken brawls in Osnabrück triples (Petersen 1856:68). In 1557, the council of Nuremberg complains about the number of wounds daily inflicted due to drunkenness. Around this time, the city is carting carry away the drunken people lying in the streets (Janssen 1910, 15:409).

PEASANTS/FESTIVE DRINKING. Elector Augustus of Saxony complains in 1557 about peasants among whom "it has become the fashion...at the high festivals, such as Christmas and Whitsuntide, to begin their jollifications on the eve of the festival and to carry them on til morning, either missing the church services altogether or else coming tipsy to church and sleeping all the time like hogs" (quoted Janssen 1910, 16:114).

The preacher Erasmus Sarcerius complains in 1551 about Sunday and festive drinking in towns and villages: "On no other day is there so much impropriety, wantonness, scandal, vice, villany, godlessness, as on the day of the Lord. In the morning, especially in the towns, people sit in the public houses drinking brandy. In the village also many people begin early in the morning with wine and beer....Easter is celebrated with inordinate eating and drinking." On Whitsunday eve, drinking lasts late into the night until everyone is intoxicated (quoted Janssen 1910, 16:34-35).

NOBLES/DRINKING. In 1555, Erasmus Sarcerius complains that "several lords and nobles, also some of the councillors in towns" are the chief promoters of the increasing love of drink (quoted Janssen 1910, 15:410).

SPIRITS PRAISED, MEDICINAL USE (GESNER). The Swiss physician Conrad Gesner (or Evonymous Philiater, 1516-1565), called "the Pliny of Germany," discusses the history and methods of distillation in his De remediis secretis (1552, 1557), which is quickly translated into several languages. He places special emphasis on water cooling, which was ignored by Brunschwig (Forbes 1970:120). He praises the therapeutic virtues of aqua vitae, gives recipes for many different spirits, and specifies the specific ailment for which each recipe should be employed. Many of the recipes deal with preserving or purifying wines, indicating the inferior quality of much of the wine available (Simon 1948:132-133; Younger 1966:326, 331-333). As translated into English by Peter Morwyng (Treasure of Evonymous, 1559:83), Gesner asserts: "It is good for them that have the falling sickness if they drink it. It cureth the palsy....It sharpeneth the wit, it restoreth memory. It maketh men merry and preserveth youth....It is marvelous profitable for frantic men and such as be melancholy. It expelleth poison."

SPIRITS CONSUMPTION. One of the advantages of distilling which Gesner (see above) observes is that even bad wine can produce good spirits. He also notes that the demand for aqua vitae is so great that even good wine is being distilled (Simon 1948:133-132). The above-quoted comments of Erasmus Sarcerius indicate that brandy drinking is still isolated in the towns.

BRANDY DRINKING CONTROLS. After the 1524 prohibition in Hesse against brandy fails to stop "inordinate brandy drinking," a more severe ordinance in 1559 orders that "no more drinking bouts [are] to be held either by innkeepers, or by burghers, peasants, nobles or commoners, and that brandy [is] only to be sold to men and women who [are] ill and infirm." In Bavaria, it is decreed in 1553 that "no one should drink more than two pfennigs worth of brandy per day," and it is forbidden to make brandy out of "wheat, barley and suchlike grain," since such drinks are highly injurious to the people (Janssen 1910, 15:418).

IRELAND

WHISKEY CONSUMPTION; DISTILLING CONTROLS. The Irish Parliament in 1556 requires that manufacturers of "aqua vitae" obtain a license from the Lord Deputy on pain of imprisonment and a four-pound fine (no license fee is mentioned). Peers, gentlemen of property, and borough freemen are permitted to distill without this license. The statute states that aqua vitae is "a drink nothing profitable to be daily drunken and used, now made universally throughout the realm of Ireland" (quoted Morewood 1838:619). Apparently, whiskey drinking has become so widespread that legislation to curb it is warranted, although no evidence exists that distilleries are abundant. It is likely that small stills run by innkeepers for local needs are common, and that large households do their own distilling (McGuire 1973:92). This statute is followed by a number of other acts in the last half of the 16th century. (French [1884:143] claims that the act greatly reduced spirit use, but provides no evidence.)

ITALY

TEMPERANCE CITED; CROSSCULTURAL COMPARISONS (CELLINI; DELLA CASA). Benvenuto Cellini's (1500-1571) *Autobiography* provides evidence of Italian sobriety. It contains only two instances of drunkenness, both involving Germans (books 1.97, 2.18). Cellini also expresses surprise when a Frenchman orders wine to accompany their conversation, with the explanation that it is the French custom to drink before discussing anything (2.31). Cellini considers this custom unnecessary (Jellinek 1976b:4).

Similarly, Giovanni della Casa's *Galateo* (1558), a Florentine book of courtesy and manners, advises that "inviting people to drink repeatedly is not one of our habits and we describe it by means of a foreign word: *fare brindisi*, a reprehensible custom which one should not adopt. I thank Heaven that, among the scourges that have come from beyond the Alps, this one, which is the worst, has not gained a footing here; I mean to consider it amusing and even estimable to get drunk" (quoted Lucas-Dubreton 1960:118-119; see also della Casa 1558/1969:115-117).

Around this same time, an English visitor is amazed by the sobriety he finds in Florence, where it is considered the mark of a true epicurean to have a good time without "swilling and gormandizing." He also observes that the medicinal benefits of wine in moderation are greatly praised (Lucas-Debreton 1960:117, 119).

MEDICINAL USE, SPIRITS (MATTHIOLUS). In his widely read commentary on Dioscorides (*Commentarrii in VI libros Pedacii Disocorides*, 1554), the Venetian pharmacist Peter Matthiolus (Mathiole) praises distilled spirits and recommends drinking a spoonful a day to fortify memory and sight and to give vivacity and spirit (Le Grand d'Aussy 1782, 3:66).

NETHERLANDS

BACKGROUND DEVELOPMENTS: WAR OF INDEPENDENCE. In the 16th century, the Low Countries are inherited from the dukes of Burgundy by the Hapsburg Emperor Charles V (1519-1558), and then by his son Philip II of Spain (1556-1598). In 1566, 200 nobles (both Catholic and Protestant) from various provinces establish a league to check "foreign" or Spanish influences, inaugurating a long period of revolt, sometimes called the Eighty Years War. In 1579, the seven northern provinces, led by Holland and Zeeland, form the Union of Utrecht and in 1581 declare their independence as the United Provinces of the Netherlands, more commonly known as the Dutch Republic or Holland, due to the latter province's dominance in the union. (For simplicity, the designation Holland is sometimes used in this chronology beginning in the 16th century in reference to the Dutch Republic.)

TAXATION. In conflict with their Hapsburg rulers, the Low Countries rely increasingly on wine taxes for public finance, preferring to tax luxuries rather than necessities. An unprecedented rise in the price of wine results, bringing about a rapid decline in consumption during the second half of the century. Owing to this rising cost and declining consumption, wine tax revenues decline (Craeybeckx 1958:15-16).

LOWER CLASSES/DRINKING (BRABANT). Representatives of Brussels complain that the high imposts on grain, meat, and beer have created unnecessary hardship for the common man, for whom these goods are more indispensable for nutrition than for the rich. Life has become so costly that "the common people are so pressed that several have a dearth of bread and nothing to drink but water." Craeybeckx (1958:41) sees this as an example of how the economy affected fluctuations in the use of water as a beverage.

BEER CONSUMPTION. The Spaniards accompanying Crown Prince Philip on his official visit to the Netherlands in 1549 are astonished to see the quantities of beer consumed. An average of three pints a day is probably consumed by adults, at least among the rich (Parker 1977:20).

WINE CONSUMPTION. By the 16th century, economic decline in the southern Netherlands brings a fall in wine drinking even in once-prosperous Bruges. The average annual wine consumption for the area is only 20 liters per person. However, because of its use by the wealthy, the Low Countries continue to be an important market for French wines. In the mid-century, wine represents 53% of the value of all merchandise imported from France and 40% of all traffic in goods between the two countries (Craeybeckx 1958:39, 43).

SPIRITS CONSUMPTION. The abuse of medicinal spirits is spreading. The Dutch physician Laevinius Lemnius (d. 568) greatly praises aqua vitae as a medicine: "No liquor, that is ministered into any use to man's body, is either lighter or more piercing, or more preserveth and defendeth all things from corruption." However, he complains that "the use of aqua vitae has grown so common in Nether Germany and Flanders that freelier than is profitable to health, they take and drink it" (quoted Roueché 1963:174; Simon 1948:130).

German soldiers serving in the Dutch army during the revolt against Spain are also said to have popularized spirit drinking in the Netherlands (Forbes 1970:104).

DISTILLING: INTRODUCTION. Phillip Hermanni's popular Constelyc Distilleerboec (1552) introduces distillation "in its modern form" into Holland (Forbes 1970:105). Strongly influenced by Braunschweig and Gesner, Hermanni gives the earliest recipe in the Dutch language of a spirit distilled from wine and mixed with juniper, which he recommends for all kinds of diseases (ibid., pp. 159-160). However, it is not until the mid-17th century that the recipe for a juniper-flavored grain spirit--what becomes known as gin--is formulated.

FESTIVE DRINKING. One of Prince Philip's Spanish companions reports that banquets sometimes last from lunch until late into the night. The guests become so drunk that they cannot move, and they urinate on the floor rather than going outside. This he finds surprising given the meticulous cleanliness of Dutch women (Parker 1977:20).

In several of his pictures, Pieter Bruegel (Brueghel) the Elder (1525-1569) portrays festive drinking in Flanders, revealing that an "enormous amount of carousing" goes on (Foote et al. 1972:117). In "The Fight between Carnival and Lent" (1559), he captures the tension and nervous excitement created by the contrast between Shrove Tuesday and Ash Wednesday. In the front of the painting, the enormous figure of Carnival is seated on a beer barrel, wearing a large pie cocked over one eye. Around him full-bodied people are portrayed in a state of intoxication and gaiety. Lent and his followers are portrayed as thin and gloomy (Henisch 1976:38-39). In the "Land of Cockaigne" (1567), he portrays three men lying on an idyllic hillside surfeited with food and drink, a vision of paradise and abandon painted in the bleak war years. Roehl (1976:116-117) observes that in his depiction of men, women, and children gorging themselves on food and drink, Bruegel captures the real spirit of medieval revelry in which, because of undernourishment, being robust and hefty symbolized well-being and security. Other portrayals of festive drinking are "The Peasant Dance" and the "Peasant Wedding."

RUSSIA

TAVERN (KABAK) MONOPOLY INTRODUCED; REVENUE GENERATION. Ivan IV (Ivan the Terrible; 1533-1584) introduces a *kabak* (tavern) in Moscow for the exclusive use of his guards, friends, and courtiers, as a place where they can congregate to eat and drink. When Ivan realizes in 1552 that considerable revenue can be derived from the kabaks, he gives orders to close existing taverns, where drink and food are served, and to open up "tsar kabaks," in which only drink is sold. Furthermore, vodka soon replaces other beverages served there. Peasants are not allowed to make their own brew; they must set up a kabak in their village, be responsible for running it, and pay taxes on it. Vodka is distilled in government plants situated near the kabaks or is supplied to them by government contractors or by concessioners. The kabaks are run according to one of two systems: as a concession, or "on faith"--that is, by someone elected by the people.

A tax rate is fixed for each individual saloon, based on the income realized the previous year, the size of the concession, and other circumstances. The main rule in determining the quota is that the profits should exceed those of the previous year. At first, the operation of the kabaks is confirmed by the tsar as a privilege to courtiers, but later, liquor licenses are sold to the highest bidder.

In order to earn a livelihood, the saloonkeeper is obliged to overcharge or to reduce the quality of the beverages sold. When first elected, a saloonkeeper must pay for the vodka supplied by the government; he thus begins by getting into debt (Pryzhov 1868; Efron 1955:493-495; Cherrington 1929, 5:2331-2332).

SCOTLAND

WHISKY SALES CONTROLS. In Edinburgh, there are a number of convictions in 1550 for breaking the privilege granted to the city's surgeon-barbers to make and sell whisky (Ross 1970:3-4). This suggests that the monopoly is profitable and that recreational drinking is spreading. One Bess Campbell (her name indicating origins in the western Highlands) is told to "desist and cease from making aqua vitae within the burgh in time coming" without the surgeon-barbers' permission (Wilson 1973:35; Ross 1970:3-4).

EXPORT BAN. A threatened famine in 1555 leads the Scottish Parliament to forbid the export of victuals, including ale and aqua vitae.

SWEDEN

DISTILLING CONTROLS. Gustavus I (1523-1560) takes measures to stop spirit production, which became common under Erik XIV and John II (Forbes 1970:97).

BEER CONSUMPTION. One of Gustavus's sons complains that his soldiers are allotted only a tankard of about 2.75 quarts of beer a day, which is not enough "to help themselves to their subsistence." Through his intervention, the ration is raised to 1 1/2 tankards a day and two on Sundays, equal to about 1,600 quarts a year. There are many instances of "a far higher consumption" (Heckscher 1954:21).

Beer consumption may be 40 times higher than in modern Sweden. In 1573, beer consumption at the courts amounts to 1,700 to 1,800 calories a day, out of a total of 4,500 to 5,500 calories; this compares with an average of 26 calories from beer out of a total of 4,400 in 1912-13. Agricultural servants receive about 835 calories out of 3,500 from beer (Heckscher 1954:69-70, who comments: "It is probable that the lords constantly reveled in food and drink in a way which was beyond the reach of the peasants except at special celebrations. The high mortality and generally poor health were certainly related to the gluttony of the age").

DRINKING INFLUENCES: DIET. Such high levels of beer consumption are related to the degree to which foods are salted and spiced to make them palatable and storable (Heckscher 1954:69-70).

BEER CHARACTERISTICS. Beer is often so unpalatable, if not totally soured after storage, that it has to be heavily seasoned with spices (Heckscher 1954:22-22).

1560

EUROPE

CHURCH CONCERNS: POPULAR CULTURE (COUNCIL OF TRENT). The Council of Trent holds its last and most decisive session in 1562-1563, issuing a number of decrees for the reform of popular culture as part of the Catholic effort to counter Protestantism. The council declares that "the celebration of saints and the visitation of relics [must not] be perverted by the people into boisterous festivities and drunkenness, as if the festivals in honour of the saint are to be celebrated with revelry and with no sense of decency." Throughout Europe in this decade, synods and provincial councils and leading Catholic reformers declare themselves enemies of taverns, plays, and drunken festivals, especially Carnival (Burke 1978:220).

ENGLAND

BACKGROUND DEVELOPMENTS: ELIZABETHAN ERA. Elizabeth I becomes queen of England in 1558 and reigns until 1603. Under her reign, the religious strifes, government debts, and military defeats of the past are reduced or ended. With the defeat of the Spanish, English rises to the level of a first-rate power and an era of national pride, peace, and commercial prosperity begins. Among the upper classes, the late 16th and early 17th centuries are a period during which prodigious conspicuous consumption is viewed as a symbol of rank and status. Contributing to this "extraordinary temptation to extravagance" among the nobility and gentry are (1) the desire to imitate the lifestyle of Renaissance Italy; (2) the competition generated by the *nouveaux riches* to demonstrate their prominence through material goods; (3) the unprecedented mobility of landed society as more and more new families become enriched; (4) the need to keep up with the pomp of royal service at court and the competition for royal favor; and (6) the medieval tradition of noble openhandedness and moral obligation, which puts generosity and display before thrift and economy (Stone 1967:86, 249, 264-265).

CONCERNS: DRINKING. In the early part of Elizabeth's reign (to about 1580), there is little attempt in dramatic literature to condemn drinking or to moralize about it; there is a paucity of references to drink. But nondramatic literature advocates temperance more frequently and angrily than in the first half of the century, especially after 1580 (Williams 1969:107-122, 192-193, 483-484). The Statute of Artificers (1563) and the act of 1597-1598 for the maintenance of husbandry and tillage include idleness, drunkenness, and unlawful games among the "lewd practices and conditions of life" which interfere with agricultural or industrial production (Hill 1964:125).

ALEHOUSES AND CRIME. Thomas Harman (*A Caveat or Warning for Common Cursitors, vulgarly called Vagabonds*, 1566) expresses the fear of many that tippling houses are often full-time headquarters for professional gangs of criminals and actually encourage the spread of crime: "For what thing doth chiefly cause these rowsey rakehells thus to continue and daily increase? Surely a number of wicked persons that keep tippling houses in all shires where they have succor and relief; and whatsoever they bring, they are sure to receive money for the same; for they sell good pennyworth" (quoted Judges 1930:65).

ALEHOUSE PREVALENCE. In the 400 parishes of Essex, between 600 and 800 licensed houses exist, or about 1 1/2 to 2 houses per parish (Emmison 1970:215).

UPPER CLASSES/DRINKING. The normal aristocractic household diet consists of huge quantities of meat and bread "washed down in oceans of beer and wine," causing moralists to attack banqueting as a scandalous waste of money. Aristocratic household accounts for the Elizabethan and Stuart periods indicate a beer allowance of 5-8 pints per person per day, and 750-1,250 gallons of wine a year for a household of between 50 to 100 people. Along with guests and their attendants, often 200 people are fed a night. The "cautious" Lord Burghley (William Cecil) supplies 1,000 gallons of wine for the celebration of his daughter's marriage in 1582 (Stone 1967:255-257, who comments that "no wonder foreigners were struck by the English capacity for beer-drinking").

SPIRITS PRAISED; DISTILLING. Housewives are encouraged in home distillation by Peter Morwyng's translation of Conrad Gesner (1550/Germany), published in 1559 as Treasure of Evonymus and reissued with a modified title in 1565 (Forbes 1970:121). In 1576, George Baker translates another version as The Newe Jewel of Health (Simon 1948:133). In his preface, Baker expresses his delight in offering this book for the benefit of the country. This new jewell (i.e., aqua vitae) "wyll make the blynde to see, and the lame to walke,...the weake to become strong, and the olde crooked age appeare yong and lustye." This work is reissued in 1599 under the title The Practice of the New and Old Phisicke.

Most early English spirits are made at home. By Elizabeth's time, the art is practiced by amateurs as well as professionals. The still room is becoming an important part of many households, where other things besides spirituous medicines are distilled (Younger 1966:326, 331-333).

FRANCE

ECONOMIC VALUE, WINE. In his analysis of the causes for the contemporary price inflation, which he attributes to the abundance of gold, Jean Bodin ("Reply to the Paradoxes of Malestroit," 1568) cites wine as one of the primary sources of France's wealth: "The English, the Scotch, and all the people of Norway, Sweden, Denmark, and the Baltic coast, who have an infinity of mines, dig the metals out of the center of the earth to buy our wines [and other products, especially salt]" (quoted Ross and McLaughlin 1953:204).

GERMANY

INEBRIETY CONDEMNED (FRIEDRICH). Protestant preacher Matthew Friedrich writes a treatise on the devil of drunkenness (Wider den Saufteufel, 1562). He complains of "drunken sots" who say that drinking is not a sin because it is not forbidden by God and who allege that "they are never more fervent in prayer than when they are intoxicated" (quoted Janssen 1910, 15:388-389; see also Janssen 1907, 12:323).

PROTESTANTS/DRINKING CONDEMNED (ANDREA). In 1568, James Andrea, the Catholic chancellor of the University at Tübingen, links drunkenness with Protestantism: "The vice of drunkenness has now, for the first time in the memory of man, become common everywhere; our dear forefathers under the papacy...never allowed drunkards and the wine bibers to hold public posts; they were shunned and fled from at all weddings and social gatherings; street boys ran after them and marked them as useless, godless people who were not wanted anywhere; now, on the contrary, drunkenness is no longer regarded as a disgrace either among high or low classes" (quoted Janssen 1910, 15:387-388).

IRELAND

INEBRIETY OF O'NEIL. According to Raphael Holinshed (1587/1808, 6:331), Elizabeth's Irish opponent Shane O'Neil (d. 1567) is a copious drinker of wine and of "Uskebagh or Aqua vite of that country." Sometimes he drinks to such excess that his attendants bury him in the earth to his chin: "That for the quenching of the heat of the bodie, which by that meanes was most extremelie inflamed and distempered, he was eftsoones conveied (as the report was) into a deepe pit and standing upright in the same, the earth was cast round about him up to the hard chin, and there he did remaine untill such time as his bodie was recovered to some temperature....and in the end his pride joined with wealth, drunkennesse, and insolencie, he began to be a tyrant."

POLAND

BEER CONSUMPTION. At a Polish manor, the average daily consumption of beer is three liters a head (Minchinton 1976:124).

SCOTLAND

WHISKY CONSUMPTION. Henry Darnley, the second husband of Mary Queen of Scots, reportedly gets the French ambassador drunk on composita in 1565. This is the earliest specific mention in Scotland of inebriety caused by spirit drinking (Wilson 1973:36).

1570

ENGLAND

ALEHOUSE, TAVERN PREVALENCE. A census of the number of retail outlets in England and Wales in 1577 is made for the Privy Council, the first serious attempt to survey all types of drinking houses (Clark 1983:41; Everitt 1973:93). It is estimated that there are 17,367 alehouses, 1,991 inns, and 401 taverns, for a total of 19,759, or about one license for every 187 persons (Monckton 1966:101-104; Monckton 1969:39-40, who notes that this compares with about one license for every 657 people today). (Bretherton [1931:155] says the census uncovered 14,000 alehouses, 1,600 inns, and 400 taverns [16,347 total], but this excluded London and several populous counties. He estimates the total for the nation without Wales at 25,000.)

REVENUE GENERATION. Based on the 1577 census of drink outlets, Elizabeth orders each licensed alehouse to pay 2s6d toward the repair of Dover harbor. This levy is typical of the occasional use of licenses to raise revenues (Bretherton 1931:156; Firebaugh 1924:260).

BEER CHARACTERISTICS. Late in the century, a chorus of complaints arise about strong brews and competition among brewers to produce the strongest drink (Clark 1983:99). There are several indications in this period that the strength of beers has increased. Queen Elizabeth criticizes brewers in 1559 for no longer producing both "single" and "double" beer as they were doing at the beginning of the century, but only a very strong "double-double" beer and orders them to brew each week as much of the single as the double.

William Harrison (1587/1968:247; see 1580/England) observes that in the weekly town markets "there is such heady ale and beer in most of them as for the mightiness thereof among such as seek it out is commonlie called huffcap, the mad dog, father-whoreson, angels'-food, dragons'-milk, go-by-the-wall....It is incredible to say how our maltbugs lug [suck] at this liquor."

Several factors probably account for this increasing strength, including a change of taste. Brewers prefer stronger beers because they can increase their profit margin (Monckton 1966:106-107; Monckton 1969:43). Previously, during the reign of Henry VIII, brewers had sought to raise the price of single beer, which was restricted by the assize to half the price of double beer (Hackwood 1909:95). Hops also makes possible stronger beers. Thomas Tusser (*Five Hundred Pointes of Good Husbandrie*, 1574:51 [ch. 39, July]) writes: "The hop from his profit I thus doo exalt,/ it strengthened drinke, and it favoreth malt./ And being well brewed, long kept it will last,/ and drawing abide, if ye drawe not too fast." (This is part of his lesson on "when and where to plant a good hop-yard.") Finally, beers may be getting stronger because of competition for customers among rising numbers of victuallers (Clark 1983:99).

INEBRIETY CONDEMNED (GASCOIGNE). Around 1575, Southampton jurors denounce the extent to which "the notorious sinne of drunkenness" has become common in the town and the excessive number of alehouses (Hearnshaw 1906:94).

George Gascoigne writes *A Delicate Diet for Daintie Mouthde Droonkards* (1576), subtitled: "Wherein the fowle abuse of common Carousing, and Quaffing with heartie draughtes, is honestly admonished" (Gascoigne 1576/1969:451-471). Gascoigne declares that "general proposition to be "that all Droonkardes are Beastes," and he quotes an epistle of St. Augustine warning that one should "feare Drookennesse, as you feare the spit of Hell" (pp. 456, 457). On his own age and nation, Gascoigne writes: "We shall find by too true experience, that we doo so much exceede al those that have gone before us....And in this accusation, I doo not onely summon the Germaines (who of auncient tyme have beene the conntnuall Wardens of the Droonkards fraternity and corporation,) bit I do also cyte to appear our new fangled Englyshe men." Now, no one can dine without being exhorted to "Quaffing, Carowsing, and tossing of Pots." He laments that "our gentrie, and the better sorts of our people" are so taken to this practice; he then turns to attacking the plethora of fashionable, strong beverages (pp. 465-467). The tract becomes a model for later temperance advocates (Younger 1966:305).

DRINKER LIABILITY. In the case of Reniger v. Fogossa (1571), it is declared: "If a person that is drunk kills another, this shall be felony, and he shall be hanged for it, and yet he did it through ignorance, for when he was drunk he had no understanding nor memory; but in as much as that ignorance was occasioned by his own act and folly, and he might have avoided it, he shall not be privileged thereby" (quoted Holdsworth 1926:441).

FRANCE

ATTITUDES (MONTAIGNE). Humanist and skeptic Michel de Montaigne (1533-1592), in his essay *Of Drunkenness* (1573-1574), views wine drinking as one of the great pleasures of life, which is not surprising as his homeland is Bordeaux. He provides evidence of excessive drinking among the nobility, but he does not perceive drunkenness as a widespread problem in France. Indeed, at times he seems to suggest that the French are too temperate as a whole. He begins: "Now drunkenness (*ivrognerie*)...seems to me a gross and brutish vice....There are some [vices] that involve knowledge, diligence, valor, prudence, skill and subtlety; this one is all bodily and earthy. Thus the grossest nation in existence today [Germany?] is the only one that holds it in esteem. The other vices affect the understanding; this one overturns it, and stuns the body....the worst condition of man is when he loses knowledge and control of himself." Although drunkenness is a vice, his own "taste and constitution" more than his reason are hostile to it; it is a "weak and stupid" vice, but less hurtful and mischievous than other vices which "clash more directly with society in general." He is critical of German swallowing of wines, but also considers drinking after the "French fashion, at the two meals and in moderation" too restrictive. Drinking needs "more time and more application. The ancients spent whole nights in this practice." He tells of a great lord "who, without any effort and in the course of his ordinary meals, used to drink scarcely fewer than 10 quarts of wine, and showed himself on leaving only too wise and circumspect, at the expense of our affairs." He adds: "The pleasure we want to reckon on for the course of our life should occupy more space in it, but people should not prolong the pleasure of drinking beyond their thirst." He personally does not "set store by drinking except after eating" (Montaigne 1574/1958:245, 247-248; on Montaigne's attitudes, see Brennan 1981:217-218 and Jellinek 1946b).

In his *Of the Education of Children* (1579-1580), Montaigne recommends that a young gentleman must be able to fit into any social milieu in which he finds himself. If among a group of heavy drinkers, he should be capable of keeping pace. This does not mean indulging in prolonged debauchery, but one should know the "mechanics thereof" and then avoid it by his own will (Wiley 1954:90). He observes: "The philosophers themselves do not think it praiseworthy in

Callisthenes to have lost the good graces of his master Alexander the Great by refusing to keep pace with him in drinking. Even in dissipation I want him to outdo his comrades in vigor and endurance; and I want him to refrain from doing evil, not for lack of power or knowledge, but for lack of will." He then quotes Seneca, "There is a great difference between not wishing to do evil and not knowing how" (Montaigne 1580/1958:123).

TAVERN CONTROLS. The Parisian bourgeoisie in 1576 attempt to reduce the number of taverns and cabarets in Paris, to control their recruitment more rigorously, and to break the community of interests that link them to the vine growers of the city's suburbs. A royal edict of 1577, promulgated at the request of the Estates General, seeks to put in order the operation of hostels, cabarets, and taverns by reducing the superfluous number of drink outlets and choosing for their operators people of better living, character, and conversation. Taverners must buy letters of royal permission to open their shops. The law criticizes retailers who open their basements into taverns and sell wine from barrels. Restrictions are placed on the purchase of wines from other than the city's designated wine sellers (except the wines of bourgeois proprietors). The royal government and the city of Paris attempt through the Old Regime to erect a fiscal barrier which will make it practically impossible to sell lower priced wines to workers living in the capital (Dion 1959:483-484, 490, 497; Bercé 1974:296).

In 1579, a parlementary decree reiterates that no one, of whatever rank, quality, or condition, can haunt and frequent inns (hostelleries), taverns, and cabarets in the city where they dwell. Permission to drink there is given only to travellers and strangers (as first stipulated by Louis IX in 1256). However, purchases can be made to drink off-premises. It is evident that this interdiction is frequently transgressed (Dion 1959:487).

BRANDY CONSUMPTION. Montaigne fails to mention brandy in any of his discussions on drinking, indicating that it is still not important as a beverage (Jellinek 1976b:77).

MEDICINAL USE, SPIRITS (LIEBAULT). Jean Liébault (d. 1596) provides one of the first detailed descriptions of distilling apparatus in French. His Quatre livres des secrets de médicine et de la philosophie chimique (Four Books of Secrets of Medicine and Philosophical Chemistry, 1573), along with the French translation of Gesner (see 1550/Germany), which inspires Liébault, is the oldest full handbook on distillation in French. His goal is to "give apothecaries a taste of distilling and stimulate them to be more and more careful in preparing their medicines." He views aqua vitae as a veritable panacea and recommends that it be drunk twice monthly. This work, along with L'agriculture et maison rustique, written with his father-in-law Charles Estienne in 1547, provides evidence that spirits are still viewed as medicines and that there is little distilling done by vignerons (Forbes 1970:145; Claudian 1970:19; Le Grand d'Aussy 1782, 3:67-68).

GERMANY

INEBRIETY PREVALENCE, CONCERNS. As the century progresses, the complaints of preachers about drunkenness multiply. Most preachers place drunkenness in the broader context of widespread overindulgence and extravagant display, and of imitation of upper-class behavior.

One preacher sermonizes in 1573: "If anyone should want to describe the life of the burghers and peasants of our time he must begin with the inordinate, extravagant display in dress and jewelry of all sorts...and then next speak of the bestial gorging and drinking, of the inhuman carousals and drinking bouts which go on in town and country after the examples of the princes and lords, as it were in the highest seat of government" (quoted Janssen 1910, 15:354). It is not unusual for Carnival participants to "pour liquid down their throats as down sluices, not stopping until they drive out their senses" (quoted Janssen 1910, 15:406).

Thomas Rorarius of Giessen comments in 1572 on the ruin and misery that has descended on Germany, including the soaring prices of food and drink and the deterioration of their quality. He goes on to complain of the idleness of the people and concludes that the reason for the general poverty is that "nobody is content with his own position." According to Polycarpus Leiser in 1571, among the causes for contemporary misery are that "as though the world were out of its senses, there is splendor and luxury in dress, in every class beyond its means, and no less extravagance in gluttonous eating and drinking, as if people were bound to throw away whatever they have in their hands" (quoted Janssen 1910, 15:498, 500, 502).

NOBLES/DRINKING. Several complaints are raised about upper-class inebriety and its bad influence on society in general. In 1565, Nicholas Selnekker, the Lutheran preacher to the court of Augustus of Saxony, describes nobles as "for the most part...disgusting gorgers and drinkers." Another preacher asserts in 1573 that when the lords do issue laws and ordinances "forbidding so much drinking and tasting," little good is done: "The people laugh at it....One hears them say, 'The rulers themselves are lying ill in bed and they want to cure others! Let them begin first at home'" (quoted Janssen 1910, 15:410-411). Selnekker attacks his fellow preachers for being ridden with the very vices that they should be the first to punish, including drunkenness (Janssen 1910, 16:116). The superintendent of the Voigtland complains (c. 1575) that even government officials are guilty of drinking "themselves poor and ill, and into hell in the bargain, rather than helping to maintain authority and preserve manly dignity" (quoted Janssen 1910, 15:344, 391).

DRINK TRADE. Around 1573, a preacher places partial responsibility on the nobility for the rise in drunkenness because "they actually encourage drinking by the erection of breweries, distilleries, and taverns" and because "they want to sell a great deal and receive plenty of duty and excise" (Janssen 1910, 15:410-411).

PROTESTANTS/DRINKING CONDEMNED. A Protestant pamphlet (1579) attacks "the vice of inordinate drinking [that] reigns so powerfully at those courts which call themselves evangelical" (quoted Janssen 1910, 15:229-230). (See also the comments of James Andrea, 1550/Germany.)

DRINKING PRACTICES (MONTAIGNE). Montaigne (1574/1958:247) observes: "The Germans will drink almost any wine with equal pleasure. Their aim is to swallow rather than to taste." They are thus better drinkers because "their pleasure is much more plentiful and ready at hand." In his travel journal he notes that in south Germany he was never invited to any drinking bouts "except out of courtesy." He characterizes inn keepers in Germany as "vainglorious, choleric, and heavy drinkers; but...they are neither traitors nor thieves" (Montaigne 1581/1958:892, 893).

BRANDY SALES CONTROLS. In 1574, in Berlin brandy can still be sold only in apothecary shops. Following the evident failure of the ordinances of 1524 and 1559, the town council of Grunberg in Hesse orders in 1579 that "henceforth brandy drinking shall not be allowed either before or during service time" (Janssen 1910, 15:418-419).

DRINKING CONTROLS. Legislation against drunkenness is passed in Prussia in 1577 and by the imperial Diet of 1578.

PLEDGING. In 1573, an Order of the Golden Ring is founded in Westphalia in which members must pledge to drink no healths or Godspeeds under penalty of paying a gold gulden to the poor and losing their membership ring. Few emulate the society (Krucke 1909:23).

ESTEEM OF BEER. In 1568, Abraham Werner advocates drinking beer instead of wine to prolong life, noting that only those people who drink only beer live the longest without becoming ill. Heinrich Knaust's *Fünff Bücher, von der göttlichen und edlee Gabe, der Philosophischen, hochthewren und wunderbaren Kunst, Bier zu brawen* (Five Books on the Divine and Noble Privilege of Brewing, 1573) becomes the most popular book on the subject. Knaust maintains that beer brewing is a divine right and "a philosophical, faithful, and wonderful art," and claims to have tasted 150 kinds of beer which he describes in detail (Loffler 1909:9-12).

DISTILLERY PREVALENCE. In 1577, mention is made of 40 home brandy stills in the town of Zittau; by the turn of the century, Zwickau has 34 and Frankfurt-on-the-Oder has 80 (Janssen 1910, 15:419-420).

ITALY

WINE DILUTION. Mathematician and physician Girolamo Cardano describes in his autobiography (*De vita propria liber*, 1575) how he adds double or more the amount of water to six ounces of wine with supper (quoted Ross and McLaughlin 1953:516).

Montaigne (1581/1958:930) records that at a dinner given by the Grand Duke of Tuscany (Francis I de' Medici), the duke and his wife were given a full glass of wine and another of water. They poured out however much wine they did not want, then filled up the rest with water, the duke adding a great deal of water; his wife, almost none. He comments: "The fault of the Germans is to use immoderately big glasses; here it is the opposite, to have them so extraordinarily small."

LIQUEUR DISTILLING. Montaigne (1581/1958:917) tells of a group of Italian monks who neither say mass nor preach, but pride themselves on being both excellent distillers of orange-flower liqueurs and similar waters.

NETHERLANDS

DISTILLING. Lucius Bols establishes a distillery at Schiedam near Amsterdam in 1575, and probably becomes the first to produce genever commercially (Doxat 1971:98). Bols is the oldest of the existing Dutch distillers.

SCOTLAND

DISTILLING BAN. In 1579, the Scottish Parliament again (as in 1555) enacts a law prohibiting the making of aqua vitae for most of the year due to the scarcity of barley and a threat of famine and the "great quantity of malt consumed in all parts of this realm in the making of aqua vitae" (Ross 1970:3). The act, however, makes one large exception to this prohibition: it excepts "earls, lords, barons, and gentlemen for their own use" (Robb 1951:35; Wilson 1973:37).

CONCERNS: POPULAR CULTURE. As part of John Knox's Presbyterian reformation, a sustained attack is directed against Church celebrations, wakes, and other festivals, and singing and dancing in general. Respectable Scots begin to withdraw from popular culture and the tavern (Burke 1978:217, 279).

SWITZERLAND

WINE PRICE CONTROLS. An agreement between Bern, Fribourg, and Solothurn in 1574 includes controls over the rising price of grains and wine (Jellinek 1976b:74).

HEALTH CONCERNS (PLATTER). In his *Medical Practice*, Felix Platter (see 1550 France), now a professor in Basel, discusses "alcoholism" and warns of the danger of excessive alcohol use in a section entitled "Imbecilitas, Consternatio, and Alienato" (Salasa 1977:15; Jellinek 1976b:82).

1580

ENGLAND

INEBRIETY PREVALENCE. In the last decades of Queen Elizabeth's reign, particularly after the intervention in the Dutch wars (1585) and the defeat of the Spanish Armada (1588), drunkenness becomes a national problem. Williams (1969) observes that England seems to embark on a prolonged drinking spree that lasts until 1645. Most writers of the day lament that inebriety has increased markedly. Camden, Nashe, and Shakespeare, among others, blame this on Dutch influences, but it is also clearly rooted in a variety of major domestic changes occuring since at least the mid-century.

(HARRISON). One of the few contemporaries to allude to sobriety is William Harrison, in his *Historical Description of England* (1577, 2nd. ed. 1587), written as the introductory book to Holinshed's *Chronicles*. A clergyman with deep moral commitments, Harrison refers to the moderate eating and drinking at the tables of "the honourable and wiser sort," among whom he includes "the nobleman, merchant, and frugal artificer." To avoid drunkenness, they do not eat salt meat. But he does note that the "meaner sort of husbandmen" and country inhabitants are "now and then" given over to "surfeiting and drunkenness, which they rather fall into for want of heed-taking than willfully following or delighting in those errors of set mind and purpose." Harrison speculates that they are "overtaken when they come up to such banquets" because they live at home "with hard and pinching diet and small drinks, and some of them having scarce enough of that." This suggests declining food consumption is making drunkenness more common. He also refers to "aleknights so much addicted thereunto that they will not cease from morrow...until they defile themselves" (Harrison 1587/1968:132, 139).

ALEHOUSES, INEBRIETY CONDEMNED (STUBBES). Attacks on drunkenness and alehouses intensify as Puritanism spreads, as reforming enthusiasm is channeled increasingly into a demand for a reformation of manners, and as shifting attitudes toward the poor give rise to the establishment of the Tudor Poor Laws (Wrightson 1981:16-18).

The Puritan Philip Stubbes (or Stubs) (*The Anatomy of Abuses*, 1583/1972) asserts that the London public houses are crowded at all times with inveterate drunkards who care for nothing except obtaining the best ale. He devotes one section to the gluttony and the "beastly vice" of drunkenness: "Every countrey, citie, towne, village and other, hath abundance of alehouses, taverns, and Innes, which are so fraughted with mault-wormes night and day, that you would wunder to se them. You shall have them there sitting at the wine and goodale all the day long, yea all the night too, peradventure a whole seek togither, so long as any money is left, willing, gulling, and carowsing from one to another, til never a one can speak a redy word. Then...how they strut and stammer, stagger and relle too and fro like madmen, some vomiting, sprewing, and disgorging their filthie stomacks....But they will say, that god ordained wines and strong drinks to cheer the hart and sustain the body; therefore it is lawful to use them to that end." He goes on to attack the great harm drunkenness does to one's mind and body and to provide evidence that the drunkard is a "brute beast" who "in his drunkenness killeth his friend, revileth his lover, discloseth secrets and regardeth no man: he either expelleth all feare of god out of his minde, all love of his friends and kinfolkes, all remembrance of honestie, civilitie and humanitie: so that I will not feare to call drunkards beasts."

CHURCH-ALES CONDEMNED (STUBBES). In another section, Stubbes attacks church-ales, at which the ales or beers are very strong, and when they are put up for sale, "well is he that can get the soonest to it, and spend the most at it, for he is counted the godliest man of all the rest, and the most in God's favour, because it is spent upon his church forsooth." The next section attacks wakes, events of "such gluttony, such drunkennesse...as the like was never seen." Many people spend more money on wakes than they do the rest of the year, causing debt and ruin.

FESTIVE DRINKING. In a discussion of how conditions in England have improved since the establishment of the Church of England, Harrison (1587/1968:36) asserts that the number of holy and festival days have declined and with them "the superfluous numbers of idle wakes, guilds, fraternities, church-ales, help-ales, and soul-ales, also called dirge-ales, with the heathenish rioting at bride-ales, are well diminished and laid aside."

DRINKING INFLUENCES: THE DUTCH (CAMDEN). In 1585, Elizabeth sends 5,000 troops to the Netherlands in support of the country's revolt against Spanish rule. The troops are said to bring home a taste for brandy (see below) and a habit of excessive drinking. Historian William Camden (Brittania, 1586) writes: "The English, who had hitherto, of all the Northern nations, shown themselves the least addicted to immoderate drinking, and been commended for their sobriety, first learned in these wars with the Netherlands to swallow a large quantity of intoxicating liquor, and to destroy their own health by drinking that of others" (quoted Jellinek 1943:468). In the 1630s, it also is claimed that the English soldiers learned of this custom of drinking healths while stationed in the Netherlands (Pepys 1971, 5:172n.4).

BRANDY CONSUMPTION. It has been asserted that about 1585 the British troops in the Low Countries also acquired a taste for brandy and that upon their return the troops spread this taste among the lower classes, who can no longer afford the increased cost of wine due to the loss of Gascony (Simon 1907, 2:249; Simon 1948:33). By 1593, brandy is allegedly supplanting beer (Roueche 1963:174). However, no evidence is supplied to corroborate these assertions, and until at least the 1660s spirit drinking is still a very limited practice and almost purely for medicinal purposes.

BEER CONSUMPTION. Historian and country clergyman William Harrison (see above) brews 200 gallons of beer every month in his household at a cost of 20 shillings. Although we know nothing about the size of his household, this indicates that daily consumption is high and that prices are low (Thomas 1971:18).

BEER PRODUCTION, TAXATION. There are now 26 breweries in London (Wright-St. Clair 1962:512). A national tax on ale and beer is proposed for the first time (Monckton 1966:104-105).

ATTITUDES: TEMPERANCE, WINE (SPENSER). Edmund Spenser (1552-1599) advocates temperance in the second book of his great work The Faerie Queene, which contains the legend of Sir Guyon, "in whom great rule of Temperance goodly doth appear." The book illustrates the virtues and happy effects of temperance and the ill effects of intemperance. But Spenser also speaks of the inspiring power of wine. Early in his career, he declared that the poet is utterly unable to write great verse unless he is heated by copious cups (Legouis

1926:6) There is no evidence that he changed his view when he later became the "grave teacher of morality." In his poem "Epithalamion," he recommends that on wedding days every guest should get drunk: "Pour out the wine without restraint or stay,/ Pour not by cups but by the bellyful" (quoted French 1884:156).

FRANCE

TAVERN CONTROLS LIBERALIZED. In October 1587, Henry III (1574-1589) permits the sellers of wine and the tavern and cabaret operators of Paris to form a guild. He also ignores the regulations limiting tavern drinking by residents that he had solemnly confirmed in 1579 (see 1570/France). Banded together in defense of their common interests, the guild members expand the selling of wine and the number of outlets. Since by this time the respect and confidence of Parisians in their degenerate king has declined greatly, this liberalization possibly seeks to restore his popularity by giving to the citizens, who scorn the old restrictions, a greater liberality to frequent taverns and cabarets. By this, he also gains the support of the merchants who sell wine (Dion 1959:488). (In 1588, Henry is driven from Paris by an uprising engineered by the Catholic Holy League, which opposes the concessions he granted to the Huguenots; a year later, he is assassinated.)

FESTIVE DRINKING. At the Carnival in Romans in 1580, the Partridge King declares that Romans was to turn itself into a land of Cockaigne, a favorite them of Provençal folktales in which wine flowed like water. In keeping with the reversal of everyday order that characterizes carnival festivities, he further decrees that foodsellers, taverners and innkeepers much sell expensive items cheaply and cheap items expensively. The carnival itself is characterized by wild dionysiac demonstrations and masquerades (Le Roy Ladurie 1980:189-190, 195, 307).

GERMANY

FESTIVE DRINKING. In Saxony, a Church ordinance of 1580 complains that bridal parties were arriving at wedding ceremonies in a state of drunkenness (Janssen 1910, 15:403).

SUMPTUARY LAWS. New sumptuary laws in Berlin emphasize the need to conserve domestic resources and maintain social distinctions. These are dominant concerns for the next two centuries. Four social classes are established, with different standards and levels of expenditures allowed for each (Janssen 1910, 25:403; Dorwart 1971:31, 36-37).

PEASANTS/DRINKING. According the Basel professor Sebastian Munster, writing in 1588, the life of most German peasants is so miserable that "water and whey are almost their only beverages" (quoted Janssen 1910, 15:171).

IRELAND

WHISKEY CONSUMPTION, CONTROLS. The English government in Ireland issues a directive in 1580 that in Munster martial law applies to "idle persons,...aiders of rebels...[and] makers of aqua vitae." Apparently, spirit drinking is viewed as a cause of unrest. Within the year, there are complaints about the smuggling of "great quantities of aqua vitae and wines" into the area. In 1584, the Lord Deputy receives a request that the "statute for making aqua vitae be put in execution" because the drink "sets the Irishry a madinge, and breeds much mischiefs." There has yet to develop a national preference for spirits over wine and beer (McGuire 1973:93).

MEDICINAL USE, SPIRITS. According to R. Stanihurst's Description of Ireland (in Holinshed's Chronicles, 1587/1808, 6:8), the Irish consume aqua vitae to remedy the many ills they suffer from the wet and cold climate. He then quotes from a treatise describing three different strengths (simplex, composita, and perfectissima) and the many therapeutic benefits of simplex, which he recommends be consumed before and after eating meat: "Beinge moderatelie taken (saith he), it sloweth age, it strengtheneth youth; it helpeth digestion; it cutteth flegme; it abandoneth melancholie; it relisheth the heart, it lighteneth the mind, it quickeneth the the spirites; it cureth the hydropsie; it healeth the strangurie; it pounceth the stone; it repelleth gravel; it puffeth awaie ventosite; it keepeth and preserveth the head from whirling, the eies from dazeling, the toong from lisping, the mouth from mafflyng, the teethe from chattering, the throte from ratling, the weason from stifling, the stomach from wampling, the harte from swelling, the bellie from wirtching, the guts from rumbling, the hands from shivering, and the sinewes from shrinking, the veines from crumpling, the bones from soaking...And trulie it is a soveraigne liquor, if it be orderie taken."

NETHERLANDS

WINE TRADE. The expanding Baltic grain trade and the development of relations with France is probably responsible for a rise in the popularity of French wines in the Baltic and a decline in the demand for Rhine wines. In 1580, Dutch merchants handle 50% of the French wine imported into the Baltic; Dutch sailors carry 58% (Houtte 1977:193).

GRAIN DISTILLING. By the end of the century, distilling is gaining importance. Grain spirits from the Low Countries come into favor with customers around the White Sea, although French wine and brandy interests attack the new product. In 1588, C.J. Coolhaes (Van seeckere seer costelycke wateren) warns against them, alleging that unsound corn is often used and that badly distilled spirits are adulterated with sugar and honey (Forbes 1970:104-105, 109).

RUSSIA

KABAK PREVALENCE. In 1550, there was only one kabak in Moscow. In the late 1580s, Giles Fletcher, ambassador from Queen Elizabeth to the Russian court, observes: "In every great towne of his [the tsar's] Realme he hath a Cabak or drinking house, where is sold aquavitae (which they call Russewine), mead, beire, etc." (Fletcher 1966:44).

SCOTLAND

INEBRIETY PREVALENCE. Harrison (1587/1968:124) observes that the Scottish people "far exceed us [the English] in overmuch and distemperate gormandize." He cites as one of his sources Hector Boetius or Boece (*Scotorum historiae*, Paris, 1527), whom he translates for Holinshed's *Chronicles*. Boetius observes that "excessive feeding, and greedie abuse of wine" cause among some Scots "long sicknesse and languishing greefes." Furthermore, Boetius complains that inebriety in general has increased: "But how far we in these present daies are swarved from the vertues and temperance of our elders, I believe there is no man so eloquent, nor indued with such utterance, as that he is able sufficientlie to expresse. For whereas they gave their minds to dowghtinesse, we applie our selves to droonkennes...and so is the case now altered with us, that he which can devoure and drink most, is the noblest man and most honest companion" (Boetius 1587/1808, 5:2, 22, 26).

WHISKY CONSUMPTION. Boetius (see above) also asserts that elders are "in meate and drink sober" and that when Scots overeat they use "a king of Aquavite void of all spice, and onelie consisting of such hearbes and roots as grew in their owne gardens, otherwise their common drink was ale" (Boetius 1587/1808, 5:23).

1590

ENGLAND

ALEHOUSE PREVALENCE. Despite the licensing laws of 1552 and 1553, alehouses and taverns are still plentiful. In 1591, a report of the queen's council dealing with conditions in Lancashire and Cheshire complains that alehouses are innumerable and that the licensing laws are unexecuted, "whereby toleration of drunkenness, unlawful games, and other abuses follow." The streets and alehouses are so crowded during the hours of church service that there is no one in church except the curate and his clerk (quoted Bretherton 1931:154; Monckton 1966:104).

WINE CONSUMPTION. In the late century, the nobility and gentry appear to greatly expand their consumption of wine. William Cecil writes: "England now consumes four times as much wine as formerly" (quoted Dowell 1888, 4:114-115). Many of these are stronger wines from Spain and Greece.

INEBRIETY CONDEMNED (NASHE, BACON, RALEIGH). In many of his works, Thomas Nashe (1567-1601) makes frequent reference to inebriety and drinking habits, which he indicates have recently gotten worse. Although he approves of the use of alcohol and accepts its medicinal value, he attacks the drunkenness of the age (Jellinek 1944:462-464). His *Pierce Pennilesse, His Supplication to the Divell* (1592) contains a "Complaint of Drunkennesse," in which he attacks the recent increase in inebriety and blames this on the Dutch wars: "Let me discend to superfluitie in drinke: a sinne, that ever since we have mixt our selves with the Low-countries, is counted honourable: but before we knew their lingering warres, was held in the highest degree of hatred that might be. Then, if we had seene a man goe wallowing in the streetes, or lie sleeping under the boord, we would have spet at him as a toade, and cald him foule drunken swine, and warnd al our friends out of his company: now, he is no body that cannot drinke *super nagulum*, carouse the Hunters hoope, quaffe *upsey feze crosse*, with healthes, cloves, mumpes, frolickes, and a thousand such dominiering inventions. He is reputed a pesaunt and a boore that wil not take his licour profoundly" (Nashe 1592/1958:204-205). He identifies the Germans and low Dutch as the great drinkers in Europe, notes that entire books and other "generall rules and injunctions, as good as printed precepts" exist which set forth the "private lawes amongst drunkards," and goes on to describe the eight kinds of drunkenness he has seen in England (p. 207).

Francis Bacon (1561-1626) writes: "All the crimes on the earth do not destroy so many of the human race, nor alienate so much property, as drunkenness" (quoted French 1884:155).

Sir Walter Raleigh (1552-1618) devotes the ninth chapter of his *Instructions to his Sonne, and to Posterity* (1632:81-93) to "What Inconveniences Happen to such as delight in Wine." He writes (p. 81): "Take especial care that thou delight not in wine, for there was not any man that came to honour or preferment that loved it; for it transformeth a man into a beast, decayeth health, poisoneth the breath, destroyeth natural heat, brings a man's stomach to an artificial heat, deformeth the face, rotteth the teeth, and, to conclude, maketh a man contemptible, soon old, and despised of all wise and worthy men; hated in thy servants, in thyself, and companions; for it is a bewitching and infectious vice" (also quoted French 1884:151).

INEBRIETY PREVALENCE. Widespread deaths in Cranbrook, Kent, are diagnosed in the parish register as divine judgment for the town's sins, especially "that vice of drunkenness which did abound here" (quoted Thomas 1971:86).

INEBRIETY OF MARLOWE. Playwright Christopher Marlowe dies in a tavern brawl while in the middle of a drinking bout.

CHURCH-ALES BANNED. With Queen Elizabeth's support, justices assembled at Bridgewater order the total suppression of church-ales in 1596. It is more than a century, however, before they finally come to an end. Of all the money-making ale feasts, church-ales seem to survive the longest (see 1630/England) (Monckton 1966:108-109; Samuelson 1878:136).

DRINK RATIONS. A gallon of beer per day per man is rationed in the royal navy (Sutherland 1969:16). Soldiers appear to receive two-thirds of a gallon. The Bury House of Corrections issues one pint (Minchinton 1976:128-129).

FRANCE

BACKGROUND DEVELOPMENTS: BOURBON DYNASTY. Following the assassination of Henry III, the last Valois king, Henry of Navarre becomes King Henry IV (1589-1610) and begins the Bourbon line. With the Edict of Nantes (1598), Henry puts an end to the religious wars that have ravaged France since 1562.

INEBRIETY CONDEMNED (LAFFEMAS). The end of the century has been identified as a period in which drinking and inebriety increased markedly. According to Dion (1959:488), it is not until the end of the 16th century that "any trace of the profound and general anxiety which the word alcoholism causes in us today" is evident in France. Much of the evidence for this assertion is derived from Barthelemy de Laffemas. One of the principal economic councillors of King Henry IV, Laffemas, as early as 1596, denounces the drunkenness that has descended upon the French people over the past 30 years or so (in *Source de plusieurs abus et monopoles qui se sont glissez et coulez sur le peuple de France depuis trente ans ou environ*). In a discussion of the problems besetting the country, he emphasizes that intoxication very often ruins households and families. This change he blames on the recent extension of the number of taverns and their indiscipline (Brennan 1981:217; Brennan 1984). In 1600, he again attributes this calamity to Henry III for having passed measures in favor of sellers of wine and outlet operators and for allowing the "good and saintly ordinances" of the past to be abandoned and to go into confusion, despite any official declaration to the contrary (in *Le quatriesme advertissement du commerce faict sur le debvoir de l'aumosne des pauvres*). He asserts that due to the passage of several laws in favor of wine sellers, especially the Paris law of 1587, great fortunes are now being made in the trade and the old laws are being ignored. A quarter of Paris is covered with wine outlets. However, the tavern restrictions are still applied in many provinces where the king's new tolerance has not yet become known (Dion 1959:488-490).

His views on drunkenness contrast markedly with those of Montaigne (see 1570/France) and may reflect the growth of mercantilistic concerns over the effect of urban drinking on the economy as much as any major increase in inebriety itself. Coming from a poor Huguenot family, Laffemas views the wealth of France as a matter of special pride, and throughout his career he complains that the great riches of the country are being squandered through bad management, laziness, and carelessness and that France is too dependent on other countries for her goods. He condemns drunkenness and laziness as not only evils in themselves, but as causes of France's widespread poverty. He recommends a severe reduction in the number of taverns and cabarets, which he believes serve no purpose but to make the poor poorer. Confirmed drunkards should be pilloried. The idle should be put to work (Cole 1939, 1:28, 32, 37-39).

GERMANY

TOASTING CONDEMNED. Tübingen professor John Sigwart attacks the drinking of toasts in 1599, complaining that "habitual topers are not satisfied with the wine which is in front of them" and fight drinking duels (Janssen 1910, 15:393-396).

DRINKING CONTESTS, INEBRIETY CONDEMNED (CAMERARIUS; ALBERTINUS). Complaints continue, particulary among preachers, about widespread inebriety and its adverse effects, including early deaths and infirmities. In a discussion "Of Drunkenness, and of the Evils that come Thereby," the jurist Philip Camerarius tells of a drinking tournament during a noble's wedding party in which the prize was won by a man who drank 18 measures of wine (the sixth part of a pipe) in a few hours. Camerarius comments: "And though such detestable and drunken trickes are not to be tolerated in a well governed State or Kingdom, (much lesse among those that are called Christian, and that it is greatly to be feared that they will draw down a sudden destruction upon Germanie) yet for all that a great many, not of the common sort, but even of the principall and cheefest of the countrie, thinks that it is a great honour to them, and they are esteemed valorous, when they have strength enough to emptie many goblets of wine" (Camerarius 1599/1621:373; see also Janssen 1910, 15:347n. 1, who gives examples of other such contests).

In De conviviis (1598), Aegidius Albertinus, ducal secretary in Catholic Bavaria and the most popular author of the Bavarian Counter-Reformation, attacks the nobility for their role in the spread of drunkenness: "Respectable sobriety has gone out of fashion everywhere and in all classes; wholesome moderation has little place; and fuddling and wine drinking has [sic] grown into necessity and habit which cannot possibly be overcome; for those who ought to punish and put a stop to it are sick with the same disease; yea verily the law-makers are the first to become law-breakers....hence it is no wonder that the subjects do likewise." The Germans, he concludes, "do nothing but eat and drink, sing and dance." Many have "drunk up all their patrimony." He also attacks toasting and health drinking and observes (in Christi Königreiche [The Kingdom of Christ], 1618) that "far more people die from overeating than through war or the sword" (quoted Janssen 1910, 15:346, 393-394, 396, 423n.2). The pessimistic Albertinus is notable for regarding all possible human weaknesses not simply as follies but as evils (Pascal 1968:103).

Lazarus von Schwendi (Ermahnung an die frommen Teutschen [Admonition to the pious Germans]) attacks overeating and overdrinking: "Gluttony and drunkenness have grown/ To honour, and as common have become,/ As though we had but these pursuits alone./ We see thereby the German nation/ Sinking into degradation..../ Of us the maxim old is true:/ 'Drink slays far more than warriors do'" (quoted Janssen 1910, 15:423).

NOBLES/DRINKING. The theologian and preacher Cyriacus Spangenberg (Adelsspiegel [Mirror of Aristocracy], 1594) complains that "so much eating and drinking" was occurring that it seems "as if people were deliberately bent on stifling and destorying nature." Among other criticisms of the nobility, including their oppression of the peasants, he cites their drunkenness: "The majority of nobles are addicted to drink. They often have to sell or mortgage a mill, an ale-house, a pond, a carriage, often even a whole village in order to get enough liquor to drown themselves in. And they are not satisfied with drinking themselves to their hearts' content, but they compel others, often with curses, to drink with them interminably." He also complains of their cupidity and extortion in the sale of beer and wine, how they turn themselves into merchants and shopkeepers and monopolize local trade. This is attributed to their meaningless life: "The young nobles have no other occupation than sleeping til high noon, and the rest of the day loafing about idly, flirting with the women, or playing with the dogs, and then drinking half through the night; next to this, all their thoughts are taken up with idiotic dressing and adornment" (quoted Janssen 1910, 15:332-333, 342-343, 346, 351).

The author of another tract (Von den vielen Anzeichen so uns den nahe be vortschen den Schrelichen jüngsten Tag verkündigen [Of the Many Signs of the Coming of the Last Judgement], 1593) complains of the "incalculable expenses" spent on eating and drinking by the nobility while others are "hungering and starving." The life of the majority of princes, it alleges, is "made up of inordinate eating and drinking, innumerable and lengthy festivities, pomp and luxury of dress and adornment." This has been "taken as a model by nobles, burgher and peasants, so that, as was plain to all beholders, the one sought to outstrip the other" (quoted Janssen 1910, 15:262-263, 332). (See also the comments of Albertinus above.)

DISTILLING CONTROLS, TAXATION. In 1595, the Berlin town council begins to tax brandy distillers. In 1604, the council of Frankfurt/Oder decrees that the 80 stills in the town be reduced to 14 (Janssen 1910, 15:419-420).

ITALY

BRANDY TARIFF. "Acquavite" appears in the customs tariffs of Venice in 1596 (Braudel 1981:243).

SPAIN

TEMPERANCE CITED. Barceloneans are known for their sobriety. It is said that in the dives of the port cities one encounters 100 prostitutes, but not a single drunkard (Le Roy Ladurie 1974:64).

The Seventeenth Century

1600

EUROPE

BACKGROUND DEVELOPMENTS. All of Europe must contend with economic crisis, bitter class wars, and a strengthening of state power to curb vagabondage and to keep the poor in their place. All states are forced to deal with the problem of poverty, which has become too vast for private charity administered by churches. Although poverty is not new, the problem is augmented in the course of the 17th century by population growth, price inflation, the substitution of complete seasonal unemployment for medieval underemployment, the tendency of great households to deal with inflation by cutting down on staffs, and the concentration of paupers in large cities by manufacturing (Hill 1964:272-279).

DISTILLING: EXPANSION; ECONOMIC VALUE. Stagnation in the evolution of distilling apparatus sets in, and the number of good technical books declines, but technological applications of distillation expand and the distilling trade is one of the most prosperous of the period. Distilling especially flourishes in Holland (see below). In the early century, most distillers' guilds are founded. By the late century, the domestic or private distiller has typically been replaced by the commercial manufacturer, though this commercial operation may be hardly more than an organization of many home distillers. There is as yet no standard of quality for spirits as there is for beer (Forbes 1970:107, 185-188).

Distilling is integrated into the rural economy. Not only does it utilize surplus corn but one of its by-products is mash or dregs, which constitute an indispensible supplement to winder fodder of cattle, producing a higher milk yield (Glamann 1977:201-202).

SPIRITS CONSUMPTION. Dutch merchants and sailors play an important role in the promotion of spirits consumption throughout western Europe. They promote wine distillation on the Atlantic coast and popularize all types of spirits in northern Europe. Engaged in the largest wine-trade of the period, they are the first to come to grips with the many problems posed by the transportation and preservation of wines. Trade in spirits is one of the main solutions. The addition of brandy gives even the most feeble wines new body. As they are more valuable than the same volume of wine, transportation costs are correspondingly less (Braudel 1981:232, 244).

Rum made from West Indian sugar cane becomes popular in England, Holland, and the English colonies in America. Brandy plays a more important role in northern Europe than in the Mediterranean lands. In continental Europe, brandy is made from cider (calvados), pears, plums, and cherries (kirsch). There is still considerable resistance to grain spirits, but expanding cereal production and trade in north Europe gradually overcome this antagonism. In the mid-century, gin is developed in Holland.

REVENUES. As part of the attempt to deal with the problem of rising prices and expenditures common to all western European governments at this time, many authorities begin to turn to alcohol (and soon to tobacco and coffee) a source for needed revenues. For example, King James I of England attempts to establish an alehouse patent in 1606 and a malt tax in 1614.

MEDICINAL USE AND INTOXICATION. Wine, both medicated and pure, has one of its greatest periods of medical popularity in Europe (Lucia 1963a:8). Seventeenth-century doctors still advise everyone to get periodically drunk for the sake of health: on three occasions (1643, 1658, 1665), this is the subject of prize-winning essays at the University of Paris. Each winner advises getting drunk at least twice a month "to strengthen the gastric juice" (Forbes 1970:187-188). (See also 1660/Europe, 1680/England.)

CLERICS/DRINKING. During the century, some monks addicted to wine are required to drink wine from antimony goblets, which causes nausea, in the hope of creating an aversion to the drink (Peters 1906:140-141).

DENMARK

BEER CONSUMPTION. The drinking of beer, especially from Germany, is enormous among the entire population. It is estimated that in the 16th century "giants could manage 20 potter or quarts a day." Four quarts a day seems to have been usual for adult laborers and sailors (Glamann 1962:129). The visit of King Christian IV to the court of James I of England (his brother-in-law) in 1606 is the occasion for a notorious drunken feast (see 1600/England) (French 1884:158).

ENGLAND

BACKGROUND DEVELOPMENTS. James I (1603-1625) becomes the first of the Stuart kings of England. Real wages reach their lowest point during his reign and taxation falls heavily on the poor. Concerns grow even greater that widespread vagrancy might lead to social disorder, especially since England (and European nations in general) has no police force. Thus, Stuart authorities aim at stabilizing the existing class structure and preventing disorder (Hill 1961:24-29; Walzer 1971:313).

INEBRIETY PREVALENCE. Under James I, the trend toward increased drinking seems to continue among all classes. Attacks on inebriety are directed at both the court and the lower classes, particularly vagrants and idlers. The literature and laws of the period "all reveal that drinking [is] a major factor in the leisure time of an energetic and creative peoples" (Williams 1969:407). Debates over whether drunkenness is a particular vice of the English are common. For example, in Shakespeare's *Othello* (act 2, scene 3), Iago says he learned a drinking song in England "where (indeed) they are most potent in potting. Your Dane, your Germans, and your swag-bellied Hollander,--Drink, oh!--are nothing to your English....Why he [your Englishman] drinks you with facility your Dane dead drunk; he sweats not to overthrow your Almain, he gives your Hollander a vomit ere the next bottle can be filled." In some literary works, drinking is treated with candor and humor, and without moralizing. But more and more writers and clerics attack the vice, with the haranguing of tipplers climaxing in the 1630s (Williams 1969:477).

LOWER CLASSES/DRINKING. Government declarations against drunkenness repeatedly attribute it especially to the "worst and inferiour people" (quoted Clark 1983:109). "Drink for many [of the poor] must have served primarily as an anaesthetic against a harsh, oppressive world" (ibid., p. 111). The lower orders are the primary alehouse patrons, and on the whole the evidence suggests several factors that are contributing to an increase in drunkenness among them (see below).

UPPER CLASSES, KING/DRINKING. Although the concerns about drunkenness are primarily directed at the lower classes, inebriety is as common, if not more so, among the nobility and gentry. Indeed, according to Thomas (1971:19), it is in the 17th century that the lord replaces the beggar as proverbially the drunkest member of the community. Viscount Conway summarizes the prerogatives of many country squires of the period as follows: "We eat and drink and rise up to play and this is to live like a gentleman, for what is a gentleman but his pleasure" (quoted Walzer 1971:252). However, King James himself is abstemious in drinking. The one known instance of excessive drinking at the court occurs during the visit of the King Christian IV of Denmark (Simon 1909, 3:16). As Sir John Harrington describes the court visit: "Those whom I never could get to taste good liquor now follow the fashion and wallow in beastly delights. The ladies abandon sobriety, and are seen to roll about in intoxication. In good sooth, the Parliament did kindly to provide his Majestie so seasonably with money, for there hath been no lack of good living; shews, sights, and banquetings from morn to eve....I neer did see such lack of good order, discretion, and sobriety, as I have now done....I do often say (but not aloud) that the Danes have again conquered the Britains, for I see no man, or woman either, that can now command himself or herself" (quoted Juniper 1933:152-153; Samuelson 1878:149).

DRINKING INFLUENCES. Although William Camden and Thomas Nashe claimed that drunkenness had increased in the late-16th century because of the influence of the Low Countries (see 1580/England, 1550/England), several other possible contributing factors have been identified. These include (1) the spread of the commercial brewing of hopped beers, which are better tasting and stronger than traditional ale, and also cheaper; (2) increased imports of stronger wines; (3) the growing numbers of alehouses and the expansion of their role as community centers; (4) the fact that people are spending more time in alehouses due to rising unemployment and the expansion of services alehouses provide for the lower orders; (5) the turn to alcohol as a cheap dietary supplement, which also enables the growing numbers of the poor to forget their physical and mental discomforts; lower food intake in this period also means that men can get drunk quickly; and (6) the "disruption of traditional communal sanctions as a result of economic and social change," which results in a "shift away from conventional drinking at home towards alehouse drinking" (Clark 1978:59). Protestant and Puritan efforts to purge the church of its role as a popular community center may also aggravate the drink problem since the people turn to the alehouse as a replacement. One result of this may be an increase in toasting and pledging as traditional neighborliness is transmitted into a new-style drinking camaraderie and as elaborate toasting rituals replace simple pledges (ibid., p. 61).

DRINKING CONCERNS. Although the evidence suggests that drunkenness has increased, several developments in society also independently increase the concerns over drinking. Most notably, under the new English Poor Law of 1601 (which remains in effect with amendments until 1834) support of the poor falls primarily on property-owners of the parish. This focuses greater attention on the socioeconomic effects of drinking. These concerns are reinforced by the rise of Puritanism and middle-class business values.

PURITANS. "Puritan" is a name given to those who want to "purify" the Anglican Church of all its Catholic practices, but apart from this they hold a multitude of quite divergent opinions. They are heavily influenced by Covenant theology and Calvinist discipline. English Puritans, with their firm conviction that God is served by honest labor for the good of the community, make diligence and the avoidance of idleness, and all social evils caused by idleness such as drunkenness, moral issues upon which one's very salvation depends. The cardinal virtues of 17th-century Puritans are self-denial (discipline), sobriety, and frugality. Puritan treatises repeatedly emphasize the calamities that overcome drunkards because of their unthriftiness (e.g., see the treatises of Richard Rawlidge under 1620 and Richard Younge under 1630). Reflecting the Puritan concern over idleness is the appearance of A Treatise of the Vocations (1603), in which William Perkins summarizes current ideas about the duty of all men to labor in their personal calling for success in both this world and the next (Wright 1935:171).

As in the case of Luther and Calvin, the attitudes of English Puritans toward drinking do not differ substantially from those of the medieval Catholic Church. Drunkenness is considered a moral offense, but alcohol itself is believed to have been created by God for man's benefit and enjoyment in moderate use. This is evident in the attitudes of John Downame (see 1610/England). During the 1620s, the disapproval of the Puritan faction seems to be directed more furiously against masques, dancing, and play acting than against indulgence in drink (Younger 1966:305). While Puritans loudly denounce drinking on Sundays, this is not so much an attack on drinking per se than it is one part of a larger Sabbatarian effort to preserve the Lord's day as a regular day of rest and meditation during which neither work nor recreational activities should interfere (Hill 1964:146, 165). Under the Puritan Commonwealth (see 1650/England), however, a greater effort is made to enforce existing laws against drunkenness, Sunday drinking, and alehouses. A focus on deterrence rather than cure and a lack of sympathy for the drunkard are among the features which differentiate the 17th-century Puritans from 19th-century temperance advocates (Harrison 1971:115).

ANGLICANS. While much of the concern over drunkenness is expressed in Puritan writings, Archbishop Laud and the Anglican Church condemn it as well, especially tippling during the hours of divine service (Wright 1935:193-194).

CONCERNS: ALEHOUSES. Complaints over alehouses reach a peak in the first half of the century, particularly among the upper and middle classes and among Puritans. They are blamed for increasing drunkenness and attacked as a major source of danger to public order. Partly these complaints are due to rising numbers. Despite the controls established in 1552, alehouse licensing is still lax and the number of alehouses has grown to an estimated 14,000 (Monckton 1966:112). (See 1550/England for a discussion of the reasons for this increase.) In some areas as many as half of the alehouses may still be unlicensed (Clark 1978:49). The complaints are also "fuelled by a fear, sometimes bordering on the hysterical, that alehouses were being transformed into the strongholds of a populist world which aimed to overthrow established, respectable society" (Clark 1983:145, who observes [p. 159] that fears about the alehouses as a revolutionary centers were exaggerated). Because England is a country without a professional police force, depending for the maintenance of order upon the old system of watch and ward, the alehouses present a serious problem. Furthermore, although the numbers of vagrants and alehouses are both symptoms of broad social changes, the government tends to view the alehouses as the source of idleness (Hudson 1933:6-7).

Roberts (1980:46) comments that "the most striking feature of the seventeenth-century alehouse was its capacity to provoke sustained atacks from the political nation; and yet the attitude of the 'government,' broadly conceived, in London and in the localities of southern England and elsewhere, was never less than ambivalent." This ambivalence is due to questions about the degree of central control the government should exercise, and even its capabilities in terms of finances and manpower to excercise central control. It also is due to desires to explore the alehouse as a source of revenue, which critics view as an indefensible encroachment on local privileges. Thus an "uneasy tolerance" of the alehouse alternates with "outbursts of vigorous suppression."

While the proliferation of alehouses undoubtedly contributes to an increase in drinking, the alehouses are probably not as great a threat to public order as many contemporaries allege. Most criminal activity related to them is "amateur, small-scale, and sporadic" (Clark 1978:60). Nor are they centers of political or religious radicalism. However, they do serve as a focus and symbol of many traditional values which are in conflict with rising Puritan concerns and the attitudes of the commercial sector (ibid., pp. 65-68).

On the reputation and function of alehouses in Tudor and early Stuart England, Bretherton (1931:199-200) writes: "[I]t was during this period that the habit [drunkenness] gained its first real hold upon the nation...and the village alehouse became the rallying point not only of the tippler and the drunkard but...of all the forces which were consciously or unconsciously anti-social in their action. But much of the character of disreputability which gathered round the alehouse and its frequenters was the result of the actions of the authorities themselves. Because drinking and gaming were placed by the law under the same shadow as robbery and theft, honest persons whose pleasures harmed few but themselves seemed to be forced into a furtive association with the scum and rascality of the nation...; while the failure to enforce by law an unpopular standard of private conduct tended to bring the law and legal authorities themselves into contempt."

<u>ALEHOUSE PATRONS</u>. Although alehouse patrons have always been predominately lower class, as the century progresses the "better sort" increasingly tend to withdraw from alehouse sociability (Wrightson 1981:7). Patrons are generally a "mixed bag of characters," but regular patrons are principally craftsmen, laborers, and servants, with a regular contingent of vagrants and laboring poor, with quite a few close to destitution. The largest single group may be young people who have finished craft training but have not yet set up their own businesses (Clark 1983:127-129, 132).

<u>CONTROLS: DRINKING, ALEHOUSES</u>. In the first years of James I's reign, concerns over alehouses and drunkenness lead to regulation of both. Early Stuart drink legislation has among its objectives to (1) prevent waste of grain; (2) prevent the poor from squandering their limited money on drink and becoming a burden on property owners; and (3) further the maintenance of good order. The laws are rooted in the belief that the king as *parens patriae* has the duty to care for the nourishment, education, and moral upbringing of his "children." This belief is given added reinforcement by a strongly Puritan Parliament with a "lively sense of responsibility for the moral welfare of others." "Thus circumstances combined with the spirit of the time to urge upon all branches of the government a greater effort to regulate the manufacture and consumption of beer, and to give to that effort a distinctly social purpose" (Hudson 1933:5-7).

ALEHOUSE CONTROLS (1604). Parliament passes an act in 1604 "to restrain the inordinate Haunting and Tippling in Inns, Ale-houses, and other Victualling-houses" because their purpose is to be rest places for travelers, not places for entertainment and the "Harbouring of lewd and idle People to spend and consume their Money and Time in lewd and drunken Manner." The act establishes a ten shilling fine on proprietors for not keeping order, with imprisonment for nonpayment. The act implements the earlier licensing act of 1552, which required a common alehouse owner to post bond binding him to maintain good order (Baird 1944b:143-144).

INEBRIETY CRIMINALIZED (1606). In 1606, the first act aimed directly against drunkenness is passed because "the loathsome and odious Sin of Drunkenness is of late grown into common Use within this Realm, being the Root and Foundation of many other enormous Sins." The act imposes a fine of 3s4d upon the tippler and 5s upon whoever should be found drunk; those unable to pay are placed in the stocks--tipplers for four hours; drunkards for six. Second offenders are required to post bond of two sureties of 10 pounds conditioned upon good behavior. The act also attempts to conserve food by setting the price of beer so low that brewers who comply with the law have to use as little malt (made from barley) as possible in order to make a profit (Longmate 1968:7; Hudson 1933:9; Baird 1944b:143, 146). The act remains in effect until 1872.

DRINKER LIABILITY. The courts modify the 16th-century rule that drunkenness does not excuse a crime by exculpating habitual drunkards if their mental deterioration has progressed so far as to amount to "insanity" (Bonnie 1975:27).

CHURCH ATTITUDES. The dean of Christ Church (Henry Aldrich) composes a "bibulous epigram" reflecting the acceptance of drinking by the Anglican church: "If all be true that I do think,/ There are five reasons we should drink;/ Good wine; a friend; or being dry;/ Or lest we should be by-and-by;/ Or any other reason why" (quoted Ellis 1956:12).

KING'S ATTITUDES: INEBRIETY, ALCOHOL REVENUES. King James I himself probably is averse to the drink trade. He is a moderate drinker (see above) and in his Counterblaste to Tobacco (1604), he describes smoking as a "branche of the sinne of drunkennesse, which is the roote of all sinnes." In rejecting the belief in the medicinal properties of tobacco, James observes: "And so doe olde drunkards thinke they prolong their days, by their swinelike diet, but never remember howe many die drowned in drinke before they be halfe old." Furthermore, many habitual smokers cannot give it up, "no more than an olde drunkards can abide to be long sober" (James I 1604/1869:106, 108-109). In 1616, he attacks the proliferation of alehouses, and during his reign legislation is passed establishing stricter controls over alehouses, taverns, and drunkenness. But his aversion to alcohol does not prevent him from utilizing it for revenues with the alehouse patent of 1606 and the malt tax of 1614.

REVENUES: ALEHOUSE PATENT (1606). James I, in order to increase government revenues, grants a patent to collect from alehouse keepers all fines and forfeitures of their recognizances and the right to compound with and pardon offenders. The introduction of this patent creates an uproar, and the matter is taken to the Commons where, in 1606, it is declared a "grievance." In 1618, a similar patent is granted, but after three years the Commons once again refuses to uphold the patent and it is brought to an end (Monckton 1966:112).

SPIRITS CONSUMPTION, PRODUCTION. Until the reign of Charles I (1625-1649), there is no control over domestic distilling; anyone is free to distill anything they please. This appears to result in the production of "much inferior and injurious spirit" (Simon 1948:135-136). Throughout the century, the most commonly available and consumed spirit is French brandy, which is imported in ever-rising quantities in spite of intervals during which all trade with France is prohibited. However, until around 1670 brandy drinking remains limited, usually is consumed for medicinal purposes, and causes little concern.

SACK CONSUMPTION. Sack consumption (see 1500/England) has risen considerably. When Drake raids Cadiz in 1578, he brings back 3,000 butts of Jerez wine, probably glutting the English market with cheap sack. After peace with Spain in 1604, it becomes even more available. In an ordinance for the royal household, James I comments on the rising popularity of sack wines. The text indicates that sack was consumed for medicinal purposes before recreational use began and that one of the complaints against its consumption was the loss of revenue to Spain to pay for its import: "Whereas in the times past Spanish wines, called Sacke, were little or no whit used in our Court, and in late years though not of ordinary allowance, it was thought convenient that such noblemen and women, and other of account, as had diet in the court, upon their necessities of sickness or otherwise, might have a bowle or glass of Sacke....We understand that within these late years it is used as a common drinke and served at meals as an ordinary to every mean officer, contrary to all order, using it rather for wantonnesse and surfeitting than for necessity to a great wasteful expense" (quoted Allen 1961:172; see also French 1884:170).

BREWING. Over the century, many of the great breweries of the future are formed. As product demand increases with population growth, breweries expand in size to meet domestic and overseas markets (Monckton 1966:111).

ATTITUDES (SHAKESPEARE). The growth of concerns over drinking is reflected in the plays of William Shakespeare (1564-1616). In his late-16th-century histories and comedies, Shakespeare treats drinking humorously and often without any moralizing. This treatment is most evident in the character of Falstaff. In general, Falstaff reflects the Renaissance view that wine is a good creature, a pleasant companion on all convivial occasions (Simon 1948:46; Legouis 1926:1). Falstaff is often "full of indulgence for the Bacchic spirit." Drinking appears hearty and inviting. No wickedness or tragedy results from it; he is never seen "the worse for it," but neither is wine a source of wisdom or virtue (Legouis 1926:8, 13). Falstaff says in Henry IV (part 2, act 4, scene 3): "If I had a thousand sons, the/ first human principle I would teach them should/ be, to foreswear thin potations and to addict themselves to sack." Falstaff speaks of abstinence from wine with scorn: "Nor a man cannot make him laugh; but that's no marvel; he drinks no wine." Yet even in the character of Falstaff, the great lover of sack, Shakespeare's concerns over alcohol emerge. As Prince Hal matures, he rejects the Falstaffian lifestyle, symbolizing instinct, emotions, and pleasure, in order to assume the responsibilities of organized society (Carter 1976:6-8).

In Shakespeare's later tragedies after 1600, most of his allusions to drinking are negative, and his comments on drunkenness are "bitter and moral"

(Legouis 1926:14). He illustrates the danger and shame of inebriety (Williams 1969:194-286). In Othello, drunkenness is the downfall of Cassio, and it is the villain Iago who sings its praises (see above). Hamlet calls it the curse of Denmark (act 1, scene 4).

INEBRIETY OF BEN JONSON. Among the notorious drinkers of the period is Ben Jonson (1573-1637) (French 1884:165-168). A contemporary says that "drink was one of the elements in which he liveth": "He would many times exceed in drink...then he would tumble home to bed; and when he had thoroughly perspired, then to studie." Jonson himself writes: "Truth itself doth flow in wine..../ Wine is the milk of Venus..../ Tis the true Phoebian liquor,/ Cheers the brains, makes wit the quicker" (quoted Legouis 1926:8-9). He "seems to have spent more time in taverns than out of them"; he "was the firmest of all believers in the stimulus of drink and gloried in his intemperance" (Sutherland 1969:162). In his writings, the alehouse appears as "the trysting-place of an underworld populated by gulls and vagabonds, robbers and whores, a world which though parasitical is also a mirror image of the moral sham, the trickery and hypocrisy of respectable society" (Clark 1978:48).

FRANCE

INEBRIETY PREVALENCE. Joseph Duchesne, doctor to King Henry IV (1589-1610), asserts in *Le pourtraict de la santé* (1606:347) that "licentiousness is so great, and gluttony and drunkenness so common and so practiced, that people make gods of their stomachs" (quoted Franklin 1889, 6:128). This supports Laffemas's assertion that inebriety has recently increased (see 1590/France). According to Magne (1944:197), drunkenness is "the approved vice of the 17th century." Hugon (1911:97) writes of the period: "One of the chief complaints levelled at the fashionable world by 17th-century moralists is that it wasted time and dissipated energy in lolling on sofas and imbibing enervating drinks. Drunkenness was so long established a vice that reformers fulminated against it more or less from habit."

This evidence almost exclusively concerns the upper classes, the literati, and town dwellers. Drunkenness is not seen as a major problem among the peasantry (see below), and inebriety itself is not a major social concern in France at any time during the century. In southwest France, there is evidence that drunkenness is aggravated over the century, that its hold on *les petits gens des villes* increases due to the spread of vineyards and distilling in Gascony and Angoumois and to the growth of taverns in city suburbs in defiance of municipal ordinances. On Saturdays and feast days, the laborers go to the taverns after mass and drink until night comes (Bercé 1974:221-223; see also 1620/France).

PEASANTS/DRINKING. Of 20 million people in France, 15 million are peasants, a huge majority of whom subsist chiefly on porridge, soup, bread, and drippings (Goubert 1970:23, 45). The "Monologue du bon vigneron" (1607), a poem familiar in Auxerre, indicates that even in the viticultural area of Burgundy wine drinking by the peasantry is still limited. In this poem, a vigneron boasts how he drinks only his own best wine and disdains water, which is only good to mix with soup, while his fellow vintners and their laborers drink sparingly of their own goods, which is *le monde du pais*. On the whole, 17th-century peasants, whether they work in the vineyards or not, drink so little wine that the countryside during this period offers no commercial market for a viticulture designed to produce wine solely for popular consumption (Dion 1959:472-473).

The majority of peasants of Beauvais suffer from almost continuous undernourishment and drink only very weak, diluted alcoholic beverages, such as apple or pear ciders or "bitter green wine," that quickly spoil and have no nutritional value (Goubert 1967:167).

UPPER CLASSES, BOURGEOISIE/DRINKING. The "Monologue du bon vigneron" (see above) describes the prestige associated with wine in wealthy urban circles and how town citizens seek to serve wine they have personally produced (Dion 1959:463). Olivier de Serres (1605) describes how the great towns are vacated by notables who go to their "rustic farms" to drink their wines (Lausanne 1970:54).

LITERATI/DRINKING. Throughout the 17th century, a literature of revelry or gluttony (de goinfres) flourishes (Magne 1944). Nearly all poets of the period write drinking songs and view the cabaret and drunkenness as the better part of their earthly happiness. In the early century, the fashion for cabaret drinking is set by a group of free-thinkers and "refractory spirits" who become known as libertines and esprits forts. They gather at the taverns of Paris, chiefly the Pomme le Pin, instead of the salons. They are characterized by epicureanism and a skeptical or indifferent attitude toward religion. They popularize satire, burlesque, and picaresque literature. The best of the poets, the persecuted Théophile de Viau (1590-1626), has been described as "the Corypohaeus of a band of young and well-born courtiers who defied all attempts to set bounds to the indulgence of their appetites" (Van Laun 1909:167). Boilleau is "the poet of the tavern" (Lyons 1982:9). Around 1625, the poet Saint-Amant (1593-1660), generally regarded as the initiator of the burlesque movement, haunts the taverns "seeking inspiration 'in the sweet song of orgies; in the glow of red, bloated faces'" (Leclant 1979:86). He appears as a sort of demi-god to all the tipplers (biberons) of the kingdom (Magne 1944:196). Claude-Emmanuel Chapelle, another great drinker, composes a song of praise to the cabaret Croix de Lorraine. He writes many of his poems while intoxicated in the numerous taverns where he exercises his "perpetual thirst" (Butler 1923:198). Playwright Paul Scarron (1610-1660) declares his love of the cabaret and praises its boisterous atmosphere (Magne 1944:193, 194). One poet writes when he is drunk: "I imagine I am a king and believe all are/ Under my law and empire./ But when I am far from a glass, I find myself/ stupid without a word to recite" (Magne 1944:197).

In an analysis of the theme of wine and food consumption in the poetry of Saint-Amant, Lyons (1982:40, 52) emphasizes that such consumption is significant as a social ceremony that unites the poet and the listener. Saint-Amant glorifies debauchery and waste in defiance of middle-class respect for moderation and social restraint.

FESTIVE DRINKING. In the towns of the southwest, complaints are raised that on Sundays and feast days laborers leave mass and go to an inn to get drunk all day, not leaving until night. Town festivities are the occasion of extraordinary drinking bouts. Wine is an accompaniment of all urban revolts and disturbances (Bercé 1974:222-223).

CABARET FUNCTIONS. Although more is known about English taverns and alehouses, the French cabaret probably serves similar functions (Burke 1978:110-111). Until after the mid-century, all classes of people go to the cabaret to have fun and socialize; but only the poor go to the tavern. By 1680, it is stated that only the menu peuple frequent the taverns (Dion 1959:485). The church and the cabaret are the twin poles of village life. People meet at the cabarets not only to drink but to play games, talk, conduct business, and find distraction. Cabaretiers are small-scale usurers and also "an essential figure in popular culture, a center of information...an organiser of collective rejoicing" (Bercé 1974:222, 297). Although lacking in comfort and decor, the cabarets appeal to the biberons, who gather there as places of "furious gaiety," warmth, comaraderie, free-thinking, and escape from the hardships of life (Magne 1944:192-193). The cabaret becomes the center of the social and intellectual life of the city. Moliere, La Fontaine, Racine, Chapelle, Boileau, and other writers of independent spirit go there to drink wine and chat (Leclant 1979:86).

CABARET, TAVERN CONTROLS. The law restricting the use of cabarets and taverns to visitors to a city is repeated in 1597 and 1635, and still exists at the beginning of the 18th century; but from at least the 17th century, it is no longer enforced and police efforts are directed more toward limiting hours of service than excluding local clientele entirely. Until 1666, closing hours are 7:00 p.m. in winter and 8:00 p.m. in summer; after 1666, they are 6:00 p.m. winter and 9:00 p.m. summer (Brennan 1981:74)

SALES CONTROLS. The royal government and the city of Paris attempt throughout the 17th and 18th centuries to erect a fiscal barrier which will make it practically impossible to sell lower-priced wines to workers in the city (Dion 1959:499).

TAXATION. At the beginning of the century, the duc de Sully undertakes the first effort to rationalize the collection of the aides. The aides are internal indirect taxes dating back to the Middle Ages which, in the 17th century, are levied especially on wine. By the end of the ancien régime, they are the most valuable of all indirect taxes. The most important and hated of all the aides are the sales taxes, fees, and taxes levied on the wine trade and later all other alcoholic beverages (Goubert 1970:41). The system as it develops over the century, particularly under the financial reforms of Colbert in the 1660s, has the following features. Both wholesale and retail transactions are taxed, and wine-dealers and innkeepers have to pay an annual fee. In addition, tolls are collected at various points, including the entry of goods into towns and cities. These are levied and collected only in the pays d'aides, a region roughly coinciding with the 14 royal generalities. Included in the aides are the entrées or octrois, the town duties that must be paid for wine or other commodities to enter a city. (Entrées is the word used when a combination of aides and octrois is collected on the goods entering a city.) The basic retail sales tax on wine levied in the pays d'aides are the quatrième and the huitième. Both are subject to a sous pour livre surtax. In four generalities (Paris, Amiens, Chalons, and Soissons) called the pays de gros, there is also levied a droit de gros, a wholesale tax consisting of an ad valorem 5% tax upon the sales prices of each muid of wine sold or resold at wholesale. An additional fixed augmentation tax is placed on each muid of wine, regardless of sale price. Other areas only pay the gros.

In levying these taxes, distinctions are made between vin d'achat and vin de cru. The former is commercial wine bought and sold in open market by regular vintners. The latter is wine produced on the estates of privileged groups and consumed or sold by the owner of the establishment. In general, a bourgeois of a city is privileged to import his vin de cru free in whole or part of the gros and the augmentation (see 1630 for Paris). Such wine is presumably for personal use. In some cities (Orleans, Lyons) a bourgeois need not pay either tax. Everywhere the bourgeois is permitted to sell vin de cru from his own house at retail free of some part of the huitième. However, nowhere is there a clear definition of what constitutes the bourgeoisie, and the General Farms attempt to restrict the legal membership; nor is there a precise definition of vin de cru. The clergy and the nobility also are exempt from paying these taxes on their own wines when making wholesale sales.

Evasion of the system is common, but not to the extent of other taxes. The most common offense becomes selling without a license (muchepot), but the penalty is only the payment of a 100 livres fine and confiscation of the wine in question. Smuggling is the chief problem facing the administration of the entrées. All levels of the population engage in using every conceivable trick to avoid tax payment, especially as the sous pour livre surtax increases after 1705 (Matthews 1958:145-167).

A distinction is also drawn between wine sold à pot at establishments offering no other fare and à assiette where food is served. The latter pay an additional livre in the huitième. In general, cabarets are considered the former; taverns and inns, the latter. Wine selling licenses are called annuels. Separate licensing is required for each category of merchandising (wholesale or retail) and for each beverage type sold.

WINE, SPIRITS TRADE. Beginning around the mid-16th century, the Atlantic coast undergoes "what might be called a peaceful Dutch invasion" (Bizière 1972:126). This invasion accelerates after the Dutch declare their independence. Although the Dutch come to southern Europe primarily in search of salt, wine becomes the next most important article of trade. Unlike the medieval Flemish traders, the Dutch establish residence or representatives in the French ports, causing the wine trade to expand greatly, especially after the wars of independence end. Dutch merchants settle in the towns, monopolize local trade, subsidize vinegrowers, and even become landlords themselves. This trade influences viticulture in several ways. First, it promotes its spread in the southern regions where the climate is most suitable, and particularly in those costal regions with heavy-yielding vineyards such as those of Aunis (coastal Charante), whose weak, white wines were consumed only locally prior to Dutch arrival. Aunis wines become an important mainstay of the Dutch trade. The Dutch are more interested in quantity than quality since they prefer strong, sweet wines and have to fortify them with spirits to both strengthen and preserve them. Second, the Dutch trade forces down the production of more northern vineyards. Third, inland vineyards (Burgundy, Orléans) are compelled to produce better wines if they are to survive. Fourth, in the regions of Dutch trade, the distillation of brandy increasingly utilizes a large proportion of wine production in order to fortify and preserve the wines for shipment and save shipping space (Bizière 1972:126-127; Dion 1959:431).

SPIRITS CONSUMPTION, PRODUCTION, CHARACTERISTICS. Brandy is still largely a medicine, but in certain regions it already has become popular as a beverage (Delamain 1935:17), and Sully complains about the money being spent on feasts and luxuries such as liqueurs (Le Grand d'Aussy 1782, 3:76). In the course of the century, however, consumption and production increase. It is from the second half of the century, especially its later years, that the spread of spirit drinking in France is dated.

Brandy production particularly spreads in areas under the influence of the Dutch wine trade (see above); distilling shifts from a craft of the apothecaries utilizing only a small part of the vignoble to a peasant occupation utilizing a considerable part of it. In 1604, Bergerac sends for "certain Flemish men in order to make a great quantity of brandy and to these ends to build their ovens." In 1619, the East India Company purchases some coniak in London for their ships. The bourgeois of La Rochelle complain in 1624 that their market is being invaded by the inferior products of viticulture: brandy, vinegar, and spoiled wines. It appears that already in the first quarter of the century brandy production is important in Aunis (Dion 1959:427, 440, 443-435). (On cognac distilling in Charente, see 1630/France.)

In his book on agriculture for the small French farmer Olivier de Serres (1605) gives recipes for the distillation of plants and herbs but does not mention brandy, indicating that at least in Languedoc it is still considered a product of the apothecary or distiller, not the vigneron (Forbes 1970:145).

As elsewhere in Europe, the 17th century in France is a period of experimentation to improve the taste of spirits, but the quality of liqueurs is still less important than quantity and cheapness. Apparatus are modified to increase the speed of the process, and methods are developed to hide bad tastes under pleasant perfumes, an art which comes from Italy (Delamain 1935:27-28).

VITICULTURE; WINE PRODUCTION. The rise in wine consumption from the reign of Henry IV (1589-1610), especially in the large cities, makes it more difficult to prevent the progress of popular viticulture, which is menacing to the privileged classes (Dion 1959:491).

Regional specialization in viticulture is slowly spreading. In 17th-century Burgundy, there are still just 11 places where it engages the whole population. Wine, like corn, continues to be largely produced and consumed on the spot, even in regions of poor soil and climate, where a season good enough to yield a vintage produces "at best a miserable rot-gut." Wine production continues in Normandy and Flanders until the 16th century. This conservatism is due to uncertainty of communications and to local demand for wine as a common beverage and in liturgy (Bloch 1966:27).

ESTEEM OF WINE. De Serres (1605) elaborates on the earlier attitudes of Chretien Estienne (see 1570/France) on the running of a 16th-century manor house, but gives more complete advice on the cultivation of vineyards, "since it is necessary to drink in order to live, and since wine is the most general and salutary of beverages." He recommends two or three cellars to hold wine (Wiley 1954:123).

CHARTREUSE DEVELOPED. The formula for a medicinal liqueur that becomes green Chartreuse--which is made from 130 different herbs, plants, and spices--is given to the Carthusian monks at the monastery of La Grande Chartreuse by an unknown donor about 1605. According to various accounts, the donor is either a French alchemist or the Marshal d'Estrées. However, little interest is shown in the formula for 150 years (Garfinkel 1978; Emerson 1908, 2:173; Seward 1979:155-157).

GERMANY

BACKGROUND DEVELOPMENTS. The fundamental event in German life during the first half of the century is the Thirty Years War (1618-1638), which is fought almost wholly on the territories of the Holy Roman Empire. A whole generation of Germans grows up knowing nothing but fighting. During the course of the war, some areas suffer a population loss as high as 70%, although other areas are hardly touched by the war. Overall, it appears that the German territories as a whole lose about 40% of its rural and 33% of its urban population. By the Treaty of Westphalia, which ends the war, Germany is composed of over 300 states which owe only nominal allegiance to the Holy Roman Emperor. German prosperity declines, and the area remains poor until the 19th century. The landed nobility manages to keep its privileged status at the expense of the educated and the prosperous burghers, who are rising in importance in other European nations. The 17th century is a period of decay for the commercial classes and towns. The social structure becomes dominated by the landowning class and a depressed peasantry. "The essential feature of German life from the late seventeenth century to [at least] the age of Napoleon...is a turning in on itself, an excessive preoccupation with local, even parochial, concerns" (Sagarra 1977:3-7, 13, 35, quote at 13).

INEBRIETY PREVALENCE. During the century, concerns over inebriety seem to decline markedly in Germany. Hauffen (1889:490) argues that the specialized drinking literature disappears and drinking once again becomes just another vice which is addressed in the more general moral tracts because the problem has dissipated. But it is unclear whether consumption itself declines. Little research has been done on drinking in this period. Jacob (1935:91-92) asserts that a high level of drinking continued because of the desire of the people to drown their sorrows and because of the late introduction of coffee into Germany.

COURTS/DRINKING. State budgets sustain a staggering round of courtly social activities designed to emphasize the might and magnificence of the prince. Elaborate festivities gradually become an end in themselves, "distracting the courtiers from their artificial world, where the enemy was boredom" (Sagarra 1977:30). At the banquet concluding the peace negotiations with the Swedes in 1648, red and white wine flow from the mouth of a stone lion for six hours. Many festivals are characterized by an underlying barbarity; many are drinking and feasting orgies (ibid., p. 33).

TAVERN DRINKING. An Augsburg preacher named M. Volcius laments that the inns and alehouses are day and night "full of people who do nothing but gorge and drink and swill, with such screaming and riotous behaviour that respectable people passing along in the streets have enough of it....Poor peple who have nothing behind or before them spend their days in drinking. Before the kreuzer has been fully earned, it has already been drunk away in the alehouse" (quoted Janssen 1910, 16:127-128).

TEMPERANCE ORDER. An Order of Temperance is founded at Hesse in 1600, restricting its members to seven glasses of wine at a meal not more than twice a day. Members also have to pledge to abstain from "full guzzling" for two years after their initiation (Krücke 1909:26).

HEALTH CONCERNS. Like Sebastian Franck before (see 1530/Germany), numerous preachers at the turn of century complain of the illness and early deaths caused by excessive drinking (Janssen 1910, 15:421-422).

DRINKING CONTROLS. Duke Maximilian of Bavaria promotes the Catholic Reformation by prohibiting excessive eating and drinking, as well as masquerades, short dresses, magic, and "shameful" language at weddings (Burke 1978:221).

IRELAND

INEBRIETY PREVALENCE. Throughout the early modern period, frequent references are made to drunkenness as a national weakness, although there is some evidence that the situation improves in the late 16th century. A contemporary asserts in the early century that "drunkenness is the only curse of the country." One Father Hollywood in 1605 ascribes the introduction of the "foul habit of drinking" to English soldiers (O'Brien 1919:15; MacLysaght 1950:69).

DISTILLING (BUSHMILLS). Bushmills, the oldest commercial whiskey distillery in the world, is established under the first Crown distillery license in Bushmills, Ulster, in 1608 (Tuohy 1981:7). The English Crown's deputy in Ulster grants the license to himself.

WHISKEY CONSUMPTION. Already whiskey is being exported (O'Brien 1919:84). Queen Elizabeth is said to have fancied a drop of Irish whiskey at a time when her English subjects deplored the barbarous Scotch whisky. Sir Walter Raleigh, stopping at Ireland in 1617, is given the "supreme gift" of a 32-gallon cask of the Earl of Court's home-distilled "uisca beatha." John Marston (The Malcontent, 1604) identifies "uisca beatha" as the defining characteristic of the Irishman (Tuohy 1981:7; Murphy 1979:142).

ITALY

WINE CONSUMPTION. Wine consumption increases in Rome throughout most of the century, chiefly due to more use by the common people, for whom it provides a cheap form of caloric compensation for a marked decline in the amount of meat and grain available, "a decrease of which they must have been the principal victims." Production in the Roman vineyards increases 50% between 1630 and 1660 and doubles from 1630 to 1710. From 1600 to 1800, the average amount of wine consumed is quite large--about 200 liters (53 gallons) per capita per year--but usually the wine is of low alcoholic content (Revel 1979:45).

NETHERLANDS

BACKGROUND DEVELOPMENTS: INDEPENDENCE. The Twelve Years' Truce (1609) in the Dutch war of independence partitions the Low Countries into the Dutch Netherlands (the seven northern provinces) and the Spanish Netherlands (the ten southern provinces). War is resumed in 1621, but in 1648 the Treaty of Westphalia confirms the independence of the United Provinces of the Netherlands.

TRADE. Since the soil in Holland produces almost nothing except dairy products, the Dutch import everything and become the great merchants and carriers of early modern Europe. They are ideally suited for this role because the country is situated at the junctures of the maritime routes linking the north and east and the south and west. Although the Dutch had already built up a large merchant fleet in the 16th century, it is in the 17th century that independence and the defeat of Spanish naval power brings about their rise to prominence. The Dutch East India Company is founded in 1602 (Zumthor 1962:277-278; Houtte 1977:195).

WINE TRADE. In 1618, the wine trade is Rotterdam's most profitable activity (Zumthor 1962:174). The trade in French wine into the Baltic expands greatly after 1580, with its greatest growth between 1623 and 1657. Most of this wine originates in Charente and is carried by Dutch merchants. Dutch traders on the south French coast take the poor, light white wines of Aunis, fortify them with spirits, syrups, or southern wines so that they will travel better, and sell them in northern Europe at enormous profit. This promotes the taste for spirits and strong wines throughout Europe (Bizière 1972:126-129). (On the role of the Dutch in the spread of spirits consumption, see 1600/Europe.)

SPIRITS PRODUCTION. Stimulated by favorable trade conditions and sufficient capital, the distilling industry becomes a national asset, particularly after 1650. From 1600, Rotterdam and Weesp are operating large-scale distilleries which are very profitable; they often are combined with pig breeding since the animals feed on the waste products (Zumthor 1962:174). Around 1630, distilling begins at Schiedam (Kellenbenz 1977:538). Two general developments contribute to this expansion. First, fuel prices decline and the distillers learn how to produce their own yeast. Second, the growing Baltic corn trade provides an ample supply of grain, and this trade is wholly in Dutch hands. In 1617, the corn exchange opens in Amsterdam and becomes the financial center of the corn trade until the end of the 18th century. This expanding trade makes the production of corn spirits more feasible (Forbes 1970:104-106, 188).

BEVERAGES CONSUMED. Beer is the national drink, even though the wealthy have a taste for wine (see below). Beer is drunk at every meal and between meals. Two types exist: double (very strong) and simple. National consumption is enormous. Five-and-a-half million gallons of beer are consumed in 1600 by the customers of Haarlem taverns alone. Although most towns have a brewers' guild, beer must still be imported from England and Germany to meet consumer demand. Wine drinking increases in proportion to the consumer's social eminence. Wine constitutes one of the very few imported commodities of which large quantities are reserved for domestic use. Spirit drinking is limited to the court and the wealthy. Very little water is drunk; it is even disdained by the poor (Zumthor 1962:72-73, 174). Although the Dutch play a large role in the spirits trade in the century, little evidence exists regarding domestic spirits consumption; certainly, it does not appear to be widespread or at least cause much concern.

INEBRIETY PREVALENCE. Although the Dutch are known for their simplicity, thrift, cleanliness, modesty, and moderation, temperance in drink is not counted among their virtues (Huizinga 1968:62). In England, John Marston (*The Malcontent*, 1604) identifies drunkenness as the defining characteristic of the Dutchman (Murphy 1979:142). In general, the Dutch bourgeois distrusts anyone who drinks less than he does and considers it rude if all his guests do not get drunk. Women drink as much as men; even young girls drink regularly through the day, giving them an inflated appearance (Zumthor 1962:172-173).

DRINKING PRACTICES. What most astonishes foreigners is "the systematic, organised character" of Dutch drunkenness. The many rules and ceremonies for getting drunk repel the visiting Frenchman Théophile de Viau, who has himself a reputation as a drinker (Zumthor 1962:172). Brereton (1844:28) describes a custom in the 1630s at which single people desiring to marry go to the Delph church. If a woman likes a man, "they go out and drink, and then in their cups they treat of portions, etc., and if all things concur, suddenly married."

FESTIVE DRINKING. The daily Dutch diet is frugal, but on special occasions "incredible quantities of food and drink" are consumed in compensation for the "parsimonious household economy." These festivities help "break down momentarily the solid barriers of family exclusiveness, and thus [constitute] fundamental manifestations of social relationship," serving as "a kind of periodical renewal of communal interests." Those who are too poor to afford much food on special occasions just drink more (Zumthor 1962:172-173). St. Martin's Day (November 11) is a favorite festival in which the patron saint of innkeepers' feast day is celebrated by everyone drinking themselves stupid (Foote et al. 1972:117).

BANQUETS; BANQUET CONTROLS. Banquets are preceded by as many toasts as guests, and drinks are served in large glasses that astonish foreigners. Guild banquets are proverbial for their extravagance. One guild of surgeons forbids its colleagues to become indecently drunk at banquets and orders them to carry home anyone who falls under the table (Zumthor 1962:170-173). (See the comments of William Temple, 1670/Netherlands.)

THE KERMIS. One of the major festivals is the *kermis* (or *kermesse*), a yearly communal gathering held by every town and village, "a periodic expression of collective instincts," lasting a week or two. Begun as a church fair (*kerke messe*), it has gradually merged into a twice-yearly gathering of villagers in a "communal explosion" of joy, drinking, and marketing (Foote et al. 1972:118). Over the century, its atmosphere degenerates into a "drunken spree serving to liberate all the vulgar passions normally held in check." The festival rarely finishes without some disorder. A social cleavage develops as "respectable" (generally Calvinist bourgeois) folk become more spectators than participants (Zumthor 1962:188-190). But on the whole Calvinism has little influence on this tradition (see below).

TAVERN LIFE. At public houses, people rarely drink alone; they are always in groups, toasting and singing. Tavern employees are generally considered dishonest. Still, the entire male population and many lower-class women frequent the taverns "assiduously." Several attempts are made over the century to control the taverns (Zumthor 1962:174; see 1630/Netherlands).

CALVINIST ATTITUDES, INFLUENCES. By 1600, Calvinism has come to predominate in the Netherlands. It entered the Low Countries from France around the mid-16th century and became a powerful inspiration during the wars of liberation in the hope of establishing an independent theocracy. In the great assembly of 1651, the Reformed Calvinist Church achieves the status and power of a state church, and its strictures strongly influence life in general. Dutch Calvinism extolls "sobriety of expression and reserve," is humorless, and condemns "all exuberant forms of spontaneity." Calvinist preachers issue vehement protests against improper individual conduct and public life, such as wearing long hair and jewelry, going to the theater, dancing, excessive drinking, and, at mid-century, coffee and tea drinking. This gives Dutch society a "rather austere appearance" (Zumthor 1962:81-82).

Although the rise of Calvinism increases the criticism of popular drinking festivals, its influence is limited: "The more stern and extreme Calvinist theologians...might thunder against many popular amusements, but it was...more to mark their disapproval than because they seriously expected the authorities to take action against them. The regents would not contemplate allowing the Church to be as rigorous as in Geneva. Most Saint's days were abolished but Saint Nicholas continued to call on 6 December....The kermeeses were still held in every town and village....William III, unlikely as it might seem, was particularly fond of the one at The Hague, where even princes and magistrates might mingle with the crowd" (Haley 1972:162-163). In general, Dutch ministers are far less successful in their efforts to reclaim Sunday from traditional popular recreations than their English counterparts, causing Puritan visitors to be shocked by Dutch activities on Sundays (Hill 1964:206-209, who believes that this was because the Dutch did not feel it was as necessary to prevent people from working on Sunday).

DRINKING CONTROLS. Amsterdam possesses two institutions for the confinement of "antisocial" men and women. Husbands can have their wives incarcerated for misconduct or drunkenness. A drunkard might be banned from all taverns for three years (Zumthor 1962:252, 254).

BREWING. Haarlem and Delft are the two centers of the beer-brewing industry in Holland, but competition is increasing rapidly. By century's end, most major towns have established their own brewers. Already in Delft, where a third of the town's craftsmen were brewers in the late 16th century, 29 breweries have closed by 1600. By the mid-century, the industry has practically come to an end in Delft (Zumthor 1962:305).

RUSSIA

VODKA INTRODUCED TO SIBERIA. Vodka is introduced by Moscow to Siberian natives living to the northeast (Perm, Pechora, and Ugra) following their subjugation in the 16th and 17th centuries. These peoples have traditionally used fermented beverages only (beer and kumys [fermented mare's milk]). Their first contact with vodka is disastrous. When the Samoyeds bring their annual taxes to the frontier trading center, the Moscow official treats them to drinks, making them drunk. The Samoyed then "drink up the goods they had brought as taxes, their dog teams, all their possessions." Officials appointed to Siberian outposts are supplied with ample quantities of vodka for "native tax expenses." By the end of the 17th century, kabaks have been established in all the main centers of Siberia. Home brewing and trade in native beverages are suppressed, and tribesmen are required to buy their beverages at the kabaks (Pryzhov 1868; Efron 1955:497-498).

CONTROL EFFECTS. Saloon keepers and special saloon spies are allowed to supervise the social and home life of the people, entering houses and searching to see that there is no bootlegging, which would undermine the profits of the saloon--and the revenue of the state. Every market day, the town criers reiterate the decree that no one may keep any beverages or distill alcohol. All this is to little effect. The people continue to brew beer, distill vodka, and run secret taverns. They avoid the kabaks, which are known as places where drunkards gather (Pryzhov 1868; Efron 1955:496).

SCOTLAND

BACKGROUND DEVELOPMENTS. Upon the death of Queen Elizabeth of England, the crown passes to King James VI of Scotland, who also becomes James I of England in 1603.

INEBRIETY PREVALENCE; DRINKING PRACTICES. At the turn of the 17th century, the bard MacVurich describes how he was "twenty times drunk every day" when he visited Chief Rory MacLeod at Dunvegan Castle (Ross 1970:22-23). This is not a novel or unusual episode. In most of the early poems of Highland society, feasting and drinking appear as heroic activities mentioned in the same breath as mundane work. Convivial drinking binges occur frequently, but there is as yet no preoccupation with the nature or psychological effects of drink, except for references to quality and purity. There are no detailed descriptions of what actually constitutes drunkenness. All these marks of true drinking songs seem absent from Gaelic tradition earlier than the 18th century (Ross 1970:27-28).

IMPORT CONTROLS: STATUTES OF ICOLMKILL. The Statutes of Icolmkill (Iona) are accepted under extreme duress by the chiefs of the Western Isles in 1609. These statutes are concerned with the establishment of the Church of Scotland and the improvement of morality, especially control over the great drinking bouts of the inhabitants of the Western Isles. The fifth statute, concerned with drinking, states that one of the special causes of the great poverty, cruelty, and barbarity which occur on the isles has been the extraordinary drinking of strong wines and spirits brought in by mainland merchants and local traffickers. Therefore, the acts forbid the importation of wine or aqua vitae (brandy) to the isles, punishable by confiscation and heavy fines: 40 pounds for the first offense, 100 for the second, and forfeiture of one's entire possessions for the third. However, barons and gentlemen are allowed to purchase wines from the Scottish Lowlands, and the domestic distillation of aqua vitae for personal use is permitted. The exception given the upper classes is probably designed to recognize the large burden of hospitality which they shoulder. It also appears that the great problem is wine drinking, and not whisky, whose distillation is still permitted (MacDonald 1914:28-29; Ross 1970:24-26).

DISTILLING PROMOTED. The Statutes of Icolmkill and subsequent restrictions on wine may have encouraged the spread of distillation among the Hebrideans and then throughout the Highlands to the east (MacDonald 1914:32; Ross 1970:26).

SPAIN

BRANDY CONSUMPTION. No evidence exists of brandy consumption in Spain before the seventeenth century (Braudel 1981:243).

SWITZERLAND

BREWING. Itinerant brewers from Germany give impetus to brewing in Switzerland. The industry spreads from urban to rural areas (Jellinek 1976b:77).

SPIRITS CONSUMPTION. The recreational use of distilled spirits in Switzerland probably begins in the late 16th century (Jellinek 1976b:78).

CONTROLS: DISTILLING, DRINKING. An increase in heavy drinking during the early years of the 1600s takes place in the German region of Bern, and the local authorities are ordered to restrain the practice. The excessive drinking of *kirshwasser* (distilled cherries) prompts a Bernese decree of 1619 that prohibits distillation from cherries, except by physicians. The decree is renewed in 1674. An ordinance of 1620 attempts to achieve moderation by limiting the amount served (beverage unspecified) by innkeepers with the evening meal to 1.5 liters per guest, an indication that the amounts consumed were considerable (Jellinek 1976b:78, 82).

DRINK AND POVERTY. In 1612, a Lausanne memoir complains of drunkenness being the cause of the impoverishment of many families in the Vaud (Jellinek 1976b:82).

1610

EUROPE

TEA TRADE BEGINS. The first cargoes of tea from the Orient arrive at the Dutch East India Company in Amsterdam in 1609, but supplies in Europe are limited until the early 18th century (Braudel 1981:250).

DRINKING PRACTICES (MORYSON). The English traveler Fynes Moryson (Itinerary, 1617) describes the drinking habits of the various nations he visited between 1605 and 1617. (His comments are quoted below under the individual countries.)

ENGLAND

ALEHOUSE, TAVERN PREVALENCE. In 1618, London magistrates still speak of "the multitude of alehouses and victualling houses within this city increasing daily." By the end of the decade, there are probably 3,000 in London alone, many in the poorer areas (Clark 1978:50). London taverns, limited to 40 in 1552, have increased to 400 by 1618 (Monckton 1966:120). In the mid-1620s, John Winthrop estimates that there are 20,000 drink shops in England and Wales, or two per parish (Bretherton 1931:152).

INEBRIETY CONDEMNED (DOWNAME; YOUNG). The Puritan John Downame writes a "Disswasion from the Sin of Drunkenness" as the second book of his Foure Treatises (1609:77-127). He recognizes that alcohol does have legitimate uses, including for friendship and "for honest delight," but he attacks inebriety and its recent increase in England: "For who seeth not that many of our people of late, are so unmeasurablie addicted to this vice, that they seem to contend with the Germanes themselves, spending the greatest part of their time in carousing, as though they did not drinke to live, but lived to drinke." Of themselves, wine and "strong drink" are "the good creatures of God, which to the pure are pure." But "the first cup is for quenching the thirst; the second for delight to make a cheerrefull heart; the third for voluptuousnesse, and the fourth for madnesse" (pp. 79-82; see also Harrison 1971:89, 423n.18). Reflecting the concerns about the association of alehouse drinking and social disorder, he writes: "When a drunkard is seated upon the ale-bench and has got himself between the cup and the wall, he presently becomes a reprover of magistrates and repiner against the best established government" (quoted Clark 1983:145, citing the 1613 edition, p. 88).

Thomas Young writes a book entitled England's Bane, or the Description of Drunkennesse (1617) because "no Nation is more polluted with this capital sinne" than England. It is a development recent in origin. Prior to the Dutch wars, "drunkennesse was held in the highest degree of hatred that might be amongst us....now it is grown for a custome and the fashion of our age." Now so much time is spent in carousing that it is as if the purpose of life is to drink. Indeed, "there are in London drinking schooles: so that drunkennesse is professed with us a liberall Arte and Science." He blames this on the proliferation of alehouses and the practices of pledging and toasting (section F3). He laments the abuse that now is being made of "God's gift" and constantly refers to drunkenness as a "monster." Nature never sent a "more deadlier plague" (E2). It is "a vice which stirreth up lust, griefe, anger, and madnesse, extinguisheth the memory, opinion and understanding, maketh a man the picture of a beast" (D1). The drunkard is "the annoyance of modestie, the trouble of civiltie, the spoile of wealth, the destruction of Reason, he is only the Brewars agent, the Alehouses benefactor, the Beggars companion, the Constables trouble" (F2).

TEMPERANCE CITED, DRINKING PRACTICES (MORYSON). Fynes Moryson maintains that drunkenness is not uncommon among certain sectors, but the English in general are a temperate people. He also indicates that toasting is not excessive: "For the point of drinking, the English at a Feast will drinke two or three healths in remembrance of speciall friends, or respected honourable persons, and in our time some Gentlemen and Commanders from the warres of Netherland brought in the custome of the Germans large garaussing, but this custome is in our time also in good measure left. Likewise in some private Gentlemens houses, and with some Captaines and Souldiers, and with the vulgar sort of Citizens and Artisans, large and intemperate drinking is used; but in generall the greater and better part of the English, hold all excess blameworthy, and drunkenness a reprochfull vice. Clownes and vulgar men onely use large drinking of Beere or Ale...but Gentlemen garrawse onely in Wine" (3.3.152, Moryson 1617/1908, 4:176).

DRINKING CONCERNS (BOOK OF SPORTS). In the face of Puritan Sabbatarianism, James I promulgates a "Declaration...concerning Lawfull Sports to be Used," better known as the "Book of Lawful Sports" (1618), incurring the rath of the Puritans (Hill 1961:85; Hill 1964:195; Malcolmson 1973:11-12, 14). James begins by explaining that he is issuing the declaration because he recently had to rebuke Puritans in Lancashire who were interfering with the people's "lawfull recreations, and honest exercises upon Sundayes and other Holy dayes, after the afternoone Sermon or Service." He therefore orders that after divine service the people should not be disturbed in their exercise of lawful recreation. He justifies this action in part on the belief that if people are denied their traditional Sunday sports, they will otherwise go to alehouses and indulge in "filthy tiplings and drunkennesse" and "idle and discontented speeches."

The book is reissued and expanded in 1633 by Charles I, who adds the following defense of wakes: "Our express will and pleasure is, that these Feasts, with others, shall be observed, and that our Justices of the Peace, in their serveral divisions, should look to it, both that all disorders there may be prevented or punished" (quoted Malcolmson 1973:11).

The Puritan opposition to this book has been taken as an example of how they were more concerned about Sabbath breaking than with drinking per se (Bainton 1945:53-54).

ALEHOUSE CONTROLS. On the authorization of James I, in 1616 the Privy Council orders justices to pay greater attention to regulating licenses and to suppressing unlicensed and unneeded alehouses. In a speech to the council, James denounces the alehouses as centers of robbery and murder (Hudson 1933:10, 11).

TAXATION: MALT LEVY (1614). In 1614, a levy is placed upon malt for the first time. Many brewers attempt to evade the tax. In 1621, the brewers present a complaint to Parliament, but their demands remain unsatisfied until 1624 when Parliament declares all patents issued by royal prerogative to be unlawful (Monckton 1966:114).

REVENUE GENERATION: INN PATENT ESTABLISHED. In contrast to alehouses, inns (about 2,000 at this time) do not come under the licensing power of justices of the peace. In 1617, James grants a patent to license inns and to receive the proceeds of inn penalties. The penalties and license fees are to be paid to the Exchequer save for a fixed sum for the holders of the patent. A year later he also grants a patent for the collection of alehouse fines. The inn patent is seriously abused; some 1,200 inn licenses are granted indiscriminately. In 1621, Parliament condemns and rescinds both patents. Authority over both the inns and alehouses returns to the justices. The early Stuarts make no further effort to combine central control with revenue generation (Monckton 1966:112; Bretherton 1931:156-158).

FRANCE

TEMPERANCE OF KING. King Louis XIII (1610-1643) is a very moderate drinker, diluting his wine with water (Franklin 1889, 6:129).

TEMPERANCE, WINE DILUTION CITED (MORYSON). Moryson describes the French as being more temperate than the English: "Drunkenness is reprochfull among the French, and the greater parte drinke water mingled with wine, and alwies French wines, not Sacke or Spanish wines (which are sold as Phisicke onely by Apothecaries)....Yet Marriners, Souldiers and many of the common sort used to drinke Perry and Syder to very drunkenness, yea, I have seene many drink wine with like intemperance....Women for the most part, and virgins alwaies...use to drinke water, except it be in the Provinces yeelding Perry and Syder, which all sorts use to drinke without exception. And at Paris I remember to have seene a poore woman to beg a cup of water, which being given her, she drunke it off, and went away merily, as if she had received a good almes" (3.3.136, Moryson 1617/1908, 4:142).

TAVERN CONTROLS AND REVENUES. A decree of the Royal Council (1613) asserts that fidelity to the ancient ordinances restricting taverns has brought much inconvenience to the public and caused a loss of revenues (aides). Thus, taverners are allowed to sell to whomever they want (Dion 1959:489).

GERMANY

INEBRIETY PREVALENCE (MORYSON; SOMMER). Moryson makes frequent reference to excessive drinking among the Germans. "The diet of the Germans is simple, and very modest, if you set aside their intemperate drinking....They use many sawces, and commonly sharpe, and such as comfort the stomacke offended with excessive drinking: For which cause in upper Germany the first draught commonly is of wormewood wine....those that take any journey, commonly in the morning drinke a little Brant wein" (3.2.82). "Let the Germans pardon me to speake freely, that in my opinion they are no lesse excessive in eating, than in drinking....to their drinking they can prescribe no meane nor end....[At a nuptial feast] there was endless drinking, whole barrels of Wine being brought into the Stoave...which we so plied, as after two howers, no man in the company was in case to give account next morning, what he did, said, or saw, after that time. To nourish this drinking, they use to eate salt meats....Salt thus pleaseth their pallat, because it makes the same dry, and provokes the appetite of drinking" (3.2.83). "And to say truth, the Germans are in high excesse subject to this vice of drinking, scarce noted with any other nationall vice....In Saxony, when the gates of the Cities are to be shut, while they that dwell in the subburbs, passing out, doe reele from one side of the streete to the other....For howsoever the richer sort hide this intemperance for the most part, by keeping at home, surely the vulgar yeeld this daily spectacle" (3.2.86). He is told by a physician that "many Germans dying suddenly upon excesse of drinking, were ordinarily (for hiding the shame) given out to die of the falling sickeness" (3.2.87). For all the legislation against drunkenness, he has never seen a prince free from the vice (3.2.89) (Moryson 1617/1908, 4:24-25, 28-29, 34, 37, 40-41).

The preacher John Sommer (Olorinus Variscus) observes in his _Geldtklage_ (1614:195-196) "that other nations say to drink freely is to Germanise." He emphasizes that drunkenness has come to permeate almost any occasion in which people gather: "Not to speak of the orgies at weddings and christenings, they have now invented so many excuses for convivial gatherings that it is impossible to describe them all. Neither Christmas nor Easter, Whitsuntide or Ascension Day can be kept in a Christian manner, unless Bacchus is worshipped at the same time....Side by side with all these high festivals there are the special eating and drinking festivals....The dead themselves cannot get out of the clutches of Bacchus until the surviving relatives, friends and neighbors have sung them a requiem from cans and glasses, with the juices of grape and barley oozing out of their eyes" (quoted Janssen 1910, 15:397-398). John Sommer does suggest that officials often try to keep peasants from drinking beer by telling them it is a sin and unhealthy (Janssen 1910, 15:168).

DRINKING PRACTICES (BEINHAUS). As part of an attack on the number of indecent books being printed, the preacher K. Beinhaus complains in 1617 of the numbers of books about "drinking and carousing, with which Germany [is] quite full and so to say congested." He continues: "Books of that sort are read widely and eagerly, and it makes no difference whether they are for or against drinking; for people only want to hear about drinking because they indulge in it daily, and if anything fresh is said about drunkards and the art of drinking, they are all agog to hear it, and they only laugh at those who preach and write against drunkenness" (quoted Janssen 1907, 12:193, 211).

BREWING: ECONOMIC VALUE. Moryson observes: "The trade of brewing is more commodious among the Germans, then any other trafficke. So as at Torg...onely the Senate hath the priviledge to sell the same by small measures (as also to sell wine), and in the rest of lower Germany, as onely the Senate buies and selles wine, so the chiefe Citizens by turnes brew beare, admitting troopes of poore people into their houses to drinke it out. As the gaine of brewing is great, so princes raise great impositions from it" (3.2.88, Moryson 1617/1908, 4:40).

IRELAND

INEBRIETY PREVALENCE (MORYSON). According to Fynes Moryson, "the mere Irish are excessively given to drunkenness" (quoted MacLysagth 1950:69).

WHISKY PRAISED (MORYSON). Moryson observes that "the Irish Aquavitae, vulgarly called Usquebagh, is held the best in the World of that kind; which is made also in England, but nothing so good as that which is brought out of Ireland" (3.3.162, Moryson 1617/1908, 4:197).

ALE CONSUMPTION, SALES (MORYSON). Hopped beer is still not made, but there is extensive brewing of strong ales, "great quantities of which [are] consumed" (O'Brien 1919:83). Moryson writes in his Itinerary that the selling of ale in Dublin "outstrecheth all [other commodities] that I have spoken of." It is "a quotidian commodity that hath vent in every house in the town every day in the week, and every hour in the day, and every minute in the hour. There is no merchandise so vendible, and is the very marrow of the commonwealth of Dublin" (quoted O'Brien 1919:83-84).

WOMEN/DRINKING (MORYSON). Moryson also observes that the English-Irish drink all alcoholic beverages in excess; of women, he writes: "I have in part seene, and often heard from other experience, that some Gentlewomen were so free in this excesse, as they would kneeling upon the knee, and otherwise garausse health after health with men; not to speak of the wives of Irish Lords, or to reffer to the due place, who often drinke till they be drunken" (3.3.162, Moryson 1617/1908, 4:197-98). The "wilde Irish" only drink alcohol when they come to a market town, and then they drink wine and whisky until they are drunk as beggars (3.3.163, Moryson 1617/1908, 4:200).

NETHERLANDS

TAVERN DRINKING; BEVERAGE PRICES (MORYSON). Speaking of his experiences in the inns of the Dutch Republic, Moryson warns that "huge impositions" are placed on wine, often exceeding the value of the actual wine. Even if one drinks only cheaper beer, the traveller "must understand that his companions drinke largely, and be he never so sober in diet, yet his purse must pay a share for their intemperance." However, although the Dutch pass the time after supper by drinking around the fire "upon the common charge," they do "not use these night drinkings so frequently nor with such excesse, as the Germans doe" (3.2.98, Moryson 1617/1908, 4:61).

SCOTLAND

WHISKEY CONSUMPTION; FUNERAL DRINKING. A description of the funeral of a Highland laird in 1618 includes one of the earliest references to the consumption of whisky ("uiskie"). Officers sent by the king in 1622 are given the best entertainment that the season and the country allow; they lack not in wine or "aquavitie." In Banffshire, that region in the Highlands where today most of the malt distilleries are located, a man is accused in 1614 of breaking into a house, committing an assault, and knocking over some "aquavitae" (Wilson 1973:37). That Moryson (see below) does not mention whisky indicates that, for Scotland as a whole, recreational whisky drinking is not common.

INEBRIETY PREVALENCE (MORYSON). Moryson asserts that the Scots are far more excessive in their drinking than the English. Even the Scots "of the better sort...spend great part of the night in drinking, not onely wine, but even beere" (3.3.156, Moryson 1617/1908, 4:184-185).

WINE DRINKING CONTROLS. James VI has a hard time enforcing the Statutes of Icolmkill, including the provisions concerning the consumption of wine. Therefore, an act is adopted in 1616 imposing a fine of 20 pounds for the purchase or drinking of wine in the Western Isles and the mainland immediately adjacent. Shippers are warned in 1622 not to send or carry any wine to the isles other than the permitted maximum quantities, under pain of confiscation (MacDonald 1914:31-32; Gunn 1935:34; Ross 1970:26).

CONTROLS: TAVERNS, CURFEW. A statute of 1617 provides that all persons who are convicted of "drunkenness or haunting of taverns and alehouses after ten hours at night or at any time of day except in time of travel or for ordinary refreshments" are to be fined three pounds for the first offense or, if unable to pay, jailed for six hours. The penalties are increased for subsequent offenses (*Scottish Law Review* 1943:86).

SWITZERLAND

DRINKING CONTROLS. Moryson observes that "if any man often offend in Drunkenness, he is imprisoned, and may drink no Wine for a yere, til he have procured pardon of the publike Senate....generally the Switzers drinke as stiffly, as those of the upper part of Germany" (3.4.263, Moryson 1617/1908, 4:413-414).

AMERICA (BRITISH COLONIAL)

VIRGINIA

ATTITUDES; DRINKING CONTROLS. The Virginia colonists continue the English idea that alcoholic beverages are a part of man's natural food and drink and therefore good. The ordinary consumption of liquor is not seen as a subject for legislation. No person is prohibited from drinking any quantity desired, and efforts are made to insure that a sufficient supply is available. However, in 1619 a statute is passed prohibiting drunkenness, which is seen as an offense in itself, a "loathsome sinne." Penalties range from a reprimand by the minister (first privately, then publicly) to imprisonment. These penalties are soon replaced by a fine, usually of five shillings or the equivalent. The legislature's concern is mainly for the lower classes, since the fines "would obviously deter no well-to-do man." In the administration of the laws, the tendency in the course of the century is "to rely less on punishment and more on direct control by regulating selling." The purpose of alcohol regulation is to control excessive drinking and drunkenness, not to restrict the everyday use of alcohol (Pearson and Hendricks 1967:5-6).

1620

ENGLAND

COURT, UPPER CLASSES/DRINKING. King Charles I (1625-1649) is a moderate drinker who only takes a glass of wine mixed with water at dinner. However, the court continues to have a reputation for being "venal and bibulous" (Francis 1972:56-57).

At the same time, the conspicuous consumption that had characterized the nobility and gentry in the late 16th and early 17th centuries begins to decline. Caroline noblemen become more economical, in part due to changing values and codes of behavior, improved economic conditions, and a reduction in social competition and pressure. Many are deeply affected by the rising tide of Puritan propaganda against waste, extravagance, gambling, and drinking. The English nobility begin to free themselves from the medieval preoccupation with publicity and open handedness according to which a man's status was judged by the number of his attendants and the scale of his hospitality. A reassessment of values places less emphasis on self-publicity and display and more on privacy, personal luxury, aesthetic qualities, and respect for the individual (Stone 1967:88, 266).

CONCERNS: ALEHOUSES. With the collapse of the alehouse patent in 1621 (see 1610/England), the campaign against alehouses accelerates in both town and countryside (e.g., see the views of Rawlidge below). Justices require licensed victuallers to post substantial sureties; unlicensed victuallers are suppressed, with persistent offenders jailed. These efforts are assisted by the growing power of the larger brewers, who seek to curb competition from petty brewers and brewing victuallers (Clark 1978:69-71).

INEBRIETY, ALEHOUSES CONDEMNED (RAWLIDGE; BURTON. According to the self-described simple "Mechanicall Man" Richard Rawlidge (*A Monster late Found Out and Discovered, or the Scourging of Tiplers*, 1628:24), the offense of inebriety has never been so great. Men are daily seen "realing and vomiting, yea like swine, tumbling in the open streets." He attributes this in large part to the proliferation of alehouses, which were once few in number, but "now every street [is] replenished" with them. He considers alehouses as one of the most imfamous "Receptacles of all manner of baseness and ludenesse" (p. 2). These have spread in part because the preachers have inveighed against public sportings on Sundays. "What then followed? Why sure alehouse haunting...so that the people would have their meetings either publicly with pastimes abroad or else privately in drunken alehouses" (pp .12-13, quoted Clark 1983:152). Inebriety has also spread because the laws are unenforced and few complain because the fines go to the poor and the informer receives nothing (p. 25). Tippling, he warns, results in poverty, distress, and the damnation of God.

Similarly, a political economist in 1627 writes that "with grief I speak [of drunkenness], the taverns, ale-houses, and the very streets are so full of drunkards in all parts of this kingdom, that by the sight of them is better known what this detestable and odious vice is than by any definition whatsoever." He defines the "most monstrous vice" of drunkenness as "the privation of orderly motion and understanding" (quoted French 1884:188).

In discussing the "Quantity of Diet" as a cause of melancholy, Robert Burton (1621/1854:142-143 [section 2.2.2]) complains that there is "immoderate drinking in every place," that men are flocking to the taverns, and that "'Tis now come to that pass, that he is no gentleman, a very milk-sop, that will not drink, fit for no company." "Gluttony kills more than the sword," he laments; we now "luxuriate and rage" in "surfeiting and drunkenness."

ATTITUDES; MEDICINAL USE (BURTON). Robert Burton (see above) refers to wine and women as the "two main plagues, and common dotages of human kind." "Wine is a cut of madness which causes sorrow, poverty, shame and disgrace" (1.2.3.13, Burton 1621/1854:181-182). Drunkards' children often suffer from melancholy (1.2.1.6, p. 134). But he does see virtue in moderate wine consumption, although he drinks none himself. He does not hold a high opinion of the medical properties of wine (Sutherland 1969:176), but he recommends a cup of wine or spirits if soberly and opportunely used as a cure for melancholy and to drive away care (2.5.1.5, p. 410; see also French 1884:180-181).

DRINKING INFLUENCES (WINTHROP). John Winthrop, a pious country gentleman who later becomes governor of Massachusetts Bay Colony, writes a memorandum on the reasons why inebriety persists and drafts an "Act for the (utter abolishinge) preventing of Drunkenness and (the preventing) of the great waste of Graine within this Kingdom" (Bretherton 1931:167, 171; Walzer 1971:228, who considers Winthrop's views on drunkenness and its cure "typical of the Puritan gentry").

CONTROLS CRITICIZED (WINTHROP). Regarding the formal indictment proceedings for drunkenness under the 1606 act, Winthrop asserts that "custom is so dilatorye and Chargable to such as prosecute the offenders" that many offenses are "lett slippe" or "not prosecuted with effect" (quoted Bretherton 1931:168).

BEER CHARACTERISTICS. Winthrop argues that the strictest laws against drunkenness will not help unless the strength of beer and ale is reduced to a level wholesome to the body, "it being now the com(mon) practice for everyone to strive to exceed the other in the strength of their beere, that they may drawe the more Customers to them: and this is one chiefe cause that Innes and Alehouses are so much frequented" (quoted Bretherton 1931:170-171).

TOASTING CONDEMNED (PRYNNE). Puritans disapprove of the pledging of healths far more than drinking itself. The Puritan William Prynne writes Healthe's Sickness (1628), dedicated to Charles I, to prove "The Drinking and Pledging of Healthes to be Sinfull and utterly Unlawful unto Christians." He considers this practice to be an "idle, foolish, heathenish and hellish Ceremony" which causes men to drink more than they would otherwise. Health drinking is "the very mother and nurse of Drunkennesse"; if it were banished, drunkenness "would quickly vanish and grow out of use." Of all the traps laid by Satan for man's souls, few are "more Dangerous, Hurtful and generally Pernicious." He further indicates that the spread of the practice is a recent occurrence, that it has "of late defiled, and overspread our Nation," or at least its popularity has recently increased "mightily." Among the causes for this he lists the negligence of justices (Sections:A, B3-B4, 1-2, 12). Barnaby Rich (The Irish Hubbub, or the English Hue and Crie, 1619) also attacks the custom of toasting (French 1884:179, who quotes a long passage describing toasting rituals.)

Clark (1978:64; 1983:156) sees the replacement of simple pledges by elaborate toasting rituals and conventions as a product of the transmutation of traditional neighborliness into "new-style drinking camaraderie" and the ritualization of drinking in the alehouse. The drinking of healths, he argues, appears to have been fully established among the general population only in the late 16th century in imitation of upper-class habits. French (1884:190) considers "the prevailing habit of toasting" to be an important cause for the widespread drunkenness of the period.

CONTROLS: DRINKING, ALEHOUSES. In 1623, the acts of 1604 and 1606 are strengthened by a statute which (1) extends the tippling provisions to all persons, including travelers; (2) reduces the evidence required for conviction of drunkenness or tippling from two witnesses to one (or confession); and (3) provides that the proprietor of any alehouse who permits the commission of either offense within his place of business should "for the Space of three Years next ensuing...be utterly disabled to keep any such Alehouse" (quoted Baird 1944b:147).

During the reign of Charles I, a series of laws are passed that are designed to decrease excessive drinking. A 1625 statute provides that the penalty for permitting tippling or drunkenness be imposed upon innkeepers as well as upon alehouse keepers. Subsequent statutes bring sellers of wine and spirituous liquor within the licensing program (Baird 1944b:157).

DISTILLING INFLUENCES. The success of Dutch spirit imports in the early century show the English how profitable the trade can be. The domestic distilling industry gains ground on brewing (Forbes 1970:195, who observes that distilleries cost only 500-5,000 pounds to build, whereas a brewery costs 2,000-10,000, the same figures are cited by Campbell 1747/1969:332-333).

FRANCE

DISTILLING CONTROLS. The first state control of distilling occurs in 1619; grain spirits are forbidden. Official manufacturing privileges have been limited to apothecaries and spice merchants, but apothecaries and distillers are combined into one guild by ordinances of 1624 and 1634 (Forbes 1970:102).

ATTITUDES: TAVERNS, WINE SELLERS. In successfully opposing the establishment of wine merchants as a seventh major guild in Paris, the guild of mercers asserts: "These people whose only purpose is to be and to fulfill an abject and servile profession...their cabarets and taverns have always been treated in a scornful manner to such a point that laws have forbidden bourgeois and inhabitants to set foot in them....Artifice, fraud and deceit are the inseparable characteristics of wine merchants [and yet] they dare claim their commerce necessary to the State. To make such a claim they must needs have been drunk on the precious liquor which they themselves sell, and the fire of such a liquor must have given them a fit" (quoted Ranum 1968:175).

ARTISANS/DRINKING. In Saint-Emilion (southwest France), it is complained that "the greatest part of artisans are so debauched that they don't leave the taverns night or day, where they unduely dispense their possessions and all that they win by their labor and industry. This causes the greatest part of families to perish of hunger. After they are completely drunk from wine, they begin an infinity of insolences and blasphemies against the holy name of God, and once night has come and they depart from the cabarets, they come to steal from the houses" (Bercé 1974:222).

GERMANY

BEVERAGES CONSUMED; PRICES. Due to the Thirty Years War, taxation of all industrial products, especially beer, increases greatly. Everywhere complaints are raised about the high cost and poor quality of beer, which is frequently adulterated. The war causes commercial brewing in Lubeck to decline markedly (Arnold 1911:276-277, 299). Rising prices for beer and wine contribute to an increase in spirits consumption. The foreign soldiers in the imperial armies will not drink German beer, but prefer distilled liquor (Cherrington 1926, 3:1092).

DRINKING CONTROLS. An Austrian ordinance of 1624 stipulates that young officers invited to an archduke's table should "present themselves in clean uniform, not...arrive half drunk, not...drink after every mouthful,...wipe moustache and mouth clean before drinking, not...lick the fingers, not...spit in the plate, not...wipe the nose on the tablecloth, not...gulp drink like animals" (quoted Braudel 1981:207).

SUNDAY DRINKING CONTROLS. A Württemberg territorial ordinance of 1621 demands a fine for drinking on Sundays and holidays (Stolleis 1981:103).

GRAIN SPIRITS DISTILLING. New centers of grain spirits production have been established at Wernigerode and Magdeburg. There is still very strong public opinion against the abuse of cereals for manufacturing alcohol; distilling is banned in several towns (Forbes 1970:103).

SCOTLAND

DRINKING CONTROLS. The newly reformed Scottish Church is an ascetic Presbyterian denomination; Catholicism still predominates in the Highlands, where much of the whisky is produced and drunk. In 1625, the Scottish Church launches a campaign against drunkenness that is so effective, tavern keepers complain that "their trade was broken, the people had become so sober" (Cherrington 1929, 5:2384).

SWITZERLAND

CONTROLS: DRINKING, TAVERNS. Zürich passes ordinances in 1628 "in these last troublesome times in order to promote a penitent Christian life and honorable conduct." One ordinance provides that, as intemperate and superfluous eating and drinking desecrates the Sabbath, taverns and guild houses must be closed on Sundays except to strangers; furthermore, giving dinners, walking in the fields, and holding weddings are prohibited. Certain carnival customs are banned as heathen superstitions. It is further stipulated: "In the city at an evening drink...the courses must be called for about 5:00...and persons must not sit together longer than til about 6:00, and at that time they must leave, and must not stop at any other place on the way home to sit down and drink again, but everyone must betake himself to his own house and when at home, all further evening drinks and 'night caps' are forbidden." Drinking brandy and strong liquors before noon is forbidden except to strangers from outside the city. Wine cannot be sold before noon by tavern keepers. No person can compel another to drink, and treating is forbidden. In 1650, these ordinances are sharpened in detail. Going to taverns for eating and drinking on Saturday afternoons is banned to allow people to attend Saturday evening services (Vincent 1898:365-366).

AMERICA (BRITISH COLONIAL)

VIRGINIA

INEBRIETY PREVALENCE. In 1622, Sir Frances Wyatt, governor of the Virginia Company, is advised by the London headquarters to effect a "speedie redress" of the enormous excess of drinking, the "cry whereof cannot but have gone to heaven, since the infamie hath spread itself to all that have heard the name Virginia" (Krout 1925:3; Pearson and Hendricks 1967:6).

1630

EUROPE

DISTILLING: GRAIN SPIRITS. The Dutch physician and chemist Angelo Sala (Hydrelaeologia, 1639) asserts that everywhere in northern Europe brandy is now made from cereals and describes the common procedures for producing it (Forbes 1970:162).

ATTITUDES; DRINKING PRACTICES (HOWELL). Royal historiographer and member of Parliament James Howell (1594-1666) writes a letter (dated October 1634) reflecting the current "general lay knowledge of alcoholic beverages" and their use throughout the world (Mendelsohn 1963:6, who reprints the letter in full, pp. 7-16; see also Howell 1635/1753:365-374). He begins by observing, "I do not know or hear of any Nation, that hath Water only for their Drink."

The use of milk and "strong-waters of all kinds" in Ireland is unique. "The prime [of these waters] is usquebaugh which cannot be made any where in that perfection: and whereas we drink it here in aquavitae measures, it goes down there by Beer glass-fulls." He complains that the Scots, because they are a "useful confederate to France," have first choice of Bordeaux wines (Howell 1635/1753:367, 371).

Spain and Italy are the two countries "freest from excess of Drinking." Germany, Holland, and Greece are the countries where excess drinking is most prevalent. But in the wine-producing areas of the Palatinate, it is held "a great part of incivility for Maidens to drink Wine until they are married" (ibid., pp. 372-374). (On his attitudes toward ale and wine, see 1630/England.)

DENMARK

FESTIVE DRINKING. James Howell, in a letter from Hamburg dated October 1632, relates that at a banquet given by the king of Denmark for the earl of Leicester, the king called for 35 toasts for various Christian princes; all the court officials got drunk and the king had to be carried away in his chair (Howell 1635/1753:249; also quoted French 1884:158; see also Glamann 1962:129).

ENGLAND

ALEHOUSES: PREVALENCE, CONCERNS. Despite efforts at restriction, complaints continue about the increasing number of alehouses. Concerns about alehouses are exacerbated in the 1630s by concerns over food shortages and the diversion of too much grain to brewing and over social distress. As many as 30,000 alehouses may exist, compared to 18,000-19,000 in 1577 (Clark 1978:50). Thomas Dekker and Ben Jonson portray the alehouse as "the trysting place of an underworld populated by gulls and vagabonds, robbers and whores, a world which though parasitical is also a mirror immage of the moral shame, the trickery and hypocrisy of respectable society" (ibid., p. 48). Dekker (English Villanies, 1638) asserts that in London "a whole street is in some place but a continued alehouse" (quoted Clark 1983:39). At about the same time, Lord Chancellor Coventry considers them the "greatest pests in the Kingdom," a source of drunkenness, thievery, and disorder (quoted Monckton 1966:120). Puritan

Christopher Hudson attacks alehouses in 1631 as "nests of Satan where the owls of impity lurk and where all evil is hatched, and the bellows of intemperance and incontinence blown up" (quoted Clark 1978:47).

INEBRIETY PREVALENCE. The preacher R. Younge (1638:336-339) attacks the prevalence of inebriety in England: "Indeed, heretofore they [drunkards] were as rare as Wolves; but now they are as common as Hogs." Whereas once they drank only at night, now they drink openly in the day. No street in England is without a drinking house. "Yea, if England plies her liquor so fast as she begins, Germany is like to loose her Charter, for drunkennesse....[W]hat bowsing and quassing, and whiffing and healthing, is there on every bench: and what reeling and staggering in our Streets: what drinking by the yard." He describes drunkards lying in the fields where the town fair is held. Furthermore, he sees little hope for improvement: "In all probability, this infectious vice of drunken good-fellowship is like to stick by this Nation, for...a common blot is held no staine."

DRINKING AND THE PLAGUE. Younge (1638:246-248) remarks that even the plague did not bring about a lessoning of drinking: "I have myself seen when the Bills [of mortality] were at the highest, even bearers who had little respite from carrying dead corpses to their graves and many others of the like rank go reeling in the streets." He who was a drunkard before is a drunkard still, and the taverns were never so full.

INEBRIETY CONDEMNED (YOUNGE; TAYLOR). Younge's massive The Drunkards Character (1638) traces to drunkenness every conceivable vice and social evil, including idleness and poverty. A drunkard deprives himself of his sense and reason; drunkenness is a "voluntary madness, a temporary forfeiture of the wits." Although wine is "a good creature of God," drunkenness is a sin against the laws of God, grace, nature, and all nations (Younge 1638:13, 20, 28, 33).

Jeremy Taylor, chaplain to King Charles, vigorously attacks drunkenness but prescribes "faith, hope, and charity," rather than abstinence, as the best remedy (quoted Harrison 1971:88). In a "Sermon on Christian Prudence," he emphasizes that "temperance hath an effect on the understanding, and makes the reason sober, and the will orderly, and the affections regular, and does things beside and beyond their natural and proper efficacy." In a his "Funeral Sermon for the Countess of Carbery," he laments that "we pour in drink and let out life;...we strangle ourselves with our own intemperance;...and we quench our souls with drunkenness" (quoted French 1884:202).

SUNDAY DRINKING CONTROLS. In Exeter (1627) and Somerset (1631), the chief justices issue orders prohibiting all ales and wakes on Sunday. Jealous of infringements on the Church's prerogatives, Archbishop Laud persuades King Charles to revoke the Somerset order. The king issues a national declaration reaffirming James I's "Book of Lawful Sports" (see 1610/England), emphasizing that "feasts of the dedications of the churches...shall be observed" (Monckton 1966:126-127). These actions are taken in an atmosphere of struggle between Puritans and Anglicans. One bishop writes to Archbishop Laud: "If the people should not have their lawful recreation upon Sundays after evening prayer, they would go either to tippling houses, and there talk of matters of church and state, or else into conventicles [secret Puritan meetings]" (quoted Hudson 1933:16, who dates the Somerset order at 1632).

TAXATION: BEER LEVY. In 1628, King Charles dismisses Parliament; in 1637, in need of money, he orders a tax on brewers. Charles summons Parliament again in

1639. Either upon his initiative or under Parliamentary pressure, the beer tax is abolished. During the two-year period, the duty amounts to 12,000 pounds, but only 3,000 are paid, "a clear indication of the unpopularity of the tax and of the brewers' unwillingness to pay it" (Monckton 1966:115-116).

DISTILLERS COMPANY ESTABLISHED. To protect the public from the inferior and injurious spirits flooding the market, Charles I, under the recommendation of his physician (Sir Theodore de Mayern), grants to the Distillers Company the first charter to establish controls over who can distill and beverage quality. The queen's medical advisor becomes the head of the company, and Mayern prepares a series of regulations to protect the public from unwholesome spirits in 1639. The quality of the company's product is said to be horrible, and there is no evidence that the establishment of this company leads to any increase in spirit drinking. The monopoly covers London, Westminister, and the surrounding areas to a radius of 21 miles (Simon 1948:136-136; Mendelsohn 1963:82; Younger 1966:326; Watney 1976:12).

The publication by the company of a book entitled The Distiller of London is one of the earliest examples of the use of the word "distiller" in the restricted sense of one who extracts alcoholic spirit by distillation rather than the earlier, general meaning of "a distiller of water" (Forbes 1970:194-195).

BRANDY. The older terms for alcoholic spirits (e.g., "aqua vitae") begin to be displaced by the words "brandwine," "brandewine," and "brandy wine," reflecting the Dutch influence (Forbes 1970:102).

ATTITUDES: ALE, WINE (HOWELL; TAYLOR). In his letter of October 1634, James Howell (1635/1753:366) writes regarding "noble Ale" that "there is no Liquor that more increaseth the radical moisture, and preserves the natural heat, which are the two pillars that support the life of man." Regarding wine, he quotes the adage (p. 73): "Good wine makes good blood, good blood causeth good humours, good humours cause good thoughts, good thoughts bring forth good works, good works carry a man to heaven; ergo, good wine carrieth a Man to Heaven."

In Drinke and Welcome (1637), devoted to the "Famous Historie" of the drinks consumed in Great Britain and Ireland, John Taylor (1580-1653) praises the virtues of moderate ale drinking. He calls English ale "the unparaleled liquor" (section A3). "It is a seale of many a good Bargaine. The Physitian will commend it; the Lawyer will defend it, it neither hurts, nor kils, any but those that abuse it unmeasurably and beyond bearing; It doth good to as many as take it rightly" (B2-B3). Beer is an "upstart and a foreigner"; "I shall not need to wet my pen much with the name of it" (B3).

MEDICINAL USE; HEALTH CONCERNS. Dr. J. Hart (Diet of the Diseased), (1633) attacks the tendency of people to reach for a bottle of wine at the slightests ailment due to the recommendations of their physicians. He further believes that spirits should be only taken medicinally and only on rare occasions (Simon 1909, 3:401, 408).

SACK CONSUMPTION. Taylor (1637:B3) refers to the "mighty power" of sack, the abuse of which "daily is in universall use." He dates the rise of sack consumption from the mid-16th century.

WINE CONSUMPTION. Taylor (1637) writes the following on wine: "For now our Land is overflowne with wine:/ With such a Deluge, or an Inundation/ As hath besotted and halfdrownd our Nation./ Some that are scarce worth 40 pence a yeare/ Will scarcely make a meale with Ale or Beere."

FRANCE

TAVERNS: REPUTATION, PATRONS. By the end of the reign of Louis XIII (d. 1643), taverns have come to be considered more vulgar than cabarets, to the point that in a 1634 update of a list compiled by Francis I in 1544, cabaretiers have replaced taverniers for the privilege of following the court and providing for it (Dion 1959:485).

TAX PROTESTS: CROQUANTS. During the peasant revolts known as the "Croquants," wine taxes are a frequent target of protest. These revolts occcurred under the first Bourban kings, Henry IV and Louis XIII, in the late 16th and early 17th centuries, particularly in the 1630s, a period of great famine and poverty throughout France. In their wine protests, they are joined by cabaret and tavern keepers. Casks of wine are often pierced and consumed by rioters; tax collectors are harassed for making wine too expensive. Riots frequently take place during feast days, during which the surveillance over taverns is increased (Berce 1974:585-586). As noted below, tax protests also contribute to the introduction of distilling in Angoumois.

COGNAC DISTILLING BEGINS. Dion (1959:443-446) shows that brandy already was being exported from La Rochelle in Aunis in the later 16th and early 17th centuries (see 1550 and 1600/France). However, according to some accounts, the distilling of cognac, as the brandy made from white wine produced in the two departments of Charente becomes known, does not become important in the province of Aunis until the 1620s and in the provinces of Saintonge and Angoumois until the 1630s. It is in Angoumois that the city of Cognac is located, the soil around which produces the finest cognacs.

Delamain (1935:15-18) lists several reasons why distilling did not begin earlier: (1) wine was plentiful and much of it was exported to other French provinces and abroad; (2) there was no local demand as the climate was mild and a tradition of wine drinking existed; (3) the misery of the wars of religion had depleted the Protestant population (especially in Saintonge), which might have been attracted to this new, costly, and yet uncertain industry (Delamain 1935:15-18).

According to Père Arcère's Histoire de la ville de La Rochelle et du pays d'Aulnis (1756-1759, 2:467), distilling becomes common in Aunis around 1622 when over-planting and production for the Dutch wine trade leads to a stockpiling of wine, much of which is carelessly made and does not keep. Thus, a few producers decide to reduce their wine's bulk and improve its storage properties by distilling it (Butler 1926:147). Finding a ready market for it, they then plant vineyards where none existed before, producing a poor wine fit more for distilling than for drinking (Butler 1926:147; Simon 1926:32-33; Delamain 1935:15, 19).

Distilling seems to have first spread inland into Angoumois in the 1630s in response to distress over poor wine sales due to the excessive duties charged on wines in transit down the Charente River. The neighboring provinces of Aunis and Basse-Saintonge, closer to the sea, were not subject to the same charges and were able to sell their wines cheaper. In 1636, as part of the revolt of the Croquants, the peasants of Angoumois demand that the imposts on wine be lowered, threatening to turn their wines into vinegar and spirits. Significantly, this is considered only as a last resort. No future is seen in distilling (Delamain 1935:18). The tariffs established in 1664 provide further impetus for expanding the industry, as does the ready availability of wood in the countryside for stoking the stills. Early Charente brandy is rarely either aged or blended, and tastes like a "raw young spirit." The industry is small and unsophisticated. Some makers perfume the brandy with herbs and flowers to hide its bad taste (Younger 1966:329).

DISTILLERS GUILD. In 1637, distillers and spirits sellers are combined into a separate organization under the name Distillateurs en l'Art de Chimie et Vendeurs d'Eau-de-Vie, and the use of certain substances in making distillates is restricted, including beer and cider (Delamain 1935:16-17). The combining of producers and sellers into one guild shows that distillation is still a home industry (Forbes 1970:102). But the formation of the guild itself indicates brandy production is becoming more important. In 1639, all manufacturers of chemical products by distilling are grouped together. Government regulations are similar to those regarding beer (Emerson 1908, 2:126).

BOURGEOIS SALES PRIVILEGES. An Arret du Conseil of June 20, 1633, allows a bourgeois or nobleman living in Paris the privilege of selling wine on his Parisian property that was made from vines on his land (Dion 1959:482; Brennan 1984). Those who reside in the city of Bordeaux also have the privilege of entering their wine into the city tax free. Peasants have to sell their wine in the countryside for considerably less than the "privileged," even though it came from the same vineyards. Frequent complaints are raised about this injustice (Forster 1961:26).

WINE CONSUMPTION. An anonymous contemporary report on the population and consumption patterns of Paris estimates that in 1637 about 160 quarts of wine were consumed per Parisian per year: 240,000 muids a year for a population of about 415,000 (Ranum 1968:176-178).

CABARET LICENSING. Facing the expense of the war with Spain and desiring to reap financial benefits from the proliferation of cabarets, an annual licensing fee is established in 1632. Its enforcement is uneven throughout the provinces, but in some areas efforts at execution provoke rioting (Berce 1974:296). In Bordeaux the tax is notoriously unpopular.

IRELAND

TAVERN LICENSING. Licensing of publicans begins later in Ireland than in England. The first liquor licensing act is passed in 1634 because excessive drinking has become a national misfortune. The preamble states that the law is necessary because "it is found by daily experience that many mischiefs and inconveniences do arise from the excessive number of alehouses." Its goal is to "redress these inconveniences and [to] reduce those needless multitudes of alehouses to a fewer number, to more fit persons, and to more convenient places." The act stipulates that no one can sell beer by retail without an annual license approved by the county commissioners, who also determine the number of licenses to be granted, judge the character of the applicant, and decide on the suitability of the accommodations for travelers. The license holder is required to enter into recognizances for maintaining good conduct on his premises and must pay a license duty of 5s6d.

This act becomes the pattern for all subsequent liquor retailing laws in Ireland. In 1665, this licensing system is extended to whisky and wine. However, the law is laxly administered and illicit retailing of all alcoholic beverages continues for more than two centuries (Wilson 1940:122; McGuire 1973:96-97; Lynch and Vaizey 1960:48; Magee 1980:75; O'Brien 1919:15).

NETHERLANDS

DISTILLERS GUILD. A distillers guild is founded in Schiedam about 1631. Several families of distillers guard their "trade secrets" jealously (Forbes 1970:107).

CONTROLS: TAVERNS; SPIRITS CONSUMPTION. In 1631, Holland orders the closing of all taverns and inns during hours of worship and after nine in the evening. Serving spirits to young people is prohibited. This is only one of several ordinances governing drinking outlets over the century (Zumthor 1962:175).

W. Brereton (1635/1844:10) notes the abundance of brewers in Rotterdam and how they defeated an effort in 1632 by a religious "burgomaister" to enforce strict observance of the Sabbath. When he imposed a fine of one guilder on everyone trading or working on Sunday, the brewers united and "in a mutinous manner told the burgomaister that they would not be subject to his new laws; and hereby all quashed formerly effected, and the hoped for reformation came to nothing."

REVENUES. Brereton (1635/1884:65) also observes that a mighty revenue is derived out of the excise paid for beer and wine by the tap-houses and wine sellers of Amsterdam.

AMERICA (BRITISH COLONIAL)

MARYLAND

DRINKING CONTROLS. In 1639, Maryland defines drunkenness as "drinking with excess to the notable perturbation of any organ or sense of motion." The penalty is a fine of five shillings. The penalty for a servant found drunk is corporal punishment or confinement in the stocks for 24 hours (Krout 1925:28).

MASSACHUSETTS

BEVERAGES CONSUMED. The ship *Arabella*, which brings 700 Puritan settlers to Massachusetts Bay in 1630, lists among its supplies "42 tuns of beer, 14 tuns of water, 1 hogshead of vinegar, 2 hogsheads of cider and 4 pumps for water and beer" (Cherrington 1920:17). Upon arrival in Massachusetts Bay, colonists establish apple cultivation from English seeds. Hard cider is produced almost immediately (Dorchester 1884:117). In an effort to increase supplies of domestically produced alcohol, colonists make use of a wide variety of native crops. In 1630, someone in Massachusetts writes home to England: "If barley be wanting to make into malt,/ We must be content and think it no fault,/ For we can make liquor to sweeten our lips,/ Of pumpkins, and parsnips, and walnut-tree chips" (quoted Dorchester 1884:110).

SPIRITS CONSUMPTION. In 1630, Governor John Winthrop records in his journal that he has "observed it a common fault with our grown people that they have themselves to drink hot waters immoderately" (quoted Asbury 1950:5).

TAVERN LICENSING, CONTROLS. In 1632, a Massachusetts Bay law requires that anyone desiring to sell wine or "strong water" must obtain permission from the governor or deputy governor (Shurtleff 1853, 1:106). In 1634, Massachusetts Bay adopts the provisions of the British alehouse Licensing Act of 1604 (Krout 1925:15). In 1637, the colony's General Court complains about "much drunkennese, wast of the good creatures of God, mispence of precious time and other disorders" in taverns and inns, "whereby God is much dishonored, the profession of religion reproached, and the welfare of this commonwealth greatly impaired, and the true use of such houses (being the necessary releefe of travellers) subverted." The court declares that no person shall lodge or remain in any inn or common victualling house "longer then for their necessary occations, upon payne of twenty shillings for every offence" (quoted Shurtleff 1853, 1:213-214). Despite these restrictions, Massachusetts' government recognizes alcohol consumption and taverns as essential to life. Accordingly, in 1637 the court orders that each town license a man or woman to sell wine and "strong water" to insure that the people are supplied with proper accommodations (Krout 1925:4, 7; Levine 1978:146).

PRICE CONTROLS. The 1632 licensing act fixes the price of beer at one penny a quart (Krout 1925:15). In 1637, the General Court orders that the maximum price of "wine, or strong waters, [or] beare, or other drinke" shall be one pence per quart (Shurtleff 1853, 1:213-214).

PLEDGING BANNED. A law is enacted in 1639 forbidding the pledging of healths "because it was an occasion of much waste of the good creatures, and of many other sins as drunkennes, quarrelling, uncleanness." The law is repealed in 1645 (Dorchester 1884:111).

NEW NETHERLAND

TAVERN PREVALENCE; SALES CONTROLS. The Dutch West India Company encourages the retail sale of liquor by the people of Manhattan in their homes. In 1637, Governor Kieft complains that one-fourth of the houses in the village are "grog-shops or houses where nothing is to be got but tobacco and beer." The following year he forbids the sale of wine anywhere but at the company store (Bridenbaugh 1955:111-112; Thomann 1887:86).

VIRGINIA

DRINKING CONTROLS. In 1632, Virginia enacts the provisions of English law against drunkenness, which includes fines for tippling and drunkenness or time in the stocks for those unable to pay a fine; proprietors who permit drunkenness within their premises can be disbarred from operating an alehouse for three years (Cherrington 1920:17).

1640

ENGLAND

BACKGROUND DEVELOPMENTS: CIVIL WAR. Civil war breaks out in 1642 following the parliamentary rebellion against the king in 1640. The war ends in 1649 with the beheading of Charles I and the declaration of the Puritan-dominated Commonwealth under Oliver Cromwell.

ECCLESIASTICAL JURISDICTION. The authority of the Anglican church courts over moral offenses is abolished in 1646; henceforth, sin is distinguished from crime. Ecclesiastical jurisdiction is revived after the Restoration (1661), but never regains the same authority (Hill 1964:343; Baird 1944b:139, who dates the abolition at 1641).

DRINKING SOCIETIES: THE EVERLASTING CLUB. Although temperance is advocated by supporters of both the king and the opposition, heavy drinking continues. To this period belongs the "Everlasting Club," devoted to singing and drinking. The members divide into shifts, and as one shift staggers off duty another moves in to take its place, so that the work of celebration goes on round the clock. It is said that its members sought to "grow immortal by drinking" and that in its brief existence the club drank 30,000 bottles of ale, 1,000 hogsheads of red port, and 200 barrels of brandy. In addition, it smoked 50 tons of tobacco (French 1884:214-215; Longmate 1968:7).

TAXATION: BEER DUTIES. As armed hostilities break out between king and Parliament, the latter turns to the liquor trade for revenues, as the king had attempted to do before. A parliamentary ordinance issued in May 1643 imposes the first duty on beer. Clark (1983:178) calls this excise the most important control innovation of the revolutionary period. In 1645, hops are added to the list of taxable items. None of these financial burdens are popular, and the taxes are collected only with considerable difficulty on a month-to-month basis. Charles issues a warrant in 1645 stating that the same excise taxes will be levied by the Royalists. Because of their unpopularity, these taxes are announced as temporary (Monckton 1966:132).

BREWING. Despite the burden of taxation now placed upon the licensed beer trade and its customers, the trade continues to thrive (Monckton 1966:132; Spring and Buss 1977:567).

SPIRITS SALES. Critics charge London distillers with keeping "shop and drinking houses" where they sell drinks "so fierce and heady" that a pint "will make half a score of men and women drunk" (quoted Clark 1983:96).

FRANCE

TEMPERANCE OF LOUIS XIV. Louis XIV begins his long reign (1643-1715) as king of France and becomes the most formidable monarch in Europe. Although a prodigious eater, Louis is a temperate drinker. Until he reaches the age of 20, he drinks only water; thereafter, he drinks only moderate amounts of diluted wine. According to Saint Simon (see 1710), such temperance is not emulated by others in Louis' court, at least in the king's later years (Rolleston 1941; Younger 1966:386-387). Lewis (1957:199) asserts that the king never touched spirits. It is true that Louis never drank a distilled beverage for recreational purposes, but the statement overlooks the king's consumption of rossolis for medicinal purposes (see 1660).

TAXATION. By royal edict in 1637, all town entry duties (*octrois*) are converted into royal *octrois* and the cities are allowed to double the levy and retain the second half of the increase (Cole 1939, 1:365).

TEA CONSUMPTION. Tea is not mentioned in France until 1635 or 1636, and is first met with suspicion. A medical student writes a thesis on the therapeutic value of tea in 1648; the thesis is rejected and the copy apparently burned (Braudel 1981:250).

POLAND

DRINKING PRACTICES. The peripatetic Peter Mundy, traveling from Danzig to Warsaw in 1643, calls heavy drinking a national failing and makes the following observation of peasant drinking practices: "A Crew of them [peasants] will combine and vow one to another to Meet in such a Crooh [country inn] and not to depart thence For a sett tyme. For a wekke, a Month or a yeare, as some say, and to spend thatt tyme in drincking, druncke s, sleeping, vomiting, til they bee sober and then drincke till they bee druncke againe, soe continue using their best witts to play the beasts and to exceed them in beastly Fillthinesse" (Mundy 1647/1925:196).

Bishop Cromer (*Beschreibung des Königreiches Polen* [Description of the Polish Kingdom]) also complains of rising excesses in eating and drinking, as well as luxury in clothing. He relates that Polish festivals frequently end in bloodshed, owing to quarrels which generally begin among the servants and retainers as it is a point of honor that they partake in the drinking. The wealthier the estate, the greater the excesses (cited Mundy 1647/1925:xliii-iv).

RUSSIA

INEBRIETY PREVALENCE; DRINKING PRACTICES. Western travelers and residents report that drunkenness is nearly universal in Russia, prevalent among peasant, priest, boyar, and tsar alike. According to Adam Olearius (1647/1669), who visits Muscovy during the reign of Tsar Michael (1613-1645), a Russian seldom willingly misses a chance to drink. To be intoxicated is an essential feature of Russian hospitality. Proposing toasts that no one dare refuse, host and guests drink numerous glasses of vodka, turning their beakers upside down on their heads to prove that they are empty. The evening is considered a failure if the guests do not go home dead drunk (Massie 1980:118).

According to J. Collins, the English physician for Tsar Alexis (1645-1676), the tsar loved to see his boyars (the landed aristocracy) "handsomely fuddled"; the boyars, in turn, enjoyed getting foreign ambassadors as drunk as possible. The "common people" drank to reach a stupor of unconsciousness in order to rapidly forget the unhappy world around them (Massie 1980:118).

SCOTLAND

BACKGROUND DEVELOPMENTS: REBELLION. In 1637, rebellion breaks out in Edinburgh against attempts to impose Anglicanism in Scotland. The need to raise funds to suppress this rebellion causes Charles I to convoke the Long Parliament, which leads to civil war.

TAXATION: FIRST WHISKY EXCISE. The Scottish Parliament imposes an excise tax on spirits (1644); whisky is taxed at a rate of 2s8d per Scots pint (Robb 1951:35; Murphy 1979:9). (A Scots pint is equal to approximately one-third of a gallon.) This measure is similar to the excise imposed by the English Parliament the year before on makers and retailers of ale, beer, cider, and perry (Williams and Brake 1980:2). At this time, Parliament badly needs money because under the Solemn League and Covenant signed in September 1643, it has agreed to raise an army to help the English Parliament in its war against Charles I. The collection of the tax is very haphazard because distillation occurs mainly in private homes, there are no roads into the Highlands, the law does not establish any machinery for collecting the tax, and no excisemen are appointed (Ross 1970:6). The tax is later lowered considerably by Oliver Cromwell during the Commonwealth. (Murphy [1979:11] says it was abolished in 1655, then reimposed in 1693 and increased in 1713.)

CHURCH BANS TOASTING. The Church of Scotland prohibits the drinking of healths by its members, blaming a rise in intemperance on the practice (Cherrington 1929, 5:2384).

AMERICA (BRITISH COLONIAL)

TAXATION. Tariffs and excise taxes begin to be used throughout the colonies for revenue production. In 1644, Massachusetts establishes an excise tax on all wines sold at retail by licensed vintners. In the same year, New York levies an excise tax on beer, wine, and brandy (Cherrington 1920:20).

MASSACHUSETTS

HOPS CULTIVATION. In 1641, the cultivation of hops is introduced into Massachusetts (Cherrington 1920:19).

DRINKING CONTROLS. In 1645, the General Court orders that no licensed drink seller shall allow "any to be drunke or drinke excessively, or continue tipling above the space of halfe an hour"; both the proprietor and the offending customer are subject to a fine. The court further defines "excessive drinking of wine" as more than a pint of wine at one time (Shurtleff 1853, 1:100).

TAVERN CONTROLS. A law passed in 1645 states that a town may select a "fitt man" to sell wine under license from the General Court or the Court of Assistance and to allow wine to be drunk in his house, provided he prevent drunkenness and immoderate drinking, under pain of a five pound penalty for every offense (Shurtleff 1853, 1:279-280).

Believing that no law against drunkenness will succeed unless the cause is removed, the court orders in 1646 that wine may be sold at retail only by taverners or other licensed persons, that taverners must pay a yearly license fee, and that constables must search taverns for disorderly conduct and illegal sale of wine. Another law prohibits drinking after 9 p.m., except for travellers, deprives keepers of victualling houses of their license for three years if they are convicted three times of violating the law against drunkenness, and strengthens the procedures for license renewal (Shurtleff 1853, 2:171-173).

FUNCTIONS: DRINK AS WAGES. Noting that forcing laborers to accept wine as wages is "a great nursery or preparative to drunkenness and unlawfull tipling, occasioning the private meetings of profane persons, whereby youth is drawen aside to lewdnes," Massachusetts orders in 1645 that no laborer can be compelled to take wine in payment for his labor. The law also forbids laborers from selling wine or settling debts with wine (Shurtleff 1853, 2:100-101).

INEBRIETY PREVALENCE. Despite the drinking and tavern controls established in the mid-decade, in 1648 the General Court observes that "a greate quantity of wine is spent, and much thereof abused to excesse of drinking and unto drunkenness it selfe, notwithstanding all the hoalsome lawes provided and published for the preventing thereof" (Shurtleff 1853, 2:257).

BEER CONSUMPTION, ATTITUDES. The Massachusetts court orders in 1649 that "every victualler or ordinary taverner shall alwayes...be provided of good and wholesome beare, for the entertainment of strangers, who, for want thereof, are necessitated to too much needless expenses in wine" (Shurtleff 1853, 2:286).

NEW NETHERLAND

TAVERN CONTROLS. In 1648, Governor Peter Stuyvesant complains that "nearly the just fourth of the city of New Amsterdam consisted of brandy-shops, tobacco or beer-houses." To correct the situation, he imposes a licensing system that includes the following provisions: (1) new taverns may not be set up without permission of the governor; (2) those tavern keepers already established may continue to do business for four years and must improve their accommodations for travelers; (3) tavern licenses may not be transferred; (4) tavern keepers may not entertain persons after 9 p.m. nor sell liquor on Sunday before 3 p.m.; and (5) a permit is required to purchase liquor. In another law of this year, Stuyvesant orders that drunkards are to be arrested and punished (Thomann 1887:93-94, quotation at 94).

PLYMOUTH

INEBRIETY PREVALENCE. In 1641, William Bradford in his History of Plymouth Plantation (begun about 1630 and completed in 1651, but not published until 1856) expresses his astonishment at the increase of drunkenness in the Plymouth colony (Bradford 1651/1908:364).

VIRGINIA

ORDINARY LICENSING. In 1644, the legislature adopts a policy of selective licensing of ordinaries (places where one can sleep and eat as well as drink) as a means to regulate alcohol use. Keepers of ordinaries and "eating houses" are required to obtain licenses from the governor after "approbation of the court of the county." The act attempts to confer upon ordinaries, as distinguished from tippling houses (which serve only alcohol), the exclusive right to feed and lodge people, while withholding the right to sell liquors other than strong beer. The ordinaries continue to sell without limitation, however, and the restriction on the sale of spirits is repealed the next year (Pearson and Hendricks 1967:13-14).

PRICE CONTROLS. By 1645, the Virginia legislature is fixing the prices of all "retailing wines or strong waters." (In 1659, price regulation is transferred to the county courts.) Generally, the prices are not unreasonable compared to prices in England, especially for wine. The main purpose of this legislation is to insure a steady supply of affordable drinks and to prevent speculation. Retailers openly violate the price-fixing policy (Pearson and Hendricks 1967:8-9).

1650

EUROPE

BACKGROUND DEVELOPMENTS: ECONOMIC CONDITIONS. A century of agricultural depression begins in western Europe, milder than that of the late Middle Ages, and interrupted by violent fluctuations, but still a serious, prolonged decline. As in the 14th century, cereal prices fall; arable land is converted to pasture, viticulture, and industrial-crop growing; real wages are relatively high; rural industries (especially textiles) develop; and urban populations expand (Slicher van Bath 1963:206, 218).

CONCERNS: POPULAR CULTURE. Throughout Europe, in Protestant as well as Catholic countries, the movement to reform popular culture enters a second phase which is more concerned with morality, licentiousness, and an ethic of respectability than with theology and "superstition." In this second phase, the laity plays a larger role than the clergy and the need for reform of popular culture is justified more often by secular than religious arguments (Burke 1978:240).

COFFEE INTRODUCED. Coffee, first appears in Venice about 1615, and in Marseilles, Lyons, Paris, and other French cities in the mid 1640s. Coffee reaches London and Vienna around 1650 and Sweden in the early 1670s (Franklin 1893, 9:33; Braudel 1981:256).

DENMARK

RATIONS; BEVERAGE CHARACTERISTICS. During the latter half of the century, the Danish navy's beer ration is reduced, probably because beer is now brewed strong and brandy has been introduced (Glamann 1962:130).

ENGLAND

BACKGROUND DEVELOPMENTS: THE COMMONWEALTH (INTERREGNUM). In 1649, King Charles I is executed and the Commonwealth is established under Oliver Cromwell. From the mid-century onwards, writers begin to become less concerned about the possible dangers of popular unrest and disorder due to too many people and some begin to advocate the desirability of a large population. This is probably due to the spread of Calvinist discipline, increasing food production, and declining population growth, which remove earlier concerns over limited means of subsistence. International economic competition also makes a larger labor supply more desirable (Coleman 1956:293).

CONTROLS: DRINKING, ALEHOUSE. Cromwell enforces rigid rules against drunkenness in his New Model Army and during the Interregnum period (1649-1660) efforts are made by the Puritans to restrict intemperance and drunkenness, primarily by discouraging tippling. There is a notable focus on deterrence rather than cure and a lack of sympathy for the drunkard (Harrison 1971:115). Convicted drunkards are sometimes made to wear a "drunkard's collar," consisting of a wooden cask with holes for head and arms, in which they are made to parade in disgrace. The Constitution "Petition and Advice" (1657) declares that "all Drunkards and common Haunters of Taverns or Alehouses" cannot serve in Parliament and should be disenfranchised (Bretherton 1931:159). Some congregations excommunicate drunkards. In addition, the Commonwealth Parliament forbids the holding of all ales and merrymaking near places of worship or on the Sabbath. In 1655, Cromwell establishes a system of major-generals who work to close public houses on Sundays and fast days, limit the number of alehouses in each parish, and restrict the number of new licenses issued. So strong is the feeling against alehouses in Somerset that justices of the peace concede local option to each village (Samuelson 1878:137; Monckton 1966:132; Harrison 1971:88-89).

But on the whole, the Interregnum period is not markedly different from the early Stuarts in terms of principles or drink legislation; its focus is generally on tightening up and more actively administering the old laws against drinking, gaming, and Sabbath-breaking. Christopher Hill observes that "the idea they [Puritans] imposed a gloomy godliness on a merrie nation is a post-Restoration myth"; in fact, few Puritans had all the "kill-joy" qualities of the post-Commonwealth stereotype (Hill 1961:172; Hill 1964:25). Bretherton (1931:159) considers the theory of disenfranchisement of drunkards to be the only new idea concerning drink control made by the Puritans. Cromwell himself ridicules the notion of promoting sobriety through prohibition. The Puritan major-generals lose their financing and authority after a year, and they are far more concerned with security (i.e., suppressing centers of popular discontent) than with godliness or morality when they close disorderly alehouses. According to Simon (1909, 3:188), their primary objective was checking the spread of royalist propaganda in the taverns. In enforcing Sabbath observations, they are only putting into effect Parliamentary legislation of the 1620s which the Stuart government ignored. In fact, in the early years of civil war and the Commonwealth, the drink-related legislation is "very inadequately enforced," local authorities lose their grasp over affairs, and a "multitude of unlicensed alehouses" spring up everywhere. While a tax is placed on beer and spirits by the Long Parliament, this is purely for financial purposes. Furthermore, the Commonwealth government restrains the JPs in some areas because they suppress so many alehouses that it feared that the excise revenues would decline (Bretherton 1931:160-161).

The Webbs (1903:13) believe that "the supervision of the Privy Council, by which the Justices had been kept up to the marks, was suddenly broken by the outbreak of the Civil War. We do not find any serious attempt was made either during the Protectorate or after the Restoration, to reconstruct the centrally supervised administrative system at which the statesmen of Elizabeth and James I had evidently aimed." However, Clark (1978:69) observes that whereas the civil war may have caused some relaxation of controls on the local level, by 1650 Puritan magistrates and preachers had allied together in a campaign against the alehouses. Furthermore, the major-generals make one significant contribution to the long-term regulation of alehouses: the establishement of special provincial licensing meetings of petty sessions, known as "brewster sessions" (Clark 1978:70). Court Session Records show that "after cases of assault and the making of affiliation orders, breaches of the licensing and drunkenness laws [provide] the largest and most regular single item of business down to the Restoration" (Bretherton 1931:163).

ALES DECLINE. As a result of the Commonwealth restrictions (see above), ales do gradually lose their importance.

INEBRIETY PREVALENCE; DRINKING PRACTICES. Despite the allegations of Samuelson (1878:150) and others, there is little evidence that supporters of the king drink any more than those of Parliament or the Puritans (Bainton 1945:53). Latham (1983:104) doubts that the Puritan Revolution reduced per capita alcohol consumption and suggests that it may have even increased it due to the establishment of large standing armies. Mid-century Englishmen still assert that "the Spaniard eats, the German drinks, and the Englishman exceeds in both" (quoted Minchinton 1976:127). In a letter, an anonymous French nobleman expresses shock at the drinking customs and houses of the period. He speculates that "a good half of the inhabitants" of the country may be keepers of alehouses, which he describes as a "meaner sort of cabarets." He writes further: "But what is more deplorable, there the gentlemen sit and spend much of their time, drinking of a muddy kind of beverage, and tobacco....As for other taverns London is composed of them, where they drink Spanish wines, and oter sophisticated liquors, to that fury and intemperance, as has often amazed me to consider it....A great error undoubtedly in those who sit at the helm, to permit this scandal; to suffer so many of these taverns and occasions of intemperance....Your lordship will not believe me, that the ladies of greatest quality suffer themselves to be treated in one of these taverns, where a courtezan in other cities would scarcely vouchsafe to be entertained....Drinking is the afternoon's diversion....I have found some persons of quality whom one could not safely visit after dinner, without resolving to undergo this drink ordeal. It is esteemed as price of wit to make a man drunk." He goes on to discuss the prevalence of taverns in London and the practice of "drinking healths," which are unknown in France, except for a single glass salute to an entire table (quoted French 1884:209-210).

ALEHOUSE PREVALENCE, CHARACTERISTICS. Although evidence is incomplete as to the effects of the Interregnum period on alehouses, between the 1630s and 1650s their numbers decline by a third in Cambridge. Cutbacks also occur in other areas, mainly under the influence of the major generals. However, numbers seem to be recovering before 1660 (Clark 1983:50). A growing proportion of drink is now imbibed on alehouse premises (ibid., 94).

TEMPERANCE OF MILTON. The Puritan John Milton (1608-1674), whose early poetry is steeped in the classics, refers to the power of wine to bring inspiration, like Rabelais and Spenser before him. He writes: "Poetry loves Bacchus,/ Bacchus loves Poetry." His sonnet to Mr. Lawrence "seems redolent of Horace in his Bacchanalian moods": "What neat repast shall feast us, light and choice,/ Of Attic taste with wine, whence we may rise..../ He who of those delights can judge, and spare/ To interpose them oft, is not unwise." But in his later adult life he is known for his moderation in food and drink; he recommends a frugal life and the use of pure water (quoted Legouis 1926:6). In his great religious epic, *Paradise Lost*, written in his last years to justify God's ways to man, he describes "Belial, flown with insouciance and wine," who rules over courts and palaces. He warns, "Intemperance on the earth shall bring/ diseases dire" (quoted French 1884:203-204). In his court play *Comus*, dedicated to marriage and fertility, Bacchus is the god that "first from out the purple grape/ Crush't the sweet poyson of mis-used wine" (quoted Carter 1976:9).

ATTITUDES: BEER. In *Ale Ale-vated into the Aletitude* (1651), John Taylor still calls beer "a Dutch boorish liquor...Alien to our nation...[and] a saucy intruder in this land" (quoted Mendelsohn 1963:30, 33; see also Thomas 1971:19).

COFFEE INTRODUCED; MEDICINAL USE. One of the first drinkers of coffee and promoters of its therapeutic benefits is the famed physician William Harvey

(1578-1657); in his will he bequeaths a large bag of it to his medical colleagues with a request that they meet monthly to drink it in his memory. By the mid-17th century it is already being prescribed by doctors. Two of Harvey's pupils, Edward Pococke and Walter Rumsey, are among the men who recommend it as a panacea and a cure for drunkenness (Ellis 1956:14-15). In a preface to Rumsey's Organon salutis (1657), James Howell asserts that "this Coffee-drink hath caused a greater sobriety among the nations; for whereas formerly Apprentices and Clerks with others, used to take their mornings' draught in Ale, Beer or Wine, which by the dizziness they cause in the Brain, make many unfit for business, they use now to play the Good-fellows in this wakefull and civill drink" (quoted Ukers 1935:53-55; see also Ellis 1956:52). Later, in 1665, a pamphleteer proclaims that "coffee and Commonwealth came in together for a reformation, to make us a free and sober nation." One coffee-house handbill around 1670 reads "Art thou surfeited with gluttony or Drunkenness? Then let this be thy common Drink" (Jacob 1935:93).

However, not all opinion is so laudatory. One contemporary considers coffee "useless since it serveth neither for Nourishment nor Debauchery" (quoted Latham 1983:107).

COFFEE HOUSES. In 1650, the first coffee house appears in Oxford, where the drink gains favor among students who form the Oxford Coffee Club, the origin of the Royal Society. In 1652, Pascal Rosea, Greek servant of an English merchant who imported Turkish coffee, opens the first coffee house in London. He advertises that the drink "will prevent drowsiness and make one fit for business."

Coffee houses and coffee drinking spread quickly throughout Britain, satisfying an important public need in the day-to-day business and social life of the community. Brewers demand that the new trade be taxed, and Pascal Rosea pays an annual report tax of one thousand sixpence (Ukers 1935:38, 52; Lillywhite 1963:17; Jacob 1935:92-94; Ferguson 1975:17-18; Monckton 1966:132).

TEA INTRODUCED; MEDICINAL USE. The English East India Company begins to import Chinese tea from Holland in 1646. Tea is first served in public in 1657, drunk very weak with sugar. It is advertised as "That Excellent and by all Physicians approved, China drink," and is very expensive (Braudel 1981:250-251; Ferguson 1975:124).

FRANCE

ATTITUDES. By this time, drunkenness shares the moral opprobrium that is attached to poverty and unemployment (Brennan 1984; see also 1650/England for the criticisms a French nobleman directs at toasting, drinking in alehouses, and drinking by women in England).

SPIRITS CONSUMPTION. Brandy production and consumption are still largely limited to Dutch sailors on the Atlantic coast, the hard drinkers of Brittany, the Gironde wine growers, and "Basque fishermen who smell so of spirits and tobacco that their wives turned them out of bed" (Le Roy Ladurie 1974:225).

DISTILLATION SPREADS. Bordeaux (see below) is the main seat of French brandy production and exports to England. Up until 1660, brandy is hardly produced at all in the Midi region; at Montpellier under Louis XIII (1610-1643), brandy is still little more than an apothecary's preparation (Le Roy Ladurie 1974:225-226). Distilling in Charente is only just beginning. But in the second half of the century, helped by expanding demand, particularly by Dutch traders, distillation spreads inland, as the problem of transport is less important for spirits than for wine. Also contributing to this spread is the viticultural crisis of 1655-1690, as distillation is one way of dealing with excess supplies of wine (see 1660/France). By the end of the century the brandy of the Cognac region is well-known and is being exported as far as the West Indies (Simon 1948:135). Armagnac is also beginning to gain an international reputation. Brandy gradually comes to be made wherever the raw materials (wine and wood) are available, and navigable waterways are nearby. The poor wine from the Meuse region in Lorraine is distilled into white brandy around 1690 (and perhaps earlier) in the same way as grape marc (Braudel 1981:244).

DISTILLING CONTROLS. The Parlement of Bordeaux proclaims (1657) that all stills must be dismantled and not erected except at the outskirts of the city. Restraining local distillers is seen as necessary because they had so increased in number that many citizens complained that they were a public nuisance (Simon 1948:135). Previous concerns over stills in Bordeaux had focused on the fear of fires (see 1550/France).

ESTEEM OF WINE. In a medical thesis, Daniel Arbinet argues that of all drinks the most agreeable and healthy are the wines of Beaune, in Burgundy. This instigates a century of verbal warfare over the relative merits of the wines of Burgundy and those of Champagne (Le Grand d'Aussy 1782, 3:34ff.; Emerson 1908, 2:147). (See also 1690/France.)

REFINEMENT OF MANNERS. The Ordre des Coteaux, one of the first "Wine and Food Societies" of western Europe, is formed in France by the marquis de Saint-Evremond and his friends. Known for their fastidiousness in matters of food and wine, they make the still wines of Champagne all the rage at the court. There is increasing attention paid by the end of the 17th century to the quality and service of food and wine, though the vigor of appetite and the quantity consumed are often unimpaired (Franklin 1889, 6:137-139; Younger 1966:376-377; Forbes 1967:95; Lausanne 1970:346).

CABARET CHARACTERISTICS. Some cabarets have sinister reputations, like the one in Paris described by a Dutch visitor of the mid-century as resembling "a brigards' liar or thieves' kitchen more than a place for honest folk" (quoted Braudel 1982:68).

MEDICINAL USE, SPIRITS. While most doctors still praise the medicinal properties of liqueurs and other spirits, Guy Patin attacks the use of rossolis as a ridiculous fad (Le Grand d'Aussy 1782, 3:77; Braudel 1981:245-246).

GERMANY

BREWING. After the ravages of the Thirty Years War (which ends in 1648), beer production picks up and rises above the output of prewar times (Klemm 1855:336).

ECONOMIC VALUE: ALCOHOL INDUSTRY. Johann Rudolf Glauber, physician and alchemist, called by some contemporaries "the second Paracelsus," discusses the economic value of alcohol production. He invents a distillery furnace which produces higher temperatures than before and increases the range of distillable substances. At the end of his life, he works for the German wine industry and then establishes the most impressive laboratory in Europe in Amsterdam. He is criticized for revealing too much about alchemy (Forbes 1970:201).

STUDENTS/DRINKING. The conduct and manners of university students are a "vulgar copy of those of the German nobility." Drinking bouts as well as dueling provide cheap outlets to people impressed with the "combative spirit of military officers and their contempt for the civilian classes" (Holborn 1964:317).

CONTROLS ADVOCATED. Veit von Seckendorf (Teutscher Fürstenstaat [On the German Prince], 1656), probably the foremost political philosopher in Germany at this time, outlines a philosophy of government and theory of state. Drawing on two centuries of ad hoc legislation, he provides a blueprint for action and a catalog of laws of a police nature which the prince should implement "for the welfare and common good of the fatherland." Among other things, the suggested regulations would (1) require church attendance and forbid blasphemy, profanity, and drunkenness; (2) "promote public health by proper measures against pestilence and contagious diseases,...by adequate care of the poor, the needy, and orphans,...and by protection against the evils of brandy and tobacco"; and (3) "promote economic prosperity by encouraging the virtues of industriousness...and by discouraging idleness" (quoted Dorwart 1971:16-17).

SUMPTUARY LAWS. In 1655, wedding celebrations in the villages of Brandenburg are limited to one a day. This and other regulations indicate concerns over peasants, indulging in immoderate eating and drinking and trying to impress their neighbors (Dorwart 1971:33, 34).

NETHERLANDS

GIN INTRODUCED. About 1650 Franciscus Sylvius (or Franz de la Boe, 1614-1672), physician of the iatrochemical school at the University of Leyden, is said to have invented gin by distilling alcohol from grain and masking the taste of the raw spirit with oil of juniper berries to produce a diuretic. He calls the medicinal preparation "aqua vitae," but it becomes known as Geneva or gin (Dutch, junever; French, genièvre). Whether Sylvius actually "invented gin" is open to question. The use of aromatic ingredients to flavor spirits has long been popular, as was a juniper-flavored wine (purportedly invented by the Count de Morret, son of King Henry IV of France, who attributed to it his long life and good health) (Cherrington 1926, 3:1108, who cites Engelbert Kaempfer as the source of his information; Houtte 1977:171). But whatever Sylvius' role may have been, it is in this period that the history of gin as we know it begins (Roueché 1960:25-26). Possibly this has more to do with the major expension of grain distilling in Holland in the second half of the century (see 1670/Netherlands) than with the discovery of any specific recipe for gin. Already in 1663, Amsterdam boasts over 400 small distilleries (Kellenbenz 1977:538). One of the major influences on this expansion is the conflict with France, which results in a temporary prohibition of French brandies in 1651 and a high tariff on wine and brandy in 1672. Furthermore, much of the success of

Dutch gin is due to the method of distilling, which enables the Dutch to obtain a milder spirit than elsewhere, a procedure that is soon imitated elsewhere (Morewood 1838:455).

RUSSIA

TAVERN CONTROLS (DVOR SYSTEM); REVENUE GENERATION. It is estimated that the total number of saloons in all of Russia is 1,000. In 1651, Tsar Alexis (1645-1676) abolishes all kabak concessions in Russia, renames them kruzhechnyi dvors (literally a "cup yard"), and tries to regulate sales. The dvors are to be closed on Sundays, cannot extend credit for spirit purchases, and cannot sell to the clergy. The number of dvors is limited, as is the amount that is supposed to be sold to customers (one cup to a customer; none to known drunkards). Whatever benefits may have resulted from these reforms are largely nullified by the continued sale of vodka alone (no beer or mead) and by a requirement that each year saloonkeepers produce profits exceeding those of the previous year (Pryzhov 1868; Efron 1955:498-499).

SCOTLAND

SOCIOCULTURAL DEVELOPMENTS: THE COMMONWEALTH. The execution of the Stuart king Charles I offends many Scots, who become royalists in retaliation. Cromwell's intolerance of Catholicism leads to trouble with the Highlanders, many of whom have remained loyal to the Pope. As a result, civil war follows and Scotland is militarily subdued. In 1653, Cromwell declares that Scotland is united with the Commonwealth ruled by London.

MILITARY/DRINKING. During the civil war, Cromwell's army captures Dundee in 1651, purportedly because the residents of the city were so drunk that they made no defense (Cherrington 1929, 5:2384).

WHISKY TAXATION. In 1655, the Commonwealth Parliament lowers by 95% the excise established in 1644 to 2d per Scots quart on spirits distilled in Scotland. The excise on spirits made in Scotland from imported wine is set twice as high at 4d. The importation of aqua vitae, presumably meaning brandy, is prohibited. Thus, whisky distilling receives some protection from foreign competition. These changes probably are enacted to equalize the tax rate throughout the whole Commonwealth.

Legislation in 1657 grants various powers to government officers to gauge vessels. This is the foundation of the excise control of the manufacture of spirits (Robb 1951:35; Ross 1970:7). (Murphy [1979:9] says the 1644 spirit tax was abolished by this legislation.)

AMERICA (BRITISH COLONIAL)

CONNECTICUT

TAVERN CONTROLS; TAXATION. A Connecticut law of 1650 recognizes the "necessary use" of ordinaries and houses of common entertainment where liquor is sold, but establishes strict rules governing the conduct of proprietors and patrons. Connecticut also imposes a heavy duty on imported liquors and an excise tax on liquors manufactured in the colony. Other colonies also increasingly rely on alcohol as an important source of revenue (Krout 1925:4; Cherrington 1920:13, 22, 24).

MASSACHUSETTS

INEBRIETY PREVALENCE. Continued widespread abuse of home-manufactured alcohol leads Massachusetts to pass a law in 1654 prohibiting others (i.e., guests) to drink within one's home. The new law states "notwithstanding the great care this court hath had and the laws made to suppress that swinish sin of drunkenness, yet persons addicted to the vice find out ways to deceive the law." Still, the new law has little effect (Lender 1973:359-360; Cherrington 1920:23).

NEW ENGLAND

DRUNKENNESS DEFINED. New England colonies attempt to establish a precise definition of drunkenness in order to fix a reasonable limit on what constitutes the abuse of one of God's gifts. These definitions include the time spent in drinking, the amounts drunk, and the behavior of drinking persons. Connecticut lawmakers include in their definition of a drunkard a person "bereaved or disabled in use of his understanding, appearing in his speech or gesture." In 1646, Plymouth Colony defines a drunkard as "a person that lisps or faulters in his speech by reason of drink, or that staggers in his going to that vomitts or cannot follow his calling" (quoted Lender 1973:355).

TAVERN REPUTATION. Many of the taverns of the 17th century are required to be run by men of good character and reputation; the tavern keeper, as operator of one of the main centers of community life, enjoys "great prestige and high social standing," often being ranked above the local clergyman (Asbury 1950:8).

RUM CONSUMPTION. Beginning about 1650, rum is imported from the West Indies to the mainland colonies, its cheapness making it particularly popular among the poor. New Englanders soon begin distilling their own rum out of molasses imported from the West Indies; in 1657, the first rum distillery in Boston is established (quoted Hooker 1981:37).

NEW NETHERLAND

WINE RATIONS; DRINKING CONTROLS. An ordinance of 1656 reads: "Inasmuch as it is necessary on a long voyage to maintain regularity in eating and drinking for the preservation of health, every one on board ship should be bound to drink his ration of wine every day, without being permitted to save or sell it." The same ordinance stipulates punishment for anyone found drunk on the ship (quoted Thomann 1887:99-100).

PLYMOUTH

DRINKING CONTROLS. Plymouth passes a law disenfranchising habitual drunkards (Cherrington 1920:24).

1660

EUROPE

MEDICINAL USE AND INTOXICATION. Some of the curative and prophylactic powers attributed to wine make it a veritable fetish. Dr. Nathaniel Hodges (Liomologia) credits sherry-sack with saving his life during the London plague epidemic of 1665. He drinks the wine before and after his meals and finds "it encouraged sleep and an easy breathing through all the pores all night" (Lucia 1963a:157). During the same plague, Samuel Pepys' doctor advises him to drink wine (see 1660/England).

While in Paris, the Italian priest Locatelli (see 1660/France) gets so drunk on a strong, heady Spanish wine and a round of toasting that he is placed in bed unconscious. He observes that such a debauch, according to his doctor, "was indispensable to me once a month if I wanted to keep clear of my headaches" (which he believes are caused by heating of the liver). However, "There's nothing better or healthier than the happy mean." He also notes that beer causes "that kind of intoxication that soon passes off" (quoted Blunt 1956:171-174). (According to Franklin [1889, 6:122], incorrigible but timorous drinkers often used the excuse that getting drunk twice a month was hygienic, but by the mid-17th century the belief was being abandoned.) (See also under 1680/England.)

ENGLAND

BACKGROUND DEVELOPMENTS: RESTORATION. The Stuarts are restored to the throne under Charles II (1660-1685). A great agricultural boom begins and casual labor replaces vagabondage. English society becomes increasingly dominated by money and the commercial market. Government policy is no longer determined by aristocrats whose main economic function is consumption. At the same time, the cheap consumer-goods industry expands so that by 1700 all sectors of the population are to some extent cash customers for goods produced outside their area. Comfort and privacy increase (Hill 1961:244, 253, 266, 307-308).

The parish becomes more exlusively a local government area, whose officers regard themselves as responsible to secular rather than ecclesiastical authority.

INEBRIETY PREVALENCE. It is frequently maintained that King Charles and his court precipitated a spread in laxness in private behavior, a proliferation of taverns, and an increase in alcoholism among all ranks of society. This view is found in Rolleston (1949:70), Sutherland (1969:163), Francis (1972:61), and Williams and Brake (1980:3), among others. French (1884:226) asserts that under the Restoration "drunkenness prevailed in every rank of society, and...the king set the example." Sutherland (1969:165) writes: "During the Restoration, drinking reached a new height in popularity with literary men....[But it is] doubtful if many of them could hold a candle to the alcoholic excesses of the aristocracy. Many of them died young through over-indulgence having little else to occupy their time."

The theme that drunkenness and vice in general increased under the Restoration is already commonly asserted in the writings of late 17th- and early 18th-century moralists, particularly the members of the Societies for the Reformation of Manners (see below). The cause for this is generally attributed to the bad example set by the court and upper classes, laxity in the enforcement of drink-related legislation, and the nation's general desire to disregard the stringent moralism of the Interregnum period. According to Daniel Defoe (A Brief Case of the Distillers, 1726:17), it was under Charles II that vice really began to grip the English nation and "lewdness and all manner of debauchery arriv'd to its meridian." He particularly identifies drunkenness as a major problem (quoted Earle 1976:155).

Law enforcement does become more lax on a national level (see below), and certainly a high level of drinking prevails. Samuel Pepys' diary (see below) records his frequent encounters with inebriates in all walks of life, particularly within the army and navy, and Pepys himself struggles to overcome his tendency to drink excessively. Thompson (1974:393) has "no doubt that there was a general and sometimes exuberant revival of popular sports, wakes, rush bearings, and rituals" in Restoration England. But it is unclear whether within the court or the country as a whole drinking increased or decreased under the late Stuarts. Mendelsohn (1963:3-4, 29) maintains that Restoration drunkenness has been exaggerated: "The profligacy of the court of Charles II has justly or otherwise become proverbial, but excessive drinking does not seem to have been the dominant feature"; England is "a country of copious drinking, but in general not of drunkenness"; overall, the decade of the 1660s appears relatively temperate compared to the 18th century. As for the king himself, although his lasciviousness is amply documented and he is known to get drunk occasionally, in general he is cautious about his food and drink (Rolleston 1944:70-71; Mendelsohn 1963:3; Francis 1972:61; Simon 1909, 3:111). Soon after he reaches London, Charles issues a proclamation against debauchery, including excessive tippling and drinking of healths (see below).

DRINKING CONTROLS: PROCLAMATION AGAINST DEBAUCHERY. On May 30, 1660, King Charles issues a proclamation against "vitious, debauch'd, and profane persons" who "spend their time in taverns, tippling-houses, and debauches, giving no other evidence of their affection to us but in drinking our health and inveighing against all others who are not of their own dissolute temper." As a remedy, he advocates moral persuasion rather than laws. He expresses his hopes that "all persons of honour, or in place and authority, will so far assist us in discountenancing such men, that their discretion and shame will persuade them to reform what their conscience will not....[What] laws cannot well provide against...may, by the example and severity of virtuous men, be easily discountenanced and by degrees suppressed" (quoted French 1884:234). Samuel Pepys asserts (June 4, 1660) that the proclamation gave "great satisfaction to all" (Pepys 1669/1970, 1:169).

In August 1663, the king renews the proclamation along with a new proclamation on the observation of the Lord's day. He requires that all attend Sunday church services, that travel be restricted on that day, and that "no person shall tipple or drink in any inn, tavern, alehouse, or victualling house, or sit idle, or play openly during the time of divine service." He requires that this second proclamation be read by every minister once a month for six months and requires that mayors, sheriffs, and JPs be "very diligent and strict in discovering and punishing" those who act contrary to the proclamation.

SUNDAY DRINKING. The king calls for the preservation of the Sabbath (see above) and does not republish the "Book of Sports." Foreigners continue to comment on the strictness of English Sunday observances (e.g., Jouvin de Rochefort in 1675 and Henri Misson in 1697 see 1670/England and 1690/England). Nevertheless, an "almost immediate change" occurs in Sunday observances, and a tendency to reject Puritan taboos is evident (Baker 1931:126-129). For example, on May 12, 1661, Pepys tells of drinking on Sundays at an inn. Apart from going to church, Sunday appears to become less a day of rest and meditation than of excursions, entertainment, and going to inns and pleasure gardens.

ALEHOUSE PREVALENCE. In the post-Restortion era, the rate of growth in the number of alehouses appears to slow (Clark 1983:55, 59).

CONCERNS: ALEHOUSES, INEBRIETY. During the Restoration (1660-1683) concerns about drunkenness and alehouses seem to dissipate. Despite the king's proclamations, the government of the Restoration largely abandons any effort to control alehouses and drinking (Bretherton 1931:161). No new drinking legislation is enacted, and the enforcement of existing laws on the national level is less vigorous. The decline in the concern over alehouses is especially evident. This has been credited to developments in agricultural production and the marketing of grain which erode government anxiety over food supplies, to greater demographic stability, to rising real incomes which check the progress of rural impoverishment, to the discrediting of the reforming enthusiasm under the Commonwealth, to the kings' fear of engendering the hostility of the local gentry, and to a fear of declining revenues from taxes on beer and spirits (Wrightson 1981:21; Bretherton 1931:161-162). The central administration of the licensing laws grows more lax, partial, and inefficient as the Restoration progresses, and in granting licenses more stress begins to be placed on outward conformity in matters of politics and religion than on good moral character and orderly behavior. Another problem is that enforcing licensing and other drinking controls is difficult because the JPs rely largely on unpaid officers (Bretherton 1931:167, 176).

The Webbs (1903:15-16) assert that by the end of the century, "a period of extreme laxness" in licensing sets in; practically anyone can obtain a license to run a public house, the "superfluous number" of which multiplies. They trace to origins of this laxness to the break down of centralized administrative control over the JPs during the Civil War and to the desire of the JPs to collect more fees.

However, Clark (1983:179) argues that the picture regarding licensing is "much less permissive" than the Webbs portray and that by the end of the turn of the century (see 1700/England) many developments continue to contribute to a sense that the alehouses are under control. This in itself may have been responsible for a decline in legislative action against them. County governments continue the administrative advances in alehouse controls begun during the Interregnum. By the 1670s, brewster sessions are adopted in most counties (Clark 1978:70).

SPIRITS CONSUMPTION. Consumption of brandy (mostly imported from France) rises, and rum is introduced (Longmate 1968:7; Samuelson 1878:151; French 1884:221, 233-234). But spirits consumption is still limited. In Samuel Pepys' diary (see below), only two references to brandy or aqua vitae occur, indicating that he was not a regular spirit drinker (Mendelsohn 1963:82). On March 3, 1660, Pepys records that he "drank a great deal of strong water, more than ever I did in [my] life at one time before" (Pepys 1669/1970, 1:76; see also entry for March 5, 1668). He makes no reference to whisky.

BEER TAXATION. In the spring of 1660, the new Parliament temporarily continues the excise duties imposed by the Commonwealth and orders brewers to pay all duty in arrears. This action comes as a shock to the brewing trade, which thought that the end of the Cromwellian government would also be the end of the beer duty. This causes strong feelings of "disappointment and resentment" among brewers, who become enmeshed in a net of bureaucratic control; but after a few years the excise revenues become an accepted part of life and soon begin to be manipulated according to the fiscal requirements of the day (Monckton 1966:11-119). One effect of this fiscal change is to encourage the development of larger commercial brewers since this simplifies supervision and tax collection.

In exchange for the king, giving up certain feudal rights and claims, two acts are passed granting to Charles "for life" the duties of 1/3d a barrel on strong beer and 3d a barrel on small, and 3s a barrel on imported beer. This "temporary" excise granted to Charles II is continued to James II and later to William and Mary (Monckton 1966:118).

<u>PEPYS/DRINKING</u>. In his diary kept between 1659 and 1669, Samuel Pepys frequently makes reference to drinking customs. Pepys is at times a heavy drinker; he refers to drinking five or six glasses of wine at a time, often in the afternoon (e.g., see September 17, 1662 [Pepys 1669/1970, 3:199-200]. But he is not a dissolute drunkard (Mendelsohn 1963:2). He repeatedly expresses his disapproval of drunkenness, complains about drinking too much, and tries to drink less heavily and less often, even making several vows to abstain from wines, apparently for only limited times (February 8, March 9, 1660; February 24, November 18, December 31, 1661; June 28, September 17, 1662 [Pepys 1669/1970, 1:46, 84, 2:42, 215, 242; 3:125, 199-200]). For example, on March 9, 1660, becoming "overheated with drink," he promises "to drink no strong drink this week." Frequently he makes reference to the ill effects he suffers from drinking, which he calls "my great folly" (August 9, December 2, December 27, 1660 [1:218, 307, 321]). At the same time, life without alcoholic beverages seems inconceivable to him.

During 1662, he prides himself on the progress he makes in avoiding drink and he frequently credits better mental and physical health, as well as business success, to his new temperance. On January 26 he writes: "But thanks to God, since my leaving drinking of wine, I do find myself much better and do mind my business better and do spend less money, and less time lost in idle company." On June 12, he expresses "great wonder" at the ease with which he "past the whole dinner without drinking a drop of wine." On June 28, he records: "My mind is now in a wonderful condition of quiet and content, more than every in all my life....and I find it to be the very effect of my late oaths against wine and plays; which, if God please, I will keep constant in" (3:18, 107, 125). In other instances he expresses guilt over having indulged in even a little wine and describes his attempts to drink only moderately (April 7, June 17, June 23, July 3, July 31, August 11, 1662; June 15, July 19, July 30, 1663 [3:61, 112, 119, 130, 151, 163; 4:186, 235, 254]). On another occasion (September 1665), he resumes drinking during a plague on the advice of his doctor. In 1665, when he hears a rumor that he has gone to gambling and drinking, he attributes this to jealousy of his temperate ways (6:243).

The diary makes few references to ale, probably because malt liquor is considered less an alcoholic beverage than an essential part of the diet. Any controversy over it centers on quality and price (Mendelsohn 1963:4-5, 29). (See below for Pepys' attitudes toward toasting and drinking in the navy.)

NAVY/DRINKING. Pepys also plays an important role in the campaign against drunkenness in the Royal Navy, "in which it was exceedingly rampant" (Rolleston 1944:70).

TOASTING. Pepys describes how on the accession of Charles to the throne, the people of London rejoiced and drank to his health on their knees in the street (May 1-2, 1660 [Pepys 1669/1970, 1:121-122]). On the issue of the drinking of healths, Pepys praises Prynne's "discourse proving the drinking and pledging of healths to be sinfull and utterly unlawful unto Christians" (see 1620/England) (June 6, 1665 [5:172]).

ATTITUDES: WAGES AND DRINK. Thomas Manly (Usurie at Six Per Cent, 1669) complains that higher wages bring higher drinking: "The men have just so much the more to spend in tipple, and remain now poorer than when their wages were less....They work so much the fewer days by how much the more they exact in wages" (quoted Coleman 1956:290-291).

UPPER-CLASS WOMEN/DRINKING (MADEIRA). The marriage of the Portuguese Catherine of Braganza to Charles II establishes the fashion of upper-class English women having a glass of Madeira wine in the morning. This ritual lasts "well into Victorian times" (Sutherland 1969:86).

COFFEE HOUSE LICENSING. All coffee houses are required to be licensed with a fine of five pounds for every month's violation of the law. They are placed under close government supervision (Ukers 1935:55).

TEA CONSUMPTION, PRICE. Tea becomes a fashionable drink but it is still prohibitively expensive (Monckton 1966:132-133).

FRANCE

DRINKING PRACTICES; WOMEN/DRINKING (LOCATELLI). The Italian priest Sebastiano Locatelli makes several observations about drinking in France during his trip to Paris in the mid-decade. In the south of France, he sings gay provencal drinking songs; in one favorite song the poet asserts he has been everywhere, seen everything, but has never come across anything better than drink. Locatelli's female companions in Paris never drink wine, "for in France, one of the greatest insults one can address to a woman is to tell her that her breath stinks of wine." However, at Lyons, he is astonished by the great number of debits de vin and claims that there they drink more wine than in a dozen cities in Italy. He praises the good wines of Burgundy and attributes the Burgundian plumpness to "drinking so much" (quoted Blunt 1956:105-106, 146, 184, 208-211). Locatelli's observations on Lyons can be taken as an example of the difference between the high level of wine consumption in the cities and the low level in the country (Dion 1959:473, 488).

TAXATION. Throughout the reign of Louis XIV (1643-1715) taxes on wine rise. As part of his financial reforms, Jean Baptiste Colbert rationalizes, centralizes, and raises the aides as part of a single general tax farm. The object of the general farm is to follow each cask of wine from the time it is laid down in the winery to the time it is decanted and sold by the bottle or glass (Cole 1939, 1:305). New tariffs are established in 1664 on drinks made from brandy, wine, or water, such as liqueurs and syrups (Forbes 1970:196).

VITICULTURAL CRISIS. Similar to conditions in the late Middle Ages, as cereal prices decline (see 1650/Europe) the vineyard acreage begins to expand, especially after 1720 (Slicher van Bath 1863:217). However, between 1655 and 1690, the southern vintages experience a period of crisis caused by declining prices, a drop in production, and eventually "an appalling contraction in the wine trade." Wine is the most seriously affected of all crops during the general agricultural depression. By 1660, the price structure is no longer renumerative. In Languedoc, all wine growing villages shut themselves off, restrict any wine or grape imports, consume barrels of local piquette in their homes and taverns in the hope that the price will again rise, and begin to distill their surpluses (Le Roy Ladurie 1974:225). (On the growth of distilling, see below.)

SPIRITS: USER CHARACTERISTICS. Spirits drinkers are mostly poor people who are content with "cheap spirits, a little rude and without great finesse" (Delamain 1935:29). But the drinking of liqueurs is spreading among the wealthy. About 1666, the abbot of Choisy writes of a dinner he attended, after which each guest had un petit coup de rossoli (Franklin 1889, 6:146-147; Younger 1966:386-387).

DISTILLING: LANGUEDOC, CHARENTE. Widespread distilling begins in Languedoc at the very bottom of the viticultural depression around 1663-1664, when the first references to brandy distillation occur in the areas of Béziers and Lunel. By the end of the 1670s, Sète has become a center of liquor exports (Le Roy Ladurie 1974:226).

A droit de sortie charged upon every tun of white wine shipped from La Rochelle is continually raised over a century and a half, beginning in 1664. Such export taxes cannot be borne by the light, white wines of the Charentes, which are of no particular merit except their low price. Increasingly, the unsaleable wine is turned into brandy (Simon 1926:32).

ATTITUDES (COLBERT). Colbert supports a reduction in drunkenness since he feels that it is antithetical to productive labor, but he encourages the export trade in wine to help the French economy (Cole 1939, 1:361, 412). Colbert also recommends that industries not be established in towns which have vineyards, "for wines are very great hindrances to work" (quoted Braudel and Spooner 1967:409).

TOASTING. On June 19, 1663, Pepys (1669/1970, 4:189) tells how "Mr. Moore showed us the French manner when a health is drunk, to bow to him that drunk to you, and then apply yourself to him whose lady's health is drunk, and then to the person that you drink to; which I never knew before, but it seems is now the fashion."

COFFEE. The first coffee house in Marseilles meets opposition from the local vintners (Jacob 1935:103).

GERMANY

DRINKING CONTROLS, CONCERNS. Saxony-Gotha passes an ordinance in 1666 stipulating a fine for public drunkenness and goes on to describe the negative aspects of drunkenness with regards to the loss of public and family responsibility (Stolleis 1981:102, 103).

BEER SALES CONTROLS. A 1661 tax edict issued by Elector John George forbids that home-brewed beer should "on any account be sold or publicly provided" (quoted Jacob 1935:89).

DRINKING PRACTICES, ATTITUDES (GRIMMELSHAUSEN). John Jacob Christof von Grimmelshausen publishes The Adventures of Simplicissimus (1669), the great plebian novel of 17th-century Germany. The novel provides a intimate look at daily life during the Thirty Years War, including drinking habits. The first German novel to deal with contemporary events in everyday language, it is an instant success. The story is the autobiography of Simplex Simplicissimus, an infant who loses his parents and is brought up by peasants and then by a hermit in the forest.

POOR/DRINKING. Having spent his life in poverty, Simpliccisimus is not accustomed to drinking alcholic beverages. When he enters into the service of a lord, the grooms try to get him drunk by using pledges with sack and malmsey, but as the grooms themselves are humble and are not used to such "heavenly Nector," they too get drunk Grimmelshausen 1669/1963:56).

UPPER CLASSES/DRINKING. Simplex describes at length one of his lords' banquets at which all arrive sober "but after the first three or four drinks 'to your health' things got livelier....I watched the guests eat like pigs, drink like cattle, behave like asses, and finally throw up like sick dogs." They pour down noble wines by the bucketful. He finds the sight amazing as he "had no knowledge of the effects of wine and drunkenness." He refers to "that Teutonic probity which insists on a man matching his neighbours glass for glass" and to guzzling "wine in buckets to the health of princes, friends, and sweethearts." Furthermore, "he who could last longest and drink deepest boasted of it and thought himself a splendid fellow." The banquet ends in bloodshed (1.23, pp. 44-47).

MILITARY/DRINKING. The traits he mentions to characterize a soldier's lot are "gluttony and drunkenness, hunger and thirst, whoring and sodomy, gambling and dicing, murdering and being murdered" (1.14, p. 27). As a soldier, Simplex spends all his money "with good companions on Zerbst and Hamburg beer, to both of which I became much addicted" (2.16, p. 82).

ATTITUDES. The chapter describing his lord's banquet is entitled, "His first sight of drunkards makes Simplex think/ There is more madness than health in drink." Simplex is told by a Parson that "where wine goes in, wit flies out." As he views the lord's guests "wantonly" wasting food and drink, he thinks of the homeless, starving peasants and the famine in the town (1.23, pp. 46-47). Later, plentiful food and drink are envisioned as paradise (2.26, pp. 104-105).

DRINKING INFLUENCES: DIET. That the salty food of the period contributed to high levels of alcohol consumption is verified by Simplex's observation (see above) that when he lived with the poor hermit they never used any salt or spices, since they did not have any cellar and did not want to raise a thirst (book 1.10, Grimmelshausen 1669/1963:20).

IRELAND

TAXATION: SPIRIT DUTY EVADED. By the 1660s, the distilling industry in Ireland is sufficiently established and large enough to attract the attention of the government as a source of revenue. By an act of 1661, domestic spirits are subject to a duty of 4d per gallon. The same act establishes the Excise

Commission, with authority to appoint gaugers and searchers to collect the duty. Evasion of the spirit duty is widespread, and the government continually strengthens the law to try to bring in the lost revenue (McGuire 1973:97-98; McGuffin 1978:9).

LICENSING. An act of 1665 brings spirit and wine retailing under the licensing system created in 1634 for beer. The license is in effect for three years. Since the commissioners earn a fee of 1s for each license granted, they are inclined to approve as many licenses as possible, thus undermining the intent of the law to restrict spirit drinking (McGuire 1973:101).

SPIRIT CONSUMPTION. Around this time, the drinking of whiskey and other aqua vitae is said to be extraordinary. Imported brandy reportedly becomes more popular than whiskey (O'Brien 1919:15, 144).

RUSSIA

TAVERN CONTROLS LIBERALIZED: KABAKS REINSTATED. The kabak concession system is reinstated in 1663 after Tsar Alexis' reforms (see 1650/Russia) are found to result in too great a loss of revenues (Cherrington 1929, 5:2332; Efron 1955:499).

SCOTLAND

TAXATION. Duties are levied on all imported beers and spirits in 1661; domestic Scotch whisky is exempted from direct taxes (until 1693). If any district fails to meet its quota of an annual grant which must be paid to King Charles II, the district must make up the difference by a tax on malt (Robb 1951:35; Ross 1970:7). This is the beginning of the malt tax, which is made a general imposition in 1681 and continues, on and off, until the 19th century.

Between 1661 and 1823, there are some 30 changes made in whisky duties. This indicates that "the policy was unclear and the system to control and raise revenue from whisky was not well designed" (Robb 1951:42). (Considerable confusion surrounds taxation policy during this period; see MacDonald 1914:39; Sutherland 1969:60.)

DRINKING CONTROLS. Penalties are established in 1661 for cursing, swearing, and drinking to excess. Penalties are graded according to rank: nobles are fined 20 pounds (Scots), while yeomen are fined only 2 pounds. Ministers are fined a fifth of their stipends. Anyone unable to pay is to be "exemplarly punished in his body" (*Scottish Law Review* 1943:86).

IMPORT CONTROLS. The Scottish Parliament in 1663 prohibits the importation of foreign aqua vitae, possibly to encourage the local distilling of spirits from imported wines, etc., which are taxed. In 1672, imports of all foreign spirits are ordered diverted to Edinburgh, probably to make the collection of the import duties due easier (*Lancet* 1905:240).

AMERICA (BRITISH COLONIAL)

MARYLAND

INNS, LIQUOR TRADE ENCOURAGED. In 1662, Maryland passes a law to encourage the establishment of inns, with special licenses granting a monopoly of the sale of liquors in a district. Like other colonies, Maryland seeks to encourage innkeeping, brewing, distilling, and trade (Krout 1925:6).

NEW YORK

SALES CONTROLS. In 1665, New Netherland becomes an English colony (New York). The new laws for the colony include provisions to regulate the drink trade. Brewers of beer for sale must be qualfied in the art of brewing; customers who purchase beer of inferior quality may recover damages from the brewer who sold the beer; liquor retailers must present a certificate of good behavior in order to obtain a license, which must be renewed annually; the price of strong beer is set at two pence a quart; and licensed drink sellers must not permit excessive drinking and are responsible for maintaining order in their establishments (Thomann 1887:103-104).

VIRGINIA

TAVERN CONTROLS. In 1661, the 1644 ordinary licensing act is extended to all tippling houses, where, according to the preamble, "many disorders and riotts" occur. The act places responsibility for any disorder on the retailers not only through the requirement of a license but also by bonding them to maintain good order and to sell at legal prices. These licenses need not be renewed and can be revoked. The new act removes all distinctions between tippling houses and ordinaries. Subsequently, the number of places selling alcohol greatly increases (Pearson and Hendricks 1967:14-15).

In 1668, the Virginia legislature directs the county courts "to take speciall care for the suppressing and restraint of the exorbitant numbers of ordinaries and tippling houses in their respective counties, and not to permitt in any county more than one or two, and those neare the court house, and noe more, unles in publique places...where they may be necessary for the accommodation of travelers." The ordinaries and tippling houses have become havens for "a sort of loose and carelesse persons." In this statute, and in one of 1671, rates are set in detail. The point is made that although drink places were "set up for a private gain," public convenience must be considered in permitting them. The restrictions do little to improve behavior at ordinaries and tippling houses (Pearson and Hendricks 1967:15-16).

DRINKING DEBT CONTROLS. As part of its price-fixing efforts, the Virginia legislature stipulates in 1666 that drinking debts may be collected only if the drinks were charged at legal prices, the person charged knew the price of the drinks, and the suit was begun during his lifetime. The law is intended to prevent the collection of huge debts created while the debtor is on a spree or to prevent suits brought into court when the debtor is no longer alive to challenge their accuracy (Pearson and Hendricks 1967:12).

1670

ENGLAND

BRANDY CONSUMPTION, CONCERNS. Evidence that brandy consumption is becoming common in England appears for the first time. Complaints are raised about Londoners on holiday drinking too much brandy. Trade sources indicate a large rise in French brandy imports coinciding with the increased production in south France (Francis 1972:74-75). A petition is presented to Parliament in 1673 to prohibit the drinking of brandy, rum, coffee, and tea on the grounds that they are a threat to the brewing trade and agricultural interests and, unlike beer or ale, a hazard to the national health. Much of the hostility towards these drinks is rooted in concerns over declining revenue from beer. The petition alleges that "brandy is now become common and sold in every little ale-house" and recommends that it be prohibited in order to "prevent the destruction of His Majesty's subjects many of whom have been killed by drinking thereof, it not agreeing with their constitution" (quoted Simon 1948:137; Monckton 1966:134).

BRANDY IMPORTS BANNED. Retaliating for a French tariff, in 1677 England prohibits the importation of French brandy out of concern that home distillation is suffering. This results in the spread of home manufacturing of brandy, aqua vitae, and other spirits from wine and malted cereal. The prohibition is soon lifted, and then reimposed in 1690 (Dowell 1888, 4:163).

DRINKING PRACTICES; INEBRIETY PREVALENCE (CHAMBERLAYNE). Edward Chamberlayne discusses drinking habits in his *Angliae Notitia, or the Present State of England*, an immensely popular description of the country by an Englishman first published in 1669. He writes that the national vices of the English were historically "gluttony and the effects of lasciviousness," including the consumption of great quantities of French wine. He blames the introduction of the "foul vice of drunkenness" on the High Dutch, adding: "This vice of late was more, though at present so much, that some persons and those of quality may not be safely visited in the afternoon, without running the hazard of excessive drinking of healths (whereby in a short time twice as much liquor is consumed as by the Dutch, who sip and prate) and in some places it is esteemed a piece of wit to make a man drunk, for which purpose some swilling insipid trencher buffon is always at hand" (quoted from the 1674 edition by Ranum and Ranum 1972:152).

DRINKING PRACTICES (ROCHEFORT). In the mid-decade, the Frenchman A. Jouvin de Rochefort (*Le voyaguer d'Europe*, 1675, 3:451) asserts that there is no other kingdom where Sunday is observed more than in England, and that "no kind of business is transacted in England without the intervention of pots of beer," which he considers the best in Europe. During a talk with a Cambridge clergyman, he was required to drink two or three pots of beer. He also complains about the English custom of toasting and about the habit of the daughters of innkeepers who sit at the table with the guests and drink as much as the males (quoted French 1884:210, 224; see also Baker 1931:138; Ellis 1956:13; Flandrin 1983:69).

DRINK RATIONS. Each soldier sailing to Virginia in 1676 is given a gallon of brandy as a three-month allowance. A new victualling contract for the navy established in 1678 (in part under Pepys' influence) gives every seaman a daily allowance of a gallon of beer, or, if south of Lisbon, a quart of wine or a half quart of brandy. This indicates increasing familiarity with brandy in England (Mendelsohn 1663:84).

COFFEE: CONCERNS. Of all the new drinks, it is coffee that evokes the most controversy at this time; much of this controversy is actually rooted in concerns over the coffee houses, which in government circles are viewed as a burden on the economy and as centers of republican sedition. This is first reflected in the 1673 petition against all the new drinks (see above). In 1674, a "Women's Petition Against Coffee" complains that coffee makes men unfruitful, causes domestic disorders, and interferes with business (women are not allowed in the coffee houses). Coffee houses are called the "great enemies of industry" because of the idle time and money men spend there. Supporters praise the coffee house for providing cheaper, healthier, more sober, and more civil diversion than the alehouse (Ukers 1935:66-67; Jacob 1935:94-95, 135; Ellis 1956:88).

MEDICINAL USE. In one of the first attempts to do justice to both sides of the coffee question in England, Dr. Thomas Willis (*Pharmaceutica rationalis*, 1674), a distinguished Oxford physician, admits that there are times that he sends his patients to the coffee house rather than to the apothecary, but he warns that, at best, coffee is a somewhat risky beverage. It can cause languor and even paralysis; it may "attack the heart and cause tremblings in the limbs." On the other hand, it may, if judiciously used, have a marvelous benefit: "Being daily drunk it wonderfully clears and enlightens each part of the Soul and disperses all the clouds of every Function" (quoted Ukers 1935:54-55; see also Ellis 1956:16-17).

COFFEE HOUSES: CONCERNS, CHARACTERISTICS. Following the Great Fire of 1666 many new and larger coffee houses are opened in London; by 1669 some are already serving beer. They begin to draw drunks from taverns and alehouses after closing hours who want to sober up, but also cause considerable disruption. In order to maintain order, some coffee houses post a rhymed set of rules stipulating that all customers must maintain order, that fines will be imposed for swearing, quarreling, and gambling, and that conversation must avoid sacred things, profanation of the Scriptures, and criticism of the state (Ukers 1935:56-57).

BAN ATTEMPTED. In December 1675, Charles II issues a proclamation for the suppression of all coffee houses in London. He refers to them as "the great resort of Idle and disaffected persons" and charges that their growing numbers have "produced very evil and dangerous effects." Most notably, "in such Houses...divers False, Malitious, and Scandalous Reports are devised and spread abroad, to the Defamation of his Majestie's Government." The immediate public uproar and discontent is so intense that, within a month, a new proclamation announces that, subject to certain restrictions and safeguards, the coffee houses can remain open. Heavy taxes are imposed instead (Ukers 1935:68; Lillywhite 1963:18; Ellis 1956:92). Following this controversy, the great expansion of the coffee houses occurs.

FRANCE

SPIRITS: MEDICINAL USE (CHARAS). Chemist and physician Mayse Charas devotes several chapters to distillation in his *Pharmacopée royale Galénique et chymique* (1676). He draws special attention to absinthe and other distillates of herbs, and lavishes praise on their medicinal value (Forbes 1970:206; Delamain 1935:30).

DISTILLING EXPANDS. The pharmacopoeia of Charas (see above) indicates that the production of good wine spirits is still considered a matter of much patience and that spirits are not yet prepared in France in great quantities (Forbes 1970:206). Other evidence indicates that in the mid-decade spirits production and consumption begin to increase in France. Distilling begin to expand greatly for the first time in both the Midi and Charente (see 1660/France). In other developments, two Scots establish a distillery in Bordeaux and begin exporting more brandy than anyone else (Simon 1948:1350). Furthermore, the industry has become large enough to warrant establishing sales privileges and controls (see below).

CONTROLS: DISTILLING, SALES. In 1674, sales of spirits are restricted to apothecaries, spice merchants, and vinegar sellers. In 1676, distillers and spirits sellers are united with limonade sellers and the entire distilling trade is combined into one organization. The "Statuts et Ordinances pour La Communauté des Maitres Limonadiers, Marchands d'Eau-de-Vie...de Paris" establishes some degree of monopoly over the making, buying, and selling of spirits, either wholesale or retail. Two hundred and fifty masters are authorized (Forbes 1970:195; Younger 1966:329; Saint-Germain 1965:297-298). The privilege to make and sell spirits becomes such a good source of income that the corporations are frequently dissolved and new corporations are created with more members (1704, 1706, 1713). The *limonadiers* are given the sole privilege of selling spirits, coffee, and limonade from tables (Le Grand d'Aussy 1782, 3:78, 91; Gottschalk 1948, 2:227). Among the spirits allowed to be sold on the streets of Paris is rossolis, which often becomes used as a generic name for any liqueur based on brandy, sugar, and herbs (Younger 1966:386-387; Le Grand d'Aussy 1782, 3:77; Franklin 1889, 6:146; Leclant 1979:90).

According to an edict of 1678 authorizing street sales of spirits, "poor sellers of spirits, possessing it and desiring to sell in retail in small measures, in the streets on tables and stools...are permitted to have portable hoods under which to put their display to avoid the injuries of weather" (Le Grand d'Aussy 1782, 3:68; Delamain 1935:28-29). An ordinance of 1680 establishes that those who have purchased spirits *à pot et à pinte* (roughly in a jug and in a quart container) need not pay the supplementary duties to resale *a porte-col* (meaning itinerants selling from portable barrels worn around the neck) and on street corners. The number of people selling spirits in Paris grows over the next years. They are described as hucksters (*regrattiers*) "who each day in the morning and when the stores begin to open and the workers and artisans go to begin work, establish their small stores at corners of the streets or walk around the city carrying a *cabaret*, bottles, glasses and measures, in a small box hung from their neck" (quoted Delamain 1935:29, who seems to make a typographical error in dating the 1680 ordinance as 1660). The issuance of frequent regulations for the spirits trade (in 1678, 1693, 1699, etc.) indicates that the use of spirits is becoming more common among the street people (Delamain 1935:29).

ATTITUDES: LIQUEUR DRINKING. A letter of Madame de Sévigné dated 1674 indicates that liqueurs are regarded by some as a luxury from which devout persons abstain (Le Grand d'Aussy 1782, 3:78; Franklin 1889, 6:147).

INEBRIETY PREVALENCE; SPIRITS CONSUMPTION. According to Le Roy Ladurie (1974:226), the slump in the wine trade and the subsequent rise in distilling has begun to fundamentally alter drinking patterns and the level of inebriety in France by the mid-1670s: "Brandy-based alcoholism annexed itself to the wine-based alcoholism of the age of Louis XIII....In the towns, the rich, with their marc brandy, no longer held a monopoly on drunkenness. By 1675 the workers, too, were sampling eau-de-vie and walnut brandy in small doses. In the streets, hawkers were selling spirits....By 1680 even in the sober Midi the bishops were fulminating against the village tavern that was the ruin of the countryside."

FESTIVE DRINKING. Local church authorities in the diocese of Alet complain that only 100 out of 600 communicants attend Easter services; the others refuse absolution in order to drink and dance at the parish feast (Bercé 1976:151).

CABARETS: PREVALENCE. There are an estimated 1,847 cabarets in Paris in 1670 (Saint Germain 1965:293).

CONTROLS: HOURS LIBERALIZED. Because of a loss of receipts to the treasury, by a declaration of the Royal Council the hours during which cabaret keepers are prohibited from selling wine to local residents are reduced in 1670 to only the hours of divine service on Sundays and *fêtes* (Dion 1959:489).

JUSTICES, VILLAGES/DRINKING. The Parlement of Paris complains in January 1672 that in certain rural villages (*bourgades rurales*), justices encourage intoxication by their own example, having no shame in going to cabarets and even rendering justice there (Dion 1959:490).

SPARKLING CHAMPAGNE INTRODUCED. Over a 20-year period (c. 1668-1690), Dom Perignon, the cellarer of the abbey of Hautvillers, develops the first sparkling champagne as we now know it. Much legend surrounds Perignon and his achievements, but it is generally accepted that by inventing (or at least recognizing the possibilities of) the cork stopper, and by taking advantage of the new stronger glass wine bottles, Perignon is able to bottle the wines of Marne in a state of secondary fermentation so that they remain permanently sparkling. He then begins to perfect the technique of producing clear white champagne wines and an improved technique of blending. Perignon probably only succeeds in making a *crémant* wine without as much sparkle as the fully sparkling *vin mousseaux*. It takes more than a century to develop fully the technique of producing true champagne, especially to learn how to prevent bottles from exploding. Champagne does not gain general popularity until the late 18th century, but it is popularized in the early 18th century by the duc d'Orléans, and later by king Louis XV (Younger 1966:345-346; Forbes 1967:39-41, 113-131). By 1700 it is fetching double the price of the best old still wines of champagne (Seward 1979:142).

COFFEE: AVAILABILITY, SALES. By 1671, coffee is sold publically in several shops; in 1672, the first Parisian coffee house opens. When the "master *limonadiers*, brandy sellers" are given bylaws in 1676, article III grants them the "right to compound and sell...coffee in the bean, as powder, and as a drink" (Jacob 1935:120-121; Leclant 1979:88-90).

MEDICINAL USE. An anonymous treatise on coffee drinking heralds coffee as a marvelous remedy (De l'usage du caphé, du thé, et du chocolate, written by Jacob Spon in 1671, but also credited to T. Dufour). However, other doctors and public opinion holds that coffee is an antiaphrodisiac and a "eunuch's drink" (Braudel 1981:257; Le Grand d'Aussy 1782, 3:100, 115-117).

CHOCOLATE: ATTITUDES. In a letter to her daughter written about 1671, madame de Sévigné retracts previous recommendation that chocolate is a nourishing drink, illustrating how quickly fashions in drink change: "I wish to tell you, my dear child, that chocolate does not enjoy the favour it once had in my estimation....All who once sang its praises to me now find fault with it. They curse it, they accuse it of causing all ills to which we are heir. It is the source of all vapours and palpitations. It gratifies you for a time, and then suddenly kindles in you a fever which brings you to your death" (quoted Hugon 1911:97).

GERMANY

COFFEE INTRODUCED. In 1675, limited coffee use appears at the court of Frederick William, the elector of Brandenburg. Most likely, it was introduced by his Dutch physician, Cornelius Bentekuh, who believes consumption of tea, coffee, and chocolate prolongs life by accelerating the circulation (Ukers 1935:41; Jacob 1935:62).

IRELAND

ALEHOUSE PREVALENCE. Writing in 1672, William Petty (Political Anatomy of Ireland, London, 1691:13-14) observes: "In Dublin where there are about 4,000 families, there are at one time 1,180 ale-houses and 91 public brew houses." He believes that there are too many people involved in the ale trade, but does not infer that too much is drunk. Rather he recommends that the same amount of drink be sold less wastefully (also quoted O'Brien 1919:139, 196).

BREWING; TAX PROTEST. In 1672 an English traveler comments on "the incomparable beer and ale, which runs as freely as water in Ireland" (quoted O'Brien 1919:196). In 1676, the brewers petition Parliament regarding the oppressive manner of the excise, beginning a long conflict, "which would not have been worth the Government's while to engage in, were not the revenue involved considerable." In addition to large public brewers, there are still many small-scale brewers who own their own little taverns (O'Brien 1919:196).

NETHERLANDS

INEBRIETY PREVALENCE (TEMPLE). In 1673, William Temple (1673/1814, 1:142-143) asserts that the upper classes in the United Provinces seldom drink except at banquets; nevertheless, there is not a single Dutchman who does not get drunk at least once during his life. In their austere existence, they know only one joy and real luxury--alcohol, without which they "would otherwise seem poor and wretched in their real wealth." He also speculates that it is the quality of the Dutch air which inclines them to drink (Zumthor 1962:173).

<u>DISTILLING INCREASES</u>. Strict prohibitive import duties (due to war) on French and German spirits and wines (in 1664, 1668, 1670, 1671) make possible the great expansion of the distilling industry in Holland. Gin has been previously produced for local consumption only and had to overcome the preference for brandy and wine over corn spirit. The new high duties end brandy's command of the market (Forbes 1970:189). A further impetus to the spread of the distilling trade is the decline in herring fishing.

In the 1660s, Schiedam distilleries are already "producing considerable quantities of hollands" (Zumthor 1962:174). Now the city becomes the center of commercial distilling and the home of the first Dutch guild of distillers. There wages are low and good transportation is available. The town council leaves distilling unhampered by any ordinance restricting its manufacture and sale; the city is also well suited for the importation of cheap grain and the exportation of cheap gin overseas. By 1700, no fewer than 77 stills in Schiedam are owned by 34 distillers. The distilleries are relatively small. During the latter decades of the century distillation also develops in Rotterdam, which had few distillers in the 16th century. In 1674, Rotterdam has 56 private distillers (Forbes 1970:106, 190; Houtte 1977:255).

<u>GIN PRICES</u>. In the mid-decade, Schiedam hollands (as gin is called) is worth wholesale about 36 florins a cask of 30 gallons; the same quantity of Bordeaux brandy costs about 70 florins (Zumthor 1962:332 n.7).

<u>MILITARY/GIN CONSUMPTION</u>. The armies of the Dutch Republic spread a taste for gin drinking when they fight in the southern provinces against Louis XIV in the last decades of the century (Houtte 1977:255). Daniel Defoe (<u>A Brief Case of the Distillers</u>, 1726:24) says that during these wars the Dutch armies brought Geneva drinking into Flanders where it spread among the other confederate armies, including England's. He also asserts that because gin was at first expensive, its consumption was limited to the generals and officers. Several Frenchmen also blame the spread of spirit drinking in Paris on these wars (see 1690/France).

RUSSIA

<u>INEBRIETY PREVALENCE</u>. A Serbian Catholic priest, Urii Krizhanich, writing during his exile in Siberia between 1660 and 1675, describes widespread drunkenness throughout Russia, which he blames on the kabaks: "You can travel over the whole wide world and nowhere will you find such horrible, disgusting, filthy drunkenness as here in Russia. The reason for this is the tavern monopoly, or the kabaks" (quoted Efron 1955:499-500).

<u>FESTIVE DRINKING</u>. S. Collins (<u>The Present State of Russia</u>, 1671:22) observes that during the last week of Carnival, Russians "drink as if they were never to drink more" (quoted Burke 1978:183).

<u>TAVERNS: CONTROLS; ECONOMIC VALUE</u>. In 1676, the 14-year-old Tsar Fyodor Alexeyevich (1676-1682) abolishes the kabak system, but he soon reinstates it when too much revenue is lost (Efron 1955:499).

AMERICA (BRITISH COLONIAL)

MASSACHUSETTS

DRUNKENNESS ATTACKED (MATHER). Puritan minister Increase Mather publishes a sermon, "Wo to Drunkards" (1673), in which he praises alcohol but warns against its waste or abuse: "Drink is in itself a good creature of God, and to be received with thankfuleness, but the abuse of drink is from Satan; the wine is from God, but the Drunkard is from the Devil." Mather defines a drunkard as "a lover of wine." He thus distinguishes between someone who is "merely drunken" and a drunkard: "He that abhors the sin of Drunkenness, yet may be overtaken with it, and so drunken; but that one Act is not enough to denominate him a Drunkard." Habitual drunkenness is a kind of madness. Man should partake of wine, but a man should not drink "more than is good for him" (quoted Lender 1973:353; Levine 1978:148-150; Rorabaugh 1979:30).

ALCOHOL AND WAGES. Despite the law of 1645 forbidding payment of wages in drink (see 1640/America), laborers continue to demand alcohol as a condition of employment. An act of 1672 refers to "sundry and frequent complaints of oppression by excessive wages of workmen on those they work for, by demanding an allowance of liquor or wine every day over and above their wages, without which...many refuse to work." In order to end this practice, the General Court reenacts the law of 1645 (quoted Thomann 1887:24; Conroy 1984).

LICENSING. In 1679, the Massachusetts court observes that on training days and other civil occasions strong drinks are sold by people without a license to do so, with the result that "many people, both English and Indians, that come to such meetings, as well as souldiers, committ many disorders of drunkenness, fighting, neglect of duty, etc." To prevent such abuse, the court orders that no one may sell "any wine, strong liquor, cider, or any other inebriating drinkes, excepting beere of a penny a quart," unless he has a license from two magistrates or the chief military officer (quoted Shurtleff 1853, 5:211).

NEW ENGLAND

CONCERNS. Beginning with Increase Mather (see above under Massachusetts), and continuing in the works of Cotton Mather (see 1700/America), Samuel Danforth (see 1700/America), Thomas Foxcraft (see 1720/America), and Jonathan Edwards (see 1750/America), to name only a few, "we can see the seeds of a modern view of habitual drunkenness, as well as the absolute limits to which colonial and Puritan thought could go on the question" (Levine 1978:147).

VIRGINIA

TAVERN, ORDINARY CONTROLS. In June 1676, during a rebellion against the royal governor, the Virginia Assembly, finding that "the many ordinaries in severall parts of the county are very prejudiciall," includes a provision in Bacon's Laws that "no ordinaries, ale houses, or other tipling houses whatsoever" be operated except at Richmond and at the York River ferries, and these can sell only beer and cider. The law never goes into effect, however (quoted Pearson and Hendricks 1967:15-16).

In 1677, a new law for "Regulating Ordinaries, and the Prices of Liquor" stipulates that licenses for ordinaries must be granted by the county courts. These licenses, which carry full liquor-selling privileges, are limited to two per county except the one where the General Court meets. License holders must provide accommodations for travlers. The law makes no mention of tippling houses, which continue to do business with or without license (Pearson and Hendricks 1967:16-17).

1680

ENGLAND

CONCERNS: INEBRIETY. Several literary attacks against drunkenness are published in the 1680s and 1690s, employing arguments similar to those of the 19th-century temperance movement. These include M. Scrivener's *A Treatise against Drunkenness* (1680); *The Way to Make All People Rich* (1685); and W. Asheton's *Discourse against Drunkenness* (1692, see 1690/England) (Harrison 1971:422-423 n.13, 14).

SPIRITS CONSUMPTION. Spirits consumption officially reaches 527,000 gallons a year in 1685 (Webb and Webb 1903:38). In 1684, exciseman Charles Davenant observes that at Canterbury "more brandy [is] drunk...than is entered [for the excise] in all the country," indicating a considerable amount of smuggling and illicit production. He also observes that an "abundance of brandy" exists in Wiltshire (quoted Clark 1984). In the preface to *A Treatise of Lewisham (but vulgarly called Dulwich) Wells in Kent* (1681), the physician John Peters complains that on the pretense of drinking the waters there, Londoners engage in "great Prophaneness" and also drink "an excessive quantity of Brandy (that Bane of English Men) or other strong Liquors, thereby many of them becoming greatly prejudiced in their Health" (quoted Baker 1931:137). Dr. Thomas Tryon (*Health's Grand Preservative*, 1682) says that "of late years" brandy, rum, and rack have "become common drinks amongst many as Beere and Ale." Their consumption, he warns, is pernicious and hurtful to one's health. He particularly attacks the practice of brandy drinking among women at the slightest disorder of the stomach (quoted Simon 1909, 3:397-398). This indicates that brandy is still closely associated with medicine, at least among women.

BEER: CONSUMPTION. Beer is the principal drink of everyone, and consumption rises throughout the decade. According to Latham (1983:104), this increase cannot be accounted for by any other cause than a rise in individual consumption. For 1684, duty is charged on 6,318,000 barrels of commercially-produced beer in England and Wales. This amounts to a per capita consumption of probably 40 gallons of beer a year or a pint a day, for a population of little over 5 million, indicating that the per capita consumption is higher than in modern times (Thomas 1971:18). Gregory King (1696/1936:40-41) estimates that an additional 70% of the total must be allowed for beer brewed privately on which excise is not charged. In 1688, a total of about 12,400,000 barrels of beer are produced for both commercial and private consumption (King 1696/1936:41; Williams and Brake 1980:3). In 1689, the consumption of beer reaches a high of 104 gallons per person annually (2.3 pints per day) for the total population (Spring and Buss 1977:567, who consider this high consumption not so much an indication of widespread drunkenness as of the fact that beer is safer to drink than water). (See also Earle 1976:155.)

CHARACTERISTICS. In a debate on a new victualling house law in 1680, it is stated that ale has become as strong as wine and can be burnt like sack. These exceptionally strong ales are given many unique names, such as huff-cap, nipitatum, and Pharaoh (Simon 1948:150; Monckton 1966:123). Weak beer is still common.

TAXATION. The strong feelings against excise revenues from beer begin to subside, and they are accepted as part of the normal pattern of life. In 1684, the "farming" of duties by contractors ceases, and the responsibility for tax collection passes to the revenue department. The duty now begins to be manipulated according to the fiscal requirements of the day (Monckton 1966:119).

MEDICINAL INTOXICATION. One lord writes in 1688, "In old Age...I am of the opinion that a little Drunkenness, discreetly used, may well contribute to our Health of Body as Tranquility of Soul" (quoted Sutherland 1969:177).

COFFEE CONSUMPTION. Between 1680 and 1730, London drinks more coffee than any other city in the world. In the last years of the century, coffee is said to have had a perceptible influence in diminishing drunkenness among the upper classes (Jacob 1935:89, 128-129; Lillywhite 1963:22).

COFFEE HOUSE FUNCTIONS. Coffee men are again summoned before the council for writing seditious news in 1679, but for the rest of the century coffee houses grow and prosper; as schools of conversation, they are referred to as "Penny Universities," since the entrance fee is only a penny. John Houghton, a Cambridge professor, compares the coffee house to the university and declares that a man may pick up more useful knowledge there than from reading books for a whole month (Ellis 1956:24, 28-29). Coffee houses are used to assist in the daily life of the city: banking, shipping, insurance, stock exchange, commerce, and trade are all conducted there. The business day usually begins and ends in coffee houses. To attract and retain customers, coffeemen keep on hand a wide variety of newspapers, bulletins, price lists, auction notices, etc. Various trades and interests congregate at particular houses, paving the way to the establishment of future exchanges (Ukers 1935:69; Lillywhite 1963:18, 20-21; Ellis 1956).

FRANCE

YOUTHS/SPIRIT CONSUMPTION. Jean de La Bruyère (Les Caractères, 1688/1929:212, 427). expresses his exasperation at young people at court for so abusing wine that they find it tasteless and seek to awaken their extinguished tastes by drinking eaux-de-vie and all the liqueurs plus violentes. According to La Bruyere, "Liqueurs" is a word that should be "foreign and unintelligble"; they are only an incentive to luxury and gluttony (also quoted Delamain 1935:30).

TAVERN CHARACTERISTICS. The government grants the taverners' request to provide tables and seating for customers in return for paying the higher tax rate charged the cabarets. The distinction between the two essentially disappears (Le Grand d'Aussy 1782, 2:358; Brennan 1981:28; Saint Germain 1962:292).

TAXATION. By 1682, the aides amount to 22 million livres compared to 5 million in 1661 (Cole 1939, 1:305).

CAFES APPEAR. In 1686, a limonadier named Procopio opens the first true café in Paris (Café de Procope). Wishing to differentiate it from the cabarets and early coffee houses, he appoints it luxuriously with mirrors, crystal, and marble tables, and he sells a wide variety of liqueurs, sherberts, ices and other confections in addition to coffee (Leclant 1979:90; Fosca 1934:8). An instant success, he spawns many imitators. The cafés come to cater to the elite and intelligentsia, and become the gathering place of actors, authors, dramatists, musicians, and, later, revolutionaries. The fashionable people overcome their initial antagonism to coffee (Ukers 1935:41; Uribe 1954:17).

GERMANY

COFFEE CONSUMPTION. The first coffee house is opened in Vienna. Coffee, tea, and chocolate quickly win over the fancy of the upper classes throughout Germany (Klemm 1856:337, who believes that this helped to reduce the enormous consumption of alcoholic beverages).

SCOTLAND

TAXATION: MALT TAX REGULARIZED. The malt tax, which had been collected locally in an irregular fashion since 1661, is made a general imposition and collected regularly every year after 1681. It remains in effect until 1695 (Robb 1951:36). The malt tax is intended to apply to brewers and the spirit tax to distillers, but because Scotch whisky is made from malt, the tax places a double burden on its manufacturers.

AMERICA (BRITISH COLONIAL)

MASSACHUSETTS

TAVERNS: CONTROLS, PREVALENCE. In 1681, the General Court orders that the selectmen of the towns must "approve of all persons to be licensed before license be granted to any of them by the County Court," with all licenses to be renewed annually. Under this law, the court also reduces the number of licenses that can be issued in each of the towns of the colony (quoted Shurtleff 1853, 5:305). One effect of the reduction of licenses is to encourage the proliferation of unlicensed houses, particularly in Boston. The court is obliged to progressively increase the number of licenses. By the end of the century, 74 licenses are allowed for Boston, compared with 24 in 1681, although the population remains relatively stable (Conroy 1984). In order to suppress the numerous unlicensed inns and taverns in the colony, in 1684 the General Court directs tithingmen, grand jury men, and constables to search out "all disordered houses that do, contrary to law, retaile wine, ale, beere, cider liquors, etc., without license" (quoted Shurtleff 1853, 5:448).

PENNSYLVANIA

DRINKING CONTROLS; TAXATION. The first assembly convened by William Penn in 1682 passes a law ordering punishment for persons convicted of drunkenness and for retailers who permit excessive drinking in their houses. The law also prohibits the drinking of healths and fixes the price of ale and strong beer at two pence per quart. In the following year, the assembly imposes an import duty of two pence on each gallon of spirits and wine and one pence on each gallon of cider. Beer is not mentioned, suggesting its importance as a necessity of every day life (Thomann 1887:145-146).

1690

EUROPE

MARC BRANDY INTRODUCED. The distillation of grape marc (the skins, pips, and stalks of grapes after they have been pressed and wine made) into a strong, crude brandy begins about 1690 in the wooded areas of Lorraine. In contrast to wine brandy, which needs low heat, marc brandy requires a high temperature and therefore large quantities of wood. The practice spreads to Burgundy and all of Italy, where the drink is known as grappa (Braudel 1981:246; Gottschalk 1948, 2:147).

ENGLAND

BACKGROUND DEVELOPMENTS: GLORIOUS REVOLUTION. The "Glorious Revolution" of 1688 dethrones King James II, and the Protestants William III of Orange and Mary (James' daughter) begin their reign (1689-1702).

DISTILLATION CONTROLS LIBERALIZED. Motivated by hostility to France (which harbors exiled James II) and by a desire to encourage the use of grain in domestic distilling, King William prohibits the importation of foreign wines and spirits and issues charters to divert surplus English corn to the production of "good and wholesome" spirits. In 1690, an "Act for the Encouraging of the Distillation of Brandy and Spirits from Corn" permits anyone to distill spirits made from English corn upon payment of a small duty and upon giving 10 days notice to the excise. As a further inducement to spirits production, William breaks the monopoly of the London Distillers Guild, ending control over the trade and enabling anyone to become a distiller providing the excise duty is paid (French 1884:245, 246; Coffey 1966:673-674; Inglis 1975:62).

SPIRITS PRODUCTION, CONSUMPTION. The distillery business thrives owing to the 1690 act. By 1696, the annual production of distilled liquors, mostly gin, rises to a million gallons (Webb and Webb 1903:38). William himself is a native of Holland and helps establish gin as a fashionable and patriotic drink. Consuming it becomes a sign of being Protestant. Hampton court is even described as a "gin temple" (Watney 1976:12). Charles Davenant (1695:138) complains that for a long while brandy imports have been rising; this benefits neither the economy nor the health of the nation. Doctors say "it extinquishes natural Heat and Appetite." The drinking of it has become a "growing Vice among the common People, and may, in time, prevail as much as Opium with the Turks, to which many attribute the Scarcity of People in the East." He recommends that "there is no way to suppress the use of it so certain, as to lay such a high Duty, as it may be worth no Man's while to take it, but for Medicine" (see Zirker 1966:84).

BEER: TAXATION. Beginning in 1689, a series of increases to the beer duty occur to help finance the wars with France. After another increase in 1690, the duty on beer has tripled from the rate in 1643, from 6 pence on small beer and 2 shillings on strong beer to 1/6d and 6/6d, respectively. In 1694, it is raised again (Spring and Buss 1977:567, 569; Monckton 1966:119-120). Davenant (1695:45-47) observes that each year since 1689 alcohol revenues have fallen, a decline chiefly due to the increase in duties to pay for war. This has caused a number of victuallers to leave the trade and many private families to brew their own drink.

CONSUMPTION. Official consumption figures drop from a peak of 104 gallons per capita in 1689 to about 75 gallons in 1700. Actual consumption probably remains nearer the former figure, since the higher duties prompt people to brew their own beer, which is not subject to tax (Spring and Buss 1977:567-568). Contributing to a decline in beer consumption are the increasing availability of a cheaper and more potent alternative in gin, the economic depression, the high price of bread, wartime dispersion of trade, and the poor state of currency. The Reformation of Manners movement may also have exerted some influence, but this is at best of only minor importance (Latham 1983:104; Bahlman 1957; Earle 1976:155).

CONCERNS: DRINK AND POVERTY (DUNNING). Richard Dunning (Bread for the Poor; or, a Method shewing how the Poor may be Maintained, 1698:1), in discussing high parish expenses for maintaining the poor, writes: "'Tis generally observ'd, That not only more Ale and Brandy is sold than formerly in single Ale houses and Brandy-shops, but the number of such Houses and Shops are also increased, that the Money spent in Ale and Brandy, in small Country-shops and Ale-houses, amounts to a vast and almost incredible sum." Furthermore, he asserts that the poor will only drink the costliest and strongest ale.

Dunning reflects the view of many contemporaries, including John Locke, that the poor spend their time, "of which they had defrauded the State, in profligate living, in luxury, and particularly in brandy drinking" (Marshall 1926:34).

CONCERNS: ALEHOUSES (LOCKE). In a 1697 royal commission report on, among other things, methods for employing the poor, John Locke recommends suppressing superfluous brandy shops and alehouses (Zirker 1966:84). However, on the whole concerns over alehouses have diminished considerably compared to fifty years previously. Although the Societies for the Reformation of Manners attack the alehouses, it is not one of their principal concerns (see below). William Asheton writes A Discourse against Drunkenness in 1692 without any mention of alehouses (Clark 1983:186). Considerable evidence exists that by the end of the century there is a growing sense that the alehouses are under control (see 1700/England).

CONCERNS: MANNERS AND MORALITY. About 1690, many among the upper-classes and the educated, believing that morals have declined since the Restoration and fearing that divine retribution is imminent, organize a movement designed to reform English manners and morals. In 1690-1691, with royal support the first Societies for the Reformation of Manners appear. In a decade, there are over 20 such societies in London alone. Membership comprises "emminent merchants, fulltime philanthropists, and a halo of higher clergy" (Porter 1982:311). Various causes for England's moral decline are given, but all reformers agree it is rooted in the nonenforcement of existing laws.

Their chief methods are propaganda and the organization of a system of informing and prosecution through which the existing laws are enforced even without the active assistance of local authorities (Bretherton 1931:178). They supplement church courts by laying prosecutions before civil courts for vice offenses. They pay informants to seek out people guilty of such vices as swearing, cursing, prostitution, and drunkenness and to initiate prosecutions against them. Although they condemn drunkenness and alehouses in general, neither are singled out for special attention. The condemnation of alehouses is "fairly muted" (Clark 1983:39, 186-187). More concern is directed toward the suppression of Sabbath breaking, sexual immorality, and attendance at theaters and fairs. In practice they only attack Sunday drinking and they disdain any intention of enforcing the laws regarding closing hours or suppressing tippling on weekdays (Bretherton 1931:179). Sabbath breaking is viewed as the most deadly of sins; it is their primary consideration and the subject of most of their prosecutions. They are never concerned with prosecuting drunkards in general, but only those who get intoxicated on Sundays or drink during the time of divine service. They thus stand in an intellectual tradition running from pre-Civil War Puritans to 18th-century Methodism (Curtis and Speck 1976:57-59).

Although they do not just prosecute the poor and they seek to end corruption at the court as well, the societies are attacked by contemporaries and posterity for their methods, especially the use of paid informants, and for being an upper-class attempt to prosecute the sins of the lower orders, especially as these societies are clearly more concerned with public than with private vice. Moreover, they fail to enlist much enthusiasm among the gentry and even the magistrates. At least one justice is inspired to extralegal efforts, but many other justices laugh at the proclamation, drink and swear with the worst offenders, and refuse to prosecute (Curtis and Speck 1976:49). Prosecutions decline in the early 18th century; in 1738, the last accounts and sermons of the reformation socities are published, but by then most of the societies had already disbanded (Bahlman 1957:66). (See 1700/England for more criticisms of this movement.)

DRINKING PRACTICES: SUNDAY DRINKING, TOASTING (MISSION). Henri Misson de Valbourg (Mémoires et observations faites par un voyageur en Angleterre, 1697) observes that the English do not differ from the French in doctrine regarding the observation of the Sabbath but they do observe it more scrupulously (Baker 1931:138). He also comments on the great amount of toasting that occurs. To drink at a table without toasting is considered discourteous (French 1884:248).

PORT CONSUMPTION BEGINS. The popularity of port wine begins as the triangle trade with Portugal makes it more available. At first, the word is used for all wines from Portugal (Younger 1966:379).

FRANCE

INEBRIETY PREVALENCE (LISTER). The English physician Martin Lister (Journey to Paris in the Year 1698) refers to increased drinking in this "once sober nation." To a large extent, he places responsibility for this change on the rise in spirit drinking among all classes, which he in part views as the result of Louis XIV's wars against the Dutch (see below). He also comments on the innumerable alehouses and the "very many publick coffee-houses" which also serve strong waters, adding, "I wonder at the great change in this sober nation in this particular, but luxury, like a whirlpool, draws into it the extravagance of other people" (Lister 1699/1972:230-231, also quoted Duby and Mandrou 1964:225, 230).

By the end of the century, Saint Germain (1962:63, 148) asserts that plays and public entertainments are often interrupted or disturbed by drunkards. Sergeants of the watch cannot pay their bowmen on days they work for fear that they will get drunk.

TAXATION. Lister observes that "the tax upon wines is so great that whereas before the war they drank them at retail at 5d. the quart, they now sell them at 15d. the quart, and dearer." Possibly the rising wine tax has contributed to the spread of spirit drinking. He also comments on the amount of beer brewed in Paris (quoted Duby and Mandrou 1964:225, 230).

SPIRITS CONSUMPTION. It is only at the end of the 17th century that brandy, despite its expanding role in the drink trade, fully looses its quasi-alchemical associations (Younger 1966:331). Often, use is still limited to medicinal purposes. When Louis XIV's immense appetite begins to impair his stomach, his physician Fagon prepares a rossolis for him which Louis drinks at dessert and which becomes known as le rossolis du Roi (Le Grand d'Aussy 1782, 3:77; Gottschalk 1948, 2:148). But a taste for liqueurs has become well developed among the wealthy. It is a development that is very slow but very clear (see below). Until this time, spirits are largely the drinks of the poor (Delamain 1935:29). Audiger (1692) (see below) illustrates that sweet liqueurs have "found their way into drinking habits" (Braudel 1981:246).

Lister (1699/1972:230-231) observes a marked increase in spirit consumption in Paris since his last visit. In general, "there is no feasting" without the drinking of all kinds of strong waters, particularly ratafias (liqueurs), a custom which he had not seen previously. Nantes brandy, once only the morning draught of porters, is now valued as one of the best wine spirits in Europe. These and other "strong waters of wines" form part of the dessert of all great feasts and are consumed freely. This, Lister observes, is a new custom he had not previously seen.

MILITARY/DRINKING. Lister (1699/1972:23) attributes the rise in spirit drinking to the influence of wars and the army: "But it was the long war [War of the League of Augsburg, 1688-1697] that has introduced them; the nobility and gentry suffering much in those tedious campaigns, applied themselves to these liquors to support the difficulties and fatigues of weather and watching; and at their return to Paris, introduced them to their tables."

Several other contemporaries also point to the influence of the wars waged by Louis XIV against the Dutch in spreading the popularity of brandy and other spirits among the military, the wealthy, and the general civilian population, at least in Paris. Prior to this time the major markets for southern brandy were still foreign. According to a report prepared in Angoumois in 1731, immense quantities of cognac were consumed in the army hospitals of the great garrisons the king maintained in Flanders (Delamain 1935:39-40). This may account for the soldiers' use of the word "brandevin" as a slang term for spirits instead of the official "eau-de-vie" (Forbes 1970:196). (See also Defoe's comments on the influence of the Dutch wars under 1670/Netherlands.)

ADVERSE EFFECTS. According to Lister (1699/1972:230, 231) the once lean Parisians are now fat as a result of daily spirit drinking. He also wonders what adverse effects spirit drinking may have "on some tender and more delicate constitutions, and weak and feeble brains," and whether it may be one of the causes of "so many sudden deaths as have been observed of late."

DISTILLING. Audiger's La maison réglée (1692/1700:193-297), a book of household advice for the "solid bourgeoisie" written by a limonadier, includes a chapter on "The True Way to make all kinds of Liqueurs in the Italian Fashion," which gives advice and recipes for their preparation. Audiger also provides a history of his early efforts to establish himself in Paris as a maker of liqueurs (having learned the art in Italy). He complains that now anyone can become a spirits vendor by paying only 50 ecus and that the trade has been infiltrated by men without experience from the worst elements of society (pp. 211-212).

SPIRITS EXPORTS, ECONOMIC VALUE: BRITANNY; LANGUEDOC. Wine and brandy make up a great part of Breton commerce. From Nantes, 7,000 pipes of eau-de-vie are exported in 1697 (Le Grand d'Aussy 1782, 3:27). By the end of the century, brandy distilling has become the salvation of the wine-growing industry of Languedoc. With the help of a few poor vintages, it contributes to pegging the prices of wine to wheat prices for the first time in 35 years. When a good harvest threatens to drive prices down in 1699, half the wine passed through Sète is distilled into brandy for export (Le Roy Ladurie 1974:227).

CONTROLS: PRODUCTION, SALES. In September 1693, a decree of the Royal Council forbids making brandy from corn (which it says should be put to better use). The ban indicates that some brandy is still made from cereals (Younger 1966:329). The ban is repealed in 1694. On March 13, 1699, the sale of all other eaux-de-vie except that made from wine, apples, or pears is again banned (Claudian 1970:14). The ordinances of 1693 and 1699 also ban retail spirit sales in small measure if not *à porte-col* at street corners. Lack of compliance is punishable with confiscation and a fine of 1,000 livres (Delamain 1935:29, 34). This ban is reiterated often.

WINE CONSUMPTION. According to Audiger (1692/1700:24, 27, 73), a bachelor *seigneur* must have at least 37 servants. Each servant gets a free pint of wine everyday, and the butler who keeps the key to the cellar need only account for 265 out of every 280-pint barrel. At a table of twelve, each person should be given a pint of wine for each meal in a house of quality. It appears that one of the important qualifications for servants is that they not be drunkards (see pp. 52, 93; see also Lewis 1957:196-197).

ATTITUDES. Mathier Fournier, a medical student in Beaune, Burgundy, asserts in 1696 that the wines of Rheims are conducive to catarrh, gout, and other diseases of the nervous system. An uproar results in Rheims, where Giles Culotteau lauds the purity, healthfulness, and superiority of the wines of Champagne. This sparks a counterattack on the wines of Rheims by M. de Salins, dean of the faculty of medicine of Beaune (Emerson 1908, 2:148-149). (On the Burgundy-Champagne rivalry, see 1650/France).

TOASTING. Henry Misson (see 1690 England) observes that the English custom of toasting is almost abolished among French people of any distinction (French 1884:248). However, Pepys' diary mentions toasting in the French manner in the 1660s (see 1660/France).

DRINK HOUSES. It is said that at the end of the 17th century in France one goes to a tavern to drink to excess, to a cabaret to drink, and to a café to talk (Butler 1923:362).

COFFEE MONOPOLY ESTABLISHED. In 1692, Louis XIV grants the first coffee monopoly in order to assure a fixed revenue to help pay for his war expenses. The concession proves of little value because a simultaneous edict sets the price of coffee so high that consumption drops. The monopoly requests that the price be lowered, but the excessive price has already ruined the monopoly (Jacob 1935:128).

GERMANY

BRANDY CONSUMPTION, CONTROLS. Bishop Ernest August of Osnabruck complains in 1691 that brandy is no longer a medicine for the common man, but has become a daily drink, and that there are drinking rooms for its consumption. Shortly thereafter, decrees against it are issued, to be read in the churches the first Sunday after Trinity, are issued (Cherrington 1926, 3:1092).

INEBRIETY PREVALENCE. F.W. Leibniz, returning from Italy in 1690, finds that Germans no longer are so "crazy and full of beer" (quoted Jacob 1935:93).

QUALITY CONTROLS. Following earlier (1603, 1676) actions Cologne passes a law that forbids forbids the use of bottom fermentation in brewing and forces the brewers to take an oath that they will prepare their beer, as of old, from good malt, good cereals, and good hops well boiled, and that they will pitch it with top-yeast, rather than with bottom yeast, raw wort, or other noxious herbs, no matter of whatever name (Cherrington 1925, 1:406).

IRELAND

INEBRIETY PREVALENCE. O'Brien (1919:126-127) and MacLysaght (1950:68, 72, 163) believe that the standard of drinking and of morality has improved in rural Ireland since the 1630s, although wakes are a considerable problem. John Stevens writes in his journal in 1690 regarding the "meaner people" that "they drink for the most part water, sometimes coloured with milk; beer or ale they seldom taste unless they sell something considerable in a market town" (quoted O'Brien 1919:140). A Catholic observes in 1683 that the Irish of County Wexford "are not inclined to debauchery or excessively addicted to the use of any liquor" (quoted MacLysaght 1950:54). In a pamphlet with "the most scurrilous anti-Irish views" (A Brief Description of Ireland, 1692), it is asserted that drinking is not one of the Irish national vices (MacLysaght 1950:70). Another anti-Irish travel account (A Trip to Ireland, 1699) also states that "drinking is not so much their vice as some of their neighbouring nations" (quoted MacLysaght 1950:70-71). However, other sources do complain about the high level of inebriety, such as Richard Lawrence's Interest of Ireland in its Trade (1682), and John Dunton's Dublin Scuffle (1698:222), which asserts that drunkenness "is now become a feminine vice" (O'Brien 1919:126; MacLysaght 1950:28, 72).

NETHERLANDS

MILITARY/GIN CONSUMPTION. Gin begins to replace beer as the general beverage of the Dutch army and navy. As use increases, heavy taxes are placed on it. By the end of the 18th century, it is traded on the Amsterdam exchange (Forbes 1970:192).

PEASANTS/DRINKING. At Rumegies, a village near Valenciennes, Hainaut, a priest complains that rich peasants sacrifice everything to luxury of dress and can be found with "unheard of insolence frequenting taverns every Sunday" (quoted Braudel 1981:311).

RUSSIA

DRINK IMPORTS. The Russians import beer from the Dutch and wine from the French. They have to pay custom duties, which make their products expensive, but supplies are plentiful (Massie 1981:125).

SCOTLAND

BACKGROUND DEVELOPMENTS. After the Glorious Revolution (1688), James II flees to France, where he and his descendants (sometimes known as the Pretenders) are maintained as the British monarchs for more than half a century. Their supporters in Britain are known as Jacobites (from Jacob, the Latin version of James). Their strongest adherents are among the Catholics of the Scottish Highlands, who rebel in support of James in 1715 and 1745.

TAXATION: WHISKEY, MALT. Scotch whisky's exemption from the 1661 excise tax on spirits is removed in 1693, and an excise of 2s per Scots pint is imposed (Robb 1951:36; Ross 1970:7; Murphy 1979:10). The excise collected on spirits not made of malt distilled and sold within the kingdom is 3s per pint. Imported spirits are taxed at 6s per pint. Foreign beer is taxed heavily at 13s per barrel (*Lancet* 1905:240). The malt tax is repealed in 1695 and the spirit tax is raised to 8s per gallon. A year later, the malt tax is reimposed.

TAX EXEMPTION: FERINTOSH. The Forbeses of Culloden are given an exemption in 1695 from the whisky tax in recompense for damages inflicted on them by Jacobites in retaliation for their support of the Glorious Revolution. Their whisky, called Ferintosh, remains free from duty until 1784, when the exemption is removed. Ferintosh becomes so popular that it becomes a synonym for whisky in the 18th century (MacDonald 1914:42-44; Robb 1951:36; Ross 1970:58-59; Murphy 1979:36). (On Ferintosh, see also 1780/Scotland.)

WHISKY CHARACTERISTICS. Writing of the island of Lewis in the Hebrides in 1691, Martin Martin (*A Description of the Western Islands of Scotland*, 1698) describes whisky and its role in Highland life. He observes that "the air is temperately cold and moist, and for the corrective the natives use a dose of trestarig or usquebaugh," which they are encouraged to make by the abundance of grain. He also cites the use of the four-times-distilled usquebaugh-baul "which at first taste affects all the members of the body: two spoonfuls...is a sufficient dose; and if any man exceed this, it would presently stop his breath, and endanger his life" (quoted Daiches 1978:24). This indicates that there are other spirits distilled in 17th-and 18th-century Scotland besides true malt liquor. An infusion of aromatic herbs and whisky is common in the Highlands into the second half of the 19th century (Daiches 1978:26). Yet by the 17th century malt whisky is already established as the characteristic Highland spirit. (Wilson 1973:40 says this passage is cribbed from Fynes Moryson [see 1610/Scotland] but we could not find it in Moryson's *Itinerary*.)

AMERICA (BRITISH COLONIAL)

MASSACHUSETTS

CONCERNS: INEBRIETY (MATHER). In 1694, Cotton Mather, Increase Mather's son, petitions the General Court of Massachusetts for effective regulation of ordinaries and begins a campaign against drunkenness, claiming that the horrible religious condition of the country is due to excessive tippling and that a "flood of excessive drinking" has begun to drown Christianity (Krout 1925:53).

PENNSYLVANIA

TAVERN CONTROLS: LICENSING. A licensing law passed in 1699 stipulates that no one may operate a public drinking house without a license from the governor of the province, that licenses are to be issued only to those persons recommended by the justices of the county in which the person proposes to run the drinking house, and that the county justices may suppress disorderly houses and permanently revoke the retailer's license (Thomann 1887:147-148).

VIRGINIA

DRINKING DEBTS. In 1691, the Virginia Legislature observes that because of "the unlimited credit given by ordinaries and tippling houses...to seamen and others,...they spend not only their ready money, but their wages and other goods, which should be for the support of their families." Thus, it modifies the drinking debt act of 1666 to protect such persons (see 1660/America). Legal collection of debts in excess of 300 pounds of tobacco is prohibited unless the debtor has two servants or is worth 50 pounds sterling. Severe penalties are established for drink sellers if they attempt to evade the law by charging liquors as some other goods. With slight alterations, this law is in force until 1734, after which time the credit limitation is continued but without restriction except for sailors and servants (Pearson and Hendricks 1967:13).

The Eighteenth Century

1700

EUROPE

BACKGROUND DEVELOPMENTS. On the whole, the 18th century brings economic improvement throughout Europe, though the improvement is uneven geographically, in time of occurrence, and in effects (Anderson 1961:47). Underlying the recovery is a population explosion (based on earlier marriages and a falling death rate), in the second half of the century. Among the causes for this population increase are ecological changes that are noxious to plague-bearing rats; more disciplined and less devastating warfare; better nutrition; acceptance of hardy new crops (e.g., the potato) as dietary staples (1709 is the last year of great famine in France); and better medicine and sanitation.

The effects of these demographic and economic changes are far reaching. First, the demand for goods increases, stimulating trade and expanding capitalization of both agriculture and industry. Second, the labor supply expands, leading to the development of intensive agriculture and industrial growth. Third, the combination of growing population and ballooning overseas trade shakes unprecedented numbers of Europeans out of their routine into the world of the distant market place. Fourth, the middle class becomes larger and richer, increasing the demand for quality domestic goods. Fifth, international trade focuses less on luxury goods and more on articles of mass consumption. At the same time, the luxury items of the 16th century begin to penetrate below the middle classes. Overpopulation occurs in some areas, the number of vagabonds increases, and an oversupply of wandering artisans become tinder for popular uprisings in town and countryside alike (Krieger 1970:119-122, 128). Although urbanization increases, only 3% of Europeans live in cities with a population of 100,000 or more (Burke 1978:245).

INEBRIETY PREVALENCE. Previously, the problem of drink has been "more of the order of damage due to occasional excess than to a definite dependence of large segments of the population upon alcoholic beverages" (Jellinek 1976b:83). Now, consumption is prodigious among all classes, and habitual drinking among the lower classes becomes a major problem for the first time. Forbes (1970:188) observes: "It was quite common in this period that the guests had to be borne home after dinner and kings and clergy led the dance. The pamphlets praising the virtues of alcohol as published in the sixteenth century and later bore fruit." Although drunkenness is still largely an urban problem, it begins to spread out into the countryside. As in the 16th century, this increase in drinking occurs during a time of commercial expansion (Braudel and Spooner 1967:408).

SPIRITS CONSUMPTION. Over the century, spirits consumption increases "by leaps and bounds" throughout Europe (Braudel 1981:244). Most prominently, there occurs a "gin epidemic" in London in the first half of the century.

ATTITUDES. The gin epidemic in England facilitates for the first time a campaign against the consumption of an alcoholic beverage (gin) rather than just the abuse of an otherwise beneficial substance. However, this condemnation is not elaborated into a general attack on alcohol drinking.

WAGES AND DRINKING. A common complaint is that when laborers earn large sums, they just spend more on drinking and idleness. In England and the rest of Europe, wage gains are often reflected in increased consumption of alcoholic beverages. In part this is because they are the only foodstuffs for which production can be adjusted to meet increased demand (Aymard 1979:8). Furthermore, few poor laborers possess the "bourgeois" spirit of thrift and hard work. With wages low and labor long, increasing earnings significantly is not possible for most. They prefer leisure to working harder and saving, and spend what they have quickly, in part because possessing savings would likely disqualify a family from poor relief (Porter 1982:106, discussing England).

NUTRITIONAL VALUE. Given the meager food rations of the lower classes before 1800, alcoholic beverages play a decisive role in the diet. Not until the end of the 19th century does a general transformation of food distribution take place, providing urban masses with increased meat, dairy products, fruit, and vegetables. With the average diet in revolutionary France consisting of 1800 calories a day, a liter of wine (12% alcohol) produces 700 calories for a man weighing 65 kg. Alcohol provides a supplementary source of energy for laborers, since cereal rations are minimal and meat is scarce (Aymard 1979:8, who observes that, although it is often assumed that alcohol supplied a maximum of 10% of all calories, this is "clearly underestimated").

MEDICINAL USE. Alcohol's popularity as a medicine reaches its height. It is believed to be especially important for maintaining the health of those who do heavy labor or live in damp and cold climates (Lucia 1963a:8; McCarthy and Douglass 1949:9). (For example, see under 1720/England.) In an age when almost everyone suffers from some physical pain and no anaesthetics are available, alcohol remains the best pain-killer (Porter 1982:30). By century's end, these attitudes have begun to change significantly under the impact of widespread spirit abuse and the growth of industrialization.

VINTAGE WINES ESTABLISHED. Vintage wines establish themselves at the beginning of the century. The best-known wines probably owe their reputation less to their merits than to the convenience of the routes in their vicinity and particularly to their proximity to a large town or waterway. Paris alone absorbs the 100,000 or so barrels produced in 1698 by the winemakers of Orléans (Braudel 1981:235). (See also 1750/France.)

WATER CONSUMPTION. In the course of the century, the problem of supplying drinkable water is "clearly posed and the solutions seen and sometimes achieved" (Braudel 1981:230). Despite everything, progress is slow and the lack of good drinking water still is one of the main influences on alcohol consumption.

DISTILLATION TECHNIQUES. Distillers begin searching for new base materials, especially potatoes. This poses problems since stills are not yet designed to handle thick potato washes. A demand for larger and better types of distilling apparatus grows (Forbes 1970:262-264).

DENMARK

BRANDY RATIONS. The Danish navy's brandy ration is half a pint, spread over a number of breakfasts. Breakfast is the only time it is apportioned, and no other food is served. By the end of the century, the value of the braendevin breakfast is questioned and reductions are recommended (Glamann 1962:130).

DRINKING PRACTICES, BRANDY. Among the general population, brandy is mixed with Danish beer "in the hope that the resulting mixture would have the same strong effects upon the drinker as had German beer" (Glamann 1962:130).

ENGLAND

BACKGROUND DEVELOPMENTS: GEORGIAN ENGLAND. In 1714, after the reign of Queen Anne (1702-1714), the German elector of Hanover (and great-grandson of James I) becomes King George I (1714-1727) of England. Four Hanoverian kings named George rule over England until 1830, hence the 18th century is frequently called Georgian or Hanoverian. Georgian England is a pragmatic, capitalist, materialist, market-oriented society in which the pursuit of money is given general social backing.

The urban middle class grows considerably in terms of numbers, relative affluence, and social esteem. It is increasingly praised as the backbone of English society (Corfield 1982:130-135). Often called the "middling sort," these are people who fall between the aristocracy/gentry and the poor, such as professionals, merchants, tradesmen, small manufacturers, clerks, skilled artisans, and yeoman farmers. In the second half of the century a "recognizable middle-class culture of respectability" emerges, allying itself with the elites to stamp out working-class cultural habits, customs, and pastimes. The growth in numbers and wealth of the urban "middling sort" is reflected in the growing demand for luxury goods and services and in the emergence of a consumer economy, together with the development of advertising and marketing techniques. By 1800, there are at least a half million middling families, well educated and with money for luxuries (McKendrick et al. 1982; Stone 1984:44). Addison's and Steele's Spectator (started 1711) becomes the first successful magazine consisting "entirely of essays on manners, social morality and self-improvement"; through it Addison and Steele seek to cultivate a new standard of polite behavior. The magazine finds a new market in the growing middle-class audience and teaches others how to exploit it (Plumb 1973:6; Ellis 1956:161-162).

Throughout the century, urbanites are still a minority of the population, but by the end of the century, the urban population has increased from 20% to 30%, with the increase being greatest in provincial cities (Stone 1984:46). Many contemporaries warn against what they view as "the perilous nature" of urban life and the insidious diffusion of city culture into rural England. Moralists criticize the towns for encouraging "worldliness, dissatisfaction, emulation, and desire for luxury" (Corfield 1982:2, 97).

In the towns, an apex of wealth rests on a broad base of poverty. Although lower-class families are often termed simply "the poor," and they all have to work long hours, there are many gradations of income levels between relative solvency and dire poverty. Ordinary artisans and laborers usually are not paupers.

POPULAR CULTURE. A resurgence and secularization of popular culture and the festival occur. New secular functions are added; publicans, hucksters, and entertainers encourage feasts when their customers have extra harvest earnings. As the influence of the church declines, village and town feasts are moved to Sundays and from winter to summer (Thompson 1974:392, 394). The early 18th century is a transitional period in which a robust popular culture rooted in the tightly knit, inward-looking world of country life still exists. The yearly calendar is filled with festivals and holidays that derive their meaning from an intimate connection with the rhythms of the agricultural year and the working life of the community. Popular leisure is public and gregarious, with plenty of drinking, and often punctuated with violence and excess. Elites tolerate and often patronize these festivities, in part because of their own membership in the village and in part out of their recognition of the value of these festivities for social control. Many traditional customs such as Church-ales have disappeared and others are declining as close-knit rural community life declines. The desire grows among the elite for greater discipline and order and less idleness and insobriety. By mid-century popular culture is subjected to increasing criticism (Malcolmson 1973:11, 18-19, 113-114, 118).

MORAL CONCERNS. The Societies for the Reformation of Manners continue their attempts to reduce the licentiousness of the age (see 1690/England). They bring thousands of prosecutions for vice offenses before the civil courts by 1730 (about 1,400 a year) but become ridiculed and hated in the process (Porter 1982:311-312). By 1702, their goals and tactics have become involved in heated party politics. The societies' sermons seem to many to be "superstitious and fanatical reminiscent of the oppressive enthusiasm of the puritans....The morality for which the societies strove [is] not the popular morality of the 18th century....not the neoclassical ideals of wisdom and virtue [but]...the negative virtues of not swearing, not whoring, not profaning the Lord's day" (Bahlman 1957:101). Among the critics of the societies are Defoe and Swift (see below).

INEBRIETY PREVALENCE. Prodigious quantities of alcohol are consumed by all classes throughout most of the 18th century, especially the first half. In the words of Joseph Addison in 1714, "In this thirsty Generation, the Honour falls upon him who carries off the greatest Quantity of Liquor, and knocks down the rest of the Company" (Spectator 1907, 8:47). As the century begins, not only is little stigma attached to drunkenness, it is even considered a virtue in some quarters. Defoe characterizes the "True-Born Englishman," (1701) as follows: "Good Drunken Company is their Delight,/ And what they get by Day, they spend by Night.../ In English Ale their dear Enjoyment lies,/ For which they'll starve themselves and Families.../ Slaves to the Liquor, Drudges to the Pots,/ The Mob are statesmen, and their Statesmen Sots" (quoted Humphreys 1954:20). Rudé (1971:70) comments, "Drunkenness was considered normal and satisfactory condition as much by Defoe or Sir Robert Walpole as it was later by Johnson or Wilkes." Simon (1926:xi) characterizes "the extraordinary drunken habits of rich and poor alike" as one of the outstanding features of the period. Many view the excessive drinking within all classes as a result of increasing "luxury and extravagance" (George 1925:272). Sutherland (1969:171) considers it a period "when too much leisure at one end of the social scale and too much misery at the other end drove people to the bottle." Excessive drinking is often linked to "a passion for gambling"; both are "interwoven with the fabric of society to an astonishing extent" (George 1925:272; see also Jones 1942:4-5).

LOWER CLASSES/DRINKING. Park (1983:69) asserts that the major alcohol-related development of the era is the "proletarianization of drinking," in which the masses gain greater access to alcohol due to the development of the commercial alcohol industry. Although this is a period of prosperity, work in London is very irregular and the poor are subject to a great deal of insecurity and depravity; when they do work, high wages become a means to purchase more to drink (see the discussion of gin drinking under 1720/England).

But wages are not the only influence. Among the working class, drinking customs are "interwoven with everyday life and work to an astonishing extent" (George 1925:271-272). Drinking is a part of all amusements, it occurs in the workplace, and wages are paid on Saturday nights at the alehouse, a custom that evokes repeated protests throughout the century. In almost every aspect of life, it is difficult for the average Londoner "to escape the ever-present temptation to drink" (George 1925:301).

UPPER CLASSES/DRINKING. Among the upper classes, the era is characterized by a pronounced contrast between extreme elegance and excessive intoxication and overeating, particularly in the first half of the century. The gentry drink inordinate amounts of wine, emphasizing quantity rather than quality (Madden 1967:143). Twelve-hour drinking sessions occur (Francis 1972:241). Daily consumption of two to three or more bottles of wine, principally port, is common. To gain a reputation as a "blade," one has to be at least a three-bottle man (Porter 1982:33). Some men have reputations for consuming six bottles a day, as in the cases of Richard Sheridan, Pitt the Younger, and Squire Mytton (see 1790/England). (However, bottles are smaller than today, probably about half the size, and the wine is less strong [Wain 1980:240].) Much drinking is competitive, with status conferred on the man who drinks everyone else under the table (Porter 1982:34). The Tatler (Number 19, of May 14, 1709) asserts that "rural esquires...are drunk twice a day" (quoted Younger 1966:337-339). Criticisms of upper-class drinking are contained in the works of Defoe and Swift (see below), among others.

Quinlan (1965:11) characterizes the upper-class Georgian as "preoccupied with drunken routs, foppery and seduction." Coffey (1966:681) considers it "remarkable how many of the most famous peers, statesmen, politicians, and literary men were famous for their hard drinking." To Sutherland (1969:165-167), "The preoccupation of the gentry with getting drunk was nothing short of astonishing."

EDUCATED/DRINKING. "The aristocracy of letters [is] infected, no less than that of rank" (French 1884:295). Typical of the era are Joseph Addison and Richard Steele (see below), Samuel Johnson (see 1760/England), and James Boswell (1760/Scotland). Oxford dons are famous for "their dull and deep potations" (Porter 1982:34, citing Edward Gibbon).

BEVERAGES CONSUMED; PRICES. There is "a tidy class division in drinking," with the poor consuming gin or cheap brandy; "the middling sort," porter or ale; and the wealthy, wine (Rude 1971:70). The prices of the various types of beer remain relatively stable during most of the century, but in the early century all are higher than gin, which is cheaper to produce and subject to a much lower tax--hence the preference of the poor for gin (Dowell 1888, 4:70-74; Spring and Buss 1977:569).

BEER: CONSUMPTION, ATTITUDES. At the beginning of the century, ale remains the popular drink among most Englishmen and an important source of nutrition; no other alcoholic or nonalcoholic beverage rivals its popularity with the general population (Lecky 1878, 1:478). The average household probably spends more money on ale and beer than on any other single commodity (Monckton 1966:137). At the beginning of the century, the lower classes generally drink two-penny, a beer sold at two-pence the pint, brewed at low heat and lightly fermented. Ale, beer, and two-penny are often mixed to personal taste, giving rise to the creation of porter in the 1720s (Cherrington 1925, 1:407). Over the century, however, per capita consumption of ale declines, largely owing to the popularity of cheaper gin among the poor, wine among the rich, and tea and coffee.

Nevertheless, beer remains an integral component of the British self-image. Protestantism, constitutional government, and beer are seen as a trinity of virtues opposed to Catholicism, autocracy, and wine. John Bull is portrayed "with an ale-pot in his hand and barley-corns in his hat." Throughout the century, beer is considered the patriotic drink, praised as the support of the farmer and landed interest, the promoter of public morality (as opposed to "demon gin") and the saviour of public revenues. So central is beer to British life that even tea is viewed with hostility. A common toast is "Here's a health to the brewer and God speed the plough" (Mathias 1959:xxi-xxii). (On the importance of beer-brewing to the economy, see below.)

REVENUES; ECONOMIC VALUE. Taxes from beer, malt, and hops make up as much as a fourth of the total public revenues, which accounts for much of the importance of beer in the English imagination and economy (Dowell 1888, 4:50, 65; Mathias 1959:xxiv).

BREWING INDUSTRY. The commercial brewing trade, still largely confined to the London area, gradually is organized into larger units. Meanwhile, the small brewing victualler declines. Some of the best-known firms of today--such as Bass, Watney, Worthington, and Whitbread--are founded by 1750. The appearance of large common (wholesale) brewers is partly the result of the government's policies begun in 1660, which encouraged the brewing industry to concentrate in fewer and larger plants. The introduction of porter (see 1720/England) revolutionizes the brewing industry and further stimulates its expansion and concentration. This expansion is encouraged by plentiful supplies of surplus grain and the increasing demands of a rapidly growing population. The industry itself is very suited for large-scale production within a single plant, enabling it to seize the potential of the expanding market. Thus beer becomes one of the first products to be governed by mass production and mass-marketing principles (Mathias 1979:215, 222, 226). Commercial brewing facilitates manufacture of products of higher, more consistent quality and lower cost. Both wholesale and retail prices are reduced, making it cheaper for the publican and the general population to buy beer than to make their own (Lynch and Vaizey 1960:40). In the second half of the century, new scientific principles are applied to industrial production. By its end, the barrelage of beer produced commercially has increased about 50%, and home production decreases 10%. Almost all the commercial increase is in the output of the common brewers (Mathias 1959:337, 542-543). Home brewing does not disappear until the early 19th century, but by the end of the 18th century, the rising price of malt, hops, and other ingredients has discouraged home and private brewing except among the wealthy (Cole and Postgate 1961:82; Monckton 1966:137).

The improvements made in the quality of beer produced and its keeping capacity attract attention "from far and wide"; English beer is increasingly exported (Glamann 1962:140).

ALEHOUSES: REPUTATION, CONTROLS. The alehouse is no longer regarded by elites as the headquarter of an alternative world, but more as an "informal buttress of the established order." By 1700, the alehouse appears "firmly under control in all but the largest urban centres" (Clark 1978:70-71). Contributing to this sense are: (1) the slowing down of the growth of alehouse numbers; (2) more powerful and bureaucratic mechanisms of control, including the influence of excise authorities; (3) greater magisterial supervision; (4) improvements in patron standards of living, landlord reputations, and alehouse facilities; (5) the expansion of the social functions served by the alehouse; and (6) the growing power of larger brewers which works as an informal regulatory mechanism. Alehouse keepers are subject to growing regulation by local justices, who, under authority of the 1552 licensing act, enforce tough measures against unlicensed victuallers and seek to reduce excess premises. Who can set up in trade, the size of a premise, and opening hours are all controlled. With authorities breathing down their necks, landlords regularly inform on highwaymen and thieves, and make greater efforts to maintain order. And as links with the established order become stronger, the alehouse keeper even comes to be seen as an ally to the political and religious establishment. Indeed, he even becomes the medium by which upper-class attitudes and fashions filter down the social scale. The coalition of opposition that had existed before the Civil War disintegrates as the ruling classes no longer feel threated. Complaints about alehouses do arise as concerns about the gin shop grow (Clark 1983:178-187, 232, 236-238).

Timothy Nourse (Compania Felix, 1700) blames abuses in the alehouses on the justices, because in their private lives and conduct of business they set an example of disorderliness. They even hold court sessions in inns "invested with Drinking and a Throng," and often a justice will "hold a Glass in one hand while he signs a Warrant with the other." He also accuses JPs of using their licensing powers to further their own pecuniary interests. Nourse considers that reform will not occur until the upper classes, and especially the justices, set an example and dissociate themselves from the disorder of the inns and ale houses (quoted Bretherton 1931:177-178). However, in his attacks on alehouses, Nourse is an exception to the general tolerance of the period (Clark 1983:186).

ALEHOUSE CHARACTERISTICS; PATRONS. Considerable change occurs among the alehouses and their patrons compared to the early 17th century. Because of rising affluence, alehouses become run by more respectable persons (including local officials) and become larger and more elaborate, offering a wide variety of drinks as well as food and accommodation, with game, club, and other special rooms. They are no longer the squalid backrooms and cellars usually kept by laborers and craftsmen in the 16th century; they are no longer "defined and limited by the poverty of many of [their] customers" (Clark 1983:232). Patrons are not as destitute and many drink less to forget their plight than for enjoyment as a leisure-time activity. Patron poverty has been mitigated by improved poor relief, rising employment opportunities and wages, and declining food prices; many laborers and artisans can now afford to take a day off from work to go drinking. The central core of customers is composed of skilled workers, small craftsmen, and lesser traders whose rising prosperity creates a greater demand for alehouse drinking and sociability. They help reorient the old tippling house away from its earlier preoccupation with the requirements of the poor. As alehouses become more respectable, some members of the upper and middle classes feel less inhibited about frequenting them (Clark 1983:222-227, 232).

FUNCTIONS. London life centers around the tavern, alehouse, and after the mid-century, the club. Besides serving as a place to drink and socialize, the public house provides an array of social services for the community. Coaching inns and nearby taverns act as clearing houses for the regional flow of migration into towns and serve as unofficial employment exchanges for artisans. Many trades designate a favorite alehouse as a "house of call" for news and information (Corfield 1982:105). On their part, owners set out to attract people and make the public house a center of recreation that provides a modicum of ease and comfort, a place for club meetings and public gatherings, and a site where workers can receive their wages (tavern owners pay employers for the privilege of dispensing wages). As public houses become more and more central to lower-class life, they become less and less attractive to the middle and upper classes (Clark 1984; Coffey 1966:679; George 1925:299, 302, 334-336; Harrison 1971:45-54, 325-332; Popham 1978:263-264; Rudé 1971:92)

If the alehouse was not just a place to get drunk, Park (1983:67-69) emphasizes that the increasingly important role played by the taverns contributes to the spread of drunkenness. The taverns bring cheap liquor to "new classes of workers and urban poor emerging from the dissolution of the agrarian society in England." They lose their old function as a meeting place for the whole community--both lords and peasants--in which drinking was still bounded by communal mores. With their "commercialization," they increasingly become hangouts primarily for laborers. Under the "relentless pressure of the drink suppliers," drinking reaches new heights. As a consequence, drunkenness becomes identified as a characteristic of working class life.

During this period, the role of the alehouse in old-style communal festivities may be declining as ruling elites seek to bring these festivities back within the ambit of the church and manor (Clark 1983:233; Malcolmson 1973:12, 56). This may be another reason for the decline in concerns over alehouses among elites.

TIED HOUSES. With beer production capacities expanding but per capita beer consumption falling, fierce competition develops. For survival, beer producers seek to insure their outlets by controlling public houses (the tied-house system) through direct ownership or financial backing, and by turning to advertising. Although the number of tied houses increases throughout the century, it is not until its end that brewers deliberately set out to gain control of the retail trade (Park 1983:64-65; Spring and Buss 1977:568; Mathias 1959:119-120, 375; Monckton 1966:148).

DRINKING PRACTICES. It is in the early 18th century that the celebration of St. Monday--the extension of Sunday relaxation and drinking over to Monday instead of working--becomes a "renowned custom," although the practice may date back a century. This development is probably related to the growth of artisan prosperity (Clark 1983:223).

WINE CONSUMPTION: CLARET, PORT. Claret is the preferred wine of the discriminating drinker from the late-17th century through the 18th century. It is so popular that the best wines of Bordeaux are reserved for the British trade.

The popularity of port is partly due to political reasons, a result of the Treaty of Methuen of 1703 as well as contempt for things French under the German Hanoverians. The treaty levies a prohibitive duty of 55 pounds a ton on all French wines while imposing a duty on Portuguese wines of only 7 pounds a ton. This initiates a shift in national taste from French wines and brandy to port (Drummond and Wilbraham 1939:252). But as important in port's popularity is the largely carnivorous diet of the upper classes, which "called out for a deluge of strong wines" (Younger 1966:361-378, 380). Port becomes particularly popular among the "hard-living country squires," who appreciate it "for the effective manner in which it [reduces] them to insensibility." Many a habitual port-drinker becomes "livid-coloured and bloated at an early age" and dies young (Sutherland 1969:88-89). Simon (1926:17, 52) blames the spread of spirit drinking for killing the taste for beverage wines and creating a demand for the stronger port. The 18th century is also the age of port, "most of it...a course, rough, heavy wine." At first the term "port" is used for any Portuguese wine (see 1680/England). Modern port is invented in the vineyards of Douro (Portugal) by the addition of grape brandy to the fermenting must, which stops the fermentation process so that the wine retains more sugar. This is probably first done to preserve the quality of the wine through the long voyage to England and to disguise any bad taste. Mature port also requires the development of the straight-sided bottle to allow aging on its side.

SPIRITS: SALES CONTROLS LIBERALIZED (1702). An act of 1701 requires retailers of British spirits to obtain an alehouse license; such a restriction, however, is considered harmful to the consumption of English spirits, so in 1702 Parliament repeals the act and allows distillers to operate as many retail outlets as they like. The act also permits free sale to "all other shopkeepers whose principal dealings shall be more in other goods and merchandises than in brandy or strong waters." The purpose of the act is to encourage consumption of malted corn and prevent the smuggling of foreign spirits (Monckton 1966:142). Park (1983:62) emphasizes that these measures were "deliberately instituted" by landed interests who controlled Parliament for the express purpose of encouraging the distilling of surplus grain into spirits, thereby increasing its availability and consumption. This view is reflected in the economic writings of Defoe (see 1710/England). Landlords consider grain distillation as essential because (1) surplus grain cannot be stored for a long time or transported long distances, and (2) grain surpluses are increasing with growing farm productivity due to improved agricultural techniques and the expansion of farm acreage caused by the enclosure movement.

CONTROL EFFECTS: CONSUMPTION, PRODUCTION. The gin act of 1702 results in the unlimited production and indiscriminate sale of gin, and leads to "a perfect pandemonium of drunkenness, in which the greater part of the population of the metropolis [London] seems to have participated" (Webb and Webb 1903:21-22).

DISTILLING INFLUENCES. In addition to the low duty on gin and other liberal legislative actions, several other factors account for a rise in gin distilling and retailing in this period. It is easier to become a distiller than a brewer since the established brewers discourage outsiders and the industry is being increasingly dominated by large-scale capitalist producers, especially in southern England. It is also much less expensive to become a distiller and the profits are greater (see the comments of Campbell under 1740/England). Since spirit retailers only sell drinks, they do not need elaborate premises, and, in large part as a consequence of legislative changes in 1702, they are much less controlled than alehouse or tavern keepers. In addition, after 1720 they are free of the requirement to billet troops. Thus, the spirits trade may serve as an outlet for people excluded from the regulated alehouse trade (Clark 1984).

GIN CHARACTERISTICS. Although English grain spirits are called "gin" in this period, they are technically different from Dutch "genever," since they are flavored with substances other than juniper.

ATTITUDES: INEBRIETY, CONTROLS (DEFOE). In *The Poor Man's Plea, in relation to all the Proclamations, Declarations...for a Reformation of Manners* (1698), Daniel Defoe reflects the view of many that England is thoroughly debauched and immoral (Bahlman 1957:1). He refers to drunkenness as the "well-bred vice" which began among the gentry "and from them was handed down to the poorer sort." He criticizes the view that "an honest, drunken fellow" is a praiseworthy character: "After the Restoration...drunkenness began to reign. The gentry caressed the beastly vice at such a rate that no servant was thought proper unless he could bear a quantity of wine; and to this day, when you speak well of a man, you say he is an honest, drunken fellow--as if his drunkenness was a recommendation to his honesty. Nay, so far has this custom prevailed, that the top of a gentlemanly entertainment has been to make his friend drunk, and the friend is so much reconciled to it that he takes it as the effect of his kindness" (quoted French 1884:242). Defoe protests against the class bias of vice laws and warns that society will never improve as long as it singles out the poor as victims. A rich man can "reel home through the open streets and no man take any notice of him, but if a poor man gets drunk or swears an oath, he must to the Stocks without remedy" (quoted Hill 1961:296; see also Quinlan 1965:16). He calls the moral laws and proclamations of the reformation societies "cobweb laws" that catch the little flies but let the great ones through. He condemns the whole reformation scheme on the grounds that drunken justices cannot be expected to enforce the laws against drunkenness and sober justices dare not touch the rich. Real improvement can only come about when the upper classes set a better example (Bretherton 1931:179).

In An Essay on Projects (1697), he says that moral laws never have any influence upon practice nor are magistrates fond of putting them into practice (Bahlman 1957:2, 28)

In Reformation of Manners, a Satyr (1702), he writes that drunkenness has become so epidemic in all classes that the young can hardly think it a crime. He fears that it will come to be "the nation's character" (quoted Bahlman 1957:2).

(ADDISON AND STEELE). Both Joseph Addison and Richard Steele are known for their weakness for wine. According to Samuel Johnson, Addison frequently sought comfort in drinking too much wine. At the same time, Addison praises temperance and abstinence as virtues and he writes an essay against drunkenness in the Spectator (no. 569, July 19, 1714) in which he asserts: "A drunken Man is a greater Monster than any that is to be found among all the Creatures which God has made....This vice has very fatal effects on the mind, the body, and fortune of the person who is devoted to it." There is no character "more despicable and deformed" than the drunkard, whose actions are in direct contradiction to reason (Spectator 1907, 8:47-48; see also French 1884:257-258, 262).

(SWIFT). Like Defoe, Jonathan Swift (A Project for the Advancement of Religion and Reformation of Manners, 1709/1808:275) complains that "the nation is extremely corrupted in religion and morals." Regarding drinking, he observes that "any man will tell you he intends to be drunk this evening, as he would tell you the time of day" (p. 277). Drunkenness is an example of the "pernicious vices frequent and notorious among us, that escape or elude the punishment of any law we have yet invented, or have no law at all against them" (p. 282). Insobriety is especially rife among the army and among the youth of the nobility and gentry: "It is commonly charged upon the gentlemen of the army, that the beastly vice of drinking to excess, has been lately, from their example, restored among us; which for some years before was almost dropped in England. But, whoever the introducers were, they have succeeded to a miracle; many of the young nobility and gentry are already become great proficients, and are under no manner of concern to hide their talent, but are got beyond all sense of shame, or fear of reproach" (p. 284). He characterizes drunkenness, whoring, cursing, and blasphemy as common faults of the upper classes.

As a cure he recommends that the queen declare that anyone "notoriously addicted to that, or any other vice" be denied presence to her or her ministers (p. 284). Regarding the Reformation Societies, although they began with the best of intentions, they "have dwindled into factious clubs and grown a trade to enrich little knavish informers of the meanest rank" (p. 293). He further recommends: "In order to reform the vices of this time which, as we have said, has so mighty an influence on the whole kingdom, it would be very instrumental to have a law made, that all taverns and alehouses should be obliged to dismiss their company by twelve at night, and shut up their doors; and that no woman should be suffered to enter any tavern or alehouse, upon any pretense whatsoever. It is easy to conceive what a number of ill consequences such a law would prevent; the mischiefs of quarrels, and lewdness, and thefts, and midnight brawls, the diseases of intemperance and venery, and a thousand other evils needless to mention. Nor would it be amise, if the masters of those publick houses were obliged, upon the severest penalties, to give only a proportioned quantity of drink to every company; and when he found his guests disordered with excess, to refuse them any more" (pp. 297-298).

Although Swift takes pride in his wine cellar, he is a moderate drinker who takes a daily bottle of wine for health's sake but seldom more. Many of his friends are characterized by a similar moderation (Francis 1972:151).

COFFEE HOUSE PREVALENCE. Under Queen Anne, there are nearly 500 London coffee houses. Coffee houses expand beyond the immediate vicinity of the City of London, with their vogue reaching its zenith in mid-century (Ferguson 1975:18, 20; Rudé 1971:77). The coffee house serves as a meeting place where the events of the day are discussed, papers read, and news and gossip exchanged. The most prominent representatives of English literature are coffee drinkers and frequenters of coffee houses (Jacob 1935:139). Almost every phase of public interest and activity finds its expression and outlet there. Different trades and groups develop their favorite houses, so that early they become divided along functional lines (Rudé 1971:78). (A figure of 2,000 coffee houses in London in 1715 is also cited by some authorities [Porter 1982:245; Ukers 1935:69]; the source of this descrepancy could not be determined.)

FRANCE

INEBRIETY PREVALENCE, CONCERNS. At least in Paris and other towns, drunkenness appears to increase during the last quarter of the 17th century, and some of this increase can be attributed to rising spirits consumption. But neither inebriety nor spirits consumption is viewed as a particular problem in France. The French are still among the more moderate European drinkers. Among the factors contributing to this are (1) the mixing of water with wine, although among the upper classes this appears to become less common through the century; (2) the relatively small consumption of spirits; (3) the lack of specialized spirits shops as in England and the restriction of spirits sales at a table to the cafés; and (4) the relative unimportance of such practices as drinking contests and toastings.

Although drunkeness is not considered honorable behavior in Paris in the first half of the 18th century, there is less concern over drunkenness per se than over drinking in taverns. In the complaints brought before the *commissaires*, the charge of drunkenness is rare; most complaints are over conflicts that occur within the tavern. Often wives issue complaints against their husbands for going to the taverns but without emphasizing drunkenness or alcohol. People are arrested if found drunk on the street at night, but more for being out at night than for drunkenness. No effort is made to round up drunkards during the day or night, though the police inspect cabarets and keep a watch over "suspicious persons." References in police records to drink-related disturbances are infrequent. Although Parisians drink at the cabaret several times a week and even a day, they do not drink much on a regular basis (Brennan 1981:222-224; Brennan 1984).

TAXATION: EFFECTS. By 1700, 60% of the cost of weak wine is Paris is taxes, and complaints are raised that the tax system is prejudicial to the poor (Dion 1959:502-503). In the course of the century, the taxes on the alcoholic beverage trade (droit d'aides) become increasingly onerous, widely resented, and frequently circumvented (Lachiver 1974:420). (For a description of the tax system, see 1660/France). High taxes on wine in Paris lead to the popularity of suburban cabarets called guinguettes (see below) located outside the city limits and make fraud and smuggling into the city commonplace (Kaplow 1972:12).

Fraud in the wine trade is especially common in the western part of the Ile-de-France during the century because of burgeoning wine production in the countryside alongside the existence of the ready market in Paris and the heavy taxes placed on the trade. Vine growers often lie about their production, and merchants and publicans attempt to hide purchases made illegally. Wine wholesalers are discouraged by heavy tolls at rivers and city gates; wine retailers complain about the heavy taxes levied on direct sales. Cabaret operators try to avoid taxes by refilling their wine barrels in order to give the appearance that little or no wine had been sold. On occasion, the publicans react to these heavy taxes with violent opposition (Lachiver 1974).

Sébastien Vauban asserts that although wine and brandy are among the goods that France sells abundantly to foreign countries, domestic consumption is low because of "the Greatness and Multiplicity of the Duties of the Aids and Provincial Tolls or Customs, which often exceed the Value of the Goods, such as Wine, Beer, and Cyder, and thence it is that so many have pluck'd up their Vines" and many people must groan with misery. He recommends that the aides be replaced by a simple tax on every hogshead of wine consumed, which would yield a high revenue and also act as a restraint upon the peasants who crowd the cabarets on Sundays and holidays. A distinction should still be made between what is consumed in cabarets and what is sold on the streets in pots and pints, which ought to be exempted (Vauban 1707/1708:2, 5-6, 72).

RURAL POOR/DRINKING. Poor rural women do not drink, but men often return home after a visit to the cabaret "suitably anaesthetized with cheap alcohol to the squalor of home and the hungry family" (Hufton 1974:114). Even in the viticultural regions, wine is still consumed by only "a miniscule part of the population, save on festive occasions" (Loubere 1978:93). The ordinary drink of the vineyard workers remains what it has been for centuries, a weak piquette from throwing water on the marc (Dion 1959:473; Le Grand d'Aussy 1782, 2:416; Beauroy 1976:137). Restif de la Bretonne (1933:67, 165) in his biography of his father, an agricultural laborer in Burgundy, reports that his father did not begin drinking wine until well after adulthood (age 20 in 1712), and then only in moderation. His mother drank only water, and she blushed at the very idea of wine (Dion 1959:472, 593).

URBAN POOR/DRINKING. The laboring poor, which make up about two-thirds of the population of Paris, often work 14-16 hours a day in low paying, insecure jobs on the verge of indigence, especially toward the end of the century (Kaplow 1972:52-54). Paris is characterized by a culture of poverty oriented toward immediate gratification and rooted in a communal gregariousness that serves as a safety valve for frustration. Laborers work hard and play hard; for the most part they accept their poor condition in life, but they get "roaring drunk in the guingettes [see below]...on Sundays....Like their English contemporaries, they often honored Saint Monday by continuing their libations" (Kaplow 1972:166-167, quote at 106, who observes that the poor enjoyed a tacit freedom to get drunk and let off steam in ways that were not dangerous to society). Parisian porters have a reputation for abandoning their loads to get drunk at cabarets and are closely watched by police (Kaplow 1972:42; see also Saint-Germain 1965:17).

WOMEN/DRINKING. The straight-laced and temperate Mme. de Maintenon, the second wife of Louis XIV, complains about the decline in decency and manners among the women at court, including drinking: "I cannot stand the women nowadays, their foolish and immodest way of dressing, their tobacco, wine, gluttony, coarseness and laziness. I cannot stand any of it" (quoted Gaxotte 1970:291).

One M. de Brilhac has his grandmother sent to a convent because she was spending all her money on wine and eau-de-vie, which she preferred even to food (Saint-Germain 1965:17).

MILITARY/DRINKING. The common custom of giving spirits to soldiers before battle, according to Lemery, does not produce "a bad effect" (Braudel 1981:244). Soldiers have a reputation as habitual drinkers, and the production of brandy becomes a war-time industry. As observed above (see 1690/France), several contemporaries attribute increased spirit drinking in France at this time to the influence of Louis XIV's many wars and the spread of the habit from the military to the civilian population.

WINE CONSUMPTION. In the early years of the 18th century, the average wine consumption is about 120 quarts per Parisian per year, or about 1 1/2 cups a day. This figure, less than the 160 quarts estimated in 1637, is certainly low because it does not include untaxable wine made and sold by Parisians on their own property or that sold at the guinguettes, or the massive amounts of wine smuggled into Paris to avoid the high city taxes. A more realistic per capita figure may be 170 quarts a year (150 *pintes*). Since this estimate is an average for the whole population, and since many males can only afford small amounts of wine a day, some male adults drink considerably more. The effect of such a consumption level is mitigated by the weakness of the diluted wine consumed, the consumption of much of the wine as an adjunct to hard physical labor, and the practice of spreading drinking throughout the day (Brennan 1981:212-213, 241). The wine drunk by Parisian laborers is very poor, often cut with vinegar and dyestuffs (Kaplow 1972:78).

PRICES. Throughout the reign of Louis XIV (1643-1715), the price of an average wine increases and its quality decreases; by the late years of his reign, few but the rich can afford more than modest amounts (Lewis 1957:70). A *pinte* of wine (slightly more than a quart) costs between 8 and 12 sous through most of the first half of the century, with wages at about 30 sous a day. Thus, to drink a *pinte* of wine a day would cost a worker about one-third of his daily wages. Skilled artisans or master craftsmen, who make two to three times as much money as wage laborers, probably spend about 15 per cent of their average income on wine; they also comprise the main body of tavern patrons (Brennan 1981:213-214; Brennan 1984).

CHAMPAGNE CONSUMPTION. Sparkling champagne begins to be produced during the first half of the century and is popularized under the Regency (see 1710/France). But it is a long time before the new champagne displaces the traditional red, grey, and white wines of the Rheims area (Braudel 1981:235).

SPIRITS CONSUMPTION. The popularity of spirits among the upper classes rises rapidly over the century as liqueurs are perfected and the public taste is refined. Although the brandy trade with northern countries flourishes, there is little evidence of French consumption up to this time. Evidently, this has been due to the availability of local wine, the patterns of commerce which have existed between southern France and northern Europe for centuries, and the beverage preferences of the English and Danes (Forbes 1970:196). Now, much of the brandy begins to go to Paris. In 1730, it is observed that spirit drinking, "almost unknown to our fathers," has become very common in Paris as well as in other cities of the kingdom (quoted Delamain 1935:29, 39-40). Feeding this rising demand, the brandy and liqueur industries of southern France greatly expand their production (see below).

Spirit drinking appears to occur most often in the early morning or late evening, providing instant warmth and revival. Spirit drinking does not increase the level of drunkenness. Spirits are not specifically identified in accusations of drunkenness; at the end of the century, Mercier (see 1780/France) does not cite spirits as a problem, although he describes brandy drinking and attacks cabarets and drunkenness from wine in general. This may be because in theory only cafés and merchants of eau-de-vie can sell spirits. Spirit consumption in 1714 appears to be about 1/40th that of wine (Brennan 1981:116, 240, 255n.82, n.83).

WATER CONSUMPTION. Paris has only 73 water fountains and the water is so poor that it is sometimes drunk only by the laboring poor, who over time develop a certain resistance to its ill effects (Kaplow 1972:78).

SALES CONTROLS. Since 1680, the distinction between cabaret and tavern has become blurred. Parisians use the word "cabaret" to refer to a variety of institutions that sell alcohol and offer a place with tables and chairs to drink it. The privilege of wine selling belongs only to the guild of wine merchants, which includes cabaret and tavern keepers. The master limonadier sells coffee and shares the right to sell brandy and other eaux-de-vie with the master distiller, grocer (*épicier*), and *vendeur d'eau-de-vie*. Since 1676, the guild of limonadiers has operated the cafes, which may sell a variety of liqueurs and spiced drinks in addition to coffee. Only they may serve spirits at tables. The beer sellers belong to no guild and are generally men and women who have other professions; beer may be purchased at hundreds of shops set up by these artisans (Brennan 1981:27-30).

SPIRIT SALES CONTROLS. The guild of limonadiers has grown so greatly and has become so unwieldy that the king temporarily restricts its growth in 1704. In need of more funds, in 1705 the king restores the guild with a fixed number (50) of hereditary members and requires a large payment for a new charter. In exchange, he concedes to them the new privilege of selling gin, cocoa, vanilla, and chocolate drinks. The rights of all rivals to sell spirits is affirmed, but only the limonadiers may serve them at cafe tables. The limonadiers offer a payment of 200,000 livres, but when they fail to make the full payment, the guild is suppressed again and a new corporation with 500 hereditary offices is created in 1706. Negotiations and complications involving the sales of these offices continue until 1713, when the guild is once more put on a legal footing. Thereafter, the number of cafes increases markedly throughout the century. Grocers and vinegar makers continue to dispute the cafes' right to sell and serve brandy. It appears that the issue of seating is just as important as selling brandy (Jacob 1935:178-179; Brennan 1981:29; Fosca 1934:12-13).

CABARETS: REPUTATION. Cabarets are increasingly viewed by police and elites as haunts of drunkenness and criminality; *cabaretiers* (cabaret keepers) are considered the lowest, most unsavory of employers. They are attacked as fences for stolen goods and for providing support for thieves, criminals, and other disreputable figures. They make up about one percent of those who come before the courts for individual crimes. They also are one of the major creditors to the poor, since they allow credit for drinks and thus contribute to the poor's destitution. It is through indebtedness that the transition from poverty to indigence usually occurs (Hufton 1974:31, 60, 229, 259, 265, 271, 302).

Among social and literary elites, the cabaret is seen as unfashionable, too popular and crude. As expressed in the works of Mercier (see 1780/France) and Restif de la Bretonne at the end of the century, they are sleezy, dangerous, receptacles for the dregs of the populace. Many critics also view cabaret drinking per se as a waste of time and money and a cause of men neglecting their work and families, a refuge from responsibility and authority. However, among the lower classes, cabarets are popularly viewed as an integral part of culture (see below).

CONTROLS. Although Paris police no longer attempt to restrict citizens from using the cabarets, they do attempt to restrict popular access to them by enforcing strict closing hours, by making sure they are closed during mass on Sundays and feast days, and by inspecting them at least once a week for suspicious people. Closing hours have risen three hours over the times existing as late as the 1660s. The hours are extended probably because of the difficulty enforcing earlier hours (Brennan 1981:73-76).

PATRON CHARACTERISTICS. The clientele of Parisian cabarets and many cafes is now almost entirely laboring class, male, and mature (between ages 19-40). It is still "somewhat unusual" to find women drinking there. In terms of socioeconomic levels, patrons reflect closely the overall distribution of the Parisian population. All but the richest Parisians go to the cabaret frequently to dine, relax, and drink; however, it is first of all a laboring-class institution. A few nobles attend them in the early century, but the upper classes, and after them the middle classes, increasingly abandon the cabaret. After 1731, the number of shopkeepers or master artisans among laborers also declines. In part this abandonment is because of the greater privacy offered by larger homes and the rising orientation toward family life, which make cabarets less necessary to the wealthy. It is also rooted in the rise of the cafe as the fashionable meeting place of social and literary elites. On the other hand, patrons of the cabarets do not appear to belong to the lowest levels of society, to the unemployed, indigent, immigrants. Soldiers appear to make up about 10% of patrons. But nearly 85% of customers labor with their hands or work at a trade, some as employers, but most as salaried workers. Cabarets are scattered all over the city, but are concentrated along the main streets of the laboring sections (Brennan 1981:30, 36-49, 57, 87).

FUNCTIONS. To the poor, the cabaret is an integral part of popular culture and an extension of and replacement for the home. The need for space is one of the primary requirements of cabaret patrons. Housing conditions are so poor that many use their homes only for sleeping. Farge (1979:73) says that "the cabaret seems to belong to the *petit peuple*...it violently reflects the brutality of the conditions of life and desire to live." The cabaret is popularly perceived as a respectable gathering place, a legitimate part of the community; not, as portrayed by police and other officials, as a den of thieves, a shelter for vagabonds, men of ill-repute, debauchery, and disorder. Far more than a place to drink and get drunk, cabarets are neighborhood centers for meeting and talking. Parisian cabaret life is not characterized by drinking contests or excessive glorification of drunkenness (Brennan 1981:57, 87, 282, 289, 311-312).

CRIMINALITY. The Paris police profoundly distrust the cabaret because of its openness and accessibility and the fear that it can be used for all kinds of evil. They particularly identify crime with vagabonds, vagrants, and other "suspicious people" outside the order of the city who they believe take refuge in cabarets (Brennan 1981:69-70). However, police records show that few criminals and suspicious people are arrested at cabarets, possibly because few criminals have enough money to frequent them. Most crimes are committed by starving people as acts of desperation. Although 29% of crimes of personal violence take place in cabarets, this is as much because they are the main public centers as because of drinking. These conflicts are of a communal nature, a manifestation of social tensions and frictions within the community (Brennan 1981:69-70, 76-78).

ATTENDANCE. Attendance at Paris cabarets seems to be fairly regular throughout the week, peaking on Sundays. Cabarets usually open before dawn so that laborers can meet to chat, have a cup of coffee or brandy, and a smoke. Laborers then make frequent stops there throughout the day, with business picking up late in the afternoon. A quarter of patrons drink between the hours of four and eight (most men work until six). Attendance declines 50% on Saturdays and peaks on Sundays, suggesting payment on Saturdays. On the whole, drinking appears integrated into daily life, with Sundays being special both because more money is available and because time is available to make a trip out to the suburban guinguettes (Brennan 1981:116, 118-119, 129).

PREVALENCE, SALES LEVELS. Sebastian Vauban calculates in 1707 that there are 40,000 licensed premises in the 36,000 parishes of France, with the average premise selling about 4,500 pints of wine a year, or about 12 pints a day (Vauban 1707/1708:72). This figure does not seem to include "off-license" sales but does indicate that the number of drinking outlets in France is relatively small and that sales are low, probably because of the amount of taxes put on wine by the aides system and the corruption of the system (Lewis 1957:70).

GUINGUETTES. A major problem involving cabaret drinking in Paris is the practice of lower-class men and women visiting the guinguettes outside the city gates. These are elaborate, large suburban cabarets which have vast outdoor courtyards and which offer entertainment, music, and dancing (Brennan 1981:34, 100; Farge 1979:77). They also offer cheaper drinks than the urban cabaret because the tax of "four sous admittance for one bottle which intrinsically is only worth three" is not chargeable outside of the city. They first appear around 1675, with the "guinguette" itself first occurring in print in 1700 (Dion 1959:505-511). As the taxes on wine sales inside the city rise, these suburban taverns become increasingly popular, especially on weekends. Their greatest growth seems to occur after 1750 (see the comments of Mirabeau under 1750/France and Mercier under 1780/France). People dress up to visit the guinguettes on special weekend outings. These weekend outings encourage binge drinking on Sundays. As one contemporary refrain asserts:

> Commoners, artisans, grisettes,
> All leave Paris and run to the guingettes
> Two pints for the price of a single booze

On two boat's benches, without cloth or serviettes,
You'll drink so much in these Bacchic stews
That out of your eyes the wine will ooze (quoted Braudel 1981:236)

SPIRITS CHARACTERISTICS, PRODUCTION: LIQUEURS. At the beginning of the century, nonaromatized, pure wine spirit does not have a very agreeable taste. Since improving the taste by several distillations is too costly, flavorings are used (Delamain 1935:34). Dr. Louis Lemery, of the Royal Academy of Sciences, writes (*Traité des aliments*, 1702:512): "Inflammable spirits have a slightly pungent and often empyreumatic taste....To remove the disagreeable taste, several compositions have been invented, which have been given the name of ratafia and which are nothing more than brandy or spirit of wine flavoured with a mixture of different ingredients" (quoted Braudel 1981:245; see also Le Grand d'Aussy 1782, 3:81). A great variety of poetically named liqueurs become available (e.g., *Eau nuptuale*, *Belle de nuit*, *Parfait amour*). These vary "with the inventiveness of their makers" (Younger 1966:387; see also Gottschalk 1948, 2:148). The great center for the production of these "waters" is Montpellier, near the brandy supplies of Languedoc (Braudel 1981:246; Le Grand d'Aussy 1782, 3:79). By 1729, the distillers of Languedoc are carrying on a "notable export trade" (Forbes 1970:196).

COGNAC. A frost in Charente in 1709 destroys nearly all the vines in impoverished Saintonge and leads to replanting with new vines (Folle Blanche), which become the basis of modern cognac (Younger 1966:327). These vines produce a greater quantity of juice in poor soil. As the brandy trade expands, vines replace grainlands and the focus of commercial viticulture shifts from the coast to the inland areas around the city of Cognac (Loubère 1978:36-37). Distilleries are founded in the area around Cognac which are still prominent today, including Martell in 1705, Ranson and Delamain in 1725, and Hennessy in 1765 (Simon 1926:34). In 1726, it is asserted in Angoumois that brandy from Cognac (meaning made in the province) is the best in the world (Delamain 1935:39). A similar statement is made in the *Encyclopédie* in the mid-century. By 1780, the fame of cognac is firmly established.

According to Delamain (1935:20), one reason for the expansion of brandy distilling at the time is that tastes probably have become more refined and the public is rejecting wines which had always deteriorated during long voyages but had previously been found acceptable.

VITICULTURE: EXPANSION, ECONOMIC VALUE. The increase in prices and markets during the early 18th century encourages vine-planting and an expansion in both production and viticultural knowledge around 1720. Still, "man knows little more about cultivating vines and making wine in the 18th century than in antiquity." Until the late century, the wine industry, though quite widespread geographically, operates on a small scale. Loubère (1978:210) maintains that it still "hardly an important economic force." Labrousse (1944:209) estimates that wine accounts for an average value of 800 million livres to the economy, second in value only to cereals.

In 18th-century Bordeaux, good-sized fortunes based on wine are amassed by the *noblesse de robe*, the judicial nobility whose origins are mainly bourgeois. Although the wine trade is risky, viticulture yields a much higher return on acreage than grain farming, at least three times as much, and viticulture continues to spread despite efforts of the government, which prefers land planted in grain, to stop it (Forster 1961).

COFFEE CONSUMPTION. Coffee, tea, and chocolate become the reigning luxury items of all the towns, but they do not permeate into the countryside until the end of the 19th century (Duby and Mandrou 1964:363). However, not all people are enthusiastic about the new drinks. For example, the Princess Palatine (duchesse d'Orléans) writes in a letter dated Christmas 1712: "I cannot stand either tea, or coffee, or chocolate. I cannot comprehend how people like those things. I find that tea has a taste of hay or straw, coffee of soot and lupine, and I find chocolate too sweet." What "I should like best," the German-born princess declares, is "beer soup" (quoted Duby and Mandrou 1964:363; Jacob 1935:91). (The princess is the sister-in-law of Louis XIV and mother of the future regent, Philippe, duc d'Orleans.

COFFEE HOUSES. The 18th century has been called the "Golden Age of French Cafes"; and Paris, the "Coffee-House of Europe." By 1715, there are 300 coffee houses in Paris. By the end of the century, this number triples (Butler 1923:362-363).

GERMANY

BRANDY DISTILLING. Whereas brandy distilling is largely an urban trade in the 16th and 17th centuries, in the 18th century it spreads out to the countryside.

BURGHERS/DRINKING. A high level of drinking still occurs among many German burghers as well as nobles. Bruford (1968:215-216) details the extraordinary expenses of one Augsburg citizen for beer, wine, and food in May 1715, but some industrious men drink only water.

NOBLES/DRINKING. Excessive drinking frequently occurs in the small courts of Germany (Lafue 1963:39-41, 67-69). Pollnitz (1739, 1:324-325) tells how, when visiting Heidelberg in 1719, the elector John William of the Palatinate insisted that Pollnitz see his great wine tun. The entire court accompanied Pollnitz to see it "in great Ceremony" with trumpets leading the way. Pollnitz tried to avoid the drinking party that ensued by surreptitiously pouring out some of his wine, but he was still overpowered with intoxication. He then tried to leave the wine cellar unnoticed, but he was prevented from doing so by wine cellar guards. A mock trial was then arranged at which Pollnitz was condemned to drink as long as he could swallow; although he was pardoned, he underwent "the heaviest part" of the sentence. For "some hours" he lost his powers of reasoning and speaking; finally, he had to be carried to bed.

VITICULTURE. After a century of warfare, the vineyards of southern Germany are devastated. By 1713, only 70,000 acres of land are under vines, compared with 300,000 in 1600 (Seward 1979:170).

IRELAND

BREWING. The notoriously poor quality of Irish beers begins to improve and Irish breweries are "reasonably well off" in the first half of the century, but their development is restricted by competition from the less heavily taxed and thus cheaper imported British beers, which are also of higher quality, and from the popularity of illicit spirits. Even after Irish customs duties on British beer are raised in 1741, "the competitive position of Irish brewers is inferior to that of their British rivals" because their market is so limited that they cannot take advantage of the savings from large-scale production (Lynch and Vaizey 1960:2-3).

WHISKEY CONSUMPTION. Whiskey becomes the staple drink of the peasantry, with the demand for the drink being satisfied both legally and illegally. Between 1700 and 1760, home production and imports rise dramatically, and small distilleries expand rapidly. Among the factors contributing to making whiskey the national drink are the low duty on it and its cheapness (it is less expensive than beer), the benefits derived from distilling to the Irish subsistence economy, the simplicity of its manufacture, the ease with which it can be transported, and the cash market that existed for it which helped pay for rents. The landed gentry is bitterly opposed to any attempt to restrict an industry that maintains a steady demand for grain (Lynch and Vaizey 1960:40-41, 49-50).

NETHERLANDS

BACKGROUND DEVELOPMENTS. By the treaty of Utrecht (1713), ending the War of the Spanish Succession, the Spanish Netherlands are transferred to the Austrian Hapsburgs.

TAVERN PREVALENCE. The towns and highly frequented routes of the Low Countries have an "extraordinary number of taverns." In Brussels, there is one for every twenty houses (Houtte 1977:254).

BREWING. Each village has at least one small brewery, often attached to an inn; in towns, large enterprises predominate. In many areas, the brewing trade declines, in part due to competition from coffee, tea, chocolate, and, especially, spirits, as well as from German beers. In Holland, the trade declines seriously (Forbes 1970:189). However, brewing in Louvain expands throughout the century (see 1750/Netherlands) (Houtte 1977:255).

DISTILLING: EXPANSION; ECONOMIC VALUE. A European-wide shortage of cereals causes the Dutch government to restrict the distilling industry between 1698 and 1699. Many distilleries close. Although other restrictions are threatened (e.g., in 1709), the next does not occur until 1771-1772. Although the century is a period of economic decline and unrest in the Low Countries, the Dutch distilling industry grows considerably. Production becomes concentrated around the river Maas (Schiedam, Delfshaven, Rotterdam) because of the excellent shipping facilities and the encouragement of local authorities seeking to compensate for the decreased herring industry. By 1700 there are no fewer than 77 stills for the production of raw gin in 34 distilleries in Schiedam, most of which are rather small. Specialization does set in; some distillers make only gin or brandy or liqueurs. The growing industry profits from the possibilities of export with all foreign countries with which the Dutch trade (Forbes 1970:188, 190-191).

In 1705, as a result of the low price of grain, the government of the Austrian Netherlands abandons its opposition to distilling and soon profits from the excise taxes it levies instead. Rural distilling comes to be favorably regarded by agronomists. Many peasants boil the must in their farms using more rudimentary and less expensive techniques than in Holland. The draff that is left is used to feed more and fatter animals. By 1795, east

Flanders has 181 distilleries and Brabant has 200. The Austrian government itself establishes a factory at Warneton on the French frontier so as to profit by smuggling, which amounts to 2,500 hectoliters for the year 1782 (Houtte 1977:236)

POLAND

NUTRITIONAL VALUE. The least expensive calories that Polish peasants absorb come from grain alcohol (Minchinton 1974:124).

RUSSIA

INEBRIETY OF PETER THE GREAT. Tsar Peter the Great (1682-1725), who was taught to drink heavily as an adolescent, gains a reputation during his reign for frequent and excessive drinking. He and his friends are often seen in the streets of Moscow or St. Petersburg drinking and singing bawdy songs. At his banquets, famous for their abundance of liquor, his guests are forbidden to stop drinking, despite protests that they have had too much already (Massie 1980:4; Clarkson 1969:191, 218-219). Patrick Gordon (1690/1968:169-170) records that at one feast "the Czar was so delighted that he made the Boyars, counsellors, and officers stay with him and carouse all night in the great hall. During the debauch, he took offence at something which was said, and was not pacified but with the greatest difficulty."

Charles Duclos' memoirs of regency France include an observation of Peter's drinking capacity during a visit to Paris: "He ate excessively at dinner and at supper, drank two bottles of wine at each meal, and usually one of liqueur at dessert, without counting the beer and the lemonade which he had between meals...He at times gave himself up to excesses with these men [his officers] whose consequences were best passed in silence" (Duclos 1727/1910:102)

ILLICIT TAVERNS. In some cities, the people have established their own kvas kabaks, alongside the state spirit kabaks. Kvas is a weak beer made from fermented bread. By 1705, all mash, kvas, and vinegar production has been placed in the hands of concessioners (Efron 1955:495).

SCOTLAND

BACKGROUND DEVELOPMENTS: UNITED KINGDOM CREATED. The United Kingdom of Great Britain is created in 1707. The Scots and English each preserve their own legal systems and churches (Presbyterian and Anglican, respectively), but their governments and Parliaments are merged, with London as the capital. The term "British" comes to refer both to English and Scottish people. The Articles of Union provide that malt made in Scotland will not be taxed "during the present war" (the War of the Spanish Succession).

INEBRIETY PREVALENCE. The 18th century marks "the zenith of conviviality in Edinburgh. People drink unashamedly" (McNeil 1956:53).

CONSUMPTION PATTERNS AND LEVELS. Following Parliamentary unification, English revenue officers cross the border into Scotland and write the first regular account of licit whisky manufacture in 1708. The revenues from ales and spirits are between 40,0000 and 50,0000 pounds a year (Barnard 1969:5; Robb 1951:37), but whisky represents a very small percentage of the total (1,810 pounds). In fact, accounts of drinking list wine, ale, rum, and brandy more often than whisky (MacDonald 1914:36-39).

ALE. Ale is almost certainly the most popular drink in Scotland in the early century. It is locally produced as a secondary occupation by the farm population, which is why the malt tax is so universally resented.

WINE, BRANDY. The upper classes mainly drink imported brandy, sherry, and claret (from France and Spain), of which the last is undoubtedly the most popular (MacDonald 1914:32-33; Ross 1970:4).

WHISKY. During the first half of the century, whisky is generally common only to the Highlands. In the Lowlands, whisky does not become socially acceptable until the last quarter of the century, when it is praised in the poetry of Robert Burns (see 1780/Scotland). Yet even in the Highlands, consumption is relatively limited until the mid-century.

WOMEN/DRINKING. It is estimated that Scotswomen in this period drink more than women in England. In fact, they may exceed men in this pastime (Ross 1970:46).

TAXATION: MALT. The malt tax, which had been reimposed in 1696, is again repealed in 1705. As before, its repeal is balanced by an increase in the duty on malt whisky (Robb 1951:37).

SPAIN

RUM BANNED. The king prohibits the distilling, sale, and consumption of rum. This ban is rooted in the view that rum is unwholesome; it is also viewed as a dangerous rival to Spanish brandy, whose popularity is increasing (Simon 1926:30, 35).

AMERICA (BRITISH COLONIAL)

RUM CONSUMPTION, PRODUCTION. Production of rum becomes a leading economic activity in the colonies, particularly in New England. Its cheapness makes it the most popular drink of the 18th century. The growing prominence of the rum industry and the attendant rise in rum consumption leads to a marked decline in the use of beer (Park 1984:4-5; Krout 1925:45).

CIDER CONSUMPTION, PRICES. In the early 1700s, except for a few moderate drinkers or abstainers, most people consider cider "scarcely potable until it had got hard; and the harder the better." Cider sells in the larger towns at from six to eight shillings a barrel (Asbury 1950:10).

TAVERN FUNCTIONS. During the first half of the century, the public house is a focus of community life; Americans meet there for rest and entertainment and to transact business and debate politics (Rorabaugh 1979:27).

MASSACHUSETTS

CONCERNS: INEBRIETY (MATHER). In "Sober Considerations on a Growing Flood of Iniquity" (1708), Cotton Mather declares that liquor "may not be amiss for many labouring men, especially when extreme heat, or extreme cold, endangers them in their labours." But voicing new concerns among American clergymen over increasing drunkenness, Mather takes a "less serene view" of alcohol than his father did. He affirms his father's teaching that alcohol is "a Creature of God," that spirits have nutritional and medical value, and that people may drink moderately to gain strength. But drunkenness, according to Mather, is a source of social unrest, a sign of divine affliction, and a warning of eternal damnation. He is concerned that the "Flood of RUM" will "Overwhelm all good Order among us" and that intoxication will result in increased crime, pauperism, gaming, and whoring--all threats to the class structure of New England society. "The Votaries of strong drink," he declares, "will grow numerous;...they will make a Party against every thing that is Holy, and Just, and Good" (quoted Tyrrell 1979:18-19 and Rorabaugh 1979:30-31).

PENNSYLVANIA

SPIRIT DRINKING CONTROLS. Quakers use distilled spirits in moderation. In 1706, the Pennsylvania Yearly Meeting advises Friends not to drink even small quantities of spirits at public houses (Rorabaugh 1979:37).

VIRGINIA

TAVERN CONTROLS. After 60 years, Virginia legislators become convinced that strict regulation of the number and quality of drinking places is required and that alcohol profits can be used to promote the development of places for public lodging. In 1705, tippling houses are forbidden, and the courts are instructed to license only those ordinaries as they "by their discretion shall judge" to be "convenient." The legislators are possibly influenced by the arguments and self-interest of the more important innkeepers. After 1705, the ordinary becomes a quasi-public and monopolistic institution (Pearson and Hendricks 1967:17, who comment: "There seems to be no parallel course of regulation elsewhere during this period").

1710

ENGLAND

DISTILLING: ECONOMIC VALUE (DEFOE). Daniel Defoe (Review, May 9, 1713) lauds the distilling trade for paying rents, employing people, and using excess grain produced by England's expanding commercial agriculture. He continues: "When [grain] markets are low abroad and no demands made for corn, that plenty which is other nation's blessing, is our intolerable burthen....The distilling trade is one remedy for this disaster as it helps to carry off the great quantity of corn in such a time of plenty, and it has this particular advantage, that if at any time a scarcity happens, this trade can halt for a year and not be lost entirely as in other trades it often happens to be....But in times of plenty and a moderate price of corn, the distilling of corn is one of the most essential things to support the landed interest that any branch of trade can help us to, and therefore especially to be preserved and tenderly used" (quoted George 1925:29; Watney 1976:15). (On Defoe, see also 1720/England.)

GIN PRODUCTION. By 1714, gin production reaches 2 million gallons a year, double that of 1696 (Ashton 1955:243). Under George I, who begins his reign in 1714, distilleries are "positively encouraged" (Williams and Brake 1980:3).

FRANCE

SOCIAL DEVELOPMENTS: THE REGENCY. Louis XIV dies in 1715, leaving as heir his 5-year-old great-grandson, Louis XV (1715-1774). A regency which lasts until 1723 is established under the new king's elder cousin, the duc d'Orléans.

COURT/DRINKING. There are several prominent drinkers in the French court during the last years of Louis XIV and, particularly, during the years of the Regency, including the regent himself (on the Regency, see below). Saint-Simon (1675-1755), in his Memoires (1691-1751), refers to several water drinkers, the moderation of Louis XIV and his second wife Madame de Maintenon, and their disapproval of drunkenness within the court. In his later years, the king only drinks burgundy diluted with equal parts of water. Despite the king's disapproval, considerable imbibing by others occurs within the court. Saint-Simon relates that the duchesse de Bourgogne (Burgundy) gave a supper party in 1710, attended by the duc d'Orléans and his daughter the duchesse de Berry, at which everyone became intoxicated--Mme. de Berry became so drunk that "the rest of the company were shocked beyond measure." He continues: "The after effects of the wine, both above and below, were embarrassing to witness but did not at all sober her, and the result was that she had to be returned to Versailles in a sottish condition." Fortunately, however, the king and Mme. de Maintenon do not find out about the episode. In another instance, he observes that Mme. de Berry "scarcely ever failed to drink herself into unconsciousness, rendering in all directions the wine she had swallowed; but if that were all, it was considered nothing." The early death of Mme. de Vendome at age 40 in 1718 is attributed to her love of "strong drink, of which her cupboards were full" (Saint-Simon 1751/1968, 2:90-91, 446-447, 469 and 1972, 3:166).

In other examples, he asserts that for 40 years the Grand Prior (Philipe de Vendome) never went to bed without being carried there dead drunk. The king's steward, the marquis de Livry, often got drunk in the evening, although he lived to the age of 80 in perfect health. M. de Caylus was "surfeited and stupefied for many years with wine and brandy"; the death of this drunkard is viewed as a deliverance by his wife and relatives. Lieutenant-General Vaillac "would have gone far but for wine and debauchery." The prince de Conti "sought to drown his sorrows in wine and other amusements unsuited for his age and debility." M. de Canin is "a sort of drunken savage." Lieutenant-General Ravetot is "an honest fellow who drank much wine, became very stupid and debauched [and] is finally ruined." The duc de Noailles is described as a glutton, drunkard, and debauchee (quoted Rolleston 1941:118-121).

The duchesse d'Orleans (Princess Palatine) makes several observations about excessive court drinking habits in her letters between 1699 and 1717. She asserts that drunkenness is very commom among women (August 7, 1699; April 29, 1704). Her grand-daughter, Madame de Berry (see above) drinks the strongest brandy she can get (November 18, 1717); the duchesse de Bourbon "can drink to excess without intoxicating herself, and her daughters try to follow her example, but are soon under the table" (May 21, 1716). Madame de Montespan and her eldest daughter "can drink remarkably without turning a hair." She writes: "I saw them one day drinking full glass after glass of stiff Italian wine [rossoli] and thought they would fall under the table, but they might as well have been drinking water" (December 12, 1717) (quoted Emerson 1908, 2:141; Franklin 1889, 6:131; Hopkins 1899:79).

One reason for the prevalence of inebriety may be the terrible monotony and tedium of court life (Rolleston 1941:121). Certainly this is a factor in the case of the duc d'Orléans (see below).

REGENT/DRINKING. Upon Louis' death, Philippe, duc d'Orléans, becomes the regent during the minority of Louis XV. Although not "the debauched monster of contemporary myth," the pleasure-loving regent is a notorious drinker who helps popularize sparkling champagne. His private life is characterized by "self-indulgence on a gargantuan scale...a daily routine of gluttony, drunkenness, and licentiousness" (Shennan 1979:127). Suppers organized by the regent and attended by his mistresses and those companions he calls his roués have been described as "notorious examples of collective alcoholism" (Rolleston 1941:118). According to Saint-Simon, who views him as a man who was born bored and who squandered great talents, Philippe and his companions were rebels against the sober atmosphere and tight-reigned conformity of Louis' last years (Saint-Simon 1972, 2:426-438). By 1720 the regent began "paying the penalty for his self-indulgent life. The suppers and especially the wine, invariably champagne, [began] to dull his quick mind and undermine his physique." (Saint-Simon 1968, 2:445, 456-457 see also 1972, 3:62-64).

In his Mémoires secrets, the moralist Charles Duclos (d. 1772) emphasizes that the regent and his associates regularly got drunk. Duclos describes the regent as sympathetic and human, talented and intelligent, but a man with a "taste for life which at times became grossly vicious," who "could not for long find resources in himself; dissipation, noise, debauch, were necessary to him." Many in his company had such little respect that even the duke calls them his roués (Duclos 1727/1910:8-9). At their gatherings, Duclos asserts: "Every supper was an orgy. There the wildest license reigned: filth and impiety was the foundation or seasoning of every conversation until total intoxication put the merry makers out of condition to speech and to hear. Those who could still walk withdrew and the others were carried away; and each day resembled the last. The regent, during the first hour of his levee, was still so dull, so oppressed by the fumes of the wine, that he could have been made to sign anything" (Franklin 1889, 6:130; Duclos 1727/1910:39, the same charge is repeated on p. 191; see also Saint-Simon 1751/1968, 2:457).

DISTILLERS' CORPORATION. All spirit distillers and retail-merchants are combined in a new corporation in 1713, which lasts until 1775 after which it is reconstituted a third time. The brandy distillers also obtain an order that all nonwine spirits of Brittany and Normandy must be exported to the colonies and that rum cannot be imported. In 1776, pharmaceutical chemists finally are detached from the corporation and formed into a separate guild (Le Grand d'Aussy 1782, 3:91-92; Forbes 1970:196; Gottschalk 1948, 2:237).

GERMANY

DRINKING CONTESTS. Johann Michael von Loen records a drinking contest held at a birthday banquet given for himself by the extravagant King Augustus the Strong of Saxony and Poland in 1718: "People drank deep where the king was and of a sudden there were issued strict orders that no one was to leave the garden. The Saxon courtiers had resolved to drink their Warsaw guests under the table; to show that they had stronger heads than the Polish magnates....The Poles, unused to such deep potations, were already pale as death, their heads waggling on their shoulders, their gait unsteady, so that they reeled as they walked before the king" (quoted Jacob 1935:93). Augustus has the palace guards bar guests from leaving the carousings. Neverthless, "side by side with drunkenness and greed [occurs] a general refinement of the palate among the upper classes" (Sagarra 1977:32).

RUSSIA

DISTILLING CONTROLS. In 1711, the penalty for bootlegging becomes exile to Siberia at hard labor, and for the failure to denounce it, fines. Nevertheless, bootlegging spreads steadily throughout Russia (Efron 1955:497).

SCOTLAND

MALT TAX PROTESTS. In 1713, the British Parliament places a tax on Scottish malt because English barley growers are resentful of the tax exemption given their Scottish counterparts in 1705 (see 1700/Scotland). The malt tax is declared a violation of the spirit of the Articles of Union of 1707 and an attack on Scottish independence. A slogan of the time is "Free Malt," and there are small riots in Edinburgh. As a result, the tax is not collected (Ross 1970:8).

AMERICA (BRITISH COLONIAL)

CONSUMPTION LEVEL. On the basis of imprecise figures, Rorabaugh (1979:32) estimates that per capita consumption of absolute alcohol increases from 2.7 gallons in 1710 to 3.5 gallons in 1800, with most of the increase being accounted for by spirits.

MASSACHUSETTS

TAVERN CONTROLS. In 1710, the Massachusetts assembly reduces the number of licenses for each town to one innholder and one retailer, unless the selectmen of the town decide that more licenses are needed to accommodate travelers. But about 1719, Massachusetts abandons the attempt to limit the number of licenses. In the five years before 1719, the selectmen of Boston approve 29% of the petitions for licenses, but in the five years after 1719, they approve 81% of the petitions. This new policy suggests the realization by the colony's leaders of the inability to continue enforcing Puritan morality on an increasingly diverse population.

Responding to the failure of exicse taxes to reduce the consumption of rum, the Massachusetts assembly prohibits the sale of distilled spirits in taverns. This is the last major piece of legislation in Massachusetts concerned with enforcing temperance until after the Revolutionary War (Conroy 1984).

ATTITUDES: INEBRIETY (DANFORTH). Samuel Danforth delivers a sermon against "The Woful Effects of Drunkenness" (1710), warning that "God sends many sore judgements on a people that addict themselves to intemperance in Drinking." Those that follow after Strong Drink have not the art of getting or keeping Estates lawfully." Furthermore, "They cannot be diligent in their Callings, nor careful to improve fitting Opportunities of providing for themselves, and for their families." Drunkards tend to commit "all those Sins to which they are either by Nature or Custom inclined" (quoted Levine 1978: 147, 148).

NEW HAMPSHIRE

DRINKING CONTROLS. New Hampshire prohibits the sale of liquor to drunkards and decrees that their names shall be posted in public houses (Cherrington 1920:33).

NORTH CAROLINA

CONTROLS: TAVERNS; PRICES. In 1715, the North Carolina assembly passes a comprehensive "Act Concerning Ordinaries and Tippling Houses" providing that (1) all retailers of beer, wine, or strong drink must conduct their business using English sealed measures (pints, quarts, bottles, gallons); (2) the retail price of beer and unboiled cider is to be fixed at 1s6d per gallon, with the price of other alcoholic drinks to be no more than "Cent per Cent above the first Cost of Value [wholesale price] of the Liquor sold"; and (3) a license from the governor is required to retail liquor.

DRINKING CONTROLS. In another act of 1715, the assembly makes public drunkenness a crime punishable by fine. The law refers to "the odious and loathsome Sin of Drunkenness [that] is of late grown into common use within this Province being the Root and foundation of many Enormous Sins" (Whitener 1945:1-2, 6).

VIRGINIA

SALES CONTROLS. With the disappearance of regular tippling houses in Virginia, the unauthorized sale of alcohol becomes a problem. Complaints are raised that "divers loose and disorderly persons" are "setting up booths, arbours, and stalls, at court-houses, race fields, general-musters, and other public places where, not only the looser sort of people resort, get drunk, and commit many irregularities, but servants and Negros are entertained, and encouraged to purloin their master's goods, for supporting their extravagancies." In 1710, the legislature prohibits anyone except licensed ordinaries from selling "either in houses...or any other place whatsoever." The practice, however, continues throughout the century (Pearson and Hendricks 1967:20-22; see also 1740/America).

INEBRIETY PREVALENCE (BYRD). In his diary, William Byrd observes that Virginians are often drunk at militia musters, on election days, and during quarterly court sessions. He records numerous instances of intoxication among the elite, including physicians. Although Byrd condemns public drunkenness if it leads to disorder, he is tolerant of intoxication among members of the Governor's Council and among his own servants. For Byrd, occasional drunkenness is a natural, harmless consequence of drinking (Byrd 1712/1941:53, 56, 75, 173, 218, 233-234, 270, 298, 324; Rorabaugh 1979:26-27).

1720

EUROPE

TEA CONSUMPTION. European tea consumption first becomes considerable during the 1720s when direct trade with China begins. Until now, the major part of the tea trade has been carried on via Batavia, founded by the Dutch in 1617. Chinese junks bringing their usual cargoes to Batavia also carry a small quantity of rough tea, which is the only variety that will survive the long journey (Braudel 1981:251).

ENGLAND

DISTILLING CONTROLS LIBERALIZED. The Mutiny Act of 1720 exempts from having to quarter troops all retailers who are distillers or who deal primarily in goods other than "brandy and strong waters" and who do not "suffer tippling in their houses." Such quartering is required of innkeepers, livery stable owners, victuallers, and retailers of strong waters. The act instigates the large-scale installation of stills by retailers of alcohol, compounding the adverse effects of the liberal production laws of 1690 and 1702.

GIN CONSUMPTION. Distilleries multiply and gin becomes so cheap that unrestrained indulgence prevails, particularly among the London poor (Inglis 1975:69; Blum et al. 1969:33-34). An "orgy of spirit drinking" occurs between 1720 and 1751 (George 1925:27, 29-30).

GINSHOP PREVALENCE; AVAILABILITY. By the 1720s, there exists a legion (up to 2,000) of small-scale compound distillers rectifying gin, in addition to fewer large-scale malt distillers producing raw spirits. In certain respects, spirit selling appears to take over the same functions as alehouse keeping in the 16th century, providing rudimentary victualling services for the lowest social groups and a source of income to the poor, who are now largely excluded from alehouse operation by tighter licensing requirements and greater operating expenses. Taxes and grain prices are low; sales and profits are high (Clark 1983:239-240; Clark 1984). A report of the Middlesex magistrates in 1726 estimates some 6,200 ginshops for a population of 700,000; in some parishes they constitute 1 in 10 houses. Cheap gin is "given by masters to their work people instead of wages, sold by the barbers and tobacconists, hawked on the streets,...openly exposed for sale on every market stall, forced on the maidservants and other purchasers at the chandler's shop, distributed by watermen on the Thames, vended by peddlers in the suburban lanes" (Webb and Webb 1903:21-22).

REASONS FOR USE; USER CHARACTERISTICS. While the cheapness and ready availability of gin is one reason for its popularity among the London poor, the 1720s and 1730s are a time of rapidly rising living standards for the lower classes and many contemporaries blame the gin epidemic on excessive wages and see it as a manifestation of undermining of society by luxuries. Many "gin addicts" are probably among the higher paid skilled artisans. Medick (1983:108) suggests, however, that what is far more important than high wages is the fundamentally uncertain conditions of life and unemployment, especially for the unskilled. Despite higher wages, mere survival is a continuous struggle without the old solidarity and protection of the neighborhood or guild. Under such conditions, many are inclined to obtain sporadic enjoyment from gin since this is the only "luxury" obtainable by them. Furthermore, it offers temporary respite from the ceaseless toil and uncertainty of their lives. Burton (1967:214) observes that before gin ruined the lives of the poor, "it kept them warm in winter, was cheaper than food, allayed hunger and pain, and offered escape from a life without hope." Fashion and marketing may have been as important as price in encouraging the shift from beer to gin (Clark 1984). It seems to have been especially popular among those in indoor or sedentary trades (Corfield 1982:144).

CONCERNS (MIDDLESEX MAGISTRATES). In 1721, a committee of Westminster magistrates expresses alarm at the prevalence of drunkenness in the City. Its report identifies the high number of gin shops as the principal cause of a rise in poverty, vice, and debauchery among "the inferior sort of people." These charges are repeated by another magistrates' committee in 1726 which inaugurates the campaign which eventually forces restrictive measures on the government. Although the members of Parliament are at this time not concerned about the plight of the poor, they are disturbed that gin drinking is occurring among soldiers and servants (George 1925:30-33; Inglis 1975:63).

As concerns over gin drinking appear, distillers and agricultural interests argue that spirits production is essential to the economy. (See the opinions of Defoe discussed below and under 1710.)

GIN CONTROLS: ACT OF 1729. Parliament attempts to control rising drunkenness by the Gin Act of 1729, which requires retailers to pay an excise tax of 20 pounds a year for an annual license and levies a duty of 5 shillings a gallon on gin and other compound spirits. Hawking of any kind of spirits in the street is forbidden. The act also subjects spirit dealers to the same license regulations as publicans and provides that a license to keep an inn or a victualling house or to retail spirits must be granted at Brewster Sessions only by the justices of the division within which the establishment is located (Webb and Webb 1903:24-25; French 1884:285-286).

GIN CHARACTERISTICS. Through the 1720s, some gin is heavily adulterated; some is literally poison (George 1925:30; Drummond and Wilbraham 1939:236). The Gin Act of 1729 suppresses the making of good gin: since the act applies only to gin and compound spirits, retailers evade the law by concocting plain drinks called by new names, "Parliamentary Brandy" being the most famous. Since "Parliamentary Brandy" is made from low-grade French brandy and other foreign materials, landowners and farmers complain of a decrease in the use of domestic corn by distillers (Coffey 1966:674; Dowell 1888, 4:168).

DISTILLING; ECONOMIC VALUE (DEFOE). In 1726, when gin consumption is attracting more and more concern, Daniel Defoe speaks out against any gin regulations in a pamphlet submitted to Parliament entitled A Brief Case of the Distillers, and of the Distilling Trade in England, "showing how far it is in the interest of England to encourage the said trade." He places the issue in the context of foreign competition and the economic benefits that England derives from the trade and argues against those who maintain that encouraging the trade encourages drunkenness. First, he argues that if British distilling is restricted, the Dutch will furnish their gin to London's poor and "cheat and impose upon us." Without restrictions, he foresees a day when the domestic trade will conquer "all foreign Importations" and overcome the dearness of French brandy and the corruption and fraud of the Dutch (pp. v-vi, 9, 41). The question, according to Defoe, is not one of encouraging drunkenness by encouraging distilling but one of keeping out foreign spirits; besides, no one makes the former argument about beer and wine (pp. 16-17). Second, he praises distillers for working for the public good: in consuming the produce of landed interests, tenants and farmers; in providing trade for navigators and shippers, and in providing employment for the poor (pp. 2-3). "England having now become what they properly call a Corn Country, produces much more Corn than it can consume. And if that Consumption should be lessen'd, many of those Lands must lie uncultivated, as it is manifest many Thousands of Acres did before." Third, the crown will get more revenues by encouraging distilling than by taxing it (p. 10). Fourth, he argues that production and consumption are separate issues and asserts that the problems resulting from consumption are the government's responsibility, not the distillers: "As for the excesses and Intemperance of the People, and their drinking immoderate Quantities of Malt Spirits, the Distillers are not concern'd in it at all; their Business is to prepare a Spirit wholesome and good. If the People will destroy themselves by their own Excesses, and make that Poison, which is otherwise an Antidote, 'tis the Magistrate's Business to help that, not the Distillers" (p. 11). Fifth, he suggests that the rise in spirits drinking has less to do with availability than with changing tastes and that a tendency to drink stronger drinks has developed in all classes. Thus in their spirit drinking, the poor are only following "this unhappy Humour of the Rich," who set the tone when gin was firt introduced as a gentleman's drink (pp. 46-48).

Although Defoe writes this defense of the value of distilling to the economy, several of his other writings in the second half of the decade express doubts over the wisdom of allowing unrestrained gin production and consumption. The result is often some "startling contradictions" (Earle 1976:114). (For Defoe, see also 1710/England). In his discussion of the liqueur trade in chapter 48 of The Complete English Tradesman, Defoe (1727/1841, 18:213-234) observes that the rise of malt distillers and the opening of the trade has completely changed the nature of distilling. Now the business is undertaken by "loose and disorderly persons." They "carry on their trade as if they were always drunk, keep no books but their slate....They are a collection of sinners against the people, for they break almost all the known laws of Government in the Nation." Whereas at first he boasts of the number of liquor tradesmen (pp. 213-219), he goes on to describe publicans as brokers of the devil whose numbers should be limited (p. 222), and soon thereafter he boasts of the amount the English drink (p. 223). (See also Earle 1976:114).

In Augusta Triumphans, Defoe (1728/1841, 18:32-33) discusses the abuse "among our lower sort" of "that nauseous liquor called Geneva," and calls for this abuse to be stopped. He expresses his concerns that it is poisoning the current and future population: "Those who deny an inferior class of people to be necessary to a body politic, contradict reason and experience itself, since they are most useful when industrious, and as pernicious when lazy....But now so far are common people infatuated with Geneva, that half the work is not done now as formerly. It debilitates and enervates them, and are they so strong and healthy as formerly. This accursed liquor is in itself so diuretic, it overstrains the parts of generation, and makes our common people incapable of getting such lusty children as they used to do. Add to this, that the women, by drinking it, spoil their milk, and by giving it to young children, as they foolishly do, spoil the stomach, and hinder digestion; so that in less than an age, we may expect a fine spindle-shanked generation." He provides examples proving that "there is not in nature so unhealthy a liquor as Geneva, especially as commonly sold," whereas "our own malt liquor...is most wholesome and nourishing." He reiterates these themes even more strongly in the sequel to this work: Second Thoughts are Best (Defoe 1729/1841, 18:10-11).

Commenting on the discrepancy between Defoe's attacks on drinking as the father of all vices and his championing of the drink trade, Earle (1976:113-114) emphasizes that a distinction must be drawn between Defoe's views as a moralist and his views as an economist. As a moralist he could be as virulent an observer of the evils of gin as any other critic. But when he wears his "economic hat," he takes pride in the enormous size of the trade, the economic benefits the nation derives from it, and the Englishman's drinking abilities. The high level of consumption he views as an outcome of economic prosperity. Thus, in the Complete English Tradesman, he raises the issue of whether reforming the nation's vices might not ruin the nation economically (Earle 1976:156-167).

(BLUNT). In response to the Middlesex Magistrate's complaints, the distiller Alexander Blunt, writing under the pseudonymn of E. Bockett, composes a poem in praise of "Geneva" in 1729 which he dedicates to Walpole. In the preface, He denies that inebriety or debauchery have increased due to gin or that gin is a problem. He asserts that alehouses are just as "pernicious" as ginshops and that "as long as men delight in Drunkenness," they will always find something to drink. He praises gin's superior medicinal benefits and describes the brewers as envious of gin. Gin is the "low pric'd dram" of the "little vulgar" who are only imitating the manners of the "vulgar great" in their drinking.

(MANDEVILLE). Bernard de Mandeville (1714/1729:57-61) praises the economic benefits derived from distilling, but emphasizes that they are no compensation for the loss to the people caused by gin; he compares distillers with robbers. He asserts: "Nothing is more destructive, either in regard to the Health or the Vigilance and Industry of the Poor, than the infamous Liquor...Gin, that charms the unactive, the desperate and crazy of either Sex, and makes the starving Sot behold his Rags and Nakedness with stupid Indolence....It is a fiery Lake that sets the Brain in Flame."

UPPER CLASSES/DRINKING: THE RAKES. Beginning in the 1720s, young upper-class rakes band together to indulge in drinking, gambling, street marauding, sexual license, and other vices of the time. They even organize themselves into formal and informal clubs, one of the most notorious of which is the Hell-Fire Club. In 1721, a royal proclamation is issued against "certain scandalous Clubs or Societies of young Persons who meet together, and in the most impious and blasphemous Manner...corrupt the Minds and Morals of one another" (quoted Jones 1942:41). This phenomenon reaches its peak in the 1760s, when "organized impiety, profligacy, bibulousness, and vandalism of the bucks and bloods [are] vividly called to the attention of the public" (Jones 1942:8).

PORTER INTRODUCED. Porter--a dark, heavily hopped beer of high alcoholic content--is allegedly first made by a London brewer named Ralph Harwood. Porter combines the qualities of all three types of beers currently in use, and is an outgrowth of the previous practice by publicans of mixing available types of beer to satisfy their clients' tastes. Porter is brewed specifically to keep, and its flavor actually improves with aging. As it can be kept in stock, it saves the brewers' money. The stability and efficiency of porter revolutionizes the brewing industry, making possible the use of mass-production techniques and large vats and utensils. Areas outside the local market can now be supplied as soon as transportation is available (Mathias 1959:12-13; Lynch and Vaizey 1960:39; Monckton 1966:144). Porter is "held to be a much more salutary drink than ordinary beer" and the most sensible drink for the ordinary working man (Sutherland 1969:26).

LABORERS/DRINKING PRACTICES (FRANKLIN). While an apprentice at a London printing house in the mid-decade, Benjamin Franklin, who drinks only water, observes that all the other workmen are "great Guzzlers of Beer" and call him the "Water-American." An alehouse boy is always in attendance to supply the workmen, one of whom drinks six pints in the course of the day, arguing that it keeps up his strength to work. Franklin is unsuccessful in deterring him from this "detestable Custom." When Franklin changes jobs and refuses to pay the customary drink imposition, he is ostracized by his fellow workers. Finally he has some success in convincing some of the workmen to stop "sotting with Beer all day"; those who do not stop are continually in debt to the alehouse (Franklin 1779/1964:99-101).

COFFEE HOUSE CHARACTERISTICS. More and more coffee houses are also selling alcoholic beverages, although this does not become common until after the mid-century. Defoe writes of Shrewsbury in the mid 1720s: "I found there the most coffee houses around the Town Hall that ever I saw in any town, but when you come into them they are but ale houses, only think that the name coffee house gives a better air" (quoted Ukers 1935:74 and Webb and Webb 1903:17, citing Defoe's Tour through the Whole Island of Great Britain, 1714-1727; however, this passage could not be located).

MEDICINAL USE: WINE. A Dr. Shaw publishes "The Juice of the Grape: or, Wine Preferable to Water," subtitled a treatise showing wine as "the Grand Preserver of Health and Restorer in Most Diseases" (1724). Quoting from numerous unnamed physicians, as well as from patients, the author declares that "wine, prudently used, has naturally a strong and direct Tendency to prolong Life and prevent Diseases....Wine of itself is a very wholesome Liquor" (quoted Lucia 1963a:157).

FRANCE

BACKGROUND DEVELOPMENTS. France begins to experience the population growth and economic revival that occurs throughout Europe during the century (see 1700/Europe), generating a period of prosperity that lasts until about 1775. As in the 16th century, the demand for luxury goods soars, but the 18th-century revival is more complex and influences more of society. Rural life becomes easier and more bearable, more protected from famines. Life spans and income increase; French life is flooded with riches. Although the French peasant remains timid and conservative with this new prosperity, the best situated peasants begin to participate in commercial life. Road construction escalates; urban products appear in rural markets; fairs and markets multiply. The economy becomes based on the consumption of consumer goods; urban life becomes more luxurious and comfortable. The demand for dinner and vintage wines increases, and rural drinking rises (see 1730/France) (Duby and Mandrou 1964:350-362).

DRUNKENNESS DENOUNCED. Doctors in theology at the University of Paris attack drunkenness (*yvrognerie*) as a mortal sin which should not be tolerated. They recognize that the manifestations of drunkenness range from incidental (the loss of the light of reason) to uncontrollable intoxication. Their primary concern is over those who suffer from a physical reliance so that they are "so accustomed to drink, that they fall into a swoon when they are deprived of it." These "alcoholics" should be allowed moderate drinking if they avoid drunkenness and attempt to conquer their habit (Brennan 1981:220-221; Brennan 1984).

VITICULTURE: EXPANSION, ECONOMIC VALUE. The profits of vignerons increase as demand for both inferior and vintage wines grows. Profits for the vignerons have always been higher than for the cereal grower, since the latter needs 40-50 acres to feed his family; the former, only 240 square yards (Duby and Mandrou 1964:356-359). The vineyard expansion begun in the 1670s becomes even more extensive. In all the viticultural provinces of France, an increase in wine drinking by rural peasants after the reign of Louis XIV elevates the price of inferior wines, resulting in a great increase in their production. Declining cereal prices also encourage this expansion. Thus, there are frequent protests calling for protection by proprietors of quality vines against the encroachments of popular viticulture (Dion 1959:597).

CONTROL EFFORTS. Beginning in 1720, parlements and provincial assemblies, where the influence of proprietors is supreme, promulgate or inspire a series of bans to prevent an increase of vine plantations. Complaints are raised that spreading viticulture has increased rural drinking and reduced the cultivation of needed grains, especially after the food scarcity of 1726-1727. As cereal prices decline dramatically during 1727-1731, restrictions are placed on the expansion of vineyards. Those vines newly planted since 1700 are ordered uprooted (Slicher van Bath 1963:217). A decree of 1731 for the whole kingdom denounces "this too great abundance of vines" because the vines occupy "a great quantity of lands good for grains or pasture" and because the great quantity of wine produced has destroyed the values and reputations of the vintages of many places. It is ordered that no new plantations be made, even in lands where they are already grown. In a latter case, however, authorization can be given to plant if it is proven that the "land is not more suitable for other crops." Many infringements occur, especially in the Midi (Dion 1959:598).

COFFEE CONSUMPTION, ATTITUDES. Although coffee drinking is expanding, it remains a luxury which only slowly infiltrates among the general population. When, for example, the popular hero Cartouche is about to be put to death (November 29, 1721), his "judge," who is drinking white coffee, offers him a cup. Cartouche replies that this is not his drink and that he would prefer a glass of wine with a little bread (Braudel 1981:260). Sort of a French Robin Hood, Cartouche is also said to be "sober in the taverns" (*sobre dans la guinguettes*) (quoted Kaplow 1972:140)

GERMANY

POTATO SPIRITS. J. Becker first suggests the possibility of making alcohol from potatoes; however, it is not until the end of the century that potato distilling begins (see 1770/Germany). The need to develop special stills to handle the thicker potatoe washes and the limited extent of potato cultivation retard this development (Forbes 1970:263-264).

COFFEE HOUSES (PRUSSIA). In 1721, Prussian King Frederick William I (1713-1740) grants a foreigner the privilege of conducting a coffee house in Berlin free of all rental charges. It is known as the "English coffee house."

SCOTLAND

<u>WHISKY CONSUMPTION, REVENUES</u>. In 1724, duty of 3,504 pounds is collected on a total of 145,602 gallons of spirits. As the population is a little over a million, this means that, discounting illicit production, per capita consumption is about one-seventh of a gallon a year (MacDonald 1914:39). Captain Edward Burt, stationed with the British army in the Highlands in 1724-1728 writes in his <u>Letters from a Gentleman in the North of Scotland</u> (1754) that whisky is like water to the Highlander. He continues: "Some of the Highland Gentlemen are immoderate Drinkers of Usky, even three or four Quarters at a Sitting; and in general, the People that can pay the Purchase, drink it without moderation. They say for Excuse, the country requires a great deal; but I think mistake a Habit and Custom for Necessity. They likewise pretend it does not intoxicate in the Hills as it would do in the low Country....The Collector of the Customs at Stornway in the Isle of Lewis told me, that about 120 Families drink yearly 4000 English Gallons of this Spirit, and Brandy together, although many of them are so poor they cannot afford to pay for much of either, which you know must encrease the Quantity drank by the rest, and that they frequently give to Children, of six or seven Years old, as much at a time as an ordinary Wine-glass will hold" (quoted Daiches 1978:42-43, who observes that it was incorrect that the poor drank brandy; only the wealthy did).

The increase in the cost of ale caused by enforcement of the malt tax of 1713 (see below) is credited for stimulating the spread of whisky drinking into the Lowlands for the first time, as well as increasing it in the Highlands. However, this change in beverage preference does not occur until enforcement of the malt tax increases after 1745 (see 1750/Scotland).

<u>MALT TAX OPPOSED</u>. In 1725, Walpole's administration in London decides to enforce the malt tax in Scotland. Even though it is only half the English rate, at 3d per bushel of malt, the tax is bitterly resented: it makes a popular drink expensive and places a double tax on whisky. Matters are made worse when the government appoints English excise officers to enforce the law. When the excise men enter Glasgow on June 23, they are met with opposition and riots follow. Troubles continue for much of the next month and lead to charges being brought against the magistrates of Glasgow, who refuse to convict arrested rioters (Ross 1970:9-12).

<u>ILLICIT DISTILLING</u>. Because opposition to the union of Parliaments is strongest in the whisky-producing Highlands, English revenue officers are intensely resented there. Illicit distillation becomes an act of political resistance as well as economically expedient and it immediately increases. Illicit whisky is also likely to be better than legal whisky since, to avoid the malt tax, legal distillers use cheaper corn rather than barley. There is little risk of getting caught for illicit brewing, since most of it is done in pot stills in small, remote highland areas which are easily protected from excise officers (Murphy 1979:10).

SPAIN

TAXATION. G.F. Gemelli Careri (Voyage du tour du monde, 1727, 6:387) observes that in Madrid "good wine can be drunk cheaply outside the town, because you do not pay the taxes there which amount to more than the price of the wine" (quoted Braudel 1981:236). The situation in Madrid would thus seem to be very similar to that in Paris.

AMERICA (BRITISH COLONIAL)

RUM CONSUMPTION, SALES. After 1720, the price of distilled spirits falls and the use of rum spreads. This demand brings an increase in the number of new sellers. Many small private dealers begin to sell spirits to friends and neighbors without a license (Rorabaugh 1979:34).

ATTITUDES: TAVERNS, SPIRITS, INEBRIETY (MATHER; FOXCRAFT). Cotton Mather along with 22 other ministers, writes Seasonable Advice Concerning the Tavern (1726), approving the role of public houses, but attacking the excessive use of spirits (Krout 1925:54).

Thomas Foxcraft, in "A Serious Address to Those Who Unnecessarily Frequent the Tavern" (1726), complains: "Taverns are multiply'd among us beyond the bounds of real Necessity, and even to a Falt, if not a Scandal." He also warns immoderate drinkers that they are "in danger of contracting an incurable Habit" (quoted Rorabaugh 1979:34 and Levine 1978:147, 150).

MASSACHUSETTS

TAVERNS: CONTROLS, PREVALENCE, FUNCTIONS. In Boston, liquor can be bought at one of every eight houses. To retain some control, public officials issue more permits for public houses. The increase in the number of licensed taverns between 1700 and 1730 only slightly exceeds the city's population growth, but the upper classes can no longer effectively control the larger number of public houses. Furthermore, the taverns become the center of popular agitation for a more egalitarian and less differential political order (Rorabaugh 1979:34; Conroy 1984).

ORDINATION DRINKING. During the ordination of the Reverend Edwin Jackson at Woburn, Massachusetts, the people consume, at the town's expense, 6 1/2 barrels of cider, 25 gallons of wine, 2 gallons of brandy, and 4 gallons of rum. Liquor is frequently plentiful at religious ordinations and functions and at funerals, and "the preacher was expected to lead the assault upon the jugs and bottles" (Asbury 1950:17).

NORTH CAROLINA

PRICE CONTROLS. In 1720 (see 1710/America), in order to correct abuses of the 1715 public house act, the North Carolina assembly places the determination of liquor prices in the hands of the precinct courts rather than leaving it to the (self-interested) judgment of the retailer (Whitener 1945:2).

PENNSYLVANIA

SPIRIT DRINKING CONTROLS. Quakers are forbidden to drink spirits at auctions because the practice tends to raise the price of the goods being sold (i.e., fosters bad business practices) (Rorabaugh 1979:37).

SPIRIT SALES CONTROLS. In a law of 1724, the assembly, noting that the sale of distilled liquors to iron workers is "very prejudicial and injurious" to employers, prohibits the granting of licenses to sell spirits within two miles of any iron works unless a majority of owners agree. The law explicitly exempts alehouse licenses if the keepers do not sell wine or spirits (Thomann 1887:153).

BREWING ENCOURAGED. In an effort to encourage the brewing of malt-liquor over distilling and to use up malt supplies, in 1722 the Pennsylvania assembly raises the duty on molasses and imposes a fine of 20 pounds on any common brewer or retailer who uses molasses, coarse sugar, honey, or other nonmalt material in brewing. The law also permits ale-house keepers who sell only malt-brewed beer to obtain a separate retail license at a reduced fee (Thomann 1887:151).

1730

EUROPE

TERMINOLOGY. Largely due to the influence of the Dutch doctor and chemist Herman Boerhaave, the word "alcohol" comes to mean in general use the "spirit of wine" and "spirit of fermented liquors" (Forbes 1970:107, 205-206). Boerhaave (*Elementa chemiae*) discusses distillation at length but adds nothing new to the development of technology.

COFFEE CONTROLS: ATTITUDES. J.S. Bach composes the *Coffee Cantata* (1732), inspired by the disturbances in France caused by a new coffee monopoly in 1725 and by the annoyance of German women at coffee restrictions.

ENGLAND

GIN LAWS LIBERALIZED (1733). In 1733, Parliament decides "to consider further methods for encouraging the making and exporting of home-made spirits distilled from the corn of Great Britain," and repeals the Act of 1729 (see 1720/England), which has been ineffective in curbing gin drinking in any case. A new act imposes a penalty of 10 pounds on retail sales of spirits except when sold in dwelling places.

CONTROL EFFECTS: SALES, CONSUMPTION. The effect of the 1733 law is to make every house a potential spirit shop (George 1925:33; French 1884:286). Gin is sold at every market stall, vended by peddlers. It appears that "one half of the town" is set up "to furnish poison to the other half" (Webb and Webb 1903:22). Many householders become publicans. In January 1736, a committee of Middlesex magistrates estimates that over 7,000 houses and shops in their district sell gin by retail--and even this figure they consider to be far short of the actual number.

CONCERNS: SPIRITS CONSUMPTION (WILSON; MAGISTRATES). Throughout the 1730s and 1740s the rising death rate in London is attributed to gin and causes great concern (George 1925:26; Trevelyan 1978:49-50). Numerous antigin tracts and pamphlets are published in the mid-1730s. "One of the most vitriolic and effective" (Clark 1984) is Thomas Wilson's *Distilled Liquors the Bane of the Nation* (1736). Wilson paints a bleak picture of the misery wrought by widespread spirit drinking among London's poor. He writes: "Every one who now passes thro' the Streets of this great Metropolis, and looks into the Distillers Shops...must see...a Crowd of poor ragged People, cursing and quarrelling with one another, over repeated Glasses of these destructive Liquors....[I]n one place...a Trader has a large empty Room backward, where as his wretched Guests get intoxicated, they are laid together in Heaps, promiscuously, Men, Women, and Children, till they recover their Senses, when they proceed to drink on, or, having spent all they had, go out to find wherewithal to return to the same dreadful Pursuit" (pp. v-vi). In addition, he asserts that the landed interests actually are suffering financially from the large amounts of cereals diverted to distilling (p. v). He blames gin drinking for the recent increase in the number and barbarity of murders and robberies (p. vi). One reason for such high prevalence of gin drinking is its easy availability at chandler's shops where it is procured by people who would be ashamed to enter a public house (p. vii). The second part of the work is devoted to the "malignant" effects of spirits on the body (p. 32ff). Gin is "a slow but sure poison" that will transmit its "deadly effects" on posterity so that in a generation or two, laborers will only be fit for servile offices and land cultivation. He specifically points out its dangers to the army and navy (p. 56). He recommends that the "cursed spirits" be put out of the reach of the "lower kind of people"; however, the "rich and great" must be left to their own reason and their physician's advice (quoted Webb and Webb 1903:22-23).

In February 1736, the Middlesex magistrates petition Parliament against gin selling and publish the findings of a new report. This petition sparks a debate this results in new prohibitory legislation that same year (see below). In the petition they complain that the evil of gin drinking "had for some years past greatly increased, especially among the people of inferior rank...[and] had destroyed thousands of his Majesty's subjects...[and rendered] great numbers of others...unfit for useful labor, debauched in morals, and drawn into all manner of vice and wickedness." Furthermore, "This pernicious liquor is now sold, not only by distillers and Geneva shops, but by many other persons of inferior trades, by which means journeymen, apprentices, and servants are drawn in to taste, and, by degrees, to like, approve, and immoderately drink thereof" (quoted French 1884:286, and Webb and Webb 1903:24; see also George 1925:33-34).

An anonymous Friendly Admonition to the Drinkers of Brandy and other Distilled Spirituous Liquors (1735) laments that man has found the means to extract a pernicious liquor from what God intended for refreshment. He describes all spirits as the masterpiece of the devil and warns that they are a "direct poison to human bodies." He then provides a long list of the adverse physical effects of spirit drinking (quoted French 1884:310-311; Simon 1926:37; Simon 1948:140).

GIN LAWS STRENGTHENED (1736). The 1736 petition from the Middlesex magistrates (see above) sparks the first parliamentary debate on prohibiting spirit use through high taxation. Bowing to public demand, Parliament reluctantly passes a harsh new Gin Act in 1736, which goes into effect in June. The act imposes a higher duty on all spirits of 20 shillings per gallon sold at retail, prohibits the sale of less than two gallons of spirits without payment of a license fee of 50 pounds a year, and limits licenses to keepers of established victualling houses (e.g., alehouses, inns) who have no other business interests. Magistrates and excisemen are given almost unlimited power to harass unlicensed and licensed retailers. Informers are encouraged with offers of a portion of any fines levied. Echoing the words of the Middlesex Magistrates, the preamble justifies these actions because widespread spirit drinking, especially among "people of lower or inferior rank" was leading to "the destruction of their health, rendering them unfit for labour and business, debauching their morals, and inciting them to perpetuate all manner of vices," harming not only this but future generations and tending "to the devastation of this Kingdom" (quoted Monckton 1969:63).

The act is passed with the avowed objective of making spirits so expensive that consumption will be virtually prohibited among the poor. In the parliamentary debates over the act, opponents argue that the higher duty is unfair, since the rich, who can purchase spirits at wholesale, will still indulge; they also insist that the higher duty will not work because spirits will inevitably find illegal entry channels. The supporters counter that heavy retail taxation will at least keep spirits out of the reach of persons of inferior rank, who are most likely to abuse them (Webb and Webb 1903:25-26; French 1884:286-288; Monckton 1969:63; Inglis 1975:65).

George (1925:34-35) comments: "That a landowning Parliament should attempt to suppress the home consumption of British spirits at a time when corn prices were still low suggests something of the incredible state of things." Clark (1984) suggests that while gin drinking was increasing, gin selling outside of established channels was much less widespread and disreputable than critics claimed, even in the East end where it was concentrated. Behind the law, however, was a coalition of moral reformers, Whig politicians, magistrates concerned about social disorder, London trade interests, especially brewers upset over declining beer consumption, and agricultural interests concerned about the fall in corn prices in the 1720s and seeking to encourage more brewing since beer employed more malt than gin.

CONTROL EFFECTS: SMUGGLING, BOOTLEGGING, CONSUMPTION. Gin sellers and the London mob protest the new act and violent "Gin Riots" ensue (Rudé 1959b:53-63). Widespread bootlegging and smuggling render the law impossible to enforce, even with numerous paid informers. Distillers take out wine licenses and sell gin under different names. Within a few years, it becomes evident that the act has failed to suppress excessive drinking. The act does result in a momentary decline in the duties paid on gin. For a population of 6.2 million, duty is paid on 6,120,000 gallons in 1736, of which it is believed half as much again is smuggled or obtained from illicit stills. The level falls to 4,250,000 gallons in 1737, but the difference is made up in an increase in the illicit trade. By 1740, the prosecution rate is falling, few traders are licensed, and spirit production is expanding. By 1743, annual consumption rises to 8,200,000 gallons. Respectable dealers abandon the business, which falls into the hands of the reckless and disreputable (French 1884:289-292; Dowell 1888:4, 288; Webb and Webb 1903:28; George 1925:35, 332n.37; Coffey 1966:673, 675; Clark 1984; Simon 1926:22; Kinross 1959).

In a parliamentary debate over the effect of the act in 1743, Lord Bathurst makes a spirited attack on its failings: "Everyone knows that the 1736 Act did not diminish the consumption, nor prevent the excessive use of spirituous liquors. They were not, it is true, retailed publicly and avowedly, but they were clandestinely retailed in every coffee-house and alehouse, and in many shops and private houses." Furthermore, "The perjuries of informers were...so flagrant and common, that the people thought all informations malicious, or at least, thinking themselves oppressed by the law, they looked upon every man that promoted its execution as their enemy, and therefore now began to declare war against informers....By their obstinacy they at last wearied the magistrates, and by their violence they intimidated those who might be inclined to make discoveries, so that the law...has been now for some years totally disused, nor has any man been found willing to engage in a task at once odious as endless, or to punish offences which every day multiplied, and on which the whole body of the common people, a body very formidable when united, was universally engaged" (quoted Webb and Webb 1903:28-30; George 1925:35).

Clark (1984) suggests that the 1736 act failed because the licensing fees and the harassment powers of magistrates and excisemen were so excessive that the well-to-to in the trade as well as the poor also became its victims; furthermore, many people believed the situation was not so serious as to justify such unprecedented harsh measures.

INEBRIETY OF WALPOLE. Sir Robert Walpole, who supports the effort to restrict gin drinking in 1736, is notorious for his own excesses of eating and drinking (Coffey 1966:682; Forster and Forster 1969:63-64; Madden 1967:143; French 1844:295).

INFANTS/DRINKING. William Forster (A Compendious Discourse of the Diseases of Children, 1738) complains about the "mad and imprudent Fondness of many Mothers, who do often permit their Infants to sip up Ale, Wine, and other Strong and Spiritous Liquors" (quoted Drummond and Wilbraham 1939:289).

DRINK HOUSES: PREVALENCE. By 1739, the number of public drinking places in London has risen to the "extraordinary figure" of 15,288 (about one for every 47 persons), of which 8,659 are gin shops; 5,975, alehouses; and 654, inns or wine taverns (Popham 1978:266-267).

CHARACTERISTICS (GONZALES). London drinking habits and establishments are described by Don Manuel Gonzales (London in 1731), who contrasts the large, handsome taverns where "a stranger may be furnished with wines, and excellent food of all kinds, dressed after the best manner" with the more numerous alehouses frequented by those "whose pockets will not reach a glass of wine" and who "sit promiscuously in common dirty rooms, with large fires, and clouds of tobacco." As to the drinks they serve, he writes that just "as the various kinds of beer brewed here are not to be paralleled in the world, either for quantity or quality, so the distilling of spirits is brought to such a perfection that the best of them are not easily to be distinguished from French brandy" (quoted King 1947:92).

DRINK RATIONS. In 1731, sailors are given the option of a ration of a pint of wine or half-pint of rum per day, instead of the gallon of beer established in 1590 (Sutherland 1969:16).

HYDROMETER INVENTED. John Clarke, a Londoner, invents a device to measure the amount of alcohol in spirits, or its strength. The device is demonstrated to the Royal Society in 1730 but not accepted by Parliament as an excise tool until 1787 (Murphy 1979:172).

FRANCE

PEASANTS/DRINKING. The progress of rural well-being and viticulture beginning around 1720 makes peasants accustomed for the first time to daily wine drinking, at least in areas where the climate permits viticulture. The criticism of drunkenness, which had been leveled in the preceding century almost entirely at the people of the cities, becomes in the 1730s and 1740s the first of the complaints made by employers of rural laborers in the wine-growing regions. Critics in part blame this increase in wine drinking on the expansion of viticulture. In 1731, the Estates of Burgundy complain that over the past 20 years and especially since 1720, vine-planting has been extended for the use of rural laboureurs and manouvriers employed in grain cultivation. The amount of wine consumed by les gens de la campagne has increased greatly because the peasant is now only concerned with debauchery. The same year, it is charged in Toulouse that plantings have so rapidly expanded since the beginning of the century that workers have become drunkards and libertines compared to the more sober and hard working people in les villes de fabrique far from the great vineyards (Dion 1959:594). These arguments are used to justify further restrictions on viticulture.

According to Braudel and Spooner (1967:408), in the aftermath of John Law's system, which collapsed in 1720, the peasants, who traditionally had lived soberly and made due with piquette, turned to drinking more and more wine. However, rural intoxication remains mostly occasional and never appears to be as extensive as in the cities (see the comments of Mirabeau under 1750/France, and Mercier under 1780/France). Peasants still largely drink piquette.

INEBRIETY IN BRITTANY. Although contemporaries frequently link rural prosperity with increased inebriety, the area with the worst drinking problem is rural Brittany, where poverty is endemic. Priests complain in vain against inebriated, underemployed men with family responsibilities who swill down cheap cider while their wives beg for bread (Hufton 1974:115). In 1733, the governor charges that although a great number of Bretons are industrious, in the poorest sections the inhabitants are idlers, drunkards, and boors, without any initiative (See 1927:46).

BRANDY CONSUMPTION. Much of the brandy produced in Charente is now sent to Paris and the reputation of Angoumois' brandy has become established among merchants. Therefore, a century after the Croquants rebellion and the first evidence of brandy distilling in Charente (see 1630/France), the fame of cognac is spreading (Delamain 1935:39-42).

MEDICINAL USE, SPIRITS. M. Malouin (Traité de chimie, 1735:260) believes that "spirits of wine rightly used is a sort of panacea" (quoted Braudel 1981:242).

COFFEE HOUSE PREVALENCE. There are now 380 coffee houses in Paris (Jacob 1935:184).

GERMANY

MONKS, CLERICS/DRINKING. In a letter from Würzburg dated September 1729, Pollnitz (1739, 1:184) asserts that few sovereigns in Germany serve a better table than the Benedictine monastery of Fulda, "for there is plenty of everything, particularly delicious Wines of which they tipple to such Excess that in a very little time they are not capable of distinguishing their Liquor. They are, I believe, the hardest Drinkers here in Europe." The court of the bishop-prince of Wurzburg is characterized by copious drinking, much of which is in the form of mandatory drinking of healths, which leads Pollnitz to follow "the laudable Custom of getting drunk twice a-day" (pp. 187-190, quote at 188). Of the ecclesiastical court of Bamberg, he later observes that "they drink as hard...as at Fulde and Wurtzbourg, so that it looks as if Drinking was an inseparable Function of the Ecclesiastical Courts" (p. 204). (On drinking at German ecclesiastical courts, see Lafue 1963:67-69, who observes that they often abandoned themselves to "interminable drinking bouts.")

COFFEE CONSUMPTION. Coffee drinking spreads from its beginnings in the Prussian courts to the higher circles of the bourgeoisie (Jacob 1935:149).

IRELAND

RUM CONSUMPTION, IMPORTS. Beginning in 1735, rum imported from the West Indies competes with domestic whiskey for the home market. By the 1770s, rum makes up 80% of spirits imports. Not until the end of the century does the legal production of Irish whiskey equal the level of rum imports (McGuire 1973:116).

DRINKING DEBTS. In an effort to discourage drinking on credit, Parliament in 1735 approves an act which prevents licensed retailers from collecting through the courts drink debts exceeding 1s. In 1759 debts incurred at shebeens (unlicensed premises) are rendered void (McGuire 1973:122).

NETHERLANDS

GIN PRODUCTION. Production of gin in Holland rises fourfold between 1733 and 1792 (Forbes 1970:190).

SCOTLAND

WHISKY CONSUMPTION. A traveller in 1736 attributes the Highlander's ruddy complexion and nimbleness to "aqua vitae, a malt spirit which is commonly used both as victual and drink" (quoted Sutherland 1969:61).

The Gin Act of 1736, aiming to stop the problem of widespread gin drinking in England by prohibitive taxes, exempts Scotland from its provisions. Scotland does not have a similar problem with spirit drinking. There, "whisky seems to [graduate] slowly and gently from being the drink of the lower classes to being everybody's drink. There [is] no sudden and massive appearance of cheap drink as in England" (Ross 1970:20).

TEA CONSUMPTION. Several complaints are voiced about tea drinking replacing the traditional Highland breakfast of ale, some toast, and a dram of whisky. Macintosh of Borlum writes of the 1730s: "When I came to a friend's house in the morning I used to be asked if I had my morning draught yet. I am now asked if I have yet had my tea. And in lieu of the big Quaigh with strong ale and toaste, and after a dram of good wholesome Scots spirits, there is now the tea kettle put to the fire" (quoted Ross 1970:5).

SWITZERLAND

DRINKING SOCIETY. In 1737, a society called the Golden Louse is founded in Bern. Members undertake to get drunk every day of the week. More than 50 members of the Grand Council and a fair number from the Small Council join this society (Jellinek 1976b:83).

AMERICA (BRITISH COLONIAL)

INEBRIETY PREVALENCE, CONCERNS. In the 1730s, a new style of drinking appears. Public drunkenness becomes associated with the expression of anger and hostility and with thievery, lechery, and brutality. Americans begin to perceive drunkenness itself as a major social problem. Although rationalist philosophy, commercialism, and rejection of tradition influence the changing attitudes toward spirits, the developing scientific approach to medicine and the adverse effects of alcohol have the greatest influence (see below) (Rorabaugh 1979:30).

CONCERNS: SPIRIT CONSUMPTION. In the 1720s, some physicians and scientists concluded that distilled spirits are poisonous; during the 1730s this view gains support (Rorabaugh 1979:38).

In 1736, Benjamin Franklin reprints in the Pennsylvania Gazette (July 22-August 2, 1736) an English article attacking spirits; Franklin comments, "Perhaps [the article] may have as good an effect in these Countries as it had in England. And there is as much Necessity for such a Publication here as there; for our RUM does the same Mischief in proportion, as their GENEVA" (quoted Rorabaugh 1979:30).

GEORGIA

DISTILLING BANNED. In February 1733, colonists arrive in Georgia, led by James Oglethorpe, who announces that the trustees of the colony have granted to each settler 44 gallons of beer and 65 gallons of molasses for use in brewing beer to encourage the establishment of a domestic brewing industry. However, the manufacture and importation of spirits is not permitted. This pronouncement, made on Oglethorpe's own responsibility, has no legal force, and many settlers proceed to use their molasses to make rum instead of beer. However, in 1734, the British Parliament passes an act (effective April 1735) prohibiting the "importation of rum and brandies" into Georgia. At first the law is rigidly enforced, but it quickly falls into disrepute. Moonshine stills, possibly the first ever operated in America, are set up in the back country. Rum is smuggled by boats from South Carolina and unloaded on the unguarded Georgia coast. Armed bootleggers then carry the liquor through the wilderness to the towns and villages. Illegal drink shops spring up everywhere. In a letter to London, a representative of the trustees estimates that there are as many illegal tippling houses in Georgia as there are gin shops in London. Colonial authorities find Georgians uncooperative in enforcing the law. Violators, when caught, demand trial by jury because jurors seldom return guilty verdicts. In 1739, a riot instigated by drunken ruffians frightens the Savannah magistrates into making a vigorous but unsuccessful effort to enforce the law. They report to the trustees that they are powerless to check the flow of rum. Another report alleges that many colonial officials themselves profit from the illegal trade (Asbury 1950:20-23).

MASSACHUSETTS

RUM CONSUMPTION, PRICES. The colonies become flooded with rum as a result of trade with the West Indies and domestic distillation, with a resulting drop in price. In 1738, a gallon of rum costs two shillings in Boston, compared to three shillings six pence in 1722, making it possible for a common laborer to afford to get drunk every day (Rorabaugh 1979:29).

1740

ENGLAND

GIN LAWS: TIPPLING ACT (1743). Parliament, recognizing the failure of the 1736 act and needing more revenue, particularly to pay for the War of the Austrian Succession, repeals the act in 1743 and adopts a more moderate law "written and introduced by direction of an important distiller" (Park 1983:63). The repeal of the 1736 act is also influenced by the fall of Walpole. The new legislation, known as the Tippling Act, endeavors to control the situation by annexing spirits retailing to the respectable world of the victualling house instead of suppressing it (Clark 1983:242). The law combines duties on the manufacture of spirits with the licensing of retailers at a moderate annual fee. It abolishes the high duty on retail sales, increases slightly the duty on the manufacturer of spirits, and reduces the retail license fee for taverns and alehouses to one pound a year--the latter measure being designed to encourage license holders to aid in suppressing illicit shops (Webb and Webb 1903:29-30, 33; George 1925:36; Williams 1962:134).

In support of the repeal of the 1736 act, Lord Bathurst argues that the new law will diminish consumption by controlling its sale: "We find by experience that we cannot absolutely prevent the retailing of such liquors. What then are we to do? Does not common sense point out to us the most proper method, which is to allow their being publicly retailed, but to lay such a duty upon the still head and upon licences as, without amounting to a prohibition, will make them come so dear to the consumer, that the poor will not be able to launch out into an excessive use of them" (quoted Webb and Webb 1903:28-30).

LIBERALIZED (1747). The Tippling Act reduces consumption for several years, but in 1747 distillers, pleading financial hardship and victimization by informers, petition Parliament for the right to retail spirits. Parliament complies by allowing London distillers who pay a certain amount to church and poor to take out licenses costing five pounds each, which are not granted by the justices, to sell small quantities of spirits for consumption off the premises. An increase in consumption and public drunkenness ensues (George 1925:36; Dowell 1888, 4:204).

CONTROL EFFECTS. In 1749, over 4,000 people are convincted of selling gin without a license and some 17,000 private gin shops are operated in London (Lecky 1878, 1:480).

GIN TRADE CONDEMNED. In the *London Tradesman*, R. Campbell (1747/1969:265-267) praises the distilling industry for increasing its output and improving the quality of its products, and for contributing to Crown revenues, but complains that it "has contributed to debauching the Morals, and debilitating the Strength of the Common People." From childhood to death, some people now never enjoy a sober thought. Because "private Vices are public Benefits," he sees little hope that "those many Calamities that attend national Drunkenness" will be redressed. He concludes that "the Evil arising from his [the distiller's] Trade to Individuals...over-balances all the Good he does the Public."

Campbell (1747/1969:280) particularly attacks the chandlers for contributing to the gin epidemic: "[The chandler] is partly Cheesemonger, Oil-man, Grocer, Distiller, etc. This last Article brings him in the greatest Profit, and at the same time renders him the most obnoxious Dealer in and about London. In these Shops Maid-Servants and the lower Class of Women learn the first rudiments of Gin Drinking...and load themselves with Diseases, their Families with Poverty and their Posterity with Want and Infamy."

Campbell (1747/1969:332-333) also estimates that it costs a man 2,000-10,000 pounds to become a brewer, but only 500-5,000 pounds to become a distiller, and even less to become a compound distiller. This suggests one reason so many people became distillers was that it was cheaper than becoming a brewer.

CONCERNS: GIN AND CRIME. In 1743, petitions to Parliament make frequent mention of the association between gin and crime. The lord mayor and the Corporation of London allege that spirits drinking "inflames" the poor with "rage and barbarity, and occasions frequent robberies and murders in the streets of the Metropolis." A petition states that spirits reduce the people to poverty and harden them "to the commission of crimes of the utmost enormity" (quoted French 1884:290-291). The "Proposals of the Justices of the Peace for Suppressing Street Robberies" (1744) emphasizes "the cruelties which are now exercised on the persons robbed, which before the excessive use of these liquors were unknown in this nation" (quoted Lecky 1878, 1:840).

TAVERN FUNCTIONS: HOUSES OF CALL (CAMPBELL). Campbell (1747/1969:192-193) complains about a "house of call" for tailors (an alehouse serving as a trade hiring hall). It "runs away with all their Earnings and keeps them constantly in Debt and Want...[T]he House of Call gives them credit for Victuals and Drink, while they are unemployed; this obliges the Journeymen on the other hand, to spend all the Money they earn at this House alone."

LABORERS/DRINKING PRACTICES. The Swede Pehr Kalm is impressed by the numbers of English laborers who go to victualler's to drink after work, a practice which he ascribes to custom and the wealth of the country (quoted Clark 1983:223n.4).

HEALTH CONCERNS (CHEYNE). Dr. George Cheyne, a Scotsman who at various times teaches in Leyden and works in London, develops a dietary scheme popular in the 1740s which is primarily concerned with the reduction of obesity, one of the major problems of the upper and middle classes in this period. He elaborates this scheme in several works: An Essay of Health and Long Life (1724), The English Malady (1733), An Essay on Regimen (1740), and The Natural Method of Cureing the Diseases of the Body (1742). He recommends drinking little alcohol and more milk and water, eating vegetables, and getting plently of exercise and sleep. He is convinced of the disastrous effects of overeating and drinking. He recommends that before the age of 15 and after the age of 50, no person should drink any alcoholic beverage; between the ages of 15 and 50, one should limit their consumption to moderate amounts of fermented beverages (Turner 1982:264). Because the order of the mind is affected by the order of the digestion, excessive consumption of food and dirnk is held to be irrational and incompatible with the exercise of professional duties (Turner 1982:268). In the Essay of Health, he gives 29 "General Rules for Health and Long Life" which he asserts that water is "the most natural and wholesome of all Drinks"; that "strong [i.e., wine] and spirituous Liquors freely indulged, become a certain tho' a slow Poison" and "there is no Danger in leaving them off all at once"; that no more than a pint of wine be consumed in 24 hours, in three glasses with water and three glasses without; that "the frequent use of Spirits in Drams and Cordials, is so far from curing Low-spiritedness, that it increases it, and brings on more fatal Disorders"; and that "malt Liquors (excepting clear small Beer, of a due Age) are extremely hurtful to tender and studious Persons" (Cheyne 1724/1725:74-75). Furthermore, he emphatically rejects the belief of "most People" that "the only Remedy for Gluttony is Drunkenness, or that the

Cure of a Surfeit of Meat is a Surfeit of Wine" (p. 47). He recommends that "if any Man has eat or drunk so much, as renders him unfit for the Duties and Studies of his Profession (after an Hours sitting quiet to carry on the Digestion;) he has overdone" (p. 38). In An Essay on Regimen, he calls "fermente'd and distill'd liquors" the sole cause "of all or most of the painful and excruciating Distempers that afflict Mankind; It is to it alone that our Gouts, Stones, Cancers, Fevers, high Hysterics, Lunacy and Madness are principally owing" (quoted Turner 1982:262). Cheyne is an important influence on John Wesley's views on eating and drinking (see below).

MEDICINAL USE. Despite Cheyne's warnings and the growth of concerns over gin, the consumption of fermented beverages is still generally considered a source of health and cheer. The 1741 tombstone of one lady reads: "She drank good ale, good punch, and wine/ And lived to the age of ninety nine" (quoted Porter 1982:235).

METHODIST ATTITUDES: INEBRIETY, SPIRITS (WESLEY). In 1739, John Wesley (1703-1791) begins an Evangelical revival within the Anglican church that becomes known as Methodism (Methodism does not finally separate from the Church of England until 1836). Wesley travels throughout England and preaches to the growing population of urban poor and, after 1760, to industrial workers, two populations "almost totally ignored by the established church." Wesley encourages sobriety, thrift, and hard work, and becomes "one of the most potent influences on the changing manners of the poor" (Coffey 1966:678). The general rules of the Methodist Society (1743) prohibit drunkenness and require members to avoid the selling, buying, or drinking of spirits (except in extreme necessity); however, Wesley does condone moderate beer drinking. He forbids involvement in the liquor trade because it is immoral to profit by hurting others (Quinlan 1965:34; Bainton 1945:55, who observes that it is from the Methodists and Quakers that the initiative in the later temperance crusade is derived). He is also not very enthusiastic about tea, and claims that it made his hand shake (Ferguson 1975:17).

Wesley's attitudes toward drinking are part of his concerns over general health and are influenced by the thinking of George Cheyne (see 1740/England). An ardent reader of medical literature and distrustful of both physicians and hospitals, Wesley writes a guidebook to health (Primitive Physic, 1747) so that his followers will have healthy bodies and minds as well as saved souls. The book consists of rules for retaining health and easy methods for curing most diseases. Among his recommendations are the following: (1) "Drink only water, if it agrees with your stomach; if not, good clear, small beer. Use as much exercise daily in the open air as you can without weariness." (2) "Nothing conduces more to health, than abstinence and plain food, with due labor." (3) "Strong, and more especially spirituous Liquors, are a certain, though slow, poison." (4) "Malt liquors (except clear, small beer, or small ale, of due age) are exceedingly hurtful to tender persons." (5) "Coffee and tea are extremely hurtful to persons who have weak nerves" (Wesley 1747/1960:29-30).

DRINKING IN LITERATURE (FIELDING; SMOLLETT). Drinking is a common occurrence in the English novel in the Georgian period, ranging from heavy drinking to alcoholism "in the sense of a crippling addiction to alcohol." Eighteenth-century novels are in the "picaresque" tradition, depicting the travels of heroes whose adventures frequently include copious drinking in inns and alehouses, usually with comic results, as in Smollett's Peregrine Pickle (1749). "Drinking is as typical of the genre as getting into the wrong bed, and leads just as often to loss of dignity and social confusion" (McCormick 1969:962). Novels of the period show drinking to be routine among all kinds and conditions of men; usually drunkenness is forgiven as natural. In Henry Fielding's Joseph Andrews (1742) and Tom Jones (1749), liquor is shown most often giving "comfort and sustenance like food" (ibid., p. 963). Bad behavior ascribable to drink appears more comic than reprehensible. Heavy drinking, however, is still implicitly criticized, as in Fielding's treatment of squire Western, "a coarse old-fashioned country squire" and hater of refined London manners to whom being drunk is "both traditional and manly." Each night, Fielding writes, "when he [Western] repaired to her [his wife's] bed, he was generally so drunk that he could not see." Criticisms of drunkenness become more frequent in the mid-century due to the gin epidemic, but the concept of drinking as jolly and a sign of good fellowship persists into the late century (ibid., p. 964).

COFFEE: HEALTH CONCERNS (CHEYNE). Cheyne regards coffee as a potential danger to the health, possibly causing or influencing scurvy, low spiritedness, and nervous disorders. But he does not consider its consumption necessarily either harmful or beneficial: "I have neither great praise nor bitter blame for the thing" (quoted Ukers 1935:55; Ellis 1956:17).

FRANCE

ECONOMIC VALUE: BRANDY. In 1745, les gentilhommes bouilleurs de cru complain that the greatest part of the revenue of Sointange and Angoumois comes from their wines, of which the quantity is too great to consume in the country and the quality is not good enough to transport, especially overseas. This situation necessitates their conversion to eau-de-vie.

It is estimated that 35,000-40,000 casks of brandy of 17 veltes are exported from La Rochelle. These require the distillation of 220,000-240,000 casks of wine (Delamain 1935:42).

ATTITUDES (MONTESQUIEU). In The Spirit of the Laws (1748), Montesquieu claims that a cold, northern climate promotes good health at the same time that it hinders sensations, and that the physiological constitution of the northener makes him impervious to the vices and crimes so rampant among southerners, including drinking. The drinking of strong liquors has a greater adverse effect on the body in hot southerly climates than in cold northern climates; similarly drunkenness carries with it fewer bad effects for northerners (14.2, 10, Montesquieu 1748/1951, 2:474-477, 482-484).

Discussing alcohol taxes, he observes that they are regarded as less onerous in England than in France because in the former they on placed on the producing whereas in France they fall on the consumer. He recommends that the English system be adopted (13.7, Montesquieu 1748/1951, 2:461).

IRELAND

UPPER CLASSES/DRINKING. Lord Chesterfield writes to the bishop of Waterford that "nine gentlemen in ten in Ireland are impoverished by the great quantity of claret, which, from mistaken notions of hospitality and dignity, they think it necessary should be drunk in their houses" (quoted Younger 1966:339).

NETHERLANDS

BREWERY PREVALENCE. In Holland alone, there are 100 brewers employing 1,200 workers (Kellenbenz 1977:537).

SCOTLAND

BACKGROUND DEVELOPMENTS: REBELLION (1745). For the second time (the first in 1715), the Highlands erupt in rebellion in 1745 (called the Forty-Five) against the Hanoverian crown, under the banner of Bonnie Prince Charles, son of the current Stuart Pretender James III. After some initial success, the rebellion is finally crushed in 1746 at Culloden, in the heart of the Speyside whisky country.

ILLICIT DISTILLING. With the final crushing of the Jacobites, the British are determined that rebellion will never again come down from the Highlands; they proceed to occupy the area and attempt to destroy the clan system (Lockhart 1951:10). It is now that the real problems with taxation and illicit distilling begin. Before this time, taxes were not really a serious matter; they were less severe than in England and often evaded (Daiches 1969:31; Sutherland 1969:60). Among the many officials sent to the Highlands to maintain the peace are gaugers or excisemen. The avoidance of paying excise tax on malt whisky becomes even more of a national pastime and a patriotic duty (Daiches 1969:31). Complicating law enforcement, the remote glens and hillsides which provide the best conditions for distilling malt whiskey are still largely inaccessible. A reward of five pounds is offered to anyone giving information about an illicit still, but few take advantage of the offer in good faith. Instead, when the "screw," the copper tubing of the still used to evaporate the alcohol, becomes worn out, it is usual for the still owner to report his own still and use the reward to purchase more copper tubing (Sutherland 1969:64).

DRAMBUIE INTRODUCED. Following the disaster at Culloden, Bonnie Prince Charles flees for his life, hiding out in the Highlands and the Isles until he can safely escape to France. His sojourn among the common folk of Scotland is the occasion for many tales and reminiscences, one of which concerns the origin of the liqueur, Drambuie. It seems that the drink was developed for the exiled King James and his son Prince Charles in France, out of aged Highland malt whiskies. While secretly travelling from Skye to the mainland on his way back to France, Bonnie Prince Charles gave the recipe to MacKinnon of Stathaird in thanks for his hospitality. It was called "An Dram Buideach," or "the drink that refreshes." Made only for the use of the MacKinnon family for 150 years, it begins to be produced in commercial quantities at the beginning of the 20th century, and its name Anglicized to Drambuie (Robb 1951:59).

TEA, BRANDY CONSUMPTION CONDEMNED (FORBES). Duncan Forbes (*Consideration Upon the State of the Nation*, 1730) complains that the main reason for low alcohol revenues and prices is the drinking of large quantities of foreign wine and brandy, which are relatively cheap due to low or nonexistent customs duties. In *Some Considerations of the Present State of Scotland* (1744:6-10, 30), he adds tea to his complaints. He attacks both drinks for preventing consumption of Scottish malt and liquor. Forbes complains that because of the lack of, or low duties on, imported wine and brandy since the Union, the drinking of ale and beer has been "in some Degree impaired, and the Use of home-made Spirits almost universally laid aside." But grain prices still held up because tea drinking was limited to "People of Condition." Now, the price of tea has declined and availability has increased, so that even "the lowest of the People" drink it. For every person who drank tea 15 years ago, there are now hundreds, and tea is replacing two-penny ale for breakfast, although these vices occur more in the town than in the country. Thus grain prices and excises have declined. The "ruinous Expence of Tea" has now been added to that of foreign spirits. He recommends that high duties should be placed on both and that tea should be forbidden to servants and all who cannot afford it.

SWEDEN

CONTROL EFFECTS. Linnaeus (Carl von Linne), naval physician at Stockholm, observes in the 1748 almanac that the prohibition of distillation in the 1740s results in a one-tenth decrease in crimes committed by sailors and a one-half decrease in illness (Makela 1980:134).

HEALTH CONCERNS, BRANDY (LINNAEUS). In his medical lectures at the University of Upsala, Linnaeus declares that brandy and tobacco are poisons; he stresses moderation in all things as the key to a long and healthy life (Goerke 1973:119).

AMERICA (BRITISH COLONIAL)

DRINKING CLUBS. In most colonial cities, gentlemen of "wealth, prestige, and popularity" gather together in private drinking clubs that meet in the back rooms of taverns. These clubs are modeled after similar London clubs (Rorabaugh 1979:32).

CONSUMPTION LEVELS; DRINKING PRACTICES (HAMILTON). During a trip through several colonies in 1744, Dr. Alexander Hamilton of Annapolis rates cities and regions by the quantity of spirits consumed and by their quality. He is disappointed in Philadelphia's low rate of consumption, and dislikes the poor quality of New England's beverages. On the other hand, in New York he finds that a reputation for hearty drinking is essential for admission to the best society and the drinks are of high quality. Preferring moderation to excess (not more that one bottle of wine each evening), Hamilton does deplore the efforts of New Yorkers to make him intoxicated by proposing too many toasts. In Annapolis, Hamilton belongs to a drinking club whose members "put away at each weekly meeting not less than a quart of Madeira wine apiece" (Hamilton 1744/1948:xvii, 6-7, 43, 88, 165, 193).

HEALTH CONCERNS, RUM. In the 1740s, American doctors become concerned about the spread of West Indies Dry Gripes. In 1745, Dr. Thomas Cadwalader of Philadelphia identifies rum as the cause of the disease and recommends abstinence. Rorabaugh (1979:39) considers this "a novel proposal that was contrary to traditional opinion concerning rum's healthful qualities." (The disease is now recognized to have been lead poisoning caused by drinking rum made in lead stills.)

GEORGIA

SPIRITS BAN REMOVED. Faced with popular revolt and pressured by British liquor merchants, the trustees of Georgia in 1742 abandon the prohibition law established in 1735. They permit importation of spirits and establish a system of licenses for taverns and tippling houses. The illicit manufacture and trade in rum declines, and the legal rum trade prospers. A report to the trustees a few years later asserts that drunkenness has greatly decreased (Asbury 1950:23).

NORTH CAROLINA

TAVERN CONTROLS; LICENSE REQUIREMENTS. In response to rising concern over abuses in the licensing system and over disorders occurring in taverns, in 1741 the North Carolina assembly passes legislation that abolishes tippling houses (by requiring all liquor retailers to provide food and lodging), empowers the county courts rather than the governor to grant annual licenses, requires manufacturers of liquor who wish to sell at retail to obtain a license, and enjoins keepers of ordinaries and houses of entertainment to maintain order in their establishments and to ensure that no more drinking occurs on Sunday than is necessary (Whitener 1945:6-7).

PENNSYLVANIA

TAVERN PREVALENCE, CONCERNS. In 1744, a Philadelphia grand jury complains of the enormous increase of public houses, which are estimated to cover a tenth of the city and which are blamed for a rise in poverty, the use of profane language, and a distaste for religion (Asbury 1950:24).

VIRGINIA

SALES CONTROLS. As a result of the 1710 law against unauthorized drinking establishments (see 1710/America), peddlers have appeared, especially in Yorktown, "selling liquor in small quantities, watered and at a cheap rate--to the detriment of the ordinaries and the general defeat of the law's purposes." In 1748, the legislature limits liquor retailing to keepers of ordinaries and merchants, including planters with stores. The penalty for violation is ten pounds or "twenty-one lashes, well laid on, at the public whipping post" (Pearson and Hendricks 1967:21-22).

1750

EUROPE

BACKGROUND DEVELOPMENTS. In the second half of the century, the population expansion turns into an explosion. The agricultural depression which began in 1650 comes to an end and cereal prices rise. Although there is some fear of a disastrous famine due to this population growth, the challenge is met by increases in industrial and agricultural production.

REFINEMENT OF MANNERS. In the grip of an "omnipresent luxury," food and manners are refined throughout Europe, but especially in France (see below) (Braudel 1981:206).

COFFEE CONSUMPTION. Coffee availability and consumption increase throughout the Continent as Europeans organize their own production and end their dependence on Arabic imports. Nevertheless, many people drink it only as a medicine, and its popularity is limited largely to major urban centers (Braudel 1981:258-260).

ENGLAND

BACKGROUND DEVELOPMENTS: GEORGIAN ERA PEAKS. Between 1740 and 1780, Georgian England is at its height, free from the passions of the 17th century and not yet disturbed by the future conflicts unleashed by the Industrial and French Revolutions. This is a lighthearted age of peace and complacency in which the dominance of the aristocracy is unchallenged (Trevelyan 1978:297). Especially after the peace established by the Treaty of Aix-la-Chapelle in 1748, the London middle class enjoys a period of rising prosperity and comfort.

As both a consequence and stimulus to economic growth, the population of Great Britain triples between 1750 and 1850. Around 1760, the enclosure movement quickens, ending peasant agriculture and creating a class of agricultural entrepreneurs and a large agrarian proletariat producing for the market. By the 1780s, the "take-off period" of the Industrial Revolution has begun. It is with the Industrial Revolution that the great expansion of the numbers and influence of middle-class businessmen begins (Hobsbawm 1962:46, 133, 204).

As the century progresses and the spending power of the "middling" groups in society grows, urban life also expands and becomes more sophisticated. The standard of urban living rises; towns become safer, pleasanter, more convenient and better places in which to live (Borsay 1977:582-584, 587, 590).

POPULAR CULTURE. From the mid-century, popular recreations are subjected to a "systematic and sustained attack"; by the last quarter, "customs which had once been tolerated [come] to be questioned and sometimes heatedly condemned." Underlying much of this hostility is the growing concern for effective labor discipline at a time when industry is viewed as the linchpin of England's progress. Another powerful influence is the spread of the Evangelical movement and the reemergence of moral earnestness in public life. Reformers seek to end all diversions and temptations which distract people from work: to end excess holidays, reform or eliminate wakes and fairs, and strictly regulate public houses and prevent them from offering recreational attractions (Malcolmson 1973:90-98, quotation at 89).

UPPER CLASSES/DRINKING. In the latter half of the century, drunkenness is said to be "almost universal" among the men at Cambridge University; the situation at Oxford is similar (Younger 1966:339). At mid-century, dinners usually end in "extreme intemperance" and "the high-dressed beau and the low libertine [are] similar in profligate indulgence" (Farrington 1819:67). However, some modification in drinking habits among the fashionable urban set and cultural elites is discernible as the century progresses. A new generation begins to reject the excesses of table and bed and the coarseness of manners that had characterized Robert Walpole (d. 1745) and his contemporaries: "The contrast between Sir Robert and his son, Horace, the cultivated, elegant diarist, symbolizes the gradual change in manners after 1750" (Coffey 1966:682). Early manifestations of this trend include Samuel Johnson and Joshua Reynolds (see 1760/England).

DRINKING PRACTICES: CLUB LIFE. Since 1700, a steady growth in the number of private clubs has occurred. The mid-century is the great period of English clubs; by century's end, all classes have their clubs which number in the hundreds, most of which meet at taverns. By 1750, larger alehouses have at least one club or society for which they provide a special room. These alehouse clubs are largely formed from the ranks of skilled workers, craftsmen, and small traders in a desire to emulate upper-class practices (Clark 1983:235). While eating, drinking, and conviviality are central to their existence, the clubs also perform important social functions by protecting their members from adversity, providing an elaborate system of reciprocal obligations, and promoting an outlook and standards conducive to business success. Nevertheless, there are many examples of tavern discussions escalating into brawls and of club members found in a gutter or wayside ditch in a drunken stupor (Brewer 1982:218, 223, who comments: "We are still in an age when drinking, at least of Portuguese wine and British beer, was a patriotic, convivial and socially desirable activity rather than the bane of nineteenth-century temperance societies or the evil spirit that haunts the novels of Emile Zola. Yet, despite the undeniable excesses, club life [including the rake's clubs] undoubtedly became less debauched and less vicious as the eighteenth-century proceeded").

GIN CONSUMPTION, CONCERNS. In a period of peace abroad and domestic prosperity, London's middle class becomes aroused against widespread drunkenness among the poor. (Some modern commentators feel that the alleged harmful national effects of the gin epidemic, which was largely confined to London, have been overstated; see Mathias 1959:xxv.) In 1751, the public outcry reaches a feverish pitch. A flood of books and pamphlets attack gin drinking and its consequences. Gin is seen as a major cause for crime and death in London. An anonymous tract entitled "An Address to an Eminent Person upon an Important Subject" (1751) paints an alarming picture of the "gin-sodden" urban population and condemns spirits as serving no useful purpose. The author recommends that "let the art be condemned as unlawful, let all spirituous liquors be seized and destroyed, wherever found, and the commonwealth saved at any rate" (quoted Simon 1926:23).

(HOGARTH; TUCKER). William Hogarth's print "Gin Lane" (1751) depicts the horrors of life among the London poor; the print features a gin shop with a sign quoting the common epigram, "Drunk for a penny, dead drunk for twopence, clean straw for nothing" (Coffey 1966:689). He depicts compound distillers plying their trade without controls or the intermediary of the pubs; the lives of those who sell themselves to gin is characterized by death, apathy, hunger, physical decay, poverty, and debt (Medick 1983:97). (On Hogarth's attitude towards beer drinking, see below.) Numerous petitions are presented to the House of Commons; they point out that "the common use of cheap spirits [is] destroying the people, shortening their lives, causing irreligion, idleness and disorder, and if not checked, [will] destroy the power and trade of the kingdom" (George 1925:36). Josiah Tucker (1751:1-2, 31) charges that the "pernicious effects" of the drinking of cheap spirits among the "common people" are increasing daily and spreading into the country and villages. He emphasizes that this must be stopped. He expects that with greater controls over gin people will increase their use of other alcoholic beverages, but these are not as bad as gin "because the Drinking of Spirituous Liquors is a Kind of instantaneous Drunkenness, where a Man hath no time to recollect or think, whether he has had enough or not."

DEATH RATE (MORRIS). In the 1750s, the crude death rate in London is one in twenty, much of it being attributed to alcohol-related diseases. In his *Observations on the Past Growth and Present State of London* (1751), Corbyn Morris writes: "The diminution of births set out from the time that the consumption of these liquors by the common people became enormous." Morris believes that alcohol lowers the birth rate by causing male impotence and female sterility and by rendering "such [infants] as are born meagre and sickly, and unable to pass through the first stages of life" (quoted George 1925:27; Warner and Rosett 1975:1397). But Corfield (1982:120) observes that "no substantial medical evidence have ever been adduced to support" the view that gin drinking fostered low fertility and high mortality.

CRIME (FIELDING). Henry Fielding's *An Inquiry into the Causes of the Late Increase of Robbers* (1751) denounces the evils of spirit drinking. Drawing heavily on Thomas Wilson (see 1730/England), to whom he refers, Fielding assumes that the poor turn to robbery to supply themselves with gin and warns of the dreadful consequences to future generations from the adverse physical effects of the habit. Like Defoe and Wilson, his concern is not the harm gin poses to the individual so much as to the state, by incapacitating the most economically useful (Zirker 1966:88). He attributes the widespread poverty and crime of the day to "that poison called Gin; which I have great reason to think is the principal sustenance...of more than an hundred thousand people in this metropolis. Many of these wretches there are who swallow pints of this poison within twenty-four hours" (Fielding 1751/1871:374). "Wretches are often brought before me, charged with theft and robbery, whom I am forced to confine before they are in a condition to be examined; and when they have afterwards become sober, I have plainly perceived, from the state of the case, that the *Gin* alone was the cause of the transgression, and have been sometimes sorry that I was obliged to commit them to prison" (ibid., p. 375). Like Wilson, Fielding draws a class distinction between how society should deal with drunkenness: conscience for the rich; repression for the poor (Zirker 1966:58-59, 95). He proposes that nothing less than "absolute deletion" will solve the problem. Establishment of an expensive license to sell gin will not deter the unscrupulous retailer. Possibly it should be made available only through chemist or apothecary shops, or its price should be raised "beyond the reach of the vulgar." Even these actions may not be effective, but something must be done to "palliate the evil, and lessen its immediate ill consequences, by a more effectual provision against drunkenness than any we have at present" (Fielding 1751/1871:378-379; see Zirker 1966:48).

ECONOMIC VALUE, GIN (TUCKER). In the mid-century, 7,000 out of 12,000 quarters of wheat sold in the London market in a week are turned into gin (Webb and Webb

1903:39). In countering the economic objections to gin controls that have been raised by the distilling industry and its supporters, Tucker (1751:1-2) first maintains that greater controls will not cause a loss of national revenues because revenues from beer and ale will go up and because the national economy will improve as health improves. Second, greater controls will not hurt the farmers because the distillers have exaggerated the amount of corn they use and there are other outlets for the farmers' products. Third, the drinking of foreign spirits will not increase because they cost too much.

TAVERN PREVALENCE, CHARACTERISTICS. One public house reportedly exists for every fifteen homes in the city of London and one for every eight in Westminster (Webb and Webb 1903:39). Typical of the opinion of many contemporaries is that of Tobias Smollett, who, in his History of England, speaks of the "incredible number of public-houses" to be found in the suburbs of London in 1752. These, he remarks, "continually resounded with the noise of riot and intemperance; they were the haunts of idleness, fraud and rapine, and the seminaries of drunkenness, debauchers, extravagance, and every vice incident to human nature" (Smollett 1805, 4:133). (In a discussion of 1742, Smollett [1805, 3:446-449] also surveys the events leading up to the gin act of 1736 and its repeal.)

GIN LAWS STRENGTHENED: GIN ACT (1751). In "a turning point in the social history of London," a new Gin Act reimposes the provisions of the 1743 Tippling Act (George 1925:36). It strengthens retail controls and aims to license the sale of liquor rather than to restrict it. Specifically, the act (1) increases the duty on spirits and raises the license fee to one pound a year; (2) prohibits distillers, chandlers, and grocers from retailing spirits; (3) authorizes spirit licenses to be granted to publicans who pay to church and poor and (in London) rent a tenement of at least 10 pounds a year; (4) prohibits brewers, inn keepers, distillers, and dealers in spirits from acting as justices in matters relating to distilling; (5) makes debts for spirits under 20 shillings unrecoverable at law; and (6) forbids the retailing of spirits in prisons and workhouses (George 1925:333n.41).

The major features of the act are that (1) compared to the 1736 act, it raises license fees and duties but not as excessively, nor is the excise department given special powers to harass retailers; (2) it seeks to bring gin production and sales under control by making them respectable; and (3) it separates production and distribution and limits the latter to licensed outlets (Medick 1983:106-107; Clark 1984).

DISORDERLY HOUSES ACT (1752). In order to reinforce the Gin Act, Parliament provides in the Disorderly Houses Act of 1752 that places of amusement obtain licenses from local magistrates. Until the end of the century, this act, supplemented by others, gives the local justices virtually unlimited authority to grant or refuse licenses and to confine the drink traffic within legitimate bounds through the imposition of heavy penalties on unlicensed sellers (George 1925:305; Webb and Webb 1903:38).

DISTILLING BANNED (1758-1759). Because of a dearth of corn, distillation is prohibited for two years beginning in 1758.

CONTROL EFFECTS: GIN CONSUMPTION. Following passage of the Gin Act (1751) and the Disorderly Houses Act (1752), the indiscriminate sale and consumption of spirits is brought under control. The number of retailers stagnates. At the same time, the quality of gin improves. As early as 1757, it is observed: "The lower people of late years have not drank spirituous liquors so freely as they did before the good regulations and qualifications for selling them....The additional excise has raised their price, improvements in the distillery have rendered the home-made distillations as wholesome as the imported. We do not see the hundredth part of poor wretches drunk in the streets since the said qualifications as before" (quoted George 1925:38). Gin consumption declines from 7.05 million gallons in 1751 to less than 2 million gallons by 1758 (Ashton 1955:243). In 1758, John Fielding observes that "gin has now become so expensive that good porter-beer has recovered its former pre-eminence" (quoted Medick 1983:105). On the whole, the decline continues to the end of the century.

Coffey (1966:673) observes, "The rise and decline of the gin epidemic can be related directly to taxation and legislation." There is a general consensus that, in addition to the astute provisions of the mid-century legislation and the immediate influence of the distilling ban of 1758, this consumption decline is sustained by (1) the success of the brewers with porter, which is more stable and more potent and of a higher quality than previous beer, and which, through mass production, has a lower price; and (2) rising corn prices and taxes in the second half of the century, which erode the price advantage of gin.

ATTITUDES: GIN, INEBRIETY. Eighteenth-century concerns over gin-drinking are rooted less in temperance ideology or regard for the individual than in the damage caused to the economic and social well-being of the nation (George 1925:37; Zirker 1966:88). For example, Josiah Tucker (1751) shows that the antigin movement is "entirely untouched by the spirit of the temperance reformer" (George 1925:37), in that he considers one of the evil consequence of gin drinking to be that less malt-liquor is consumed. He also pays little attention to the reasons why the poor drink, only how they can be prevented from drinking (Zirker 1966:91). Drunkenness remains a virtue among upper classes; most reform efforts are focused on unacceptable public manifestations of inebriety (Corfield 1982:144). The various laws designed to deal with the gin epidemic are explicitly intended to control the behavior of the "people of lower or inferior rank," as stated in the second Gin Act of 1736. In the Church of England's attacks on the evils of drink, it is "very much the case of one law for the rich and another for the poor" (Sutherland 1969:125).

ATTITUDES: WAGES AND DRINK (TEMPLE). William Temple (*A Vindication of Commerce and the Arts*, 1758:56-58) emphasizes that the result of low wages is industry and sobriety: "The only way to make them [laborers] temperate and industrious is to lay them under a necessity of labouring all the time they can spare from meals and sleep." High wages and job security render them "indolent and debauched" (quoted Malcolmson 1973:97). Along with Josiah Tucker, Temple is one of the proponents of a new labor discipline coming from the commercial and manufacturing regions of the West Country and London (ibid., p. 160).

BEER: CONSUMPTION. For a while the restrictions on gin lead to an increase in the consumption of beer (Mathias 1959:xxv; see also the above comments of J. Fielding). H. Jackson (*An Essay on Bread*, 1758) observes: "Beer, commonly called Porter, is almost become the universal Cordial of the Populace, especially since the necessary Period of prohibiting the Corn-Distillery [in 1758]; the Suppression presently advanced the Price of that common Poison Gin, to near three times its former Price, and the Consumption of Beer has kept pace with such advance" (quoted Drummond and Wilbraham 1939:236-237). However, the rise in consumption is not maintained. By 1830, per capita consumption has declined to half the level of 1720 (Mathias 1959:375).

CHARACTERISTICS. Jackson also complains that beer has become weaker than "a few Years ago." This is because the growing popularity of tea is forcing the brewers to brew weaker beer and to keep their prices as low as possible. This small table beer has an alcohol content as low as 2% or 3% (Drummond and Wilbraham 1939:236-237).

ESTEEM. In response to the gin epidemic, beer is praised as the support of the farmer and landed interests and as the savior of public morality and revenues. It is viewed as a necessity of life purchased by the poor and urban dwellers who have no supplies, space, or knowledge to brew their own (Mathias 1959:xxi, xxv). In 1759, the superiority of "The Beer Drinking Briton" is praised over the wine-drinking French: "Let us sing our own treasures, old England good cheer--/The profits and pleasure of stout British beer." Such attitudes are expressed in numerous ballads and sketches of the period. Whereas Hogarth's "Gin Lane" depicts the evil effects of gin on the poor, his "Beer Street" is an idealized picture showing middle-class beer drinkers "as being happy and, if a little over-fat...prosperous and contented" (King 1947:101). Here there exists a life of thrift and industry not dominated by the pursuit of mercantile gain that created Gin Lane (Medick 1983:101).

PRODUCTION. The brewing industry has become dominated by 12 giant wholesale (common) brewers, responsible for 44% of all beer produced for sale, some producing as many as 50,000 barrels annually (compared to an average 3,000 per year for all common brewers at the end of the century) (Mathias 1959:551; Park 1983:64).

VILLAGES/DRINKING; MIDDLE-CLASS ATTITUDES (TURNER). In 1755, Batista Angeleoni (John Shebbeare) remarks in his Letters on the English Nation that "good order, sobriety, and honesty" mark the village as against the "anarchy, drunkenness and thievery" of the town (quoted Humphreys 1954:31). However, the diary of shopkeeper Thomas Turner shows that "people of substance" in rural Sussex often were inebriated (Porter 1982:34). He mentions drinking constantly (Turner 1755/1979:1-5, 13-16, 20-24, 27, 29-31, 34, 40, 42-44, 48, 53-55, 58-59, 61, 65, 68-69, 73). He refers to the "many instances...we almost daily see of people's receiving hurt when in liquor" (September 27, 1758). He comments on the relationship of drinking and luxury: "I think that luxury increases so fast in this part of the nation, that people have little or no money to spare to buy what is really necessary. The too-frequent use of spirituous liquors, and the exorbitant practice of tea-drinking has corrupted the morals of people of almost every rank" (July 15, 1758). His own drinking habits are often excessive and, like Pepys a century earlier, he is always filled with remorse and vows to reform after having indulged too much. Frequently his drinking bouts occur on Sundays, which causes him particular guilt. On Sunday, February 8, 1754, he writes: "I will never drink more than four glasses of strong beer: one to toast the King's health, the second to the Royal Family, the third to all friends, and the fourth to the pleasure of the company. If there is either wine or punch, never upon any terms or persuasion to drink more than eight glasses, each glass to hold no more than half a quarter of a pint." It is a vow he rarely keeps on Sunday or other days. On Sunday, March 28, 1756, he laments having gotten drunk: "Oh! with what horrors does it fill my heart, to think I should be guilty of doing so, and on a Sunday too!." But in only the third entry after this he records he came home drunk (see also August 5, 1758; February 21, 1761; March 28, August 1, 1762.) When he does come home sober, he records it with pride (January 26, 1758; February 20, 1762). Frequently, he describes his own and others' drunkenness in terms of mad or beastly behavior: "We continued drinking like horses, as the vulgar phrase is, and singing till many of us were very drunk, and then we went to dancing, and pulling wigs, caps and hats; and thus we continued in this frantic manner, behaving more like mad people than they that profess the name of Christian" (March 7, 1758; see also August 23, 1757; June 10, June 30, 1758.) He praises an essay against drunkenness and calls it a "hateful vice...a crime almost productive of all other vice!" (June 22, 1755; January 11, 1764).

COFFEE CONSUMPTION; COFFEE HOUSE POPULARITY. In London, coffee drinking peaks in the mid-century and the public coffee houses decline as tea gains in popularity and their upper-class clientele join exclusive "clubs" and gradually move to their own houses (Younger 1966:359). Around 1760, many coffee houses are establishing themselves as private clubs and barring entry to "undesirables" or nonmembers. By the end of the 18th century there are as many clubs as there were coffee houses at the beginning of the century. Coffee houses begin to serve more wine, ale and other liquors, in what seems to be a first step towards their decay. For example, Campbell (1747/1969:281) says a coffee house keeper "is a kind of Publican" and that most sell other liquors "of which they make large profits." Gambling rather than polite conversation becomes the chief preoccupation there (Ukers 1935:71; Ferguson 1975:21; Ellis 1956:69, 172, 223-224). However, outside London, coffee houses continue to be prominent meeting places for newly formed debating societies, which reflect the development of a new popular political articulateness (Money 1971).

TEA: AVAILABILITY REVENUES. The British East India Company increases tea imports with the encouragement of the government, which benefits much more from the customs duty on tea than from those on coffee. In 1757, 4 million pounds a year are imported (Ellis 1956:239).

CONSUMPTION: AFTERNOON TEA; TEA GARDENS. Tea begins its rise as the national drink of England. Late afternoon teas become popular when the duchess of Bedford makes the habit fashionable because she eats only a small lunch and felt "a sinking feeling" around 5:00. It still is enough of a luxury that most households keep it locked up (Ferguson 1975:25-27). Still, by 1765 it is estimated that 90 families out of every 100 drink tea twice a day (Marshall 1962:15).

One contribution to the decline of the coffee houses is the transformation of the public pleasure garden into large, elaborate tea gardens in London, which provide garden walkways, dancing, bowling greens, and a variety of entertainment and concerts. The most famous are the Vauxhall Gardens and the Ranelagh. These tea gardens are popular resorts for women as well as men, contrary to the coffee houses, and while they serve coffee and other beverages, the women favor the tea (Rudé 1971:73; Ukers 1935:77-78).

CONCERNS (HANWAY). The rise of tea consumption is attacked by moralists not only as a health threat to the lower classes but as another sign of the growth of extravagant expenditures on luxuries that is undermining the nation. Among the most vocal critics is Jonas Hanway, who attaches *An Essay on Tea* to the 1757 edition of his *Journey...from Portsmouth Kingston-upon-Thames*. In it he identifies tea as the "apotheosis of luxury spending on needless extravagance by the poor" (Mathias 1979:162). He writes: "Will the sons and daughters of this happy isle for ever submit to the bondage of so tyrannical a custom as drinking tea?...It is an epidemical disease...You may see labourers who are mending the roads drinking their tea...it is even...sold out of cups to haymakers." Indeed, "the vast consumption and injurious effects of tea seemed to threaten the lives of the common people equally with gin." Violently attacking gin, he adds: "What an army has gin and tea destroyed" (quoted Marshall 1962:15; Drummond and Wilbraham 1939:243; Longmate 1968:8).

FRANCE

UPPER CLASSES/DRINKING. As manners and food become more refined and as the appreciation for vintage wines spreads, drunkenness becomes less pronounced among the upper classes (Franklin 1889, 6:132; Younger 1966:339). In 1765, Charles Duclos observes that over the past 60 years a revolution has occurred in Paris "as far as its tables, costumes and customs are concerned" (quoted Braudel 1981:206). Nevertheless, Ninon, the principal character in Voltaire's Depositaire (1767), asserts that the love of wine is a vice of the times (Franklin 1889, 6:132).

VINTAGE WINE DRINKING. With the increasing differentiation of vintages, wine develops more and more into a luxury product. By 1750, all the great vintages of today are established, and there begins to develop on a wide scale a "truly notable" wine industry (Loubère 1978:xviii). Nevertheless, J. Savory des Bruslons (Dictionnaire universel de commerce, 1759-1765, 5:1215-1216) expresses surprise that the Romans considered "the age of wines as their claim to excellence, while in France wines are thought to be stale (even those from Dijon, Nuits and Orleans, the most suitable of all for keeping) when they reach the fifth or sixth leaf [year]." Savory considers the wines of Champagne and Burgundy the best (ibid., 4:1222-1223). According to L. Caraccioli's Dictionnaire critique, pittoresque et sentencieux (1768, 3:112), the new expression sabler le vin de champagne (meaning "to toss off a drink") is fashionable among the smarter set (quoted Braudel 1981:234-236).

VITICULTURE PROSPERS. As the agricultural depression that began around 1650 ends and as cereal prices increase, restrictions on the expansion of vineyards in effect between 1727 and 1737 are repealed in 1759. Vine growing continues to experience a period of prosperity with high prices and high annual yields until around 1778 (Slicher van Bath 1963:217, 236; Forster 1960:98).

SPIRITS: PRODUCTION, PRICES. Until the end of the century, brandy production remains difficult and on the scale of a craft-type organization, but it has expanded rapidly in the first half of the century (Braudel 1981:244). The chemist Polycarpe Poncelet (La chymie de gout et de l'ordorat, 1752) describes how to blend spirits, herbs, spices, and other flavorings to create a symphony of agreeable tastes in aperitifs (Delamain 1935:33). In 1699, Sète exported 12,640 hectoliters of brandy; in 1753, 62,096; and in 1755, a record of 68,806. At the same time the price falls: 25 francs per verge (equal to 7.6 liters) in 1595; 12 in 1698; 7 in 1701; 5 in 1725. A slow increase after 1731 brings prices back to 15 francs in 1758 (Braudel 1981:244). Savory (Dictionnaire universel de commerce, 2:216-217) discusses large-scale liqueur production and merchandising in Montpellier, where merchants have set up a vast warehouse where taverners obtain supplies wholesale (Braudel 1981:246).

BRANDY CHARACTERISTICS. The average quality of brandy is called three-six and contains 79 to 80 degrees of alcohol; three-eight, at the top of the scale, is pure wine spirit with 92 to 93 degrees (Braudel 1981:244). In the Encyclopedie of Diderot and d'Alembert (see below), the city of Cognac is described as being famous for its eaux-de-vie (Delamain 1935:41).

ATTITUDES: INEBRIETY. In the Encyclopédie (1751) of Diderot and D'Alembert, the discussion of yvrognerie is "virtually copied out of Montaigne" (see 1570/France). It can thus be seen as a continuation of the tradition of tolerance toward drunkenness embodied by Montaigne alongside the contemporary intolerance of the social utilitarians like Duchesne (see 1760/France). However, the discussion does go beyond Montaigne in that drunkenness is viewed as a vice and an offense against reason. In this they share the opinion of the Paris theologians of the 1720s (see 1720/France). It is observed that "drunkenness...is not always a fault, against which it is necessary to be on one's guard; it is a breach that one makes in natural law which orders us to preserve our reason." The "least dose of an intoxicating liquor (une liqueur enivrante) suffices to destroy it [reason]" (quoted Brennan 1981:220; Brennan 1984).

Popular literature often represents drunkenness as accidental and comical, not harmless, but not very objectionable. This casual attitude is typical of the daily complaints brought before the commissaires. In Caylus' Correspondant de la guinguette, a girl is horrified when she finds the hero drunk. Caylus writes of her, "She believed drunkenness [yvrognerie] was hateful in a young man....[but] this was a simple bourgeoise who did not know the great world of Paris." Another character declares, "The man with whom I get drunk is more dear to me than wife or child" (quoted Brennan 1981:234; Brennan 1984).

STUDENTS/WINE DRINKING: RATIONS, NUTRITIONAL VALUE. At the Military School of Paris, wine rations are considerable, but vary among different groups: students, 0.25 liter; servants, 0.50 liter; supervisors, 1.291 liters; and teachers, 1.221 liters. In most colleges, wine provides the third most important source of calories (about 9%) after cereals (usually about 50%) and meat (17%-25%) (Frijhoff and Julia 1979:79).

PARISIANS/DRINKING; URBAN/RURAL COMPARISON (MIRABEAU). The marquis de Mirabeau (L'ami des hommes ou traité de la population, 1756) writes a bitter attack on urbanization, the frivolity of the nobility, the disregard for the misery of the peasantry, and other changes he sees about him. Regarding drinking, he observes that, although "it is rather commonly said that people in the country are drunkards," it is far worse in Paris: "I could establish two paradoxes in this connection: one being that the drinking which is so distasteful to water-drinkers is not harmful, and the second that, all in all...there is proportionally more drunkenness in Paris than in the country and that it is more harmful in the city. The entire population of Paris leaves the city on holidays [to go to the guinguettes, outside Paris], and even the bourgeoisie is in the habit of flocking there with the whole family, taking even young children. Half of the people come back drunk, gorged with adulterated wine, paralyzed for three days, and soon blunted for the rest of their lives. The local wine on which the peasant gorges himself does not have these dreadful aftereffects. He comes home drunk on Sunday night, that is true enough (although he is now [due to increased poverty] only too thoroughly cured of this meager luxury), but he finds his wife sober, which makes all the difference in public decency and society as a whole..., and early the next morning he goes to work" (quoted Forster and Forster 1969:80-81).

FESTIVE DRINKING; TAVERN REPUTATION (VOLTAIRE). Several passages from Voltaire's (1694-1778) Dictionnaire philosophique (1764) illustrate the prevalence of drunkenness on Sundays, holy days, and other days of nonwork. In the Catechismé du curé, he describes the Church's concern over the drunkenness that is the usual way of celebrating festival days: "You see some of them overcome by a liquid poison, their heads sunk on their knees, their hands hanging down, deaf and blind, reduced to a state much below that of animals, taken home tottering by their despairing wives, incapable of working the next day, and often ill and brutalized for the rest of their life. You see others, whom wine had converted into madmen, indulging in bloodshed, exchanging blows, and sometimes ending by murder." However, it is only on nonwork days that such behavior occurs: "It is the idleness of the festival which takes them to the public house. Debauchery and murder do not take place on working days" (Voltaire 1764/1962:149).

In his discussion of festivals, Voltaire suggests that tavern owners are responsible: "It is doubtless the publicans who have invented such a prodigious number of festivals: the religion of the peasant and artisans consists in becoming drunk on the day of a saint whom they only know by this form of worship. It is during these days of idleness and debauchery that all the crimes are committed" (quoted Rolleston 1925:184-185; Bercé 1974:221). He estimates that in a year 82 days of work are lost to Sundays (52) and festivals (30). This loss of income puts a strain on workers (Bercé 1976:154; Kaplow 1972:54 estimates 111 days of nonwork a year).

Such complaints are voiced by many, especially parish priests, under the Old Regime. It is generally asserted that the main pursuit of the poor on Sundays and holidays is going to the cabaret because they know of no other way to spend their leisure time. The parish priest actively competes with the cabaret keeper for the leisure hours of at least a minority of their male spiritual charges. "Gregarious drinking [is] for the adult male a more than acceptable alternative to mass-going, perhaps the more so because his wife and children [do] not join him" (Hufton 1974:303, 363-364). This is especially the case in the towns. Thus the church seeks repeatedly to ban Sunday openings.

DRINK HOUSE PREVALENCE (PARIS). There are about 2,000 drinking outlets in Paris in the second half of the century. One contemporary estimates that there are 1,500 wine merchants (not all of whom keep cabarets), 380 limonadiers, and 640 *épiciers* (Brennan 1981:30, 60n13; see also Cahen 1910:110).

CHARTREUSE INVENTED. Brother Gerome Maubec (a former apothecary) of La Grande Chartreuse transforms a formula known to the monks since 1605 as "an herbal elixir of long life" into a consumable potion. His work is perfected in 1764 when a fellow monk, Brother Antoine, produces an *élixir de table*, which is the present-day green Chartreuse (100 proof). Although the ingredients are secret, they are mixed with honey in a brandy base and then distilled six times, during which further ingredients are added. The mixture is then aged for up to four years. (Today, Chartreuse is still the only liqueur made by monks. Benedictine no longer has any connection with the monastic order) (Garfinkel 1978:78; Seward 1979:155-157; Emerson 1908, 2:173-174).

CAFES. Cafés flourish in the expanding Right Bank of Paris. There are a total of six on the Left Bank compared to seven hundred on the Right. The Café de la Régence, founded in 1681 in the Palais-Royal square, is especially popular. The vogue in cafes gradually lowers further the social status of the taverns. The fashion is the same in Germany, Italy, and Portugal (Braudel 1981:257-258).

GERMANY

COURT DRINKING. Life at court is heavily influenced by French models, especially in eating and drinking ceremonies. Dinner is often preceded by aperetifs, accompanied by many wines, and ended with liqueurs or brandy.

COFFEE. During the reign of Prussian King Frederick II, the "Great" (1740-1786), there are at least a dozen coffee houses in Berlin, and Frederick Meisner publishes the first comprehensive treatise on coffee, tea, and chocolate (Ukers 1935:41-42). Coffee begins to supplant flour soup and warm beer as the traditional German breakfast.

SCOTLAND

INEBRIETY PREVALENCE. McNeil (1956:20) characterizes the years from 1750 to 1900 as a period when "Scottish social life was indelibly stained with the vice of drunkenness."

ATTITUDES: LOWER CLASSES/DRINKING. Glasgow hammermen striking for shorter work hours are told that the leisure time thus created would be spent in public houses with a resulting loss of work efficiency (Thomas 1964:61).

WHISKY CONSUMPTION: HIGHLANDS. By mid-century, the distilling of whisky is "as much a part of the Highlander's way of life as the eating of his daily brose [porridge]. No matter how poor the house, the bottle of whisky to welcome or speed the guest [is] inevitable" (Sutherland 1969:60). However, according to Colonel Stewart of Garth (Sketches of the Highlands, 1822, quoted MacDonald 1914:37), it is not until 1750 that spirits of any kind are consumed as much as ale among the common people or wines and brandy among the gentry. Whisky's rise in popularity really begins around 1775.

LOWLANDS. Whisky ceases to be an almost exclusively Highland drink and becomes increasingly popular in the Lowlands. The brewing of ale diminishes as whisky increases, a development lamented by many of the parish ministers in the old Statistical Account. Several writers attribute this shift in consumption patterns to the excise tax on malt which after 1725 discouraged the brewing of weak, two-penny ale, which then was the primary drink of the lower class (Ross 1970:63). Whisky toddies made with sugar and hot water become especially popular in the Lowlands after the mid-century. Toddy is drunk by gentlemen; neat whisky is more often the poor man's drink. Claret is the preferred drink of the Edinburgh literati; enormous quantities of it are often consumed. Whisky in the Highlands has always been a classless drink; in the Lowlands, once it becomes common, "it [is] the democratic drink, the people's drink ('Freedom and whisky gang thegither!'). Later, amid the squalid horrors [of] the Industrial Revolution...[it is] often the only comfort of the destroyed and desperate [until] high taxation put it virtually beyond the reach of the common man" (Daiches 1978:42, 45).

WHISKY PRODUCTION. Officials estimate the manufacture of licit whisky at 433,811 gallons a year in 1756, compared to 50,884 gallons reported in 1708 (Barnard 1968:5; Robb 1951:37; Wilson 1973:47). Since it is unlikely that Scotland could have increased its capacity to make and drink whisky by that much in so short a time, it seems likely that some illicit whisky making became legal. However, the figure declines as additional duties are imposed in subsequent years (Cherrington 1929, 5:2384). Higher duties stimulate illicit production.

Contrary to the situation in England, spirit distilling in Scotland is mainly a family occupation. Little whisky is offered for sale before the end of the century (Simon 1926:25).

WHISKY CHARACTERISTICS. Evidence from the mid-century indicates that at least some whisky is still a flavored compound spirit made by the rectification of a distillate. For example, in an early edition (1755) of his Dictionary, Samuel Johnson defines "usquebaugh" as a "compound distilled spirit, being drawn on aromatiks." The Irish version, he asserts, "is paticularly distinguished for its pleasant and mild flavor." That made in the Scottish Highlands is "somewhat hotter" (quoted Barnard 1968:3, and Daiches 1978:25).

TAXATION: EXPORT DUTIES. By 1751, the duties on English and Scottish spirits differ so much that Parliament, prodded by English gin distillers, imposes an equalizing tax on imported Scottish spirits to stop the flow of cheaper whisky into England, where it is often redistilled into gin (Robb 1951:37; Daiches 1978:38).

ALEHOUSE CONTROLS. The first Scottish alehouse licensing act, a measure for raising revenue, establishes a license for the retail sale of ale at 20s a year. This act, contrary to all prior acts for a half century, does not exempt whisky. No one can be given a license to sell spirits who cannot produce a license to sell ale (Scottish Law Review 1943:86).

SPAIN

INEBRIETY. Drunkenness is still a vice odious to Spaniards; travelers frequently comment on their lack of inebriety. In the early 19th century, it is observed that any soldier seen drunk in the streets of Spain was usually a foreigner (a quarter of the Spanish army consisted of foreigners). Still, card playing and drinking in taverns are two vices to which artisans seem to be "irresistibly attracted"; in these pursuits they often spend their entire weekly pay on a single holiday. Many artisans also extend Sunday relaxation into Monday (Kany 1932:150, 163-164, 236).

BEVERAGES CONSUMED; TAVERN CHARACTERISTICS, CONTROLS. *Aloja* is the drink of the peasants and is the cheapest of all drinks. Spanish wines are stronger than French, but are generally diluted with water before consumption. Several new drinks are introduced from France, including the liqueur *rosoli*, popularly called *rasoliso* (roughly a drink smooth as satin). Coffee drinking spreads slower than in northern Europe.

Alcoholic beverages are primarily sold at *botillerias* (general refreshment parlors) and *alojerias* (lower-class taverns which sell only *aloja* or metheglin). Coffee houses also serve liqueurs. All drinking outlets must close at 10:00 in the winter and 11:00 in the summer. No card playing or smoking is permitted (Kany 1932:148-150).

SWITZERLAND

TAVERN CONTROLS. In Protestant Zürich, ordinances regarding drinking at times that interfere with church services become more exact and the prohibitions more decisive. People must not only go to church on Sundays but must remain there until the end of services. On Wednesday and Saturday nights, the guild houses and taverns must be closed so that there may be no hindrances to attendance at evening prayer (Vincent 1898:369).

AMERICA (BRITISH COLONIAL)

RUM PRODUCTION. By 1750, Boston, Providence, New Haven, and Philadelphia have become centers of a "burgeoning distilling industry" (Rorabaugh 1979:29).

TAVERN CHARACTERISTICS. As drinking increases and tavern keeping becomes more profitable, the issuance of tavern licenses is determined more by political and other considerations than by the social standing of the applicant, and the character of the tavern and drinkshops declines. A lower class of men gradually acquires control of the trade; dramshops and gin mills appear, "housed in rude shacks and attracting thieves, ruffians, loafers, and other ornaments of a developing underworld." Eighteenth-century taverns are usually dirty, disorderly, and lacking in satisfactory accommodations and food (Asbury 1950:8).

TAVERN CONTROLS. Having failed to reduce the number of licensed taverns, colonial authorities impose stricter laws for their regulation, including discouraging Sunday sales, requiring all taverns to provide lodging for travellers, revoking licenses if gaming is permitted on the premises, prohibiting sales to seamen, and banning a slave from buying liquor without his master's consent. These provisions, "judging from their frequent modifications and reenactments," fail to stop the erosion of upper-class authority over public drinking (Rorabaugh 1979:34).

ATTITUDES: INEBRIETY (EDWARDS). In *Freedom of the Will* (1754), theologian Jonathan Edwards uses the example of the drunkard to argue that people cannot be compelled by appetite or desire to do something against their will. "It cannot be truly said...that a drunkard, let his appetite be never so strong, cannot keep the cup from his mouth." Also, "A man never, in any instance, wills anything contrary to his desires....His Will and Desire do not run counter at all" (Edwards 1754/1966:218-219, 199). In choosing to drink or to get drunk, the drunkard is choosing his desire, his pleasure.

Levine (1978:150) comments: "That Edwards felt it necessary even to raise the question of volition with regard to the drunkard suggests that, by 1750, some people were beginning to view drunkards as individuals who had completely lost their ability to drink moderately."

CONNECTICUT

TAXATION. In 1756, Connecticut passes an act fining a town 200 pounds if it fails to appoint collectors for the liquor excise. This inaugurates a period of growth in liquor-trade taxation, resulting in increased problems with revenue collection and smuggling (Krout 1925:22).

NORTH CAROLINA

SALES CONTROLS. In 1758, slaves and indentured servants are prohibited from puchasing liquor without their master's consent (Whitener 1945:7).

1760

ENGLAND

SPIRITS: TAXATION. Finding that "the high price of spirits hath greatly contributed to the health, sobriety and industry of the common people," Parliament increases the duty on spirits in 1760.

CONSUMPTION. Following the imposition of higher spirit duties in 1760, the consumption level for the year is just over 2 million gallons; it remains at that level for 20 years (George 1925:38-39). By this time, only 12 out of the 30 London distillers existing in 1750 still survive (Simon 1926:25-26).

KING/DRINKING. George III (1760-1820), who ascends to the throne in 1760, is "a harbinger of the Victorian age" in his devotion to his family and moderate drinking habits. It is claimed he never drank a glass of wine before dinner in his life (Francis 1972:225).

SAMUEL JOHNSON/DRINKING. Samuel Johnson (1709-1784) frequently has to abstain from drinking alcohol for long periods (between 1736-1757 and 1765 to the end of his life) and is a self-described "hardened and shameless tea-drinker" (French 1884:299-300; Wain 1980:240). This periodic abstinence is rooted in his inability to drink moderately. During those periods when he does drink he performs "some creditable feats of ingestion" (Wain 1980:240), asserting to his friend and biographer James Boswell that he once drank three bottles of port "without being the worse for it" (Boswell 1791/1933, 2:187). (On Boswell see 1760/Scotland.) Johnson has been said to show many of the features of alcoholism, "notably a self-acknowledged difficulty in restricting his drinking to moderate amounts, solitary drinking, guilt concerning his alcohol intake, and periods of abstention." The years 1760-1767 are spent in a "struggle with alcohol" (Madden 1967:142). Johnson himself asserts: "I used to slink home when I had drunk too much"; he emphasizes that the goal of drinking alcoholic beverages is intoxication: "No, Sir, claret is the liquor for boys; port for men; but he who aspires to be a hero (smiling) must drink brandy....brandy will do soonest for a man what drinking can do for him" (quoted Boswell 1791/1933, 2:287, 292, see also 392-393). He asserts: "I have no objection to a man's drinking wine, if he can do it in moderation. I found myself apt to go to excess in it, and therefore...I thought it better not to return to it" (ibid., 1:65). In this "vulnerability to alcohol [Johnson is] encouraged by inclinations in the society of which he [is] a member" (Madden 1967:143). He thus becomes an advocate of restraint in drinking: "Intemperance, though it may fire the spirits for an hour, will make life short or miserable" (quoted French 1884:303). Despite his emphasis on personal temperance, Johnson does not disapprove of drunkenness per se. He feels that the harsh laws against gin were discriminatory against the poor (Sutherland 1969:125-126).

ATTITUDES: TAVERNS (JOHNSON). Samuel Johnson is an inveterate tavern haunter. He always prefers entertaining and meeting people at taverns or inns. Boswell (1791/1933, 2:661) quotes him as asserting that "there is nothing which has yet been contributed by man, by which so much happiness is produced as by a good tavern or inn." To another contemporary, he said that "a tavern chair was the throne of human felicity. As soon as I enter the door of a tavern I experience an oblivion of care and a freedom from solicitude...wine there exhilarates my spirits, and prompts those whom I most love: I dogmatise and am contradicted, and in this conflict of opinions and sentiments I find delight" (quoted Wain

1980:240 and Boswell 1791/1933, 2:661n.1; see also Porter 1982:244). However, at the very time Johnson is praising the tavern, it is declining as an upper-class institution, in large part due to the rise of the coffee house (Clark 1983:13).

JOSHUA REYNOLDS/DRINKING. The painter Joshua Reynolds is a close acquaintance of Johnson's in the 1760s, with whom he forms a conversational group called The Club in 1764. According to Farrington (1819:64-65), Reynolds and his associates help set a new standard of gentlemanly behavior and sobriety: "But it was not by the productions of his professional skill, and cultivated taste only, that Sir Joshua Reynolds attracted admiration;--his exemplary moral conduct, his amiable and well-regulated temper, the polished suavity of his manners...made his society agreeable to everyone....Such an example at the head of the arts had the happiest effect upon the members of the profession. At this time [c. 1760], a change in the manners and habits of the people of this country was beginning to take place. Public taste was improving. The coarse familiarity so common in personal intercourse was laid aside...and bacchanalian intemperance at the dinner-table was succeeded by rational cheerfulness and sober forbearance."

RAKERY DECLINES. Around 1765, the whole tradition of rakery goes into decline, although there are "a good many interesting survivals even after the turn of the century" (Jones 1942:8). A new generation of young blades appears which is less interested in street rioting and dueling than in their own clothes, manners, and superior graces. They are numerous enough "to set a new standard of behavior, one which eschewed violence, preferred tea to wine, and encouraged dallying with sin rather than embracing it as their fathers and older brothers had done." As a result, upper-class street marauding almost disappears in London by 1770, "and by the turn of the century the code of the middle class [is] gradually spreading out to encompass the aristocracy" (ibid., p. 9).

TAVERN FUNCTIONS: WAGE PAYMENTS. John Fielding (brother of Henry Fielding), in *Observations on Penal Laws* (1761, 2nd ed. 1768:107), condemns the practice of tradesmen paying their workmen at public houses, "commonly called pay-tables." The practice is "very injurious, as the men are too often kept out of their money till late on Saturday night, out of indulgence to the publican, by which means the mechanic goes home drunk and empty-handed to his family, where distress begets words then blows" (quoted George 1925:296, who could find no earlier condemnation of this practice).

RECREATION. The issue is raised whether publicans should be allowed to sponsor public sports and recreational pastimes "on the view of profit to themselves by the promotion of idleness and drinking....How often do we see the whole inhabitants of a country village drawn from their harvest work, to see cudgel playing, or a cricket match?" (quoted Porter 1982:312).

BREWING TECHNOLOGY. Knowledge of and books about brewing multiply. Michael Combrune produces *An Essay on Brewing* (1761), probably the first book on the principles of the process. G. Watkins (*Complete English Brewer*, 1768) emphasizes "the superior excellence of malt liquor above wine, in point of wholesomeness and in consideration of its being a product of our own kingdom" as opposed to the wine of "our enemies [France]" (quoted Mathias 1959:xxvi). Later, John Richardson's *Philosophical Principles of the Science of Brewing* (1784) reveals an enormous increase in the application of scientific techniques in brewing and describes the saccharometer, which measures exactly the amount of fermentable matter contained in a unit weight of malt or barley. These and similar works reflect and contribute to the expansion of commercial brewing in the second half of the century (Monckton 1966:138-139; Mathias 1979:224).

MEDICINAL USE. John Pringle (Observations on the Diseases of the Army in Camp and Garrison, 1752) states that wine and alcoholic beverages tend to suppress "putrid diseases" and are thus indispensable for the good health of the troops (Braudel 1981:244, who gives a date of 1763 for this statement in the text).

TEA AVAILABILITY. The English overtake the Dutch in volume of tea traded. Exports from Canton in 1766 are 6 million pounds (weight) on English boats. Veritable tea fleets gradually appear. The trade rises enormously (Braudel 1981:221).

FRANCE

TAXATION. A movement is underway to impose additional *sous pour livre* surtaxes on the *aides* but to reduce the areas of exemption (Matthews 1958:85).

BOUREGOIS PRIVILEGES ENDED (PARIS). In September 1759, the ancient privilege given to the citizens of Paris in 1633 to sell the wines produced on their own estates without paying taxes is ended. Henceforth excise taxes are imposed by the General Farm on all wine sold. This causes "considerable misfortune" to the poor because it eliminates one source of cheap wine. This privilege had often been abused by Parisian vineyard proprietors who presented as their own wines what in reality they had purchased elsewhere. These proprietors also sold their own wines as "imports" at higher prices (Dion 1959:482).

SURTAX RAISED. In 1760, an additional sou is added to the surtax, increasing the tax of the *aides* by 5% to 25% of value; in 1763, another is added to the entry duties of Paris; and in 1771, two more are imposed and the sou of 1763 is applied to all farmed taxes. This means that the surtax on alcoholic beverages increases from the four *sous pour livre* existing in 1718-1760 to eight in 1771 (Matthews 1958:85-86).

GUINGUETTES: POPULARITY RISES. Around 1750, Jean-Joseph Vadé (*La pipe cassée*) describes how the devotees of the *petit vin française*, the curious, and socialites go to the guinguettes in search of gaiety "To see Paris without seeing La Courtille/ Where the joyous swarm,/ Without frequenting Les Porcherons,/ The Rendez-vous of good chaps,/ It's like seeing Rome but not the Pope./ Also those who avoid nothing/ Often leaving the Luxembourg [palace]/ In order to enjoy in some suburbs/ The spectacle of the guinguette" (quoted Dion 1959:505; Vadé 1750/1831). In the 1760s, the ending of the wine privileges of the bourgeois of Paris and the increases in taxes provide further incentive to attendance at the guinguettes. It is at this time that Ramponeau opens the Tambour Royal in the faubourg (suburb) of Basse Courtille; Ramponeau's becomes the most famous of the guinguettes. He attracts patrons through an aggressive advertising campaign that emphasizes the freedom, lack of restraint, and excesses that occur there (Brennan 1981:101-102, 134; Dion 1959:505).

LABORERS/DRINKING. Except for the weekend trips to the guinguettes, poverty ensures sobriety among most Parisian laborers (Kaplow 1972:78-79).

WINE CONSUMPTION. In 1763, Tobias Smollett (1765/1979:66-67, 168) comments that the wine commonly consumed in Burgundy was so weak and thin that it would not be drunk in England. In Nice, the common people drank an inferior cheap wine. He suspects that all good wines end up in the houses of nobles or are sent abroad.

POLICE ATTITUDES (DUCHESNE). Paris magistrate Etienne Duchesne (Code de police, 1767, section 12) articulates the concerns over drinking in terms of social utility, as earlier voiced by Laffemas and Colbert. A political theorist with mercantilist influences, he identifies drinking as a cause of indigence and an offense against the work ethic. He also considers drunkenness more of a problem in viticultural areas than elsewhere: "In those places, where there is a superabundance of wine which is not exported...the low price of wine means that the common people are drawn into the cabarets...leave their work, their business, and become miserable" (quoted Brennan 1981:218).

VITICULTUAL EXPANSION: BORDEAUX. The Parlement of Bordeaux asserts in the preamble of one of its decrees in 1764 that popular well-being since 1720 has raised the price of wine and brought about plantations extraordinaires not only in this province but in all the pays vignobles (Dion 1959:596).

GRAIN SPIRITS BAN. The use of cereals or molasses for the production of alcohol spirits is again strictly forbidden in 1762, and only wine is allowed (Forbes 1970:196).

GERMANY

REVENUES: BEER, COFFEE. Frederick II discovers that more than 700,000 thalers are paid annually to foreign coffee merchants and that, as coffee drinking has increased, there has been a corresponding decrease in the brewing trade, from which he derives considerable revenue. In 1763, he tries to restrict coffee drinking by making it a luxury for people of "quality" and imposing a tax of eight silver groschen per pound of coffee (Ukers 1935:42; Jacob 1935:160).

IRELAND

DISTILLING CONTROLS. As in England, a scarcity of grain leads to a prohibition on distilling in 1758-1759. In 1760, private distilling without a license becomes a crime, turning many citizens overnight into criminals (McGuffin 1978:9; O'Brien 1918:41). Further proposals to prohibit distilling are rejected on the grounds that needed revenue would be lost (McGuire 1973:115, 123).

WHISKEY PRODUCTION, TAXATION. In 1760, spirits production is down to 225,217 gallons from 400,000 in 1758, but by 1765 it has risen to 715,475 gallons. Official records show that by 1762 both imports and home production of spirits have tripled since 1700. Illicit distilling is probably as large again as that licitly distilled. The duty on spirits is raised from 8d to 10d per gallon, but this still is so low that it constitutes government encouragment of the industry (Lynch and Vaizey 1960:49-50).

ILLICIT DISTILLING. After the temporary prohibition of distilling in 1758, the number of illicit stills multiplies and, according to O'Brien (1918:41), "the evil of drunkenness in the country greatly increased." The problem seems to have been as much due to the poor quality of the drink produced as to the level of drinking. In A Dram for Drunkards (1759), it is asserted: "A fiery spirit was distilled in all quarters, whereof the population drank so eagerly that scarce a week passed in which some did not die suddenly" (quoted O'Brien 1918:41). However, an Englishman traveling through Fermanagh in 1766 indicates that, while it is widely sold and consumed, whiskey has no adverse effects: "There are neither inns nor Alehouses on Ys road, yet almost everyhouse have for public Sale Aquavitae or Whiskey, which is greatly esteemed by ye Inhabitants, as a wholesome balsamic Diuretic; they take it here in common before their Meals....What is surprising they will drink it to Intoxication and are never sick after it neither doth it impair their health" (quoted McGuffin 1978:7).

CONSUMPTION LEVELS. Around 1765, some 5 million gallons of spirits (licit and illicit, domestic and imported), 700,000 barrels of beer, and 1.4 million gallons of wine are consumed annually by a population of about 2.5 million people (McGuire 1973:123).

CONCERNS: INEBRIETY. The high level of alcohol consumption produces much drunkenness, which distresses merchants more out of concern for property damage caused by intoxicated workers than for public morality (McGuire 1973:123). As early as the mid-1750s, William Henry is expressing his concerns over the "growing abuse of spirituous liquors," which he charges "has already nearly extirpated out of the minds of our common people all sense of religion, virtue and modesty," as well as promoted madness, criminality, idleness and the loss of several thousands of lives every year. He also complains about the numbers of drunken beggars who become unlicensed sellers in the outskirts of towns (quoted O'Brien 1918:41).

RUSSIA

ALCOHOL MONOPOLY: EFFECTS, REVENUES. In 1767, following the recommendation of a commission (appointed in 1765) to study how to increase the profitability of the alcohol monopoly to the state, Catherine the Great (1762-1796) establishes a system of farming out (otkupa). The government grants to the highest bidder the privilege of selling alcohol in a certain locality, with prices and the number of outlets determined by the government. The tax farmer receiving the contract collects the tax and turns it over to the government. In order to maximize their profits, the "farmers" encourage drinking. They soon become "local despots and political bosses," robbing the peasants, confiscating their crops in payment for debts owed the kabaks, and occasionally having them imprisoned (de Madariaga 1981:463-464; Cherrington 1929, 5:2331; Johnson 1915:113).

SCOTLAND

WHISKY CONSUMPTION. The Gentleman's Magazine (1766:211) speaks contemptuously of the common people of Scotland as "brutalised in appearance...in their lives they differ little from the brutes except in their love of spirituous liquors" (quoted Mathias 1979:151).

JAMES BOSWELL/DRINKING. James Boswell (1740-1795) makes frequent references in his journal to drinking too much and making unsuccessful resolutions to become "strictly sober" by 1774, probably under the influence of Samuel Johnson. It appears that by the age of 36, he has a drinking problem and is unable to control himself on some occasions. Around age 40, he is a daily excessive drinker, although he never drinks alone. On one occasion he is found "dead drunk" on the pavement (Rix 1975:73-74; see also French 1884:302). Although Scottish, he does not relish drinking whisky except as a morning dram to cure a hangover. He mentions only one or two occasions on which he or his companions drank gin (Coffey 1966:682).

AMERICA (BRITISH COLONIAL)

UPPER-CLASS CONCERNS: TAVERNS (ADAMS). Reflecting fears of other upper-class leaders, John Adams (1735-1826) is so concerned about the level of drunkenness that he proposes to the Braintree, Massachusetts, town meeting that the number of taverns in the town be limited; his suggestion is ridiculed. In his diary (May 29, 1760), he argues that taverns and inns, although necessary for travelers and for townspeople on public occasions, are otherwise nuisances. So many licensed taverns exist that publicans, in pursuit of profits, sell to immoral people. He laments that the licensing of tavern keepers, formerly selected from among the most worthy, has given way to licenses for the multitude. Adams blames public houses for the weakening of religious influence, the creation of political factions, and the degeneration of the political process. He complains that a clever politician can cultivate publicans and win the support of the masses, who are often a majority of the populace. "But the worst effect of all," he writes, "and which ought to make every man who has the least sense of his privileges tremble, these houses are become the nurseries of our legislators. An artful man, who has neither sense nor sentiment, may, by gaining a little sway among the rabble of a town, multiply taverns and dramshops and thereby secure the votes of taverner and retailer and all; and the multiplication of taverns will make many, who may be induced to flip and rum, to vote for any man whatever" (Adams 1804/1961, 1:128-129; Rorabaugh 1979:34, 35).

NEW YORK

TAVERN CONTROLS. Citing the great number of "mean taverns and tippling houses" that have opened recently in Dutchess County, the New York assembly directs officials of the county to specify the number of licenses to be issued, establish rules for the conduct of the drink places, and appoint commissioners to regulate the trade (Thomann 1887:123-124).

PENNSYLVANIA

SPIRITS CONSUMPTION. In his History of Pennsylvania (1798), Robert Proud describes the drink situation in Pennsylvania in the 1760s: "Malt liquor, which among the first settlers was more common, is made in some of the principal towns in small quantity. It has long been at times more or less an article of exportation from Philadelphia, and the use of tea, coffee, and chocolate, etc., is very common.

"But the liquors of the native growth and produce of the country...are, in general, but mean, or scarce and inferior, compared with the provisions, which are good and plentiful. This seems to arise, at least in part, from too much neglect in this particular, occasioned by getting rum and spirits at such exceeding low rates from the West Indies, which has rendered malt-liquor, though more wholesome and profitable for the country, less used than formerly" (quoted Thomann 1887:157).

VIRGINIA

ORDINARY CONTROLS. By several laws passed between 1705 and 1762, the legislature defines more clearly the accepted idea of an ordinary: "[It is] no place for servants or seamen or Sunday drunkards or gamblers or the impoverished local tippler, but a hospitable resort for transients with money or good names and for local men good in the courts for 20 shillings a year" (Pearson and Hendricks 1967:20).

1770

ENGLAND

BACKGROUND DEVELOPMENTS. By the end of the decade, people are becoming "healthier, longer-lived, less subject to the calamities of imprisonment for debt, and to temptations to excessive drunkenness." There has arisen a new spirit of humanity, a new command of material resources, and a "new belief in environment rather than Providence as the cause of many human ills," paralleled by the progressive revolution in manners after about 1760 (George 1925:320-321).

The consumer revolution rapidly expands in the third quarter of the century. More people are acquiring material possessions than ever before, sparking a rise in criticism over wasteful expenditures on luxuries, although more writers are praising the benefits of high wages and consumption. Wage rates rapidly increase between 1760 and 1780 (McKendrick 1982:9, 14-16, 19).

CONCERNS: ALEHOUSES. Clear signs appear among the upper classes of a renewed anxiety over alehouses as centers of disorder, idleness, and dissoluteness. These concerns are reflected in the attacks of John Scott (*Observations on the Present State of the...Poor*, 1773) and John Disney (*Thoughts on...Licensing Public Ale Houses*, 1776) (Clark 1983:254-255).

TAVERN REPUTATION (SMITH). Adam Smith observes of tavern keeping in the *Wealth of Nations*: "The keeper of an inn or tavern, who is never master of his own home, and who is exposed to the brutality of every drunkard, exercises neither a very agreeable nor a very creditable business. But there is scarce any common trade in which a small stock yields so great a profit" (Smith 1776/1950:103).

DRINKING INFLUENCES (SMITH). Adam Smith discusses various factors that serve to encourage or moderate alcohol abuse, advocating that taxes and prices should be lowered: "It is advantageous to the great body of workmen, notwithstanding, that all these [liquor] trades should be free, though this freedom may be abused in all of them, and is more likely to do so, perhaps, in some than in others. Though individuals, besides, may sometimes ruin their fortunes by an excessive consumption of fermented liquors, there seems to be no risk that a nation should do so. Though in every country there are many who spend upon such liquors more than they can afford, there are always many more who spend less. It deserves to be remarked too, that, if we consult experience, the cheapness of wine seems to be a cause, not of drunkenness, but of sobriety. The inhabitants of the wine countries are in general the soberest people in Europe....People are seldom guilty of excess in what is their daily fare....On the contrary, in the countries which, either from excessive heat or cold, produce no grapes, and where wine consequently is dear and a rarity, drunkenness is a common vice, as among the northern nations....Were the duties upon foreign wines, and the excise upon malt, beer, and ale, to be taken away all at once, it might, in the same manner, occasion in Great Britain a pretty general and temporary drunkenness among the middling and inferior ranks of people, which would probably be soon followed by a permanent and almost universal sobriety. At present drunkenness is by no means the vice of people of fashion, or of those who can easily afford the most expensive liquors" (Smith 1776/1950:456-457).

UPPER CLASSES/DRINKING. Some evidence exists that drinking has become more temperate among the upper classes, as exemplified by the behavior of Joshua Reynolds and his friends (Harrison 1971:91). As quoted above, Smith (1776/1950:457) claims that drunkenness is "by no means the vice of people of fashion." However, the upper-class English are still less temperate than their counterparts in France, as is indicated by the comments of John Moore (see 1770/France).

There are several prominent abstainers in the late century, including Erasmus Darwin, a successful provincial doctor, who asserts, "I drink only water, and am always well" (quoted Porter 1982:320, who observes that sectors of affluent society are becoming "starchy" as the age of Jane Austen approaches, exhibiting greater self-restraint and stressing sociability, good manners, and conversation). In the *Gentleman's Magazine* in 1770, Thomas Norwood lists 80 ways of calling a man drunk (Clark 1983:297; see also Porter 1982:33, who says he lists 99 ways).

FRANCE

BACKGROUND DEVELOPMENTS: ECONOMIC CRISIS. After 1760, prices begin to inflate rapidly, and around 1775, a half century of prosperity in France comes to an end. Between about 1770 and 1790, consistently poor harvests generate an economic crisis, a rising death rate, and a falling population. Between 1780 and 1800, grain prices increase 65%; wages only 22% (Labrousse 1933:89; Hufton 1974:12, 16). Compared with a base period of 1726-1741, the overall cost of living increases 45% in 1771-1789, and 62% in 1785-1789 (Lefebvre 1965:125, also citing the work of Labrousse). Industrial prices decline; unemployment and vagrancy spread; the aristocracy seeks to recoup their losses by extracting more from the peasants; poverty and dissatisfaction multiply (Duby and Mandrou 1964:370).

UPPER CLASSES/DRINKING; TEMPERANCE CITED (MOORE). The Englishman John Moore (1779, 1:325-326) lauds the sobriety of young French noblemen compared with their English counterparts: "Instead of spending their time in clubs or taverns, with people of their own age, the greater part of the young nobility pass their evenings with some private-family, or in those societies of both sexes to which they have the entrée. There the decorum due to such company restrains of course the vivacity and wantonness of their behavior and conversation; and adventures occur which interest and amuse without being followed by the nausea, languor, and remorse, which often succeed nights spent at the gaming table, or the licentiousness of tavern suppers....Even the most dissipated among them are unacquainted with the unbounded freedom of a tavern life." This, he suggests, is because most of the wealth in France is held by a very few young nobles. In England, on the other hand, there is a greater number of young men who come into greater fortunes "before they have acquired any fixed and determined taste" and squander it in gaming and tavern life out of indolence and desire for amusement.

WINE: PRICES. By the end of the decade, viticulture is in a crisis; from the highs of 1777-1778, prices fall and continue to fall until just before the revolution. The position of the small-scale vigneron is particularly acute. Contributing to this decline are overabundant wine harvests of poor quality after 1778 which slash the value of wine at the same time that grain and other prices are soaring and political conflicts (the American Revolution and a

British blockade) which cause a decline in the export trade and the economy in general, reducing the buying power of many French workers. Wine sells at a loss between 1770 and 1775 and especially between 1780 and 1785. After 1781, wine prices actually decline; they rise no more than 13%-14% between 1785 and 1789, compared to a 65% increase in cereal prices (Labrousse 1944, 1:323ff.; Hufton 1974:16; Slicher van Bath 1963:235-236; Lefebvre 1965:124-125).

CONSUMPTION. Since low-quality wine is important as a cheap foodstuff, the price tends to fall whenever grain becomes too expensive. At such times, its use could provide an alternative source of cheap calories, but the higher cost of bread and declining buying power of many laborers also reduces the amount of money available to purchase wine (Braudel 1981:237; Slicher van Bath 1963:236). The average Frenchman must allocate at least half his income on bread (Lefebvre 1965:125). Nevertheless, Restif de la Bretonne (1779/1933:67, 165), describing the limited wine drinking of his peasant father in the early century (see 1700/France), implies that in his own day wine is more commonly consumed (Dion 1959:593).

SALES. The wine merchants become one of the major guilds of Paris in 1776 (see 1620/France) (Gottschalk 1948, 2:241). Turgot establishes unrestricted transport and trade in wines throughout the kingdom in 1776 (Lausanne 1970:70).

DISTILLING PROMOTED. To encourage and improve distilling, the Société de l'Agriculture offers a prize in 1776 for the best answer to the question: "What is the most economical way of distilling brandy from wine as regards the quality and quantity obtained and the costs?" A decade later (1787), the Société Libre d'Emulation pour l'Encouragement des Arts, Métiers et Inventions Utiles holds a competition on "What are the most advantageous forms of the stills, furnaces and all the instruments used in the work of the large distilleries?" (Forbes 1970:219-223).

GERMANY

SPIRITS: CONSUMPTION, TAXATION. Observing an increase in spirit drinking in Berlin, Frederick II imposes a spirit tax in the hope that high prices will reduce consumption. He considers rum "horrible stuff" (Cherrington 1929, 5:1092).

POTATO SPIRITS INTRODUCED. By 1770, the introduction of distilling spirits from potatoes is fundamentally altering spirits production and consumption in Germany, especially in eastern areas such as Prussia. Potatoes eliminate the problems distilleries previously faced with limited amounts of raw materials (wheat, wine, fruits) due to poor weather conditions, low yearly crops, and epidemic numbers of vermin. Several factors influence this development. First, famine conditions following a series of bad corn harvests cause a major extension of potato cultivation in Germany. Second, technological developments increase the amount of output from the same quantity of raw material, reduce considerably the production time and manpower needed, and thus lower costs. In 1760, the first potato distillery opens at Mannheim. By the end of the century, potato distilleries outnumber all others and the overall level of spirit production has risen steeply and steadily. Most of the new potato spirit is consumed by poor Germans working in the industrial and agricultural sectors. In Prussia, distilleries are established in rural areas where production is not taxed as in the towns (Vogt 1984, who observes: "The production of spirits was one of the first sectors where big industries sprung up organizing the market according to its own interests"). It is also around 1770 that the word "schnapps" appears (Forbes 1970:263-264).

BEER CONSUMPTION. In many German-speaking areas, especially Prussia, beer consumption declines after the mid-century as spirits production escalates (Vogt 1984).

ATTITUDES: BEER, COFFEE (PRUSSIA). After the Seven Years War (which ends in 1763), Frederick II turns to reconstructing the financial disorder of Prussia. Both tobacco and coffee are major imports which disturb his plans. As a mercantilist, he seeks to establish a favorable trade balance of higher exports than imports. In 1777, Frederick issues a coffee and beer manifesto stating that the people, and especially soldiers, must drink beer for its nourishing qualities: "It is disgusting to notice the increase in the quantity of coffee used by my subjects, and the amount of money that goes out of the country as a consequence. Everybody is using coffee. If possible, this must be prevented. His Majesty was brought up on beer, and so were his ancestors, and his officers. Many battles have been fought and won by soldiers nourished on beer; and the King does not believe that coffee-drinking soldiers can be depended upon to endure hardship or to beat his enemies in case of the occurrence of another war" (quoted Ukers 1935:42).

IRELAND

UPPER CLASSES/DRINKING (YOUNG). Arthur Young (1779/1887:181, 186) refers to people of "considerable fortune" in Ireland as hard drinkers but on the whole observes that drinking has declined among them: "Drinking and duelling are the two charges which have long been alledged against the gentlemen of Ireland, but the change of manners that has taken place in that country is not generally known in England. Drunkenness ought no longer to be a reproach....Nor have I ever been asked to drink a single glass more than I had an inclination for; I may go further and assert that hard drinking is very rare amongst people of fortune." However, among "the class of little country gentlemen...drinking, wrangling, quarreling, fighting, ravishing, etc., etc., are found as in their native soil."

WHISKY CONSUMPTION. John Rutty (*Natural History of Dublin*, 1772:26) comments on the enormous increase in spirit consumption that has occurred since 1724 (O'Brien 1918:40).

TAXATION: SPIRITS. In an effort to reduce excessive drinking and raise revenues, which become increasingly needed over the years to pay for the wars in the American colonies and against the French, in 1775 Parliament raises the spirit duty from 10d to 1s2d a gallon. This duty continues to be increased to the end of the century. At the same time, Parliament lowers license duties. This law has two effects. First, the higher duty makes poteen (poitin) distilling, until now generally done for personal consumption only, a lucrative proposition and encourages "quite a few" distillers to go illegal (McGuffin 1978:10). Second, the lower license duty results in more retail shops, which defeats the purpose of the act. McGuire (1973:124) comments: "With more liquor shops there would follow more drunkenness and more revenue, more inefficiency in industry, more harm to the economy and more crime. It was the perennial government problem. The revenue was badly needed, so was a sober, industrious and law-abiding nation; the elusive solution was to balance the two to the least disadvantage."

BEER. Also in 1775, Parliament reduces the duty on Irish beer by 5d a barrel. This inaugurates a period of legislative encouragement of the brewing industry (Lynch and Vaizey 1960:55).

DISTILLING CONTROLS: LICENSING ACT (1779). The most important effort in the 18th century to deal with the evasion of the spirit duty is the Still License Act of 1779, which attempts to insure a minimum revenue from each still by basing the duty on the estimated number of times a still can be worked off in 28 days. A distiller has to produce a minimum number of gallons weekly, a quota many small distillers cannot meet. One effect of this duty is to drive out the smaller distillers, who are replaced by numerous illicit stills. Since the still license duty is based on a minimum amount of spirit produced in a given time period, the distiller is encouraged to produce as much spirit as possible and with little regard for its quality. There is a decline in the quality of legal whiskey, which becomes known as "Parliament whiskey." Thus even in areas where poteen is expensive, it is often preferred. In order to collect the duty on the excess spirit produced, the government continually increases the minimum amount subject to duty. The race between the distiller and the government escalates until the system becomes so unmanageable and unpopular that it is abandoned in 1823 (McGuire 1973:127-128, 131; McGuffin 1978:10-11).

TOWNLAND FINING SYSTEM (1779). To discourage illicit distilling, the Irish Parliament also passes an act in 1779 which imposes a 20 pound fine on the entire county or town in which an illicit still has been discovered (later the area is reduced to the townland or parish). Although subsequent changes make the townland fining system less arbitrary and the collection more efficient, the system remains unpopular and is of little effect in suppressing illicit distilling. Inhabitants band together to prevent the discovery of stills (McGuire 1973:397-398, 399).

BREWING. After 1750, beer consumption begins to increase as the number of better-paid town dwellers expands. Still, competition for whiskey and London porter brings Irish brewing to the verge of collapse in the early part of the decade. It is only saved by adapting itself to meet the popular taste for porter and by seeking financial advantages in the 1790s (Lynch and Vaizey 1960:41, 54).

ITALY

WINE CONSUMPTION. A sharp decline in overall wine consumption in Rome begins in the 1760s, lasting until 1820. This causes a serious economic crisis in the vineyards. The reasons for this decline are unclear, but are related to the successive wheat shortages and high cost of food characteristic of the years 1770-1800. Bread, and to some extent meat, uses up most of the buying power of the urban masses. Clearly wine is viewed as a secondary food. Though the public refuses to give up meat during periods of serious grain shortages (1764, 1772, 1779), they reduce wine-drinking without complaint (Revel 1979:46). (Compare with the similar developments under 1770/France.)

NETHERLANDS

DISTILLERY PREVALENCE. There are now 122 distilleries in Schiedam alone (Forbes 1970:190).

SCOTLAND

INEBRIETY PREVALENCE; DRINKING PRACTICES (BOSWELL). In Boswell's Journal of a Tour to the Hebrides (1785), a trip which he took in 1773 with Samuel Johnson, several references are made to the consumption of spirits, often in punch made with whisky, rum, or brandy. When Johnson first tries whisky, he asks for it by saying, "Come, let me know what it is that makes a Scotsman happy" (Boswell 1785/1936:348, see also pp. 294-295, 302-303). Boswell records that in the Hebrides, the mornings begin with a glass of whisky, called a skalk, taken as a stimulant (pp. 240-241, 282, 327). In a passage omitted in the first published version, Boswell comments on this custom, Scottish sobriety, and his own drinking propensities: "I had now got the habit of taking a scalck [skalk] or dram every morning. It really pleased me to take it. They are a very sober people, the Highlanders, though they have this practice. I always loved strong liquors. I was glad to be in a country where fashion justified tasting them. But I resolved against continuing it after leaving the isles. It would become an article of happiness to me" (pp. 240-241; later [p. 285] Boswell records Johnson reproaching him for drinking a morning dram. (Other references to drinking occur on pp. 103, 119, 173-174, 212, 222-223, 344-345; see also Murphy 1979:35-36.)

WHISKY CONSUMPTION. Tobias Smollett (1771/1929:288) contrasts Lowland and Highland drinking. In the Lowlands, a weak beer is drunk, but the Highlanders "regale themselves with whisky; a malt spirit as strong as geneva, which they swallow in great quantities, without any signs of inebriation. They are used to it from the cradle." MacDonald (1914:98) says that even in the 1770s the consumption of whisky in the Highlands is not great. But most evidence indicates that it is increasing in both the Highlands and Lowlands.

Hugo Arnot (1779:335-336) complains about "the almost total change which has taken place of the liquors used by the commons"; specifically, since 1724 the substitution of the drinking of whisky and tea for two-penny, previously "their chief if not their only drink." He finds neither of these new beverages wholesome: "Instead of malt liquor the lower class of people have betaken themselves to tea and Whisky. The first of these, to people who are not able to afford a generous diet, and liquors, cannot be deemed wholesome. The last is equally pernicious to health and morals; yet, the use of that destructive spirit is increasing among the common people of all ages and sexes, with a rapidity which threatens the most important effects upon society." In a footnote about whisky, he adds: "It is needless to descant upon the tendency of this evil. It is so important a nature as to require the hand of government to root it out, before the pernicious habit of the capital spreads itself over the country." Arnot estimates that there are as many as 400 unlicensed stills in Lowland Edinburgh alone--and only eight legal ones. There are 2,000 spirit retail houses in the county of Edinburgh (population 100,000), of which 1,850 "are destined for the entertainment of the lower class of people." To remedy this situation, he recommends increasing the tax on whisky, decreasing the tax on ale and rum, and prohibiting distilling for personal use (see also Robb 1951:27; Ross 1970:61).

DISTILLING IN LOWLANDS. In 1776, England opens its borders to the importation of Scottish spirits, presumably malt whisky. The yearly manufacture of the spirits increases (Barnard 1968:5; Cherrington 1929, 5:2384). Several large, professional distillers in Edinburgh make spirits for the English market, where it is rectified into gin (Wilson 1973:42). The government prefers the large distillers because they are "an easily controlled source of revenue" (Wilson 1973:42). Large-scale distillers in the Lowlands are bent on maximizing output through advanced production techniques, using stills with capacities up to thousands of gallons (versus 10-80 for most malt distillers).

TAVERN LIFE IN LOWLANDS. Edinburgh poet Robert Fergusson (1750-1774), who probably dies an alcoholic, describes the gaiety, squalor, and drunkenness of Lowland urban life and the conviviality of Edinburgh's taverns (Rooney et al. 1977). Although drinking plays an important role in his poetry, whisky does not, indicating that its consumption in the Lowlands is still limited among the literati. He does write "A Drink Eclogue," in which brandy and whisky conduct a dialogue, the former attacking the latter as a drink of coarse indigents. Whisky laments the snobbery which makes people prefer foreign spirits. Of whisky he writes: "And though, great god of aqua vitae!/ Who sways the empire of this city" (quoted Daiches 1978:44).

FUNERAL DRINKING. Smollett (1771/1929:292-293) describes mass drinking at funerals and weddings and the belief in the value of whisky to cold open-air life. A Highland gentleman considers it dishonorable if 100 gallons of whisky are not consumed at his grandmother's funeral at a feast for 50 people. Boswell (1785/1936:327) also describes whisky drinking as common at funerals. He laments that no spirits could be found because "a burial some days before had exhausted them."

SWEDEN

ALCOHOL MONOPOLY. In 1775, Gustavus III (1771-1792) establishes a royal monopoly over spirits production. Landlords resent the system since it interferes with the "favorable cultivation" of the land; peasants find that the monopoly makes spirits expensive and denies them the long-held right to distill their own brandy. The public in general protests the enormous revenue generated for the king by the crown distilleries (Goadby 1895:10; Gordon 1932:175). According to Morewood (1838:476), "Discontent rose so high, that even in the metropolis, it was found necessary to station guards at the Royal Brandy Factories, to prevent their destruction by an indignant populace."

AMERICA (BRITISH COLONIAL)

TEMPERANCE ADVOCATED (METHODISTS). In 1771, Francis Asbury arrives in America as John Wesley's General Assistant. A strict abstainer from hard liquor, Asbury begins to preach temperance, and in 1780 he procures the adoption by the General Conference of the Methodist Societies of a resolution disapproving the distillation of grain into liquor and urging the disownment of friends who do not renounce the practice (Asbury 1950:25).

(RUSH). Benjamin Rush, in Sermons to Gentlemen upon Temperance (1772), urges moderation in wine and strong drink. He acknowledges, however, that everyone knows people "who have made it a constant practice to get drunk almost everyday for thirty or fourty years, who, not withstanding arrive to a great age, and enjoy the same good health as those who have followed the strictest rules of temperance" (quoted Levine 1978:145-146). Rush is influenced by the medical theories taught at the Edinburgh College of Medicine, where he studied in the 1760s.

On this, Rorabaugh (1979:39) comments: "To men captivated by the Enlightenment spirit, the discovery that a too extensive consumption of alcohol produced illness and disease had two important implications. First, the information itself was a sign of progress, a sign of the triumph of rational, experimental inquiry over irrational tradition and custom. Second, the realization that alcohol was detrimental to health would be certain, it seemed, to lead men to abstinence and, as a consequence, to rationality."

SPIRITS: HEALTH CONCERNS (BENEZET). Philadelphia Quaker and reformer Anthony Benezet publishes the first American pamphlet advocating total abstinence from distilled liquor. The pamphlet, "The Mighty Destroyer Displayed" (1774), argues that spirits have an extremely destructive effect on health and morality (Krout 1925:65; Levine 1978:151). In 1775, he writes "Remarks on the Nature and Bad Effects of Spirituous Liquors." On these works, Rorabaugh (1979:37) comments: "Here for the first time we see liberty viewed in a new light, not as a man's freedom to drink unlimited quantities of alcohol but as a man's freeedom to be his own master, with the attendant responsibility to exercise self-control, moderation, and reason."

PRODUCTION CONTROLS. The First Continental Congress meets in 1774 and recommends that colonial legislatures attempt to curb "the pernicious practice" of distilling from grain (Winkler 1968:417).

AVAILABILITY. During the Revolutionary War, rum and molasses imports drop sharply because of the British blockade. Although distillers increase production of whiskey, they are unable to meet domestic demand. As a result, spirits are in short supply and expensive, and people have to do without them or resort to beer (Rorabaugh 1979:65).

JUSTICES/DRINKING. At court sessions, which are frequently held in taverns, considerable liquor is consumed. A New Hampsire court bill shows that on April 15, 1772, a judge was provided "a bowl of punch, two bottles of wine, and a mug of rum flip, at a cost of eight shillings twopence." The jury was supplied with five mugs of flip and two mugs of cider (Asbury 1950:6). (A mug varied from a pint to a half gallon.)

VIRGINIA

CONTROLS: TAVERN DRINKING. As the colonial period ends, there are no regulations in Virginia against the production and wholesaling of liquor, nor against consumption, unless it results in drunken disorder. Drinking on the premises is regulated through "selected and licensed privately owned ordinaries which must provide accommodations for transients, maintain good order, sell at prices fixed under law to freemen only and on credit only to people not of the neighborhood." Retailing for off-premise drinking is limited "to ordinaries, merchants, or even planters on the plantation, and to houses devoted solely to the selling of liquor in small quantities from door to door." Enforcement by county courts and juries is "lax at times," even though "the legislature normally offered at least a part of the fines charged for violations to those who informed on lawbreakers" (Pearson and Hendricks 1967:22).

1780

EUROPE

HEALTH CONCERNS; DISEASE SYNDROME. The dangers to health and work efficiency from drinking begin to be more widely recognized by such diverse figures as Benezet and Rush (see 1770, 1780/America), Lettsom (see 1780/England), and Trotter (see 1780/Scotland). The long-held belief that alcohol aids work performance is questioned and eventually rejected. These men mark the beginning of serious medical investigation of diseases directly associated with alcoholism (a term not yet invented) (Keller 1966:821; Bynum 1968). In the writings of Rush and Trotter, the concept of chronic drunkenness as a disease syndrome begins to emerge. However, the disease concept is largely overlooked by specialists and the public alike in favor of a moral concept (Howland and Howland 1978:39, 44).

CONSUMPTION PATTERNS. In 1781, August Ludwig Schölzer of Gottingen comments on the tremendous change that the introduction of new intoxicants, hot drinks, herbs, and spices has made on Europe since the Middle Ages: "It is indisputable that the discovery of spirits, the arrival of tobacco, sugar, coffee and tea in Europe have brought about revolutions just as great, if not greater, than the defeat of the Invincible Fleet, the Wars of the Spanish Succession, the Paris Peace, etc." (Schölzer 1781:93, quoted Medick 1983:96).

ENGLAND

BACKGROUND DEVELOPMENTS: INDUSTRIALIZATION. Hobsbawm (1962:46) refers to the years 1780-1840 as the "take-off period" of the Industrial Revolution. In the 1780s, the effects of industrialization begin to profoundly influence British life. The impact is not sudden. As late as the 1830s, only a small fraction of workers are employed in factories, but factories already are regarded as the coming mode of production (Palmer and Colton 1969:425-426). Industrialization brings about a great increase in the numbers of middle-class businessmen and the influence of an ideology of sobriety, thrift, and self-help.

CONSUMPTION PATTERNS. In the last 15 years of the century, the rate of consumption of excised commodities in mass demand, such as tobacco, soap, fabrics, and alcoholic beverages, increases more than twice as fast as the population (Mathias 1979:162).

INEBRIETY PREVALENCE. Francois Lacombe (*Tableau de Londres*, 1784) characterizes the populace of London as addicted to hangings, drunkenness, and street fighting (quoted Humphreys 1954:21). The Baron d'Archenholz (*Picture of England*, 1789, 2:110-111) comments that in London "they like [to drink] everything that is powerful and heavy." He attributes the English fondness for port wine to their attachment to strong liquors (quoted Simon 1926:50, who concurs). In the last decades of the century, the trend toward a moderation of drinking behavior and a refinement of manners continues, led by the middle classes but also increasingly evident among the upper classes. Nevertheless, intemperance continues to characterize many of the upper and lower classes.

UPPER CLASSES/DRINKING. In a letter to his wife in January 1788 from Pall Mall Court, Sir Gilbert Elliot questions how businessmen and MPs can function considering that "men of all ages drink abominably." Elliot gives as examples William Pitt the Younger, Charles James Fox, Richard Sheridan (MP, dramatist, and close friend to the Prince of Wales), and Viscount Howick Grey: "Fox drinks what I should call a great deal, though he is not reckoned to do so by his companions; Sheridan excessively, and Grey more than any of them; but it is in a much more gentlemanly way than our Scotch drunkards, and is always accompanied with lively clever conversation on subjects of importance. Pitt, I am told, drinks as much as anybody" (Elliot 1806/1874:187). Both Pitt and Fox appear drunk in the House of Commons on several occasions. Elliot also describes a dinner party disturbed by three riotously drunk young men (p. 135).

Another contemporary tells of watching the men around him at a drinking party as they fall under the table. When he tries to avoid further drinking by also going under the table himself, he finds a servant there loosening neckclothes to prevent apoplexy or suffocation (Simon 1926:46).

<u>DRINKING INFLUENCES: MODERATING TRENDS</u>. Several developments and events occur during the late decades of the 18th century which moderate drinking behavior. These include: (1) the Industrial Revolution; (2) the French Revolution; (3) the spread of Evangelicalism; (4) the spread of popular education; (5) increased tea and coffee drinking; (6) the controls over gin and the decline in its consumption since 1750; (7) a tightening of local restrictions against licensed houses beginning in 1786 and a decline in the number of alehouses; and (8) growing health concerns within the medical profession (see also the comments of King 1947:117).

The moderation of drinking habits is part of "a general stiffening of manners," which results from a new "social conservatism" between 1780 and the accession of Queen Victoria in 1837. The upper classes begin to take a more serious view of life: the gambling, masquerades, and revels of the past century lose popularity; social gatherings become more sober; and there is a noticeable rise in piety. Among the lower classes, a new concern over respectability and improved social standards begins to develop (Quinlan 1965:2-4).

<u>INDUSTRIALIZATION AND DRINKING</u>. With the spread of the Industrial Revolution, the dangers to health and efficiency from drinking become more recognized. The long-held belief that alcohol aids work performance begins to be questioned and eventually is rejected (Keller 1966:821-822). The desire to impose work discipline makes industrialists hostile to many traditional amusements and levities, especially the traditional "play" element in work--the ancient inseparability of work and recreation. These traditions are seen as fostering inefficiency and immoral idleness. A new emphasis is placed on respectable habits (Lambert 1975:293, 305).

As a working class begins to materialize at the end of the century, many of its first leaders and chroniclers call for more enlightenment, order, and sobriety in their own ranks in order to improve the conditions of working men. Many of these leaders are self-educated men like Francis Place, who raise themselves up by a self-discipline which requires that they turn their backs on the "happy-go-lucky tavern world" (Thompson 1963:57-58).

<u>EVANGELICAL, MORAL REFORM: PROCLAMATION SOCIETY</u>. Around 1785, a movement of national reform begins which lasts until the early Victorian era. This movement is launched by a handful of Evangelical Christians who are shocked at the decline of religion and morals. In 1787, William Wilberforce records in his diary his belief that "God has set before me as my object the reformation

of manners" (quoted Brown 1961:2). He helps persuade the king to issue a royal proclamation against vice, which is sent to every magistrate's bench, and he establishes the first of many Evangelical reforming institutions, the Society to Effect the Enforcement of his Majesty's Proclamation Against Vice and Immorality. In 1802, he forms a Society for the Suppression of Vice, which supersedes and goes beyond the Proclamation Society.

The primary concern of the reformers is to remove wickedness from public view and to avoid further contagion by evil example. They attempt to achieve this aim by proselytizing the ranks of the great, establishing societies for religious and moral reform, capturing the Anglican Church to their cause, and undertaking immense religious and moral propaganda. By the end of the century they have begun to win over important peers, MPs, and clergymen (Brown 1961:1-10, 18). Malcolmson (1973:101, 105) comments: "The standards of morality and propriety which were advanced by evangelicalism established more rigorous criteria for the evaluation of diverse forms of social behaviour....Customs which had previously been relatively unquestioned came to be seriously challenged....Evangelicalism had a profund impact on English society. Its voice is to be heard time and time again in the sources of the several decades after the 1790s."

Wilberforce's Vice Society in particular reflects the upper-class belief at the end of the century in the "intimate correlation between moral levity and political sedition among the lower classes." Prosecutions for drunken and lewd behavior increase markedly, and "the amusements of the poor are preached and legislated against until even the most innocuous [are] regarded in a lurid light" (Thompson 1963:402). Patronized largely by wealthy businessmen, the Vice Society is dubbed by the Reverend Sydney Smith "a corporation of informers supported by large contributions bent on suppressing not the vices of the rich but the pleasures of the poor" (quoted Porter 1982:312; on Smith's views, see also Malcolmson 1973:193). Harrison (1971:93) comments that the Vice Society "showed a zeal for prosecution, a concern with purely <u>public</u> vices and an inability to concentrate on any one of them--which disqualify it for any prominent place in temperance history."

Although drunkenness is a major concern, it is not a primary issue. Greater concern is directed toward lax sexual morality. Reformers are not opposed to drinking per se. Many Evangelical reformers are brewers, and it appears all Evangelical ladies and gentlemen drink wine (Harrison 1971:93). Even Hanah More, "the high priestess of reform" (Quinlan 1965:257), is not opposed to beer drinking by the poor worker in a suitable manner such as having a "wholesome cup of beer by his own fire-side" with his family (Brown 1961:405-406, 430). But reformers do feel that spirits drinking is not suitable for gentlemen, and they seek, with some success, to close down dram shops that sell only gin (no publican could legally sell spirits without a beer license) (Harrison 1971:93).

<u>POPULAR EDUCATION AND DRINKING</u>. Charity schools and the Sunday school movement provide the first systematic attempt to educate working people (Trevelyan 1978:315). This helps to influence drinking patterns in two ways. First, the curriculum emphasizes individual moral reform; second, it helps bring about a significant expansion of the reading public, enabling later temperance reformers to reach a larger audience (Quinlan 1965:160).

<u>HEALTH CONCERNS (LETTSOM)</u>. In 1789, John Lettsom publishes the <u>History of Some of the Effects of Hard Drinking</u>. It is one of the first serious medical investigations of the diseases directly associated with chronic drunkenness (French 1884:339; Keller 1966:821).

GIN CONSUMPTION, TAXATION. The imposition of higher duties on gin in 1782 results in a decrease in consumption to one million gallons annually. Although high corn prices have made distilling less vital to the agricultural sector, distillers complain about this decline and emphasize that smuggling has increased and revenues have been reduced as a result of the higher duties. When the gin duty is lowered in 1785, production and consumption rise and continue to increase until the mid-1790s (George 1925:39).

ALEHOUSES: CONCERNS, CONTROLS. The anxiety over alehouses that began to increase in the 1770s escalates sharply in the 1780s. Local clergy and evangelicals especially condemn alehouse drinking, as opposed to drinking in private. John Howard (An Account of the Principal Lazarettos in Europe, 1789:173) attacks "the great and increasing numbers of the alehouses that I observe...throughout the Kingdom" (quoted Clark 1983:39). Beginning around 1786, local justices throughout England adopt a policy of public house restriction and regulation. This policy is given encouragement in 1787 when the king issues the royal proclamation against vice (see above), further "prodding the justices into making fuller use of the powers they already possessed" (Monckton 1969:66). Among the social evils the JPs and the Proclamation Society seek to eliminate is the payment of wages in public houses (George 1925:305).

As described by the Webbs (1903:49), this is "the most remarkable episode in the whole history of public-house licensing in England." The actions taken include "the deliberate and systematic adoption, by benches of magistrates in different parts of the country, of such modern devices as Early Closing, Sunday Closing, the Refusal of New Licenses, the Withdrawal of Licenses from badly conducted houses, the peremptory closing of a proportion of houses in a district over-supplied with licenses, and in some remarkable instances, even the establishment of a system of Local Option or Local Veto, both as regards the opening of new public-houses and the closing of those already in existence, all without the slightest idea of compensation" (Webb and Webb 1903:49-51). The number of houses licensed to sell spirits declines markedly; hundreds are closed over the next few years.

Underlying this new magisterial onslaught are (1) the expansion of drinking at home among the middle and upper classes and among Evangelicals, (2) the onset of rapid industrialization, which contributes to concerns that the attractions of the alehouse will obstruct the interdiction of more regular, systematic work practices, (3) the fear that the number of outlets is getting out of control, especially the number of smaller establishments, which seem less susceptible to official controls, less respectable, and more prone to be meeting places of the laboring poor, and (4) mounting concerns over the problems of social distress and poor relief (Clark 1983:186-187, 256).

FUNCTIONS, PATRONS. At this time, the alehouse begins to lose part of its traditional clientele, which becomes fragmented as the drinking tastes of the well-to-do are transformed and focused around the home and private clubs, and as the social organization of the lower classes becomes more stratified. Poorer industrial workers move to dram shops and other rival establishments. The alehouse also loses its function as a center for business and social organization as competition from rival organizations grows. This causes publicans to make increased efforts to court their central group of customers: skilled workers, small craftsmen, and lesser traders. By century's end, alehouses become more spacious and congenial, with nicer parlors which serve more expensive drinks (Clark 1983:306-308, 312).

TEA CONSUMPTION, TAXATION. In the early decade, it is estimated that duty is paid on as little as one-third of the tea consumed annually (Griffiths 1967:21; Trevelyan 1978:340). Prime Minister William Pitt, in order to curtail rampant tea smuggling, reduces the import duty in 1784. Smuggling stops "almost overnight" (Ferguson 1975:24). By 1785, yearly consumption has risen to nearly 11 million pounds, compared with 6 million in 1768 (Griffiths 1967:21).

CONCERNS: LOWER CLASSES/DRINKING (TOWNSEND). Reflecting the attitudes of many in the commercial sector that "temperance and labour are the only source of happiness and wealth to individuals," Joseph Townsend (1786/1971:25, 67) criticizes the spread of drinking and idleness among the lower classes: "The reins have been held with a loose hand at a time when idleness and extravagance, the drunkenness and dissipation, with the consequent crimes and vices of the lower classes of the people, called for the most strenuous exertions of the magistrates, and the most strict execution of the laws." Those counties and provincial districts that do the least for the poor "hear the cheerful songs of industry and virtue," while those that provide for the poor "see drunkenness and idleness clothed in rags" (p. 20). "It is notorious" that the prevailing appetite of the common people is for strong drink. "When therefore, by the advance in wages, they obtain more than is sufficient for their bare subsistence, they spend the surplus at the alehouse, and neglect their business" (p. 30).

FRANCE

COURT/DRINKING. The "ordinary drinker" at the French court does not seem to be immoderate. Mostly wine is consumed, some brandy, and a small quantity of liqueurs (Younger 1966:341). Champagne is the favorite drink of King Louis XVI (1774-1792); Queen Marie Antoinette is a water-drinker (Franklin 1889, 6:137).

PARISIANS/DRINKING (MERCIER; RESTIF). In 1782, Louis-Sébastien Mércier (whom Braudel [1981:213] characterizes as "moralistic though always anxious to entertain") complains at length in his Tableau de Paris (1782-1788) about the drinking habits of Parisians. He asserts that "Paris is nightly re-entered by squadrons of drunkards staggering in from the [guinguettes in the] suburbs, rebounding from the walls." Sundays and feast days are especially a problem, and many accidents between coaches and drunkards occur. He concludes that "drinking is the source of endless disorder" and loss of family income (Mercier 1933:227-228). (See specific entries below for other information from Mercier.) Mercier's attacks on Parisian drunkenness and the cabaret are colored by his contempt for and fear of the lower classes, whom he views as primitive and violent (Brennan 1981:1).

In the discussion of "Flower Girls" in Les nuits de Paris (1788), Restif de la Bretonne (1788/1964:108-111) reflects on "the extent of luxury, which reaches its murderous hand even to the lowest classes" and undermines "sound morals" and "true happiness." He criticizes the Parisian wage earner for being overpaid, living only for the present, and squandering his wages on drink: "I reflected on the terrible drawbacks of the high cost of labor; thought of this in terms of the noxious effects on the masses, who, like savage tribes, think only of the present. If they earn enough in three days for necessities, they work for three days only and spend the other four in debauchery." He recommends reducing the price of labor so that the masses will work six days instead of three (also quoted in Forster and Forster 1969:262-265).

CROSSCULTURAL COMPARISON. John Andrews (A Comparative View of the French and English Nations, 1785:41, 65, 250) comments that the English "rabble" is better nourished and fed than the Paris "rabble," but spends most of its money on drink. The French poor save their money to keep up their physical appearance, and therefore they are better dressed (quoted Maxwell 1932:8).

VILLAGERS, PEASANTS/DRINKING. Mercier (1782/1933:227) asserts that drunkenness is less of a problem in the villages than in Paris: "A peasant or an artisan who takes a drop too much in his village is sure to get home safely enough; his wife arrives, gives him the rough edge of her tongue, and out to the pothouse he goes."

The material existence of the peasants is still quite miserable. Food is often insufficient; meat is rarely eaten. The wife of an artisan in Chatellerault declares, "Bread and soup [are eaten] several times a day, because meat is too expensive....At home wine is drunk, but on Mondays the master goes to the inn with his journeymen and drinks wine" (quoted Sée 1927:31, 176, who believes that except in sections where wine is plentiful, water is the usual beverage; see, however, the observations of Arthur Young given below).

Most vignerons still do not differ from other peasants in their standard of living. All are badly housed, partake of a monotonous diet, work hard, and die young. Wine and bread are staples of their diets. Many small growers, especially in bad years, still sell all their grapes and wine in order to make ends meet. Even in good years, the poorest people probably drink little but weak piquettes, although perhaps as much as 1-3 liters a day. Men drink most of their wine as part of their wage while working in vineyards, "without any apparent ill effect, an indication that its alcoholic content must have been low and that much was eliminated through perspiration." Many vignerons do not drink wine with meals at home except on special occasions. Some is consumed in cafés (Loubère 1978:235, 238-240).

WINE CONSUMPTION. In his account of his travels in France, Arthur Young (1794/1969:358-359) offers many comments on wine consumption. He estimates that there are 5 million acres of vineyards in France, producing wine valued at 875 million livres, with the people drinking an average 2s a head per diem. He asserts that among the poor in every part of the kingdom there is regular consumption of either wine or cider. Water is drunk only "in cases of failing crops." In addition, "an immense quantity [of wine] is consumed which is neither bought or sold."

TOULOUSE; LANGUEDOC. Young (1794/1969:358) tells of meeting laborers in Languedoc who drink three bottles of strong wine a day. By the end of 1780, the subdelegate to the intendant of Toulouse attacks the "enormous production" of vines which yield "a vile juice" and deplores increased wine-drinking, which he attributes in part to the luxury that has been introduced into all households and to population growth (quoted Forster 1960:98).

PARIS. It has been estimated that wine consumption in Paris between 1781 and 1786 is annually about 120 liters per person (Husson 1875:272; Braudel 1981:237, who observes that this "is not scandalous in itself"). In addition, 8.9 liters of beer are consumed (Husson 1875:276). Young (1794/1969:358-359) estimates that Parisians consume 4s of wine per person per day, compared to an average of 2s for France as a whole, "but this, as everyone well knows, was not the whole; for it supposes nothing for contraband, which probably was not less than one-eighth and which would make it nearly 4 1/2s a head." He later estimates that wine consumption in Paris over the last 20 years was 245,000 muids a year or 70,560,000 Paris pints or English quarts. This indicates a daily consumption of 193,315 quarts, to which he adds another 1/6th again for smuggling, to arrive at a consumption estimate of 1/3 a quart and 1/10th of that third per head per diem. For Brest in 1778, he estimates a consumption rate of 1/3 quart per head per diem for all alcoholic beverages (ibid., pp. 374-375).

WINE DILUTION. It is considered an unusual practice when Jacques-Louis Ménétra (see below) is required to mix water with wine as a cure for venereal disease (Ménétra 1780/1982:215), suggesting that wine is now being drunk pure (Brennan 1984). Arthur Young (1794/1969:237) indicates that the degree to which wine is diluted depends on the social class of the drinker, the occasion, and the quality of the wine. He also indicates that the French reputation for sobriety was somewhat exaggerated. In his entry for January 18, 1790, he writes: "I have met with persons in England, who imagine the sobriety of a French table carried to such a length, that one or two glasses of wine are all that a man can get at dinner: this is an error; your servant mixes the wine and water in what proportion you please; and large bowls of clean glasses are set before the master of the house and some friends of the family, at different parts of the table, for serving the richer and rare sorts of wines, which are drunk in this manner freely enough....[T]he common beverage is wine and water."

BRANDY CONSUMPTION: WOMEN. Mercier (1782/1933:32) complains of the new, "manly" habit of brandy drinking in the form of punch among women, who used to disdain "the reek of brandy on their breaths." While "the best-born and best-bred women" used to drink only wine, although plenty of it, "now they prefer strong spirits." The only good in this is that they are swallowing brandy "which is better for them than all these distilled liquors."

PARISIAN LABORERS. Mercier (1782/1933:77) writes that "at one in the morning 6,000 peasants arrive [in Paris], bringing the town's provision of vegetables and fruits and flowers....Then the brandy starts to flow across town counters; poor stuff, half water, but laced with long pepper. Porters and peasants toss down this liquor, the soberer of them drink wine." Parisian porters, both men and women, are accustomed to drinking brandy diluted with water and flavored with long pepper as a means of contending with the tax on wine entering Paris. The clientele of "smoking-rooms"--popular taverns where "idle" working-class smokers take their pleasure--do the same (quoted Braudel 1981:244-245). A "stentor" (a carriage-caller outside a theater) attributes his superhuman vocal powers to his preference for brandy over wine, but stentors die early "from a superfluity of grocers' liquor" (Mercier 1933:129).

COGNAC CONSUMPTION, PRODUCTION. Aged cognac is mentioned in 1783, apparently for the first time (Delamain 1935:49). By the 1780s, the amount of cognac exported from Charente has more than doubled from the level of the 1730s due to the rise in the number of firms engaging in the cultivation of grapes for distilling. By the end of the decade, eaux-de-vie de cognac has become so popular that Spanish brandies are smuggled into La Rochelle, blended with French brandy, and then reexported as cognac (Simon 1926:33-34).

BEER CONSUMPTION. Le Grand d'Aussy (1782, 2:307, 315) maintains that since beer is the drink of the poor, every difficult period sees an extension of its consumption, even in the viticultural regions. Conversely, prosperity turns beer drinkers into wine drinkers. Thus during the "disasters" of the Seven Years War (1756-1763) towns which had previously only known wine begin to drink beer, and in Champagne there are places were four breweries are set up "in the same time in a single year" (quoted Braudel 1981:239). In Paris, the number of breweries and the production level drop markedly between 1750 and 1782; the breweries decline from 40 to 23 (see also Kaplow 1972:78; Gottschalk 1948, 2:231).

GUINGUETTES, CABARETS. According to Mercier, the name of the guinguette Ramponeau, founded in 1760 in suburb of La Courtille and aggressively advertised, "is a thousand times" better known to the multitude than Voltaire's and Buffon's. "It has deserved to become celebrated in the eyes of the people, and the people are never ungrateful. It waters the thirsty populace of all the suburbs, at three and a half sous a pint: surprisingly moderate for a cabaretier and until then never seen before." At "the famous beggars' saloon" at Vaugirard, men and women dance barefoot in a tumult of dust and noise. When the guinguettes of Vaugirard and the other suburbs are full on a Sunday, "the next day you see empty barrels by the dozen in front of wine merchants' stalls. These people drink enough for the whole week" (quoted Braudel 1981:236; Dion 1959:505).

Around 1780, the abbot Mulot complains about faithless artisans who "go to drink in the guinguettes with the dregs of whoredom" and then give venereal diseases to their wives (quoted Kaplow 1972:94).

The practice of drinking at the guinguettes on Sunday seems to have increased over the course of the century, undoubtedly in response to higher taxes and greater enforcement efforts in Paris, but travel to the suburbs is still not convenient and other evidence suggests that much drinking is done in the city at neighborhood cabarets patronized steadily throughout the week (Brennan 1984).

ATTITUDES (MERCIER). By the end of the decade, the cabaret has declined appreciably in the view of elites who now consider it as a haunt of the common people (Dion 1959:485). Mercier devotes considerable attention to a critical description of the cabaret, which he calls the "receptacle for the dregs of the populace," a view he shares with Restif de la Bretonne. He holds them in disdain and regards their popular customers as wine-soaked and criminal. He tells his readers that one reason he is writing is to save the curious among his "delicate" readers from the "disagreeable sensation" of visiting a cabaret. In one description of tavern life, he records: "At other taverns I have had the occasion to see what is called 'drink a pint or quart.' A quart pitcher is on a crude wooden table...[a] leaden pitcher coated with red grime. If a fight breaks out from the effects of the adulterated wine, oaths and fists fly together." The intervention of a guard is all that prevents these scoundrels from killing each other. Such fights, he tells his readers, are common (Mercier 1782/1851:307, 310; also quoted Brennan 1981:1, 45-46, 99). He describes a troop of beggars feasting on their day's income, from which they voraciously eat for two hours. Cabarets are the home of swindlers and brigands who steal only in order to pass the rest of the night amidst women, wine, and cards (quoted Brennan 1981:78).

DRINKING ATTITUDES (MERCIER; MENETRA). Mercier (1782/1933:33) believes that "drunkenness, and the crimes due to it, should never be condoned; the old law of Pytacus, from which our own might take example, punished the drunken criminal twice over, once for his condition and once for his crime."

Author-artisan Jacques-Louis Ménétra (1780/1982:192) condemns fathers who "by their drunkenness are responsible for the fall of their children." Yet drinking is a frequent theme in his writing and he expresses pleasure in recounting his own "carousings," although he asserts that he never "sacrificed to Bacchus" except in company (Menetra 1982:259).

PEASANTS/DRINKING ATTITUDES. Everyone in the wine regions is convinced that wine is a healthy drink, a beneficial tonic for hard workers on the land. "Alcoholism," they believe, is hardly a concern among them, being rather a vice of urban dwellers, who consume distilled beverages of high alcoholic content (Loubere 1978:240).

BEVERAGE CHARACTERISTICS, PRICES. In his discussion of the evils caused by drunkenness in Paris, Mercier (1782/1933:227-228) expresses his opinion that much of the problem is due to the detestable quality of the wine. "In other countries excessive drinking is an inconvenience, true, but a passing one; here [Paris] it is a hideous thing; horrifying because wine is so adulterated with grapes from many places that it affects the digestion and therefore the reason, sooner than any other liquor can." He recommends that it would be better for the French to drink small beer as in England. "Wine-drinking darkens the skin and outlook of a nation," making the French more petulant and irritable than northern people." Parisians often "pay through the nose" for detestable wine. The beer available is "pernicious beyond belief."

TAXATION: PUBLIC PROTESTS. In 1781, an additional two sous is added to the aides surtax, resulting in a 50% increase in the actual tax rate since 1705 (Matthews 1958:85-86).

The cahiers (notebooks of grievances) submitted by the guild assembly of Rouen in the spring of 1789 demand the abolition of the aides and list as the guild's first justification the high cost of wine that results: "A man will not give a neighbor in distress a bottle of wine if he has paid an enormous duty on it when it entered [the town]....It is therefore for the entire kingdom that we demand the abolition of the aides, so that it will be possible to take care of unfortunate people and also bring all the wines and spirits one produces into town" (quoted Forster and Forster 1969:205).

Due to widespread smuggling of taxable goods into Paris, which amounts to an estimated fifth of the receipts indicated by records of entry duties, in 1783 General Farmer Lavoisier begins building a massive new wall (the seventh) around the city. The 10-foot-high wall is designed to insure proper collection of indirect taxes by incorporating areas beyond the established city gates. Parisians strongly protest the building of the wall, arguing that it will interfere with supplies of food and cheap wine. The wall encloses for the first time many taverns on the Right Bank, which must now charge higher prices for wine and suffer a loss in business. Wine merchants, tavern keepers, grape growers, and consumers work to keep contraband drinks flowing. Pipelines are placed under the barriers and sacks filled with wine are catapulted over them. Public protests beinning in 1786 succeed in suspending contruction in 1787. On July 12, 1789, two days before the storming of the Bastille, a restless mob demolishes and burns several of the new gateways and barriers. Attacks on the city barriers are led by persons known to be engaged in the contraband wine trade (Dion 1959:511-514; Kaplow 1972:14, 79; Matthews 1958:171-173; Lausanne 1970:15).

VITICULTURAL CRISIS; WINE PRICES. Generally abundant wine harvests causes wine prices to continue to fall in the first half of the decade, and the poor state of the economy in general also causes demand to decline. Prices rise again in 1788-1789, in large part due to poor harvents in 1786, 1787, and 1789, but grain prices rise even more. Peasants who sell wine and buy grain probably suffer the most. Wine prices in Angers are on the average much lower in the 1780s than in the previous decade, about two sous per pinte compared to three sous (Labrousse 1944:579-580; Tilly 1964:213-214).

Arthur Young disagrees with the predominant view in France "that the wine provinces are the poorest in all France" and that viticulture "is mischievous to the national interests" because it utilizes land and labor better put to grain cultivation. His own opinion is that France greatly benefits from its wine industry since nothing else will grow in many viticultural regions, the financial returns are higher than from arable crops, and French wine is "one of the greatest trades of export that is to be seen in Europe." The poverty of the viticultural regions results from the high level of capital needed for viticulture and from poorer vine growers having to work for their richer neighbors instead of for themselves, although wages are high (Young 1794/1929:125, 294-295; see also Young 1794/1969:99, 357-358).

INN RESTRICTIONS ADVOCATED. Parish memorials for 1789 demand that restrictions be placed on the numerous inns that exist in villages and along the roads, because they promote drunkenness, which is seen as an incentive for violence (See 1927:48).

DISTILLING EXPANDS. As in the late 17th century, the depression in viticulture encourages distilling. In Montpellier, the abundance of even extremely cheap wine is so disproportionate compared to demand that inhabitants distill 32 ships full (2.4 million gallons), most of which probably is smuggled to England (Morewood 1838:401).

DISTILLATION TECHNOLOGY. Technological developments, including continuous cooling by a double current, make brandy production less difficult and allow for improved output. The crucial changes that make it possible to distill wine in one operation come later, but near Montpellier, the "preheater" is developed. The device uses the wine about to be distilled as a cooling liquid in a warm-cooler. This process greatly increases the output of the still and its use spreads quickly (Forbes 1970:245; Braudel 1981:244).

COFFEE CONSUMPTION. Coffee drinking spreads beyond the fashionable world, largely due to greater availability and affordability. While all other prices are rising, the cost of a cup of coffee remains almost unchanged due to high production in the islands. From about 1785 to 1789, France imports 36,000 tons of coffee (half of it from Santo Domingo). Of this, 10,000 to 12,000 tons are reexported; Paris keeps about a thousand tons for its own use. Le Grand d'Aussy (1782, 3:125) observes that since 1750 coffee consumption has tripled in France. "There is no bourgeois household where you are not offered coffee, no shopkeeper, no cook, no chambermaid who does not breakfast on coffee with milk in the morning. In public markets and in certain streets and alleys in the capital, women have set themselves up selling what they call cafe au lait to the populace, that is to say poor milk coloured with coffee grounds which they buy from the kitchens of big houses" (quoted Braudel 1981:258). Unsugared cafe au lait sold on the streets to workmen costs two sous (Kaplow 1972:78). Cheap, nourishing, and flavorful, it is enormously popular. Workmen drink it in prodigious quantities, saying that it sustains them from breakfast until the evening (Mercier 1782/1933:78 and 1853:210).

Some provincial towns still do not welcome the new beverage. The Limoges bourgeois only drink coffee "as a medicine." Only certain social categories (postmasters in the north, for example) follow the fashion (Braudel 1981:260).

Moreover, the cost of coffee and sugar is still high enough that the coffee drunk by the general population must have been very weak (Kaplow 1972:78).

CAFES. In Paris, there are "well past a thousand" cafes by the end of the decade; they have become the rendezvous for men of fashion and the leisured (Brennan 1981:29). This partly accounts for the decline in the fashionability of cabarets. However, not all Frechmen approve of them. Mercier looks upon them as unfavorably as he does the cabaret, and he considers coffee and tea as destructive drinks which burn the stomach. According to Mercier (1782/1929:17), they are "the ordinary refuge to the idler and the shelter of the indigent." He complains that "our forebears frequented an inn, and they say that they were a cheerful spirit. We no longer dare go to the coffee houses, the discoloured hot water that we drink there is far more unhealthy than the generous wine which intoxicated our fathers. Gloom and bitter talk prevail in these mirrored rooms. Grumbling is heard on all sides; is it the new drink that has worked this difference?" Some men, he asserts, stay in the cafes from 10 in the morning until 11 at night (the compulsory closing time, supervised by the police). Their dinner consists of a cup of coffee with milk; their supper, Bavarian cream (a mixture of syrup, sugar, milk and sometimes tea) (quoted Braudel 1981:260).

Another contemporary writes: "I think I may safely assert that it is to the establishment of so many cafes in Paris that is due the urbanity and mildness discernible upon most faces. Before they existed, nearly everyone passed his time at the cabaret, where even business matters were discussed. Since their establishments, people assemble to hear what is going on, drinking and playing only in moderation" (quoted Ukers 1935:97).

GERMANY

SPIRIT CONSUMPTION. Germany between the Rhine and the Elbe is a brandy-consumption area. In 1783, Hamburg receives 4,000 casks of brandy from France (20,000 hectolitres). Alcohol made from grain is consumed almost exclusively beyond the Elbe and around the Baltic. In 1783, Danzig receives about 100 casks of brandy which are reshipped, after being scented with cumin, to Stockholm, another town devoted to grain-based alcohols (Braudel 1981:247).

COURTS/DRINKING. An anonymous Introduction to the Knowledge of Germany (1789) asserts that in some German courts a visitor "is initiated and purchases in a manner his freedom by submitting to drink til he has lost the use of his reason" (quoted Bruford 1968:93).

TEMPERANCE OF GOETHE. In his diary for 1779, Goethe notes how much better he was for watering down his wine ("drinking half wine"). He considers this the most beneficial change in his habits since he stopped drinking coffee, which he apparently had done a year or two earlier (Bruford 1962:173).

COFFEE MONOPOLY. For a while, after Frederick II's coffee manifesto of 1777, beer drinking increases in Prussia and coffee remains a luxury affordable only by the rich. Then coffee smuggling increases despite all attempts to stop it. Following the French precedent, Frederick creates a royal coffee monopoly in 1781.

Frederick makes exceptions to the monopoly in the cases of nobility, clergy, and government officials, to whom he issues special licenses permitting them to do their own roasting. This, in effect, confines the use of coffee to the upper class. They purchase their supplies from the government, and since the price is greatly increased, Frederick at first derives substantial revenues (Ukers 1935:42). However, general trade begins to fall off, and the import duty, increased to one thaler per pound, leads to a reduction in sales. Even though the duty is then reduced by half, the ordinary people turn instead to coffee substitutes such as barley, corn, wheat, dried figs, and chicory (Jacob 1935:161).

In order "to remove all desire and inclination for fraud," Frederick's successor, Frederick William II (1786-1797), issues an edict that reduces the tax on coffee in 1787 and ends the restrictions on roasting (Jacob 1935:162).

IRELAND

INEBRIETY PREVALENCE. In *View of the State of Ireland* (1780), it is asserted that a third of Dublin shops are spirits vendors; the healthy English porter and artisan is contrasted with "the emaciated artificer of Dublin with long and hollow visage whose trembling hands can scarcely receive the alms which his miserable appearance extorts" (quoted O'Brien 1918:41). In 1789, the Quakers petition Parliament to do something about the universal presence of drunkenness throughout Ireland (ibid, p. 42).

DISTILLING. It is asserted that the whole country is teeming with illicit stills and that in certain districts every farmer is both a spirit distiller and retailer. There are an estimated 1,000 legal stills alone in 1779, most of which are small (O'Brien 1918:42; Magee 1980:55-56).

POLAND

REVENUES. Since the mid-17th century, the monopoly on the sale of alcohol in the *propinacja* (local peasant taverns) has been the major method used by landholders to take from peasants the money revenues the peasants derive from the market economy. Under this monopolistic system, the peasants are required to buy certain amounts of grain-based alcohol as part of their obligations to landlords. As a percentage of the overall income of the royal properties, these revenues have increased dramatically from 6.4 percent in 1661 to 40.1% in 1789. This enormous increase is even more spectacular when it is taken into consideration that between 1660 and 1790 overall income from royal properties has risen considerably (Kula 1976:134-137, who concludes: "Thus the increase in overall income during the period from 1661 and 1784 is brought about to a greater extent through increased effectiveness in siphoning off peasant money than through growth in the profitability of demesne production" [p. 139]). The distillation of alcoholic beverages is critical to the royal estates in the southeast where there are enormous stacks of grains left without any possibility of being sold. According to Prince Jozef Czartoryksi, propinacja sales are the only way to increase revenues in this region. He observes that "in our country vodka distilleries could be called mints, because it is only thanks to them that we can hope to sell off our grain in years when there is no famine." Another account in 1783 claims that "the production of beer, the distillery and the *propinacja* are the heart of net revenues" (Kula 1976:137; Park 1983:70).

SCOTLAND

WHISKY CONDEMNED. A contemporary asserts in 1782 that the low cost of spirits due to an excise of only 10 pence was the main cause of the drunkenness and poverty of the lower classes: "The exceeding cheapness of spirits is the principal cause of the poverty, and especially the increase of drunkenness among the common people; it is the parent of that wretchedness, poverty, and misery, which we see continually in our streets, where we find our manufacturers and tradesmen remaining idle from Saturday to Tuesday, and frequently to Wednesday, in every week in a state of intoxication, and, when their own indolence prevents their being able to earn sufficient to support their families, we find them running to whiskey houses and entering into combinations to force an increase of wages" (quoted O'Brien 1918:42).

TAXATION. In the last year of the survey system of figuring duty (1784), the duty charged on spirits rises to 3s11d plus 15% per gallon. Duty is paid on 239,350 gallons (MacDonald 1914:49). The annual license fee on Lowland spirit stills is doubled to 3 pounds, 12 shillings in 1788.

DISTILLING CONTROLS, TAXATION: WASH ACT (1784). As part of Pitt's financial reforms, the Wash Act of 1784 establishes two systems of whisky taxation, one for the Highlands and the other for the Lowlands. The Lowlands tax is based on the quantity of wash capacity of a still at the rate of 5d a gallon. Wash is the fermented but undistilled liquor that is put through the still. In the case of whisky, the wash is made of malt and, perhaps, barley and other cereals; it is essentially ale. The act assumes that 100 gallons of wash will produce 20 gallons of spirit at from one to ten points over proof. However, no account is taken of the gravity of the wash and, since artificially saccharified washes are used in England but not in Scotland, the act is discriminatory against the Scottish distillers, especially the Lowland distillers who use large stills and thus make less spirit out of a like amount of wash than their English counterparts (Ross 1970:66; Wilson 1973:48). For the Highlands, the act imposes a license fee of 20s per gallon of still capacity per year. This is almost identical to the system adopted two years later for all of Scotland (Daiches 1978:32).

LICENSING ACT (1786). By the Scotch Distillery or Licensing Act of 1786, the Pitt government replaces the Wash Act of 1784 with a new licensing system calculated according to the capacity of the still to produce a given amount of spirits over a year. It is believed by the government that the new license system will be easier to enforce since the capacity of a still can be easily measured and its output of spirits figured for a 24-hour period and thus for a year. It is assumed that a still cannot be worked off in less than 24 hours. Wilson (1973:48) calls this system the "most foolish, unpractical and misconceived taxation process of the century."

ILLICIT DISTILLING. With the Wash Act, the "floodgates of illicit distilling" are opened (Wilson 1973:122). In response to the Licensing Act, Scottish distillers seek to maximize their production and profits by speeding up the time to work off a still in less than the 24 hours assumed. Stills of different shapes and size are tried, leading to wider, stouter stills that will boil faster than taller stills (Wilson 1973:53; Ross 1970:66-67). Soon, distillers and excise officials are engaged in a race. Every increase in the distiller's ability to produce more spirits within the parameters of the Licensing Act is met with a rise in the annual license fee. The effects of this system are hardest on small Lowland distillers who cannot afford the investment in license taxes or the cost of replacing stills now more quickly worn out. Thus they cannot produce whisky of as good a quality as before. Small Highland distillers are also driven further into moonshining, depriving the treasury of much revenue and fomenting further discontent with the excise gaugers. In addition, scarcely two stills in Scotland are paying the same rate of duty, causing great confusion (Robb 1951:39).

WHISKY PRAISED (BURNS). The first book of poems by Robert Burns, who becomes the most famous Scottish poet, is published in 1786. Burns is "perhaps the first Scottish poet to realize that whisky was becoming the national drink of Scotland," virtually identifying Scotland with whisky (Ross 1970:29). Several of the poems, written in Scottish dialect, contain references to whisky and other alcoholic drinks, especially the one titled "Scotch Drink" and "The Author's Earnest Cry and Prayer," the latter petitioning Parliament to reform the duty on whisky. Significantly, Burns is a Lowland poet, illustrating the spread of whisky drinking southward.

As part of the financial reforms of Pitt the Younger, the exemption of the Forbeses of Culloden from paying taxes on the whisky they produced on their estates is removed in 1784. Burns laments the cessation of the production of Ferintosh in a famous verse published two years later: "Thee, Ferintosh! O sadly lost!/ Scotland lament frae coast to coast!"

HEALTH CONCERNS: DISEASE SYNDROME (TROTTER). Dr. Thomas Trotter studies inebriety while he is a candidate for a medical degree at the University of Edinburgh. In his thesis of 1785, revised and published as an *Essay, Medical, Philosophical and Chemical on Drunkenness* (1804/1981:17), Trotter defines drunkenness in medical terms as "strictly speaking...a disease; produced by a remote cause, and giving birth to actions and movements in the living body that disorder the function of health." He further emphasizes that "It is to be remembered that a bodily infirmity is not the only thing to be corrected. *The habit of drunkenness is a disease of the mind*. The soul itself has received impressions that are incompatible with its reasoning powers" (p. 174, author's italics). Drunkenness also "induces" a wide variety of other diseases and physical ailments (pp. 99-132). Trotter observes that the subjects of health problems from drinking and alcohol's mode of action have been largely ignored in the medical literature (p. 111). When published in 1804, the work is widely acclaimed by physicians and some government officials (see also Hirsh 1949:231-232).

SWEDEN

MONOPOLY ABOLISHED. Faced with widespread opposition to the alcohol monopoly, Gustavus III abolishes the crown distillery monopoly in 1788 and allows any landlord who pays a small license fee to manufacture and sell spirits (Goadby 1895:10; Gordon 1932:175).

SWITZERLAND

TAVERN CONTROLS. In Basel, the drinking conditions have so deteriorated that in 1776 a commission investigates the damage caused by alcoholic excess, and two years later the mayor issues severe regulations. The increase in the number of drinking places is so conspicuous toward the end of the 18th century that the council issues a law in 1801 governing taverns and inns that forms the basis of later legislation in this field (Jellinek 1976b:82).

BRANDY SALES CONTROLS. In 1789, Bern bans the sale of spirits for both on- and off-premise consumption, except in taverns, which are permitted to sell native brandy and kirschwasser to their guests "in moderation" (Jellinek 1976b:78).

AMERICA, UNITED STATES

BACKGROUND DEVELOPMENTS: REVOLUTION ENDS. The surrender of Cornwallis on October 17, 1781, ends the American Revolution (peace signed, 1783).

INEBRIETY PREVALENCE, CONCERNS. Following the war, controls on drunkenness weaken and consumption increases, especially of whisky. Fear grows among moral leaders that the American appetite for strong drink might well endanger the future of the nation. As in other western countries, the American medical profession begins to focus more attention on the drinking problem.

HEALTH CONCERNS: DISEASE SYNDROME (RUSH). Benjamin Rush's *An Inquiry into the Effects of Ardent Spirits upon the Human Body and Mind* (1784) calls into question the routine practice of using alcohol in medical treatment and suggests less harmful remedies. In this work, Rush attacks the prevalent belief in the beneficial effects of spirits and argues that habitual drinking is a disease caused by spirits and that its cure requires abstinence from spirits. Rush insists that although an individual at first voluntarily chooses to drink, if consumed long enough spirits become a necessity, leading to a "disease of the will." According to Rush, "poverty, misery, crimes and infamy, disease and death, are all the natural and usual consequences of the intemperate use of ardent spirits." Rush insists that the effects of chronic drinking (liver disease, jaundice, consumption, epilepsy, gout, madness) may occur in the absence of drunken behavior. In place of spirits, Rush suggests the use of cider, malt liquors, wine, coffee, water, or water mixed with vinegar and sweetened with molasses (Rush 1784/1814:11, 12, 15-19). The "Inquiry" initially is published as a newspaper article and later as a pamphlet, which appears in many editions and becomes the prototype for later temperance literature (Rorabaugh 1979:40-43, 68; Levine 1978:152; Hirsh 1949:234-236).

In 1789, Rush publishes "A Moral and Physical Thermometer," which graphically displays the effects of various beverages. Water, milk, and small beer result in "Health, Wealth, Serenity of mind, Reputation, long life, and Happiness," while chronic use of gin, brandy, rum, and whiskey leads to theft, murder, suicide, madness, palsy, apoplexy, and death. The "Thermometer" becomes a prominent item in temperance literature (Rorabaugh 1979:43-44).

SPIRITS CONSUMPTION. After the war, a decline in the consumption of rum occurs because of the disappearance of foreign markets, the decline in supplies of molasses, increasing competition from other drinks, particularly whiskey, and

SPIRITS CONSUMPTION. After the war, a decline in the consumption of rum occurs because of the disappearance of foreign markets, the decline in supplies of molasses, increasing competition from other drinks, particularly whiskey, and the national resentment against relying on foreign imports. Congress imposes a duty on imported rum and molasses, which raises the price of rum in comparison with the price of whiskey, even though the latter includes an excise tax (Rorabaugh 1979:66-67).

REVENUE GENERATION; CONTROLS. In the Federalist Paper No. 12 (1788), Hamilton argues that an import duty on spirits could provide considerable revenue to the federal treasury. In addition to raising revenue, the duty might result in a reduction in spirits consumption, which would be "equally favorable to the agriculture, to the economy, to the morals, and to the health of the society" (*Federalist* 1961:95-96).

1790

EUROPE

<u>TEA CONSUMPTION</u>. Only a tiny part of western Europe--Holland and England--has taken to tea drinking on a large scale. France consumes a tenth of its own cargoes at most. Germany prefers coffee; tea is largely unknown in Spain (Braudel 1981:252).

DENMARK

<u>DRINKS CONSUMED</u>. Johan Molbech argues in 1791 that the beer-drinking ancestors of contemporary Danes lived longer and were healthier because they did not indulge in "the continual drinking of the fashionable hot, spiced and fiery drinks." Around this time, stronger kinds of beer such as porter that are produced industrially and thus more cheaply stimulate beer drinking over brandy and other rival drinks (Glamann 1962:131).

ENGLAND

<u>INEBRIETY PREVALENCE</u>. By the end of the century, many forces are at work which are beginning to reduce consumption levels. Although drunkenness is not advancing as some later temperance workers allege, it is still a major problem among all classes into the early 19th century (Bovill 1962:108). Clark (1983:298) concludes: "Overall, it seems likely that the traditional culture of heavy drinking was starting to fragment by the early 19th century, in the process becoming less closely linked to the respectable world of the public houses."

<u>UPPER CLASSES/DRINKING</u>. Heavy drinking has declined somewhat; however, this decline is less marked than that which occurs in France (Younger 1966:341). Many people still consider it the duty of a host to get guests drunk. Gentlemen still drink port after dinner until almost all are drunk under the table. The diaries of John Byng indicate that the drinking of a quart bottle of port is quite normal and that three or more bottles a day is not too taxing for a man in good health with plenty of exercise. In 1793, Byng says he daily drank three pints of port and a half pint of brandy (Francis 1972:243).

<u>INEBRIETY OF MYTTON</u>. Considered one of the prime examples of 18th-century upper-class dissolution (although he is not born until 1796), Squire Jack Mytton is said never to have been sober in the last 12 years of his short life (he dies in 1834). He begins drinking as a student at Rugby, where it is considered unmanly not to get drunk (Bovill 1962:109-110). A contemporary asserts that as an adult Mytton drinks the "almost unheard quantity" of four to six bottles of port every day: "He shaved with a bottle of it on his toilet; he worked steadily at it throughout the day, by a glass or two at a time, and at least a bottle with his luncheon." The rest he drinks after dinner and supper. By the last years of his short life, "the nearly constant state of intoxication in which he lived" makes him "somewhat insufferable to his oldest friends," and he is reduced to a state of "perfect apathy and imbecility" (Nimrod n.d.:83-87).

<u>INEBRIETY OF THE PRINCE OF WALES</u>. The Prince of Wales (b. 1770), the future King George IV (1820-1830), is a notorious drunk who has a fondness for getting

other people drunk (Bovill 1962:108). His drinking excesses "more than once nearly [prove] fatal to him" (Simon 1926:47). At his wedding, the intoxicated prince has to be held up by two dukes (French 1884:343). Before his death, however, he is a painful moral anachronism regarded with open contempt by nearly everyone (Brown 1961:21).

CONCERNS: POPULAR CULTURE. By the end of the century, the opposition to popular culture has "clearly gained the upper hand." Refinement, gentility, and the spirit of Addison, Steele, and their followers, have been "gradually absorbed by the high culture," which has dissociated itself from popular tastes (Malcolmson 1973:163).

LABORERS/DRINKING PRACTICES. Scottish temperance reformer John Dunlop (1839) describes in detail 300 drinking practices in 98 different trades existing in the British Isles in the late 18th and early 19th centuries. He complains that "on the whole, in Great Britain, we seem to be behind the more refined nations of modern Europe, in our progress of getting quit of these Barbarities; and there appears no parallel elsewhere to the multiplicity and complication of our drinking usages" (p. 261). These drinking customs seem to have their strongest hold on skilled craftsmen (Thompson 1963:243).

DRINKING INFLUENCES: FRENCH REVOLUTION. The outbreak of the "atheistic" French Revolution in 1789 and the wars with France that follow serve to reinforce the growing British conservatism in manners, including drinking behavior, and to make the British more receptive to moral reform. Edmund Burke ("Letters on a Regicide Peace") argues that the revolution has spawned a licentious system of manners, and Evangelicals vigorously renew their demand for a thorough reformation of morals and manners. The upper classes are urged to set a good example and abandon the luxury and riotous living of the past. The revolution is seen as an example for all Britains that morality and social order are interconnected (Quinlan 1965:69, 96-104).

Thompson (1963:57) observes: "The sensibility of the Victorian middle class was nurtured in the 1790s by frightened gentry who had seen miners, potters and cutlers reading Rights of Man and its foster-parents were William Wilberforce and Hannah More." Such a disposition among the propertied "reinforced the natural tendency of authority to regard taverns, fairs, any large congregations of people, as a nuisance--sources of idleness, brawls, sedition or contagion."

DRINKING INFLUENCES: EVANGELICALISM. By the end of the century, an "aggressive moral earnestness" has emerged (Malcolmson 1973:168), most noticeably in Evangelical circles and in commercial and industrial areas. In the late century, the Evangelical movement launched by John Wesley noticeably quickens. The Methodist spirit is infused into older Nonconformist sects. This revival is assisted by: (1) the British horror of French republicanism and atheism, which is seen as underscoring the need for moral reform in Britain; (2) fresh proselytizing in industrial areas; and (3) increasing literacy (Quinlan 1965:96, 104; Trevelyan 1978:435). Methodism instills in the new working class many of the values of middle-class utilitarianism; foremost, that to maintain grace one must employ a methodical discipline in every aspect of life and that labor itself is a sign of grace. Thompson (1963:375, 401) attributes the spread of Methodism among working people to direct indoctrination, to the sense of community the religion supplies in a hostile world, and to the psychic consolation and appeal of its violent emotionalism to the despairing. One of its effects is to help mediate the work-discipline of industrialism.

GIN CONSUMPTION, SALES. Although gin consumption has increased since duties were lowered in 1785, in 1796-1797 corn distilling is prohibited due to poor crops and at the end of the century gin consumption is still less than half the mid-century levels. Largely due to the local restrictions begun in the 1780s, the number of gin houses licensed to sell spirits drops nationwide from 377,000 in 1779 to 30,000 in 1799 (Quinlan 1965:43). "Thus the century [ends] on a note of greater respectability" (Monckton 1969:66).

This decline is generally credited to the combination of moderate duties and comprehensive licensing regulations; the rise in popularity of tea; the increasing price of grain in the second half of the century which increases the price of spirits; and the influence of Methodism.

SUNDAY DRINKING. A foreigner complains that the observance of Sundays is so strict in England that "a man can do nothing but go to a public house...and when there you can do nothing but drink." Because everything else is closed accept the dram shops, the English people have been left "to gloom and drink" (quoted Brown 1961:441-442).

FESTIVE DRINKING: WAKES. One Reverend Macaulay, discussing the people's attachment to wakes, observes that "with the lower sort of people, especially in the manufacturing villages, the return of the wake never fails to produce a week, at least, of idleness, intoxication and riot" (quoted Porter 1982:171).

ALEHOUSES: PREVALENCE, CONCERNS (COLQUHOUN). Patrick Colquhoun, a leader of the movement to kill unrestricted licensing, writes Observations and Facts Relating to Public Houses in the City of London (1794), which has been called "the most sober and best-informed pamphlet on the topic" (Mathias 1959:126-127). The report is submitted to the king by the Middlesex magistrates. Colquhoun (1794:2-3) estimates that there are around 6,000 public houses in London and its environs, "a very considerable proportion" of which are 4th and 5th class alehouses. This is far beyond the number which is either necessary for the community or which can be supported by it. Many people are ruined trying to operate alehouses which are poorly situated. Thus denial of licenses is "humanity in the end" (p. 12). He criticizes public houses because they harbor thieves and because now whole families attend them. Through them, the new generation of laboring people are being initiated into profligacy, vice, and immorality (pp. 15-16). "An ill-governed public-houses," he asserts, "is one of the greatest nuisances which can exist in civil society: for it spreads its poison far and wide" (p. 19). Because of indiscriminate licensing and the number of "low and unworthy people" keeping alehouses, it is no longer a respectable profession. He advocates that licenses be granted only to men of sober manners and respectability, that there should not be separate gin shops, that surplus alehouses beyond the legitimate needs of their localities be suppressed (based on an estimate that 60 families are needed to support a public house), and that all public houses which offer pay tables, idle amusement, and encourage whole families as patrons should be suppressed (pp. 18, 28, 32-35).

ALEHOUSE CONTROLS; SALE OF BEER ACT (1795). Magistrates in London, and to a lesser extent in the country, begin to look for opportunities to disapprove licenses that are up for renewal. By the Sale of Beer act of 1795, any person convicted a second time of selling without a license is barred thereafter from holding any license for the retailing of alcoholic beverages. This act marks "a vast step forward in the control of licensed premises and also an important step in the attempts to prevent excessive drinking and drunkenness" (King 1947:119).

BEER: PRODUCTION, CONSUMPTION. The 12 leading brewers now control three-fourths of the market (Mathias 1959:214). Technological improvements and more efficient management have increased the production capacities of the breweries. For the first time in any brewery in the world, Whitbread's produces over 200,000 barrels of porter in one season (Mathias 1959:xxiii).

Although the major contraction in beer drinking occurs after 1800, it starts to decline as prices increase in the 1790s, in large part due to high war-time taxation, a revival in spirits consumption and a continued increased in tea and coffee drinking. Sir Frederick Eden (1797/1928:106-107) shows that due to the high cost of beer, consumption has largely evaporated from among the diets of rural workers in the south, being replaced by tea (see below). He also notes that a "very considerable quantity is brewed at home" (Clark 1983:292-293).

TIED HOUSES. The campaign to restrict licenses, bad harvests, and war-time taxation cause breweries to fight fiercely for shares of the dwindling market. This further exacerbates the struggle among the brewers to control public houses and leads to increased use of advertising. The larger brewers seek to insure their outlets and sales by controlling public houses through direct ownership or financial backing; these "tied houses" sell beer only from the controlling brewery. Complaints against the tied-house system appear as early as 1773 (King 1947:116-117), but the spread of the system is encouraged by the fear of competition in the late century (Mathias 1959:128, 136-138; Park 1983:64-65). To attract a larger market, sales promotion and advertising of different brands of beer also occur for the first time. This development is linked to the spread of tied houses, which provide a venue for marketing (Mathias 1959:136-137). In 1794, the first publication of the Morning Advertiser, a newspaper especially for the licensed trade, also appears (it is Britain's second-oldest national daily newspaper) (Monckton 1966:153).

TEA CONSUMPTION. Eden (1797/1928:107) records that among poor Middlesex and Surrey families, "tea is not only the usual beverage, in the morning and evening, but is generally drank in large quantities at dinner" (see Drummond and Wilbraham 1939:243; Trevelyan 1978:340).

FRANCE

BACKGROUND DEVELOPMENTS: FRENCH REVOLUTION. The French Revolution ends the ancien régime in France and establishes a republic. On May 5, 1789, the Estates General is convened for the first time since 1614 in an effort to resolve the state's fiscal crisis. In June 14, a National Assembly is proclaimed in defiance of the king. On July 14, the Bastille is stormed and in August the Assembly abolishes all feudal privileges. Finally, in 1793 the king and queen are tried and executed for treason during the revolutionary wars with England and Austria.

TAXATION: ENTRY DUTIES ABOLISHED. The National Assembly abolishes the entry duties into Paris, including those on wine (Kaplow 1972:12; Matthews 1958:171-173). In August 1789, it abolishes "forever" all tax privileges and special privileges of provinces and cities (DiCorcia 1928:233). As part of its abolition of all feudal privileges, the Great Privilege or Wine Privilege in Bordeaux limiting wine trade to Bordeaux citizens is abolished as an unjust. One of the main charges leveled against the monasteries during the revolutionary period is the excessive amount of revenue they derive directly or indirectly from their vineyards (Forbes 1967:89).

GUINGUETTES. Discourses and petitions to the National Assembly in 1790-1791 emphasize that the guinguettes are necessary to the maintenance of the physical and moral health of the people, particularly in fulfilling the need for relaxation and pleasure on Sunday. The commune of Montmartre declares that "the poor of this city have a right to relaxations like the rich and don't find them but in the cabarets known as guinguettes." Representing the wine growers, Deputy Etienne Chevalier emphasizes that wine is the base of the subsistence of the multitude of workers from all the provinces who have come seeking jobs in the rapidly expanding city and who "unable to accustom themselves to the wine of the city, went in the suburbs to look in the guinguette for the wine of the wine grower, more salubrious and less expensive." There they also find cheaper food, and more agreeable and warmer surroundings than their homes, benefits which make it easier for them to bear the rigors of winter and their indigence (quoted Dion 1959:510).

TOASTING. After a group of patriots gathers at a friend's house, one writes to his absent host: "We ate like ogres and drank like Templars to the health of the Republic first, and then to your own, to that of your wife, your brother and Fréron; I counted up to thirty toasts." At one banquet, the same author tells how he drank "long and deep, and loudly hailed the Republic." He tells another friend that when he visits him they will have "at least a thousand and one healths to drink, beginning with that of the Republic" (quoted Robiquet 1964:185-186).

DRINKING PRACTICES. Mercier's account of the trial of the king and Assembly debates tells how women gathered in the boxes at the back of the hall and drank liqueurs; in the topmost galleries, the "common people" drank wine and brandy "as if they were in some low, smoke-filled tavern" (quoted Robiquet 1964:91).

IRELAND

INEBRIETY PREVALENCE. By about 1790, hard drinking appears to have declined among the gentry, but it reaches a peak among the lower classes, especially in Dublin. Employers complain that drunkenness injures business and causes many workers to be idle half the time. The high prevalence of drunkenness in Dublin is attributed to the low price and ready availability of spirits and the ease with which anyone can take out a spirit license.

Henry Grattan, the Irish parliamentary leader, estimates that every seventh house in Dublin is licensed to sell spirits and that in the country the number is even greater (O'Brien 1918:42). He asserts that spirits drinking has become "a great national evil" (Shipkey 1973:291).

The letters of "Agricola" (The Dram Shop, 1791) indicate that drunkenness among the lower classes has probably never been worse: "The present excessive use of whiskey in Ireland is the most dangerous plague which ever infected a nation; a malady that extends to every corner of the land, that sweeps away thousands in the prime of their days, and generates almost all the vices by which the poor are debilitated and deluded....Can it be supposed...that universal drunkenness does not prevail among the lower orders of the people? They alone are the consumers, and it would be impossible that such numbers could live by the vile traffic was not the consumption immense" (quoted O'Brien 1918:39).

Samuel Crumpe (1795/1968:172-273) asserts: "Drunkenness is an evil of considerable magnitude in the catalogue of national vices. It is one to which the lower Irish are peculiarly addicted, and that from which the most serious obstructions arise to their industry and employment. That vile beverage, whiskey, so cheaply purchased, and so generally diffused, affords them an easy opportunity of gratifying this destructive passion." Because of this, the Irish are "remarkably riotous." Crump (p. 193) also observes that "the misery and idleness occasioned by poverty and oppression united, is a principal source of the prevalent tendency to ebriety, and the consequent riotous feuds so remarkable among the Irish." He feels that when national wealth increases, drunkenness will decrease.

Newenham (1805:233) asserts that drunkenness is "very prevalent among the common people" of the large towns, but not in the countryside where laborers and farmers "are scarcely never seen intoxicated and never use spirits at their meals."

TAVERN FUNCTIONS. Although elites attack the whiskey shop, to the majority of the nation they are an accepted and valued part of daily life. McGuire (1973:159) comments: "The wealthy had their own private stocks and there is no evidence to show that they set a good example in sobriety. These attacks on taverns, therefore, were generally unpopular and it is not surprising that the shebeen continued to flourish."

PRODUCTION. Through the adoption of large-scale production techniques, brewing and distilling begin to undergo rapid expansion, and Ireland becomes an exporter rather than importer of alcoholic beverages (Shipkey 1973:291).

ATTITUDES; CONTROLS. In 1792, Grattan (see above) charges that the brewer has been sacrificed to the distiller and should be given preference in the tax structure. To this, Chancellor of the Exchequer Beresford responds: "It is of very little consequence to the morals of the people (if they will get drunk) what they get drunk with; it is however the duty of the legislature...to make the means of intoxication as difficult to come by as they possibly can: this can only be done by laying duties as high as the article will bear. For if you lay a duty amounting to a prohibition, or beyond a certain point, you defeat your own purpose--you indemnify the risque of the smuggler--the article so taxed is produced in greater plenty than ever--the people's morals are corrupted in a double sense by the practice of fraud and the practice of intoxication, and you collect no revenue at all" (quoted Lynch and Vaizey 1960:63).

Crumpe (1795/1968:50) believes that taxation is the best means to reduce high levels of inebriety: "Of such practices none are more injurious, to none are a poor and indolent people more inclined, than drunkenness; nor is there any, perhaps, not liable to immediate punishment, which can be more effectually checked by the proper exertions of legislative power. To this purpose statutes will avail but little; the plain and efficacious mode appears to be, taxing the materials of ebriety, whether directly or indirectly, so high, as to render the gratification of the desire extremely difficult to the lower and laborious class."

ECONOMIC VALUE: DISTILLING. A local rector comments on the importance of distilling to the peasant economy: "Even if distilling were made a felony punishable by death the poor of my parish would not desist since they had no other means of paying their rents and might well hang as starve" (quoted McGuffin 1978:15).

SPIRIT SALES; DISTILLING CONTROLS. Attempts to restrict drinking are frustrated by the landowners' desire to keep up the price of corn and by the government's need for the revenue generated by the spirit duty. Furthermore, any reduction in supply or increase in price of legal spirits is quickly made up by the illicit distiller (O'Brien 1918:38-43).

WHISKEY LICENSING ACTS (1791/1792). The licensing act of 1791 increases the retail license duty and restricts spirit retailing because, according to the preamble, "the use of spirituous liquors prevails to an immoderate excess, to the great injury of the health, industry and morals of the people; and...the laws now in being for regulating licenses...have been found insufficient." Attempts to regulate the spirit trade are resented by the common people, who view the whiskey shop as a part of daily life and not as a blight on the nation (McGuire 1973:159-160). Grocers are allowed to apply for licenses for retail spirits sales in quantities not less than one pint; this is designed to spare women having to enter public houses (Magee 1980:76).

In yet another effort to limit spirit drinking, Parliament in 1792 grants retail spirit licenses only to those innkeepers, tavern keepers, and victuallers who also sell beer and ale at reasonable prices. The sale of spirits before 1 p.m. on Sunday and before sunrise on other days is prohibited. This unpopular and feebly enforced law proves ineffective (McGuire 1973:160, 161).

WAGE AND DEBT CONTROLS. Parliament also makes it illegal to give liquor in lieu of wages and orders that wages cannot be paid out in a public house. A debt over 20 shillings incurred for the sale of spirits on credit is no longer recoverable by law (Magee 1980:76).

BREWING PROMOTED. To encourage beer over spirit drinking, Parliament in 1795 removes all duties from beer (although the malt used to make beer is still taxed). Domestic beer brewing begins to expand rapidly after 1795, due to "a conjunction of many causes" (Lynch and Vaizey 1960:3). Between 1790 and 1804, beer imports decline tremendously (Newenham 1805:233). Underlying this expansion of domestic brewing are the following: (1) the lowering of beer duties in 1795, which makes Irish beer cheaper; (2) disruption of British brewing due to restrictions, shortages, and increased transportation costs caused by the French wars; (3) domestic economic inflation after 1797, which increases demand and enables Irish brewers to expand the scale of their operations; and (4) the beginning of porter brewing. The 1795 act is probably the main reason for the emergence of brewing as one of Ireland's leading industries. It has been called "the most decisive and important single event in the whole history of Irish brewing" (Lynch and Vaizey 1960:68).

WOMEN/DRINKING. The licensing reforms of 1791 allowing grocers to retail spirits is said to foster drinking among women by increasing the ease with which drink was available to them (Magee 1980:76).

NETHERLANDS

GIN CONSUMPTION. By now gin is traded on the Amsterdam exchange and is even the object of a gambling craze. As gin drinking rises, it becomes the subject of heavy taxes (Forbes 1970:192).

DISTILLING. The more than 50% increase in population in Schiedam in the second half of the century is "probably because of the progress of the local gin industry" (Houtte 1977:231).

RUSSIA

DRINK HOUSES. After many local changes and adjustments by the end of the 18th century, Catherine the Great establishes a uniform system of alcohol concessions throughout the empire (Efron 1955:501).

SCOTLAND

LICENSING CONTROLS. Throughout the later 1780s and 1790s, licensing fees and duties are raised. This causes great damage to the Lowland distillers and to the Highland distillers who export to England. The reason for the increase in the fee is to keep up with the speed with which the distillers can work off a still (Great Britain 1834:42; Wilson 1973:53; Daiches 1978:32, 40).

DISTILLING: EFFECTS; ECONOMIC VALUE. In the old *Statistical Account of Scotland* (1793), the Reverend John Downie attacks the prevalence of whisky drinking in taverns: "The worst effect of the great plenty of spirits is, that dram shops are set up in almost every village for retail, where young and idle people convene and get drunk. The tippling huts are kept by such only as are not able to pay a fine, or procure a license. They are the greatest nuisance in the parish." A year later, the Reverend John Smith charges that the whisky industry is "extremely ruinous to the community; it consumes their means, hurts their morals, and destroys their undertakings and their health." Yet he adds: "In this place, however, very few, comparatively speaking, are given to drunkenness, as people are seldom given to excess in what is their daily fare." The danger caused by the trade is the destruction of grain and the waste of many lives (quoted Daiches 1978:35, 36-37).

A different view of distilling is taken by Alexander Simpson in 1794, who applauds the benefits that the economy and the people derive from whisky: "The small licensed whisky stills in the neighbourhood" provide "a good market for barley" and supply "us with good whisky, of a quality greatly superior to what we have from the large stills in the southern districts, as well as cheaper, and no less wholesome than foreign spirits." He concludes that these stills, "in every point of view, are a reciprocal advantage to the farmers, and the country at large" (quoted Daiches 1978:36).

The Reverend David Dunoon discusses the economic support for distilling: "I will be asked, Why then so many distilleries? For these reasons: Distilling is almost the only method of converting our victual into cash for the payment of rent and servants; and whisky may, in fact, be called our staple commodity" (quoted Daiches 1978:35). The spread of illicit distillation at this time is indicated by the statement of John Stein, a prominent Highland distiller, in testimony before a Committee on Distilleries: "The distillery is in a thousand hands. It is not confined to great towns or to regular manufacturers, but spreads itself over the whole face of the country....There are many who practice this art who are ignorant of every other" (quoted Daiches 1978:33). Barnard (1968:6) observes that distilling was promoted by "want of a ready market for the disposal of grain grown in many remote districts."

LICENSING CONTROLS (1799). Following a two-year study of the distilling situation in Scotland, in 1799 Parliament amends the Wash Act of 1786 by adding to the still-license fee a duty of 1s1 1/4d (Lowlands) and 1s4 1/4d (Highlands) on each gallon of spirits actually produced. The new duty greatly increases the cost of operating a still since each gallon of still content is capable of making many hundreds of gallons of spirits each year. The number of gallons on which duty is actually paid drops to less than half of the previous year's total and revenue collections suffer (Great Britain 1834:42).

SWEDEN

DRINKING PRACTICES, FUNCTIONS (CLARKE). During a trip through the Scandinavian countries in 1799, E.D. Clarke observes the varying drinking customs of the Swedes. He reports that meals are begun with a small appetizer and a dram of brandy, "as a whet for the appetite." Swedes assert that alcohol protects them from the illnesses from which travelers to Sweden frequently suffer (e.g., sore throats, fevers, rheumatism). "If you ask the inhabitants, whose diet consists principally of salted provisions, how they escape these disorders, they will answer, 'That they preserve their health by drinking brandy, morning and evening.' That even the most temperate adhere to this practice of dram-drinking is strictly true" (Clarke 1799/1838, 1:167-168).

AMERICA, UNITED STATES

SPIRITS CONSUMPTION. In the 1790s, whiskey begins to replace rum as the favorite drink of Americans (Rorabaugh 1979:68). Influenced by the writings of Benjamin Rush and his supporters, an increasing number of individuals, primarily from the upper-classes, cease drinking spirits; but these individuals constitute a small percentage of the general population (Rorabaugh 1979:45-47).

WHISKEY TAXATION. Alexander Hamilton proposes a financial program which includes an excise tax on domestic spirits. Subsequently, Congress approves this tax in the Revenue Act of 1791. Revenues accruing from the tax are used to pay the state debts that the federal government assumed in accordance with Hamilton's financial program. In response to the tax, widespread protests take place in the South and West, and in 1792 the law is modified by exempting "personal stills" from taxation (Miller 1960:155-156; Rorabaugh 1979:52-53).

EFFECTS: WHISKEY REBELLION. Opposition to the modified whiskey law surfaces in 1792 and is rampant in the West. Enforcement of the tax is difficult since it cannot be collected in many parts of the country, and where it can be, the costs are disproportionately high. Further, the tax leads to an increase in illegal distilling. Open rebellion takes place until 1794, when western farmers publically disrupt collection of the whiskey tax. Hamilton himself joins the army which is raised to suppress the Whiskey Rebellion. The Federal Army encounters virtually no resistance and captures few prisoners.

The tax fails because the vast majority of the population refuses to allow the upper classes to impose their standards of morality and behavior on the common man. The tax is repealed in 1802 (Miller 1960:157-159; Rorabaugh 1979:55-56).

CONSUMPTION LEVEL. By the end of the century (1800), the typical American annually drinks only about one-tenth of a gallon of wine, but annual per capita consumption of (absolute) alcohol from all sources is 3.5 gallons, almost all of which is in distilled spirits. This compares to approximately 2.7 gallons of (absolute) alcohol in 1710 (Rorabaugh 1979:8-10).

References

Every effort has been made to provide original composition or publication dates for the editions of primary sources. These dates are given within parentheses immediately following the title, followed by publication information of the edition actually consulted. In cases where the original date of composition is uncertain, or where publication occurred long after composition, the approximate date of composition is preceeded by "c."

REFERENCES

Abel, W. Stufen der Ernährung (Stages of Diet). Göttingen: Vandenhoeck and Ruprecht, 1981.

Abram, A. English Life and Manners in the Later Middle Ages. London: Routledge and Sons, 1913.

Adams, J. Diary of John Adams. (1755-1804). In: The Diary and Autobiography of John Adams. Edited by L.H. Butterfield. Vol. 1. Cambridge: Belknap Press of the Harvard University Press, 1961.

Africa, T. The Ancient World. Boston: Houghton Mifflin, 1969.

Allbut, T.C. The Historical Relations of Medicine and Surgery to the Sixteenth Century. London: Macmillan, 1905.

Allen, H. The History of Wine. New York: Horizon, 1961.

Ammianus Marcellinus. History. Translated by J. Rolfe. 3 vols. Loeb Classical Library, 1935-1939.

Anastasi, P. Alexander's alcoholism disputed in Greece. New York Times, October 14, 1980, p. C2.

Anderson, F. An Illustrated History of Herbals. New York: Columbia University Press, 1977.

Anderson, M.S. Europe in the Eighteenth Century, 1713-1783. New York: Holt, Rinehart and Winston, 1961.

Ariès, P. Centuries of Childhood. Translated by R. Baldick. New York: Knopf, 1962.

Arnald of Villanova. Arnald of Villanova's Book of Wine. (1310.) Translated by H. Sigerist. New York: Schuman's, 1943.

Arnold, J.P. Origin and History of Beer and Brewing. Chicago: Wahl-Henius Institute of Fermentology, 1911.

Arnot, H. History of Edinburgh. London: J. Murray, 1779.

Asbury, H. The Great Illusion: An Informal History of Prohibition. Garden City, New York: Doubleday, 1950.

Ashton, T.S. An Economic History of England: The Eighteenth Century. London: Methuen and Co., 1955.

Athenaeus. The Deipnosophists. Translated by C. Gulick. 7 vols. Loeb Classical Library, 1927-1941.

Audiger. La maison réglée (The Regulated Household). (1692.) 3rd ed. Paris, 1700.

Ausonius. Translated by H.G. Evelyn-White. 2 vols. Loeb Classical Library, 1919.

REFERENCES

Austin, G. Perspectives on the History of Psychoactive Substance Use. NIDA Research Issues Series, 24. Washington, D.C.: GPO, 1978.

Aymard, M. Toward the history of nutrition: Some methodological remarks. In: Forster, R., and Orest, R., eds. Food and Drink in History. Baltimore: Johns Hopkins Press, 1979. pp. 1-16.

Baader, J. Zur Geschichte des Branntweins (On the history of brandy). Anzeiger für Kunde der deutschen Vorzeit, 15:315-318, 1868.

Bahlman, D.W.R. The Moral Revolution of 1688. New Haven: Yale University Press, 1957.

Bailey, P. Leisure and Class in Victorian England. Toronto: University of Toronto Press, 1976.

Bainton, R. The churches and alcohol. Quarterly Journal of Studies on Alcohol, 6:45-58, 1945.

Bainton, R. The Reformation of the Sixteenth Century. Boston: Beacon Press, 1952.

Baird, E. The alcohol problem and the law. I. Ancient laws and customs. Quarterly Journal of Studies on Alcohol, 4:535-556, 1944a.

Baird, E. The alcohol problem and the law. II. The common law bases of modern liquor controls. Quarterly Journal of Studies on Alcohol, 5:126-161, 1944b.

Baird, E. The alcohol problem and the law. III. The beginnings of the alcoholic-beverage control laws in America. Quarterly Journal of Studies on Alcohol, 6:335; 7:110-162, 272-296, 1946.

Baker, W. The observance of Sunday. In: Lennard, R., ed. Englishmen at Rest and Play, 1558-1714. Oxford; Clarendon Press, 1931. pp. 79-144.

Bakhtin, M. Rabelais and his World. Translated by H. Iswolsky. Cambridge, Massachusetts: MIT Press, 1968.

Bales, R. Cultural differences in rates of alcoholism. Quarterly Journal of Studies on Alcohol, 6:480-499, 1946.

Balzer, R. Adventures in Wine: Legends, History, Recipes. USA: Ward Ritchie Press, 1969.

Barnard, A. Whisky Distilleries of the United Kingdom. London, 1887. Reprint London: Newton Abbot, 1969.

Barty-King, H. A Tradition of English Wine. Oxford: Oxford University Press, 1977.

Basserman-Jordan, F. von. Geschichte des Weinbaus (The History of Viniculture). Frankfurt: Anstalt, 1932.

REFERENCES

Bax, E.B. The Peasants War in Germany, 1525-1526. New York: Russell and Russell, 1968.

Beauroy, J. Vin et société a Bergerac (Wine and Society in Bergerac). Saratoga, Calif.: Anma Libri, 1976.

Becker, W.A. Charicles or Illustrations of the Private Life of the Ancient Greeks. Translated by F. Metcalfe. New York: Longmans, Green and Co., 1906.

Beech, G. A Rural Society in Medieval France: The Gâtine of Poitou in the Eleventh and Twelfth Centuries. Baltimore: Johns Hopkins Press, 1964.

Bell, D. L'ideal éthique de la royauté en France au moyen age d'après quelques moralistes de ces temps (The Ethic Ideal of Royalty in Medieval France according to Several Contemporary Moralists). Paris: Librairie Droz, 1962.

Bennett, H.S. Life on the England Manor: A Study of Peasant Conditions, 1150-1400. Cambridge: Cambridge University Press, 1969.

Beowulf. Translated by K. Crossley-Holland. London: Macmillan, 1968.

Bercé, Y.-M. Histoire des croquants (History of the Croquants). Vol. 1. Paris: Libraire Droz, 1974.

Bercé, Y.M. Fête et révolte. Paris: Hachette, 1976

Billiard, R. La vigne dans l'antiquité (The Vine in Antiquity). Lyons: Lardanchet, 1913.

Bizière, J.M. The Baltic wine trade 1563-1567. Scandinavian Economic History Review, 20:121-132, 1972.

Blanke, F. La reformation contre l'alcoolisme (The reformation against alcoholism). Société de l'histoire du protestantisme Française, 99:171-185, 1953.

Bloch, M. Feudal Society. Translated by L.A. Manyon. 2 vols. Chicago: University of Chicago Press, Phoenix Books, 1964.

Bloch, M. French Rural History. Translated by J. Sondheimer. Berkeley and Los Angeles: University of California Press, 1966.

Blum, R.H., and Associates. Society and Drugs. San Francisco: Jossey-Bass, 1969.

Blunt, W. Sebastiano. London: James Barrie, 1956.

Boccaccio, G. The Decameron. (1351.) Translated by R. Aldington. Garden City, New York: Garden City Publishing, 1949.

Boetius, H. Description of Scotland. (1527.) Translated by W. Harrison. In: Holinshed, R. Chronicles of England, Scotland, and Ireland. (2nd ed., 1587.) Vol. 5: Scotland. London, 1808. pp. 1-28.

REFERENCES

Boilleau, E. _Le livre des métiers_ (The Book of Trades). (1268.) Paris: Imprimerie Nationale, 1879.

Boissonnade, P. _Life and Work in Medieval Europe_. Translated by E. Power. New York: Knopf, 1927.

Bolte, J. Trinkerorden. _Jahrbüch des Vereins für Niederdeutsche Sprach Forschung_, 17:167-168, 1891.

Bonnie, R., and Sonnenreich, M. _Legal Aspects of Drug Dependence_. Cleveland, Ohio: CRC Press, 1975.

Born, L. The _Specula principis_ of the Carolingian Renaissance. _Revue Belge de philologie et d'histoire_, 12:583-612, 1933.

Borsay, P. The English Urban Renaissance: The development of provincial urban culture c. 1680-c. 1760. _Social History_, 5:581-603, 1977.

Boswell, J. _Life of [Samuel] Johnson_. (London, 1791.) New York: Oxford University Press, 1933.

Boswell, J. _Boswell's Journal of a Tour to the Hebrides with Samuel Johnson_. (London, 1785.) New York: Literary Guild, 1936.

Bovill, E.W. _English Country Life 1780-1830_. London: Oxford University Press, 1962.

Bradford, W. _History of Plymouth Plantation (1606-1646)_. (c. 1650.) Edited by W.T. Davis. New York: Charles Scribner's Sons, 1908.

Brant, S. _The Ship of Fools_. (1494.) Translated by E. Zeydel. New York: Columbia University Press, 1944.

Braudel, F. _Capitalism and Material Life 1400-1800_. Translated by M. Kochan. London: Weidenfeld and Nicolson, 1973.

Braudel, F. _Civilization and Capitalism, 15th-18th Century_. Vol. 1: _The Structures of Everyday Life_. Translated by M. Kochan; translation revised by S. Reynolds. London: W. Collins and Sons, 1981.

Braudel, F. _Civilization and Capitalism, 15th-18th Century_. Vol. 2: _The Wheels of Commerce_. New York: Harper and Row, 1982.

Braudel, F., and Spooner, F. Prices in Europe from 1450 to 1750. In: Rich, E., and Wilson, C., eds. _Cambridge Economic History of Europe_. Vol. 4: _The Economy of Expanding Europe in the 16th and 17th Centuries_. Cambridge: Cambridge University Press, 1967. pp. 378-486.

Braunschweig, H. _Kleine Destilierbüch_. (1500.) Translated as: _The Booke of Distyllacyon of Herbes_. London, 1527. Reprint New York: De Capo, 1973.

Brennan, T. "Cabarets and Laboring Class Communities in Eighteenth Century Paris." Ph.D. dissertation, Johns Hopkins University, 1981.

REFERENCES

Brennan, T. "Social drinking in Old Regime Paris." Paper presented at the Conference on the Social History of Alcohol, sponsored by the Alcohol Research Group, Berkeley, California, January 1984.

Brereton, W. Travels in Holland, the United Provinces, England, Scotland and Ireland, 1634-1635. (1635.) Edited by E. Hawkins. Vol. 1. London: Chetham Society, 1844.

Bretherton, R. Country inns and alehouses. In: Lennard, R., ed. Englishmen at Rest and Play. Oxford: Clarendon Press, 1931. pp. 147-201.

Brewer, J. Commercialization and politics. In: McKendrick, N.; Brewer, J.; and Plumb, J. The Birth of a Consumer Society: The Commercialization of Eighteenth-Century England. London: Europa Publications, 1982. pp. 197-262.

Bridenbaugh, C. Cities in the Wilderness: The First Century of Urban Life in America 1625-1742. 2nd ed. New York: Knopf, 1955.

Brown, F. Fathers of the Victorians: The Age of Wilberforce. Cambridge: Cambridge University Press, 1961.

Brown, W. Inebriety and its "cures" among the ancients. Quarterly Journal of Inebriety, 20:125-141, 1898.

Bruford, W.H. Culture and Society in Classical Weimar, 1775-1806. Cambridge: Cambridge University Press, 1962.

Bruford, W.H. Germany in the Eighteenth Century: The Social Background of the Literary Revival. Cambridge: Cambridge University Press, 1968.

Brunner, T. Marijuana in ancient Greece and Rome: The literary evidence. Bulletin of the History of Medicine, 47:344-355, 1973.

Bubenik, V. The world's first beer brewers. New Orient, 5:163-166, 1966.

Buhac, I. Hieronymus Bosch und Alkoholismus. Deutsche medizinische Wockenschrift, 96:1775-1776, 1971.

Burckhardt, J. The Civilization of the Renaissance. Translated by S.G.C. Middlemore. Vol. 2. New York: Harper Torchbooks, 1958.

Burke, P. Popular Culture in Early Modern Europe. New York: Harper and Row, 1978.

Burns, R. The Complete Poetical Works of Burns. Cambridge, Mass.: Houghton Mifflin Co., 1897

Burton, E. The Georgians at Home. London: Longmans, Green and Co., 1967.

Burton, R. Anatomy of Melancholy. (1st ed., 1621.) Philadelphia: J.W. Moore, 1854.

Butler, F. H. Wines and the Wine Lands of the World. London: T. Fisher, 1926.

REFERENCES

Butler, K. History of French Literature. Vol. 1. New York: E.P. Dutton, 1923.

Bynum, W. Chronic alcholism in the first half of the 19th century. Bulletin of the History of Medicine, 42:160-185, 1968.

Byrd, W. The Secret Diary of William Byrd of Westover, 1709-1712. (1712.) Edited by L.B. Wright and M. Tipling. Richmond, Va.: Dietz Press, 1941.

Caesar, G.J. On the Gallic War. Translated by H. Edwards. Loeb Classical Library, 1917.

Cahen, L. Merchands de vin et debitants de boissons à Paris au milieu du XVIIIe siècle (Wine merchants and drink outlets in Paris in the mid-18th century). Bulletin de la Société d'histoire moderne, March 6, 1910. pp. 109-111.

Calvin, J. Institutes of the Christian Religion. (1st ed., 1559). In: A Calvin Treasury. Edited by W. Keesecker. New York: Harper and Brothers, 1961.

Campbell, R. The London Tradesman. London, 1747. Reprint London: David and Clarke, 1969.

Camerarius, P. The Living Librarie. (c. 1599.) Translated by J. Malle. London, 1621.

Cantor, N. Medieval History. 2nd ed. London: Macmillan, 1969.

Carcopino, J. Daily Life in Ancient Rome. Translated by E.O. Lorimer. New Haven: Yale University Press, 1940.

Carlowitz, G.H. Von der Versuch einer Culturgeschichte des Weinbaues von der Vorzeit bis auf unsere Zeiten (The Attempt in Cultural History of Viniculture from Pre-historic Times to our Day). Leipzig: Engelmann, 1846.

Carter, E. Wine and Poetry. Davis: University of California Library, 1976.

Carus-Wilson, E.M. Medieval Merchant Venturers. London: Methuen and Co., 1967.

Cassedy, J. An early American hangover: The medical profession and intemperance. Bulletin of the History of Medicine, 50(3):405-413, 1976.

Castiglione, B. Il Cortegiano. (1528.) Translated as: Il Cortegiano, or the Courtier (A New English Version). London: Bowyer, 1724.

Catlin, G. Liquor Control. London: Butterworth, 1931.

Cato, M. (the Elder). On Agriculture. Translated by W. Hooper and H. Ash. Loeb Classical Library, 1936.

Celsus. On Medicine. Translated by W. G. Spencer. 3 vols. Loeb Classical Library, 1935-1938.

Chaucer, G. The Canterbury Tales. (c. 1400.) Baltimore: Penguin Books, 1963.

REFERENCES

Charles, R.H. The Apocrypha and Pseudoepigrapha of the Old Testament. 2 vols. Oxford, 1913.

Cherrington, E. Evolution of Prohibition in the United States of America. Westerville, Ohio: American Issues Press, 1920.

Cherrington, E., ed. Standard Encyclopedia of the Alcohol Problem. 6 vols. Westerville, Ohio: American Issues Publishing Co., 1924-1930.

Cheyne. G. An Essay of Health and Long Life. (1724.) 4th ed. London: G. Straham, 1725.

Cicero. On the Republic. Translated by C. Keyes. Loeb Classical Library, 1928.

Clapham, J.H., and Power, E., eds. The Cambridge Economic History of Europe. Vol. 1: The Agrarian Life of the Middle Ages. Cambridge: Cambridge University Press, 1944.

Clark, E.F. The "Grobianus" of Sachs and its predecessors. Journal of English and Germanic Philology, 16:390-396, 1917.

Clark, P. The alehouse and the alternative society. In: Pennington, D., and Thomas, K., eds. Puritans and Revolutionaries: Essays in Seventeenth-Century History Presented to Christopher Hill. Oxford: Oxford University Press, 1978. pp. 47-72.

Clark, P. The English Alehouse: A Social History, 1200-1830. New York: Longman, 1983.

Clark, P. "The gin mania and official controls in early Hanoverian England." Paper presented at the Conference on the Social History of Alcohol, sponsored by the Alcohol Research Group, Berkeley, California, January 1984.

Clarke, E.D. Travels in Various Countries, Including Denmark, Sweden, Norway, Lapland and Finland. (c. 1799.) 3 vols. London: T. Cadell and W. Davies, 1838.

Clarkson, J.D. A History of Russia. 2nd ed. New York: Random House, 1969.

Claudian, J. History of the usage of alcohol. In: Tremoiliers, J., ed. International Encyclopedia of Pharmacology and Therapeutics. Section 20, vol. 1. Oxford: Pergamon, 1970. pp. 3-26.

Coffey, T.G. Beer Street: Gin Lane: Some views of 18th century drinking (with 5 illustrations from William Hogarth). Quarterly Journal of Studies on Alcohol, 27:669-692, 1966.

Cohen, H. The Drunkenness of Noah. University: University of Alabama Press, 1974.

Cole, C. Colbert and the Century of French Mercantilism. Vol. 1. New York: Columbia University Press, 1939.

REFERENCES

Cole, G.D.H., and Postgate, R. The Common People 1746-1946. London: Methuen, 1961.

Coleman, D. Labour in the English economy of the seventeenth century. Economic History Review, Section 2, 8:280-296, 1956.

Collins, S. The Present State of Russia. London: John Winter, 1671.

Colquhoun, P. Observations and Facts relating to Public Houses in the City of London and Environs. London: The Middlesex Magistrates, 1794.

Columella, L. On Agriculture. Vol. 3. Translated by H.B. Ash. Loeb Classical Library, 1941.

Connell, K.H. Illicit distillation: An Irish peasant industry. Historical Studies, Papers read before the 4th Irish Conference of Historians, 3:58-91, 1962.

Conroy, D.W. "Puritans in taverns: Law and popular culture in colonial Massachusetts, 1630-1720." Paper presented at the Conference on the Social History of Alcohol, sponsored by the Alcohol Research Group, Berkeley, California, January 1984.

Contenau, G. Everyday Life in Babylon and Assyria. Translated by K.R. and A.R. Maxwell-Hyslop. New York: St. Martin's, 1954.

Corfield, P. The Impact of English Towns, 1700-1800. Oxford: Oxford University Press, 1982.

Cornwall, E. Notes on the use of alcohol in ancient times. Medical Times, 67:379-380, 1939.

Corti, C. History of Smoking. London: Harrap and Co., 1931.

Coulton, G.G. Medieval Panorama: The English Scene from Conquest to Reformation. New York: Meridian Books, 1957.

Coulton, G.G. The Medieval Village. Cambridge: Cambridge University Press, 1925.

Coupe, W.A., ed. A Sixteenth-Century German Reader. Oxford: Clarendon Press, 1972.

Crafts, W., et al. Intoxicating Drinks and Drugs, in All Lands and Times. Washington, D.C.: International Reform Bureau, 1900.

Crafts, W., et al. Intoxicants and Opium in All Lands and Times. Washington, D.C.: International Reform Bureau, 1905.

Craeybeckx, J. Un grand commerce d'importation: Les vins de France aux anciens Pays-Bas (XIIIe-XVIe siècle) (A Great Import Trade: The Wine of France to the Old Low Countries [13th-16th centuries]). Paris: SEVPEN, 1958.

REFERENCES

Crombe, A.C. Medieval and Early Modern Science. Vol. 1. Garden City, New York: Doubleday and Co., 1959.

Crothers, T.D. Inebriety in ancient Egypt and Chaldea. Cincinnati Lancet-Clinic, 50:354-358, 1903.

Crothers, T.D. A review of the history and literature of inebriety. Journal of Inebriety, 33:139-151, 1911.

Crothers, T.D. The alcohol problem as seen in ancient and modern times. American Journal of Clinical Medicine, 25:43-46, 1918.

Crumpe, S. An Essay on the Best Means of Providing Employment for the People. 2nd ed. London, 1795. (1st ed., 1793.) Reprint New York: Augustus Kelley, 1968.

Curtis, T.C., and Speck, W. The Societies for the Reformation of Manners. Literature and History, 3:45-64, 1976.

Daiches, D. Scotch Whisky--Its Past and Present. Revised ed. London: Andre Deutsh, 1978.

Darby, W.J.; Ghalioungui, P.; and Grivetti, L. Food: The Gift of Osiris. Vol. 2. London: Academic Press, 1977.

Davenant, C. An Essay upon Ways and Means of Supplying War. London: Jacob Tonson, 1695.

Davis, N.Z. Society and Culture in Early Modern France. Stanford: Stanford University Press, 1975.

Davis, R. The Rise of the Atlantic Economies. Ithaca, New York: Cornell University Press, 1973.

Defoe, D. Complete English Tradesman. (1727.) In: The Novels and Miscellaneous Works of Daniel Defoe. Vol. 18. Oxford: D.A. Talboys, 1841.

Defoe, D. Augusta Triumphans; or, the Way to make London the most Flourishing City in the Universe. (1728.) In: The Novels and Miscellaneous Works of Daniel Defoe. Vol. 18. Oxford: D.A. Talboys, 1841.

Defoe, D. Second Thoughts are Best. (1729.) In: The Novels and Miscellaneous Works of Daniel Defoe. Vol. 18. Oxford: D.A. Talboys, 1841.

Delamain, R. Histoire du cognac (History of Cognac). Paris: Librairie Stock, 1935.

della Casa, J. Galateo. (1558.) Translated as: A Treatise of the Manners and Behaviours. London, 1576. Reprint New York: Da Capo, 1969.

de Madariaga, I. Russia in the Age of Catherine the Great. London: Weidenfeld and Nicolson, 1981.

REFERENCES

Dickens, A. G. Reformation and Society in Sixteenth-Century Europe. New York: Harcourt, Brace and World, 1966.

DiCorcia, J. Bourg, bourgeois, bourgeoisie de Paris from the eleventh to the eighteenth centuries. Journal of Modern History, 50:215-233, 1978.

Diderot, M., and d'Alembert, M. Encyclopédie, ou dictionnaire raisonné des sciences des arts et des métiers, par une Société de gens de lettres. (1751-1772.) 3rd ed. Geneva: Pellet, 1778-1779.

Diethelm, O. Historical notes on the recognition of alcoholism as a medical-social problem. In: Selected Papers, International Congress on Alcohol and Alcoholism, Frankfurt-am-Main, September 6-12, 1964. Vol. 1. Ontario: Addiction Research Foundation, 1965. pp. 53-54.

Dill, S. Roman Society in the Last Century of the Western Empire. 2nd ed. London: Macmillan, 1899. Reprint New York: Meridian Books, 1958.

Dill, S. Roman Society in Gaul in the Merovingian Age. London: Macmillan, 1926.

Dimont, M. Jews, God and History. New York: New American Library, 1962.

Dinitz, S. "The Relation of the Tavern to the Drinking Phases of Alcoholics." Ph.D. dissertation, University of Wisconsin, 1951.

Dio Cassius. Dio's Roman History. Translated by E. Cary. 9 vols. Loeb Classical Library, 1924-1927.

Diodorus Siculus. Library of History. 12 vols. Translated by C. Oldfather (1-6), C. Sherman (7), B. Wille (8), R. Geer (9-10), and F. Walton (11-12). Loeb Classical Library, 1933-1957.

Dion, R. Histoire de la vigne et du vin en France des origines au XIXe siècle (History of the Vine and Wine in France from their Origins through the 19th century). Paris, 1959.

Diogenes Laertius. Lives of the Eminent Philosophers. 2 vols. Translated by R. Hicks. Loeb Classical Library, 1925.

Dioscorides. The Greek Herbal of Dioscorides. Translated by J. Goodyear in 1655. Edited by R. Gunther. Oxford: Oxford University Press, 1934.

Dollinger, P. The German Hansa. Translated by D.S. Ault and S.H. Steinberg. London: Macmillan, 1970.

Dorchester, D. The Liquor Problem in All Ages. New York: Philips and Hunt, 1884.

Dorwart, R.A. The Prussian Welfare State Before 1740. Cambridge: Harvard University Press, 1971.

Dowell, S. A History of Taxation and Taxes in England from the Earliest Times to the Year 1885. 2nd ed. Vol. 4. London: Longmans, Green and Co., 1888.

REFERENCES

Doxat, J. The World of Drinks and Drinking. New York: Drake Publishers, 1971.

Dozer, J.B. "The Tavern: Discovery of the Secular Locus of Medieval France." Ph.D. dissertation, University of California, Los Angeles, 1980.

Drummond, J.C., and Wilbraham, A. The Englishman's Food. London: Jonathan Cape, 1939.

Duby, G. The Early Growth of the European Economy: Warriors and Peasants from the 7th-12th Century. Translated by B. Clarke. Ithaca: Cornell University Press, 1974.

Duby, G. Rural Economy and Country Life in the Medieval West. Translated by C. Postan. Columbia: University of South Carolina Press, 1968.

Duby, G., and Mandrou, R. A History of French Civilization. Translated by J. Atkinson. New York: Random House, 1964.

Duclos, C. Secret memoirs of the Regency. (c. 1727.) Translated by E.J. Meras. New York: Sturgis and Walton Co., 1910.

Dunlop, J. The Philosophy of Artificial and Compulsory Drinking Usage in Great Britain and Ireland. London: Houlston and Stoneman, Paternoster Row, 1839.

Dunn, R.S. The Age of Religious Wars, 1559-1689. New York: W.W. Norton, 1970.

Earle, P. The World of Defoe. London: Weidenfield and Nicholson, 1976.

Eberlein, H.D., and Richardson, A.E. The English Inn Past and Present. Philadelphia: J.B. Lippincott Co., 1926.

Eden, Frederick. The State of the Poor. (1797.) Abridged and edited by A.G.L. Rogers. London, George Routledge and Sons, 1928.

Edwards, J. Freedom of the will. (1754.) In: Edward, J. Basic Writings. New York: New American Library, 1966. pp. 196-223.

Efron, V. The tavern and saloon in Russia. Quarterly Journal of Studies on Alcohol, 16:484-503, 1955.

Einhard. The Life of Charlemagne. (c. 830.) In: Einhard and Notker the Stammerer. Two Lives of Charlemagne. Translated by L. Thorpe. London: Penguin Books, 1969. pp. 49-90.

Elias, N. The Civilizing Process. Translated by E. Jephcott. New York: Urizen Books, 1978.

Elliot, G. Life and Letters. (c. 1806.) Edited by the countess of Minto. Vol. 1. London: Longmans, Green, and Co., 1874.

Ellis, A. The Penny Universities: A History of the Coffee-Houses. London: Secker and Warburg, 1956.

Elyot, T. The Castel of Helth. Revised. London, 1541.

Emboden, W. Dionysus as a shaman and wine as a magician drug. Journal of Psychedelic Drugs, 9:187-192, 1977.

Emerson, E.R. Beverages Past and Present. 2 vols. New York: G.P. Putnam's Sons, 1908.

Emmison, F. Elizabethan Life and Disorder. Chelmsford: Essex County Council, 1970.

Erasmus, D. Colloquy on Inns. (1523.) In: Ten Colloquies. Translated by C.R. Thompson. New York: Library of Liberal Arts, 1957. pp. 14-21.

Erasmus, D. In Praise of Folly. (1509.) In: The Essential Erasmus. Edited by J. Dolan. New York: New American Library, 1964.

Erlanger, P. Les idees et les moeurs au temps des rois (Ideas and Manners at the time of the Kings). Paris: Flammarion, 1969.

Erman, A. Life in Ancient Egypt. Translated by H.M. Tirard. London: Macmillan, 1894.

Euripides. The Bacchanals. Translated by A. Way. Loeb Classical Library, 1912.

Everitt, A. The English urban inn, 1560-1760. In: Everitt, A., ed. Perspectives in English Urban History. London: Macmillan, 1973. pp. 91-137.

Farge, A. Le vol des aliments à Paris au XVIIIe siècle (The Theft of Food in 18th-century Paris). Paris: Plon, 1974.

Farge, A. Vivre dans la rue à Paris au XVIIIe siècle (Street Life in 18th-century Paris). Paris: Ed. Fallimard/Julliard, 1979.

Farrington, J. Memoirs of the Life of Sir Joshua Reynolds. London: Cadell and Davis, 1819.

Federalist Papers. Edited by C. Rossiter. New York: Mentor Book, New American Library, 1961.

Feldman, W. Alcohol in ancient Jewish literature. British Journal of Inebriety, 24:121-124, 1926.

Ferguson, S. Drink. London: B.T. Batsford, 1975.

Ferrero, G. Characters and Events of Roman History. New York: G.P. Putnam's Sons, 1909.

Fichtenau, H. The Carolingian Empire. Translated by P. Munz. New York: Harper and Row, 1964.

Fielding, H. An Enquiry into the Causes of the Late Increase of Robbers, etc. (London, 1751.) In: Brown, J.P., ed. Works of Henry Fielding. Vol. 10. London: Bickers and Son; H. Sotheran and Co., 1871. pp. 341-483.

REFERENCES

Firebaugh, W.C. The Inns of the Middle Ages. Chicago: Pascal Covici, 1924.

Firebaugh, W.C. The Inns of Greece and Rome. Chicago, 1928. Reprint New York: Benjamin Blom, 1972.

Flandrin, J. La diversité des gouts et des pratiques alimentaires en Europe des XVIe au XVIIIe siècle (The diversity of tastes and alimentary practices in Europe from the 16th to the 18th centuries). Revue d'histoire moderne et contemporaine, 30:70-77, 1983.

Fletcher, G. Of the Russe Commonwealth. (1591.) Edited by R. Pipes and J.V.A. Fine, Jr. Cambridge: Harvard University Press, 1966.

Foote, T., et al. The World of Bruegel. New York: Time-Life Books, 1972.

Forbes, D. Considerations upon the State of the Nation, as it is Affected by the Excessive Use of Foreign Spirits. Edinburgh, 1730.

Forbes, D. Some Considerations of the Present State of Scotland. Edinburgh, 1744.

Forbes, P. Champagne: The Wine, the Land and the People. London: Victor Gollanz, 1967.

Forbes, R.J. Short History of the Art of Distillation. Leiden: E.J. Brill, 1970.

Forster, E., and Forster, R., eds. European Diet from Pre-Industrial Modern Times. New York: Harper and Row, 1976.

Forster, R. The Nobility of Toulouse in the Eighteenth Century. Baltimore: Johns Hopkins Press, 1960.

Forster, R. The noble wine producers of the Bordelais in the 18th century. Economic History Review, 14:18-33, August 1961.

Forster, R., and Forster, E., eds. European Society in the Eighteenth Century. New York: Harper and Row, 1969.

Fosca, F. Histoire des cafés de Paris (History of the Parisian Cafés). Paris: Firmin-Didot and Co., 1934.

Fowler, W.W. Festivals of the Period of the Republic. London: Macmillan, 1916.

Fox, J. The Poetry of Villon. Westport, Conn.: Greenwood Press, 1976.

Francis, A. The Wine Trade. London: Black, [1972].

Franklin, A. La vie privée d'autrefois (Private Life of Yesterday). Vol. 6: Les repas; vol. 9: Le Café, le thé et le chocolat; and vol. 23: Variétés Parisiennes. Paris: Librairie Plon, 1889, 1893, 1901.

REFERENCES

Franklin, B. *Autobiography*. (1779.) Edited by L.W. Labarree, et al. New Haven: Yale University Press, 1964.

Frazer, J.P. *The New Golden Bough*. Abridged by T.H. Gates. New York: Criterion Books, 1959.

French, R.K. *History and Virtues of Cyder*. New York: St. Martin Press, 1982.

French, R.V. *Nineteen Centuries of Drink in England: A History*. London: Longmans, Green and Co., 1884.

French, W. *Medieval Civilization as Illustrated by the Fachnachtspiele of Hans Sachs*. Baltimore: Johns Hopkins Press, 1925.

Frijhoff, W., and Julia, D. The diet in boarding schools at the end of the Ancien Regime. In: Forster, R., and Ranum, O., eds. *Food and Drink in History*. Baltimore: Johns Hopkins Press, 1979. pp. 73-85.

Garfinkel, B. La grande chartreuse. *High Times*, no. 38:77-79, October 1978.

Garrison, F.H. *An Introduction to the History of Medicine*. Philadelphia: W.B. Saunders Co., 1922.

Gascoigne, G. A delicate Diet for daintimouthde Droonkards. (London, 1576.) In: *The Complete Works of Gascoigne*. Vol. 2. Edited by J.W. Cunliffe. 1907-1910. Reprint New York: Greenwood Press, 1969. pp. 451-471.

Gaxotte, P. *The Age of Louis XIV*. Translated by M. Shaw. New York: Macmillan, 1970.

Gayre, G.R. *Wassail! In Mazers of Mead*. London: Phillimore, 1948.

George, M.D. *London Life in the Eighteenth Century*. London: Kegan, Paul, Trench, Trubner and Co., 1925.

Gervais, J. *Mémoire sur l'Angoumois*. (1731.) Edited by G. Babinet de Rencogne. Paris, 1864.

Ginzberg, L. A response to the question whether unfermented wine may be used in Jewish ceremonies. In: Schneiderman, H., ed. *American Jewish Year Book*. Vol. 25 [year 5684]. Philadelphia: Jewish Publication Society of America, 1923. pp. 401-425.

Glamann, K. Beer and brewing in pre-industrial Denmark. *Scandinavian Economic Historical Review*, 10:128-140, 1962.

Glamann, K. The changing patterns of trade. In: Rich, E., and Wilson, C., eds. *Cambridge Economic History*. Vol. 5: *The Economic Reorganization of Early Modern Europe*. Cambridge: Cambridge University Press, 1977. pp. 185-289.

Glatt, M. The English drink problem: Rise and decline through the ages. *British Journal of Addiction*, 55:51-67, 1958.

REFERENCES

Goadby, E. The Gothenburg Licensing System. London: Chapman and Hall, 1895.

Goerke, H. Linnaeus. Translated by D. Lindby. New York: Charles Scribner's Sons, 1973.

Goodenough, E.R. Jewish Symbols in the Greco-Roman Period. Vol. 6: Fish, Bread and Wine. New York: Pantheon, 1956.

Goodman of Paris: A Treatise on Moral and Domestic Economy by a Citizen of Paris. (c. 1393.) Edited by E. Power. New York: Harcourt, Brace and Co., 1928.

Gordon, L., ed. The New Crusade. Cleveland: Crusaders, 1932.

Gordon, P. Passages from the Diary of General Patrick Gordon of Auchleuchries, 1635-1690. (1690.) London: Frank Cass and Co., 1968.

Gottschalk, A. Histoire de l'alimentation et de la gastronomie depuis la prehistoire jusqu'à nos jours (History of Alimentation and Gastronomy from Antiquity to the Present). Paris: Ed. Hippocrate, 1948.

Goubert, P. The French peasantry of the seventeenth century: A regional example. In: Aston, T., ed. Crisis in Europe, 1560-1660. Garden City, New York: Doubleday, Anchor, 1967. pp. 150-176.

Goubert, P. Louis XIV and Twenty Million Frenchmen. Translated by A. Carter. London: Allen Lane/Penguin, 1970.

Great Britain. Parliament. Commission of Inquiry into the Excise Establishment. Seventh Report. Parliamentary Papers 1834 (7), XXV, pp. 1-458.

Greenfield, K.R. Sumptuary Law in Nurnberg. Baltimore: Johns Hopkins University Press, 1918.

Gregory of Tours. The History of the Franks. (c. 575.) Translated by L. Thorpe. London: Penguin Books, 1974.

Griffiths, P. The History of the Indian Tea Industry. London: Weidenfeld and Nicoholson, 1967.

Grimmelshausen, H. von. The Adventures of a Simpleton (Simplicius Simplicissimus). (1669.) Translated by W. Wallich. New York: Frederick Ungar Publishing Co., 1963.

Gulick, C.B. Modern Traits of Old Greek Life. New York: Longmans, Green and Co., 1927.

Gunn, N.M. Whisky in Scotland. London: G. Routledge and Sons, 1935.

Gusfield, J., ed. Protest, Reform, and Revolt. New York: Wiley, 1970.

Guyot, J. Etude des vignobles de France (Study of the Vineyards of France). 2nd ed. Paris: Georges Maison, 1876.

REFERENCES

Hackwood, F.W. Inns, Ales, and Drinking Customs of Old England. London: T. Fisher Unwin, 1909.

Haley, K. The Dutch in the Seventeenth Century. London: Thames and Hudson, 1972.

Hamilton, A. Gentleman's Progress, the Itinerarium. (1744.) Edited by C. Bridenbaugh. Chapel Hill: University of North Carolina Press, 1948.

Hanford, J.H. The medieval debate between wine and water. Publications of the Modern Language Association of America, 28:315-367, 1913.

Hanford, J.H., ed. Wine, Beer, and Ale, and Tobacco. By Gallobeligicus. Studies in Philology, 12:5-54, 1915.

Harrison, B. Drink and sobriety in England, 1815-1872: A critical bibliography. International Review of Social History, 12:204-276, 1967.

Harrison, B. Drink and the Victorians: The Temperance Question in England, 1815-1872. Pittsburgh: University of Pittsburgh Press, 1971.

Harrison, W. The Description of England. (2nd ed., 1587.) Edited by G. Edeler. Ithaca, New York: Cornell University Press, 1968.

Haskins, C.H. The Renaissance of the Twelfth Century. New York: World Publishing, 1957.

Hastings, C. The Theatre: Its Development in France and England. London: Duckworth, 1902.

Hatton, R. Europe in the Age of Louis XIV. London: Thomas and Hudson, 1969.

Hauffen, A. Die Trinkliterature in Deutschland bis zum Ausang des 16. Jahrhunderts (The German drink literature to the end of the 16th century). Vierteljabreschrift für Literaturegeschichte, 2:481-516, 1889.

Hearnshaw, F.J.C. A sidelight on the liquor traffic 300 years ago. British Journal of Inebriety, 4:93-98, 1906.

Heckscher, E. An Economic History of Sweden. Cambridge, Mass.: Harvard University Press, 1954.

Heers, J. Fêtes, jeux et joutes dans les sociétés d'occident à la fin du moyen âge (Festivals, Games and Matches in Western Society at the end of the Middle Ages). Paris: J. Vrin, 1971.

Heers, J. Les métiers et les fêtes "médiévalés" en France du Nord et en Angleterre (The professions and medieval festivals in northern France and England). Revue du Nord, 55:193-206, 1973.

Henisch, B. Fast and Feast: Food in Medieval Society. University Park, Penn.: Pennsylvania State University Press, 1976.

REFERENCES

Herodotus. The Persian Wars. Translated by G. Rawlinson. New York: Random House, 1942.

Herlihy, D., ed. Medieval Culture and Society. New York: Harper Torchbooks, 1968.

Hesiod. Works and Days. In: The Homeric Hymns and Homerica. Translated by H.G. Evelyn-White. Loeb Classical Library, 1954.

Hill, C. The Century of Revolution, 1603-1714. New York: W.W. Norton, 1961.

Hill, C. Society and Puritanism in Pre-Revolutionary England. London: Secker and Warburg, 1964.

Hill, C. Change and Continuity in Seventeenth Century England. London: Weidenfeld and Nicolson, 1974.

Hirsh, J. Enlightened eighteenth century views of the alcohol problem. Journal of the History of Medicine and Allied Science, 4(2):230-236, 1949.

Hirsh, J. Historical perspectives on the problem of alcoholism. Bulletin of the New York Academy of Medicine, 2nd ser., 29:961-971, 1953.

Hobsbawm, E.J. The Age of Revolution. New York: New American Library, 1962.

Holborn, H. A History of Modern Germany. Vol. 1: The Reformation; Vol. 2: 1648-1840. New York: Knopf, 1959, 1964.

Holinshed, R. Chronicles of England, Scotland, and Ireland. (2nd ed., 1587.) 6 vols. London, 1808.

Holdsworth, W.S. A History of English Law. Vol. 8. Boston: Little, Brown, 1926.

Holmes, U.T. Daily Life in the 12th Century, Based on the Observations of Alexander Nickam. Madison: University of Wisconsin Press, 1952.

Homans, G. English Villagers of the Thirteenth Century. New York: Russell and Russell, 1941.

Hooker, R. The History of Food and Drink in America. Indianapolis, Ind.: Bobbs-Merrill, 1981.

Hopkins, T. An Idler in Old France. New York: Charles Scribner's Sons, 1899.

Horace. Works. Translated by C. Smart. London: George Bell, 1888.

Horn, B.D., and Ransome, M. English Historical Documents. Vol. 10: 1714-1783. London: Eyre, 1957.

Horsch, J. The Hutterian Brethren, 1528-1931. Goshen, Ind.: Mennonite Historical Society, 1931.

REFERENCES

Houtte, J.A. An Economic History of the Low Countries. London: Weidenfeld and Nicolson, 1977.

Howell, J. Familiar Letters on Important Subjects. (c. 1635.) 10th ed. Aberdeen: Douglass and Murray, 1753.

Howland, R.W., and Howland, J.W. 200 years of drinking in the United States: Evolution of the disease concept. In: Ewing, J.A., and Rouse, B.A., eds. Drinking Alcohol in American Society: Issues and Current Research. Chicago: Nelson Hall, 1978. pp. 38-60.

Hudson, H.G. A Study of Social Regulations in England under James I and Charles I: Drink and Tobacco. Chicago: Private edition, distributed by the University of Chicago Libraries, 1933.

Hufton, O. The Poor in Eighteenth-Century France. Oxford: Clarendon Press, 1974.

Hugon, C. Social France in the XVII Century. London: Methuen and Co., 1911.

Huizinga, J. The Waning of the Middle Ages. Translated by F. Hopman. Garden City, New York: Doubleday and Co., 1954.

Huizinga, J.A. Dutch Civilization in the Seventeenth Century and other Essays. Translated by A.J. Pomerans. New York: Frederick Ungar Publishing Co., 1968.

Humphreys, A.R. The Augustan World: Life and Letters in Eighteenth Century England. London: Methuen and Co., 1954.

Husson, A. Les consommations de Paris (Consumption in Paris). 2nd ed. Paris: Libraire Hachette, 1875.

Hyams, E. Dionysus: A Social History of the Wine Vine. New York: Macmillan, 1965.

Iles, C. Early stages of English public house regulation. Economic Journal, 12:251-262, 1903.

Inglis, B. Forbidden Game. New York: Charles Scribner's Sons, 1975.

Jacob, H. Coffee. Translated by E. and C. Paul. New York: Viking, 1935.

James I, king of England. The Essayes of a Prentise, in the Divine Art of Poesie (1585); A Counterblaste to Tobacco (1604). Edited by Edward Arber. London, 1869.

James, M.K. Studies in the Medieval Wine Trade. Oxford: Clarendon Press, 1971.

Janssen, J. History of the German People. Translated by M.A. Mitchell and A.M. Christie. 15 vols. London: Kegan, Paul, French, Trubner and Co., 1905-1910.

Jastrow, M. Wine in the Pentateuchal codes. Journal of the American Oriental Society, 33:180-192, 1913.

REFERENCES

Jeanmaire, H. Dionysos: Histoire du culte de Bacchus (Dionysus: History of the Cult of Bacchus). Paris: Payot, 1970.

Jeanroy, A., and Langfors, A. Chansons satiriques et bacchiques du XIIIe siècle (Satiric and Bacchic Songs of the 13th Century). Paris: H. Champion, 1921.

Jeanselme, E. Le vin, la vigne et l'alcoolisme dans les Gaules a l'époque de l'établissement des barbares (V-Xe siècle). Bulletin de la Société Française de l'histoire de la médicine, 14:264-291, 1920.

Jeffery, C. The development of crime in early English society. Journal of Criminal Law, Criminology, and Political Science, 7:647-666, 1957.

Jellinek, E.M. A document of the Reformation period on inebriety: Sebastian Franck's "On the Horrible Vice of Drunkenness," etc. Quarterly Journal of Studies on Alcohol, 2:391-395, 1941.

Jellinek, E.M. Old Russian church views on inebriety. Quarterly Journal of Studies on Alcohol, 3:663-667, 1943.

Jellinek, E.M. The observations of the Elizabethan writer Thomas Nash on drunkenness. Quarterly Journal of Studies on Alcohol, 4:462-469, 1944.

Jellinek, E.M. A specimen of the sixteenth-century German drink literature--Obsopoeus' Art of Drinking. Quarterly Journal of Studies on Alcohol, 5:647-661, 1945.

Jellinek, E.M. The ocean cruise of the Viennese. A German poem of the thirteenth century. Quarterly Journal of Studies on Alcohol, 6:540-548, 1946a.

Jellinek, E.M. Montaigne's essay on drunkenness. Quarterly Journal of Studies on Alcohol, 7:297-304, 1946b.

Jellinek, E.M. Drinkers and alcoholics in ancient Rome. Journal of Studies on Alcohol, 37:1718-1741, 1976a.

Jellinek, E.M. Jellinek Working Papers on Drinking Patterns and Alcohol Problems. Edited by R. Popham. Toronto: Addiction Research Foundation, 1976b.

John of Salisbury. Policraticus. (1159.) Partially translated by J.B. Pike as: Frivolities of Courtiers and Footprints of Philosophers. New York: Octagon Books, 1972.

Johnson, W.E. The Liquor Problem in Russia. Westerville, Ohio: American Issues Press, 1915.

Jones, C.W., ed. Medieval Literature in Translation. New York: David McKay Co., 1950.

Jones, G. Honor in German Literature. Chapel Hill: University of North Carolina Press, 1959.

Jones, L.C. The Clubs of the Georgian Rakes. New York: Columbia University Press, 1942.

REFERENCES

Judges, A.V., ed. The Elizabethan Underworld. New York: E.P. Dutton and Co., 1930.

Juniper, W. The True Drunkard's Delight. London: Unicorn Press, 1933.

Juvenal. Satires. In: Juvenal and Perseus. Translated by G. Ramsay. Loeb Classical Library, 1918.

Kalevala, or Poems of the Kaleva District. Compiled by E. Lönnrot. Translated by F.P. Magoun, Jr. Cambridge: Harvard University Press, 1963.

Kamen, H. The Iron Century. Social Change in Europe, 1550-1660. New York: Praeger, 1972.

Kany, C. Life and Manners in Madrid, 1750-1800. Berkeley: University of California Press, 1932.

Kaplow, J. The Names of Kings. New York: Basic Books, 1972.

Kellenbenz, H. The Organization of Industrial Production. In: Rich, E., and Wilson, C., eds. The Cambridge Economic History of Europe. Vol. 5: The Economic Organization of Early Modern Europe. Cambridge: Cambridge University Press, 1977. pp. 462-548.

Keller, M. Beer and wine in ancient medicine. Quarterly Journal of Studies on Alcohol, 19:153-154, 1958.

Keller, M. Alcohol in health and disease: Some historical perspectives. Annals of the New York Academy of Sciences, 133:820-827, 1966.

Keller, M. The great Jewish drink mystery. British Journal of Addiction, 64:287-296, 1970.

Keller, M. Problems with alcohol: An historical perspective. In: Filstead, W., et al., eds. Alcohol and Alcohol Problems. Cambridge, Mass.: Ballinger, 1976. pp. 5-28.

Kelso, J. Drunkenness. New Standard Bible Dictionary. New York, 1926.

Kerling, N.J.M. Commercial Relations of Holland and Zeeland with England from the Late 13th Century to the Close of the Middle Ages. Leiden: E.J. Brill, 1954.

King, F.A. Beer Has a History. London: Hutchinson's, 1947.

King, G. Natural and Political Observations and Conclusions upon the State and Condition of England. (1696.) In: Two Tracts. Edited by G.E. Barnett. Baltimore: Johns Hopkins Press, 1936.

Kinross, Lord. Prohibition in Britain. History Today, 9:493-499, 1959.

Klemm, G. Die Getraunke. In: Allgemeine Kulturwissenschaft. Leipzig, 1855. pp. 313-350.

REFERENCES

Knowles, D. The Monastic Order in England. Cambridge: Cambridge University Press, 1941.

Koenigsberger, H.G., and Mosse, G. Europe in the Sixteenth Century. New York: Holt, Rinehart and Winston, 1968.

Korb, J.G. Diary of an Austrian Secretary of Legation at the Court of Tsar Peter the Great. (1700.) Translated and edited by the Count MacDonnell. London: Bradbury and Evans, 1863. Reprint New York: DaCapo, 1968.

Krieger, L. Kings and Philosophers, 1689-1789. New York: W.W. Norton, 1970.

Krout, J. Origins of Prohibition. New York: Knopf, 1925.

Krücke, C. Deutsche Massigkeitsbestrebungen und -vereine im Reformationszeitalter (German attempts in self-containment and self-containment edicts during the Reformation). Archiv für Kulturgeschichte, 7:13-30, 1909.

Kula, W. An Economic Theory of the Feudal System: Towards a Model of the Polish Economy, 1500-1800. London: NLB, 1976.

Labrousse, C.-E. Esquisse du mouvement des prix et des revenus en France au XVIIIe siècle (Outline of the Movement of Price and Revenues in 18th-Century France). 2 vols. Paris: Dalloz, 1933.

Labrousse, C.-E. La crise de l'économie française a la fin de l'ancien régime et au debut de la revolution (The French Economic Crisis at the End of the Old Regime and the Beginning of the Revolution). Paris: Presses Universitaires de France, 1944.

La Bruyère, J. de. The Characters. (1688.) Translated by H. Laun. London: G. Routledge and Sons, 1929.

Lachiver, M. La fraude et les fraudeurs à propos du commerce du vin dans la partie occidentale de l'Ile-de-France, au XVIIIe siècle (Fraude in the wine trade in the western part of the Ile-de-France in the 18th century). Revue d'histoire moderne et contemporaine, 21:418-444, 1974.

Lafue, P. La vie quotidienne des cours allemandes au XVIIIe siècle. Paris: Hachette, 1963.

Lambert, W. Drink and work-discipline in industrial south Wales, c. 1800-1870. Welsh History Review, 7:298-306, 1975.

Lancet, 168:240, 1905. The definition of whisky in olden times.

Landon, E. A Manual of Councils of the Holy Catholic Church. 2 vols. Edinburgh: John Grant, 1909.

Langland, W. Piers the Ploughman. (c. 1375.) Translated into modern English by J. Goodridge. Revised edition. Baltimore: Penguin Books, 1966.

Latham, R., ed. The Diary of Samuel Pepys. Vol. 10: Companion Book. Berkeley and Los Angeles: University of California Press, 1983.

REFERENCES

Laufer, B. Sino-Iranca. Chicago: Field Museum, 1919.

Laufer, B. Introduction of Tobacco into Europe. Chicago: Field Museum, 1924a.

Laufer, B. Tobacco and Its Use in Asia. Chicago: Field Museum, 1924b.

Lausanne, E. The Great Book of Wine. Great Britain: Patrick Stephens, 1970.

Leake, C. The Old Egyptian Medical Papyri. Lawrence: University of Kansas Press, 1952.

Leake, C., and Silverman, M. Alcoholic Beverages in Clinical Medicine. Chicago: Year Book Medical Publishers, 1966.

Lecky, W.E.H. A History of England in the Eighteenth Century. Vol. 1. London: Longmans, Green and Co., 1878.

Leclant, J. Coffee and cafés in Paris, 1664-1693. In: Forster, R., and Ranum, O., eds. Food and Drink in History. Baltimore: Johns Hopkins Press, 1979. pp. 86-97.

Lefebvre, G. The movement of prices and the origins of the French Revolution. In: Kaplow, J., ed. New Perspective on the French Revolution. New York: John Wiley and Sons, 1965. pp. 103-135.

Legnaro, A. Ansätze zu einer Soziologie des Rausches: Zur Sozialgeschichte von Rausch und Ekstase in Europa (Attempts concerning a sociology of drunkenness: On the social history of "being high" and "ecstasy" in Europe). In: Volger, G., ed. Rausch und Realität: Drogen im Kulturvergleich. Vol. 1. Cologne: Rautenstrauch-Joest-Museums, 1981a. pp. 52-63.

Legnaro, A. Alkoholkonsum und Verhaltenskontrolle: Bedeutungswandlungen zwischen Mittelalter und Neuzeit in Europa (Alcohol consumption and the control of behavior: Changes between medieval and modern Europe). In: Volger, G., ed. Rausch und Realität: Drogen im Kulturvergleich. Vol. 1. Cologne: Rautenstrauch-Joest-Museums, 1981b. pp. 86-97.

Le Goff, J. The town as an agent of civilisation, 1200-1500. In: Fontana Economic History of Europe. Vol. 1: The Middle Ages. New York: Barnes and Noble, 1976. pp. 71-107.

Legouis, E. The Bacchic element in Shakespeare's plays. Proceedings of the British Academy, pp. 1-20, 1926.

Le Grand d'Aussy, M. Histoire de la vie privée des Français (History of the Private Life of the French). Vols. 2 and 3. Paris, 1782.

Leibowitz, J. Studies in the history of alcoholism: II. Acute alcoholism in ancient Greek and Roman medicine. British Journal of Addiction, 62:83-86, 1967.

Lender, M. Drunkenness as an offense in early New England: A study of "Puritan" attitudes. Quarterly Journal of Studies on Alcohol, 34:353-366, 1973.

REFERENCES

Lennard, R., ed. _Englishmen at Rest and Play, 1558-1714_. Oxford: Clarendon Press, 1931.

Lennard, R. _Rural England 1086--1135_. Oxford: Clarendon Press, 1959.

Le Roy Ladurie, E. _The Peasants of Languedoc_. Translated by J. Day. Urbana: University of Illinois Press, 1974.

Le Roy Ladurie, E. _Carnival in Romans_. Translated by M. Feeney. New York: George Braziller, 1980.

Lesko, L.H. _King Tut's Wine Cellar_. Berkeley: B.C. Scribe Publications, 1977.

Levine, H. The discovery of addiction: Changing conceptions of habitual drunkenness in America. _Journal of Studies on Alcohol_, 39:143-174, 1978.

Levine, H. The vocabulary of drunkenness. _Journal of Studies on Alcohol_, 42:1038-1051, 1981.

Levron, J. _Daily Life at Versailles in the 17th and 18th Centuries_. Translated by C. Engel. New York: Macmillan, 1968.

Lewin, L. _Phantastica; Narcotic and Stimulating Drugs: Their Use and Abuse_. New York: E.P. Dutton and Co., 1964.

Lewis, W.H. _The Splendid Century: Life in the France of Louis XIV_. Garden City, New York: Doubleday, Anchor, 1957.

Lichine, A. [in collaboration with W. Massee]. _Wines of France_. New York: Alfred A. Knopf, 1951.

Lichine, A., and Fifield, W. _Lichine's Encyclopedia of Wine and Spirits_. 3rd ed., revised. London: Cassell, 1974.

Lillywhite, B. _London Coffee Houses: A Reference Book of Coffee Houses of the Seventeenth, Eighteenth, and Nineteenth Centuries_. London: Allen and Unwin, 1963.

Lister, M. _A Journey to Paris in the Year 1698_. (3rd ed. London, 1699.) Selected passages in: Ranum, V., and Ranum, P., eds. _The Century of Louis XIV_. New York: Walker and Co., 1972. pp. 214-237.

Livy, T. _The History of Rome_. Translated by W. M'Devitte. London: Bohn, 1950.

Lockhart, R.B. _Scotch: The Whisky of Scotland in Fact and Story_. London: Putnam, 1951.

Lockhart, R.B. _Scotch: The Whisky of Scotland_. Glasgow: Blackie, 1975.

Löffler, K. Die ältesten Bierbücher (The oldest beerbooks). _Archiv für Kulturgeschichte_, 7:5-12, 1909.

Longmate, N. _The Waterdrinkers: A History of Temperance_. London: Hamish and Hamilton, 1968.

REFERENCES

Loubère, L. The Red and the White: A History of Wine in France and Italy in the Nineteenth Century. Albany, New York: SUNY Press, 1978.

Lucas-Dubreton, J. Daily Life in Florence. London: George Allen and Unwin, 1960.

Luchaire, A. Social France at the Time of Philip Augustus. Translated by E.B. Krehbiel. New York: Henry Holt and Co., 1912.

Lucia, S. A History of Wine as Therapy. Philadelphia: Lippincott, 1963a.

Lucia, S. The antiquity of alcohol in diet and medicine. In: Lucia, S., ed. Alcohol and Civilization. New York: McGraw-Hill, 1963b. pp. 151-166.

Lucian of Samosata. The Works of Lucian of Samosata. Translated by F.G. and W. Fowler. 4 vols. Oxford: Clarendon Press, 1905.

Lucretius. Nature of Things. Translated by W. Leonard. New York: J. Dent, 1921.

Luther, M. "Sermon on Soberness and Moderation." (1539.) In: Luther's Works. Vol. 51: Sermons. Edited by J. Doberstein and H. Lehman. Philadelphia: Muhlenberg Press, 1959. pp. 291-299.

Luther, M. An Open Letter to the Christian Nobility. (1520.) Translated by C. Jacobs. In: Three Treatises. Philadelphia: Fortress Press, 1960. pp. 9-111.

Luther, M. "Drunkenness a Common Vice of Germans" (Table Talk No. 4917, May 16, 1540); "Drunkenness at the Saxon Court and Elsewhere" (Table Talk No. 3468, October 27, 1536). In: Luther's Works. Vol. 54: Table Talk. Edited by T. Tappert and H. Lehman. Philadelphia: Muhlenberg Press, 1967. pp. 205-206, 371-372.

Lutz, H. Viticulture and Brewing in the Ancient Orient. New York: J.C. Heinrichs, 1922.

Lyons, J.D. The Listening Voice: An Essay on the Rhetoric of Saint-Amant. Lexington, Kentucky: French Forum Publishers, 1982.

Lynch, P., and Vaizey, J. Guinness's Brewery. Cambridge: Cambridge University Press, 1960.

McCarthy, R., and Douglass, E. Alcohol and Social Responsibility. New York: Thomas Crowell, 1949.

McCormick, M. First representations of the gamma alcoholic in the English novel. Quarterly Journal of Studies on Alcohol, 30:957-980, 1969.

MacDonald, I. Smuggling in the Highlands. Inverness: Wm. Mackay and Sons, 1914.

McGuffin, J. In Praise of Poteen. Belfast: Appletree Press, 1978.

McGuire, E.B. Irish Whiskey: A History of Distilling, the Spirit Trade, and Excise Controls in Ireland. New York: Barnes and Noble, 1973.

McKendrick, H. Introduction: Commercialization and the economy. In: McKendrick, H.; Brewer, J.; and Plumb, J.H. The Birth of a Consumer Society: The Commercialization of Eighteenth-Century England. London: Europa Publications, 1982. pp. 1-96.

McKendrick, H.; Brewer, J.; and Plumb, J.H. The Birth of a Consumer Society: The Commercialization of Eighteenth-Century England. London: Europa Publications, 1982.

Mackenzie, O. A Hundred Years in the Highlands. London: Edward Arnold, 1924.

McKinlay, A. The Roman attitude toward women's drinking. Classical Bulletin, 22:14-15, 1945.

McKinlay, A. The wine element in Horace. Classical Journal, 42:161-168; 229-236, 1946/47.

McKinlay, A. Temperate Romans. Classical Weekly, 41:146-149, 1948a.

McKinlay, A. Ancient experience with intoxicating drinks: Non-classical peoples. Quarterly Journal of Studies on Alcohol, 9:388-414, 1948b.

McKinlay, A. Early Roman sobriety. Classical Bulletin, 24:52, 1948c.

McKinlay, A. Ancient experience with intoxicating drinks: Non-Attic Greek states. Quarterly Journal of Studies on Alcohol, 10:289-315, 1949a.

McKinlay, A. Roman sobriety in the later Republic. Classical Bulletin, 25:27-28, 1949b.

McKinlay, A. Bacchus as health-giver. Quarterly Journal of Studies on Alcohol, 11:230-246, 1950a.

McKinlay, A. Roman sobriety in the early Empire. Classical Bulletin, 26:31-36, 1950b.

McKinlay, A. Roman sobriety in the Comedies. Classical Outlook, 27:56-57, 1950c.

McKinlay, A. Attic temperance. Quarterly Journal of Studies on Alcohol, 12:61-102, 1951.

McKinlay, A. New light on the question of Homeric temperance. Quarterly Journal of Studies on Alcohol, 14:78-93, 1953.

MacLysaght, E. Irish Life in the Seventeenth Century. 2nd ed. Cork, Ireland: University of Cork Press, 1950.

McNeil, F. The Scots Cellar: Its Traditions and Lore. Edinburgh: Richard Paterson, 1956.

REFERENCES

McNeil, W. History Handbook of Western Civilization. Chicago: University of Chicago Press, 1953.

Madden, J.S. Samuel Johnson's alcohol problem. Medical History, 11:141-149, 1967.

Magee, M. 1000 Years of Irish Whiskey. Dublin: O'Brien Press, 1980.

Magne, E. Les plaisirs et les fêtes en France au XVIIe siècle (Pleasures and festivals in France during the 17th century). Geneva: Editions de la Frégate, 1944.

Mäkelä, K. Differential effects of restricting the supply of alcohol: Studies of a strike in Finnish liquor stores. Journal of Drug Issues, 10:131-144, 1980.

Mandeville, Bernard de. The Table of the Beers: or, Private Vices, Publick Benefits. (1st ed., 1714.) 6th ed. London: J. Tonson, 1729.

Marshall, D. The English Poor in the Eighteenth Century: A Study in Social and Administrative History. London: George Routledge and Sons, 1926.

Marshall, D. Eighteenth Century England. London: Longmans, 1962.

Martial. Epigrams. Translated by W. Ker. 2 vols. Loeb Classical Library, 1920, 1925.

Maspero, G. Life in Ancient Egypt and Assyria. New York: Frederick Unger Publishing, 1971. (Based on the 1892 English translation published by Chapman and Hall.)

Massie, R.K. Peter the Great. New York: Alfred A. Knopf, 1980.

Mathias, P. The brewing industry, temperance and politics. Historical Journal, 1:97-114, 1958.

Mathias, P. The Brewing Industry in England, 1700-1830. Cambridge: Cambridge University Press, 1959.

Mathias, P. The Transformation of England: Essays in the Economic and Social History of England in the Eighteenth Century. London: Methuen and Co., Ltd., 1979.

Matthews, G. The Royal General Farms in Eighteenth-century France. New York: Columbia University Press, 1958.

Maxwell, C. The English Traveler in France (1698-1815). London: George Routledge and Sons, 1932.

Mead, W.E. The English Medieval Feast. London: Allen and Unwin, 1931.

Medick, H. Plebian culture in the transition to capitalism. In: Samuel, R., and Steadman Jones, G., eds. Culture, Ideology and Politics: Essays in Honor of Eric Hobsbawm. London: Routledge and Kegan Paul, 1983. pp. 84-112.

REFERENCES

Mendelsohn, O. Drinking with Pepys. London: Macmillan, 1963.

Ménétra, J.L. Journal de ma vie (Journal of My Life). (c. 1780.) Edited by D. Roche. Paris: Montalba, 1982.

Mercier, L.-S. Tableau de Paris. (1782.) Edited by G. Desnoiresterres. Paris: Pagnerre/Lecou, 1853.

Mercier, L.-S. The Picture of Paris Before and After the Revolution. (Partial translation of Tableau de Paris, 1782.) London: George Routledge and Sons, 1929.

Mercier, L.-S. The Waiting City: Paris, 1782-88. (Partial translation of Tableau de Paris, 1782.) Translated and edited by H. Simpson. London: George Harrap and Co., 1933.

Miller, J.C. The Federalist Era, 1789-1801. New York: Harper and Row, Harper Torchbooks, 1960.

Minchinton, W. Patterns and structure of demand, 1500-1750. In: Fontana Economic History of Europe. Vol. 2: The Sixteenth and Seventeenth Centuries. New York: Barnes and Noble, 1976. pp. 83-176.

Miskimin, H. The Economy of Early Renaissance Europe, 1300-1460. Englewood Cliffs, N.J.: Prentice-Hall, 1969.

Modi, J. Wine among the Ancient Persians. Bombay: Bombay Gazette Steam Press, 1888.

Monckton, H.A. A History of English Ale and Beer. London: Bodley Head, 1966.

Monckton, H.A. English ale and beer in Shakespeare's time. History Today, 17:828-834, 1967.

Monckton, H.A. A History of the English Public House. London: Bodley Head, 1969.

Mone, E. Trinkgeschirre im 16. Jahrhundert und deren Verwendung ("Tools" for drinking and their use). Anzeiger für kunde der deutschen Vorzeit, 7:178-183, 1838.

Money, J. Taverns, coffee houses and clubs: Local policies and popular articulacy in the Birmingham area in the age of the American Revolution. Historical Journal, 14:15-47, 1971.

Montaigne, M. Of Drunkenness. (1573-1574.) In: The Complete Works of Montaigne. Translated by D. Frame. Stanford: Stanford University Press, 1958. pp. 244-251.

Montaigne, M. Of the Education of Children. (1579-1580.) In: The Complete Works of Montaigne. Translated by D. Frame. Stanford: Stanford University Press, 1958. pp. 106-135.

REFERENCES

Montaigne, M. Travel Journal. (1580-1581.) In: The Complete Works of Montaigne. Translated by D. Frame. Stanford: Stanford University Press, 1958. pp. 861-1039.

Montesquieu, baron de. De l'esprit des lois. (1748.) In: Oeuvres complètes. Vol. 2. Edited by R. Caillois. Paris: Librarie Gallimard, 1951.

Moore, B. Social Origins of Dictatorship and Democracy. Boston: Beacon Press, 1967.

Moore, J. View of Society and Manners in France, Switzerland, and Germany. London, Straham and Cadell, 1779.

Morewood, S. Philosophical and Statistical History of the Inventions and Customs of Ancient and Modern Nations in the Manufacture and Use of Inebriating Liquors. Dublin: Curry and Carson, 1838.

Moryson, F. An Itinerary. (1617.) 4 vols. Glasgow: James MacLehose and Sons, 1908.

Mundy, P. The Travels of Peter Mundy in Europe, Asia, 1608-1667. Vol. 1: Travels in Europe, 1639-1647. (1647.) Edited by R.C. Temple. London: Hakluyt Society, 1925.

Munz, P. Life in the Age of Charlemange. New York: Capricorn Books, 1969.

Murphy, B. The World Book of Whiskey. Chicago: Rand McNally, 1979.

Nashe, T. Pierce Pennilesse. (1592.) In: The Works of Thomas Nashe. Edited by R. McKerrow. London: A.H. Bullen, 1904-1910. Reprint, with corrections by F.P. Wilson, Oxford: Blackwell, 1958. Vol. 1. pp. 157-245.

Neuburger, M. Ein Mässigkeitsverein vor 400 Jahren (The society for self-containment 400 years ago). Mitteilung zur Geschichte von der Medizin der Naturwissenschaft, 16:118-121, 1917.

Neuer, J. "The Historical Development of Tischzuchtliteratur in Germany." Ph.D. dissertation, University of California, Los Angeles, 1970.

Newenham, T. A Statistical and Historical Inquiry into the Progress and Mgnitude of the Population of Ireland. London: C. and R. Baldwin, 1805.

Nicander. The Poems and Poetical Fragments. Translated and edited by A. Gow and A. Scholfield. Cambridge: Cambridge University Press, 1953.

Nimrod, C.J.A. Memoirs of the Life of the Late John Mytton. 2nd ed. London: Kegan, Paul, Trench and Trubner, n.d.

O'Brien, G. The Economic History of Ireland in the Eighteenth Century. Dublin: Maunsel and Co., 1918.

O'Brien, G. The Economic History of Ireland in the Seventeenth Century. Dublin: Maunsel and Co., 1919.

REFERENCES

O'Brien, G. The Economic History of Ireland from the Union to the Famine. London: Longmans, Green and Co., 1921.

O'Brien, J.M. The enigma of Alexander: The alcohol factor. Annals of Scholarship, 1(3):31-46, Summer 1980a.

O'Brien, J.M. Alexander and Dionysus: The invisible enemy. Annals of Scholarship, 1(4):83-105, Fall 1980b.

Obsopoeus, V. On the Art of Drinking. (1536.) Translated by H.F. Simpson. Quarterly Journal of Studies on Alcohol, 5:663-679, 1945.

Olearius, J.A. The Voyages and Travels of the Ambassadors Sent by Frederick Duke of Holstein to the Great Duke of Muscovy and the King of Persia. (1647.) Translated by J. Davis. London, 1665.

Olmstead, A.T. History of Assyria. New York: Charles Scribner's Sons, 1923.

Ovid. Fasti. Translated by J. Frazier. London: Macmillan, 1929.

Ovid. Heroides and Amores. Translated by G. Showerman. Loeb Classical Library, 1921.

Owst, G.R. Literature and Pulpit in Medieval England. 2nd ed. Oxford: Blackwell, 1966.

Painter, S. French Chivalry. Ithaca, New York: Cornell University Press, 1940.

Palmer, R.R., and Colton, J. A History of the Modern World. New York: Alfred A. Knopf, 1969.

Park, P. Sketches toward a political economy of drink and drinking problems: The case of 18th and 19th century England. Journal of Drug Issues, 13:57-75, 1983.

Park, P. "Material conditions and drink production in American history before Prohibition." Paper presented at the Conference on the Social History of Alcohol, sponsored by the Alcohol Research Group, Berkeley, California, January 1984.

Parker, G. The Dutch Revolt. Ithaca, New York: Cornell University Press, 1977.

Pascal, R. German Literature in the Sixteenth and Seventeenth Centuries. London: Cresset Press, 1968.

Patrick, C.H. Alcohol, Culture and Society. Durham, N.C.: Duke University Press, 1952.

Patton, W. The Laws of Fermentation and the Wines of the Ancients. New York: National Temperance Society, 1871.

Pearson, C.C., and Hendricks, J.E. Liquor and Anti-Liquor in Virginia, 1619-1919. Durham, N.C.: Duke University Press, 1967.

REFERENCES

Pepys, S. The Diary of Samuel Pepys. (1660-1669.) Edited by R. Latham and W. Mathers. 11 vols. Berkeley and Los Angeles: University of California Press, 1970-1983.

Peters, H. Pictorial History of Ancient Pharmacy. Translated and revised by W. Netter. 3rd ed. Chicago: Engelhard and Co., 1906.

Petersen, J.W. Geschichte der deutschen Nationalneigung zum Trunke (History of the German National Proclivity for Drinking). Stuttgart, 1856.

Petri, W. Social Life in Ancient Egypt. London: Constable and Co., 1923.

Petronius. The Satyricon. Translated by M. Heseltine. Loeb Classical Library, 1939.

Pfaff, F. Die Weinpreise in Rotenberg am Neckar 1545-1620 (Wine prices in Rotenberg and Neckar 1545-1620). Alemannia, 19:167-168, 1892.

Philo. On Drunkenness; On Sobriety. In: Philo. Translated by F. Colson and G. Whitaker. Vol. 3. Loeb Classical Library, 1930. pp. 308-481.

Pirenne, H. Un grand commerce d'exportation au moyen age: Les vins du France (A great export commerce in the Middle Ages: The wines of France). Annales d'histoire economique et social, 5:225-243, 1933.

Pirenne, H. Economic and Social History of Medieval Europe. Translated by I.E. Clegg. New York: Harcourt, Brace and World, 1937.

Pirenne, H. Medieval Cities. Translated by F.D. Halsey. New York: Anchor Books, 1956.

Plato. Laws. Translated by R.G. Bury. Vol. 1. Loeb Classical Library, 1926.

Plato. Symposium. In: Dialogues of Plato. Translated by M.A. Jowett. Vol. 1. New York: Random House, 1937. pp. 301-345.

Platter, F. Beloved Son Felix: The Journal of Felix Platter. (1552-1557.) Translated by S. Jennett. London: Frederick Muller, 1961.

Pliny the Elder (Plinius Secundus, C.). Natural History. Translated by H. Rackham, W. Jones, and D. Eichholz. 10 vols. Loeb Classical Library, 1938-1962.

Plumb, J.H. The Commercialization of Leisure in 18th century England. Stenton Lecture, 1972. University of Reading, England, 1973.

Plumb, J.H. Commercialization of society. In: McKendrick, H.; Brewer, J.; and Plumb, J.H. The Birth of a Consumer Society: The Commercialization of Eighteenth-Century England. London: Europa Publications, 1982. pp. 265-334.

Plutarch. Plutarch's Lives. Translated by B. Perrin. 10 vols. Loeb Classical Library, 1914-1921.

REFERENCES

Pollnitz, C.-L. Memoirs of Charles-Lewis, baron de Pollnitz. (1st ed., 1732.) 2nd ed. 2 vols. London: D. Browne, 1739.

Polo, M. The Book of Marco Polo. Edited by H. Yule. 3rd ed. 2 vols. London: John Murray, 1929.

Polybius. The Histories of Polybius. Translated by E. Shuckburgh. 2 vols. Bloomington: Indiana University Press, 1962.

Pompen, A. The English Versions of the Ship of Fools. London: Longmans, Green, 1925.

Popham, R. The social history of the tavern. In: Israel, Y., et al., eds. Research Advances in Alcohol and Drug Problems. Vol. 4. New York: Plenum Press, 1978. pp. 225-302.

Popham, R., and Schmidt, W. The effectiveness of legal measures in the prevention of alcohol problems. Addictive Diseases: An International Journal, 2:497-513, 1976.

Porter, R. English Society in the Eighteenth Century. London: Penguin, 1982.

Price, J.C. Culture and Society in the Dutch Republic During the 17th Century. London: B.T. Batsford, 1974.

Pryzhov, I. Istoriya Kabakov v Rossii v svyazi s istoriyei russkovo naroda (The History of Saloons in Russia in Relation to the History of the Russian People). Moscow, 1868.

Quinlan, M. Victorian Prelude. Hamden, Conn.: Archon Books, 1965.

Rabelais, F. The Five Books of Gargantua and Pantagruel. (1533.) Translated by J. Le Clercq. New York: Modern Library, 1936.

Ranum, O. Paris in the Age of Absolutism. New York: John Wiley and Sons, 1968.

Ranum, O., and Ranum, P., eds. The Century of Louis XIV. New York: Walker, 1972.

Rashdall, H. The Universities of Europe in the Middle Ages. Vol. 2, part 2. Oxford: Clarendon Press, 1895.

Rau, H. Arztliche Gutachter und Polizeivorschriften über den Branntwein im Mittelater (Appraisals by Doctors and Police Regulations on Brandy in the Middle Ages). Leipzig, 1914.

Raymond, I. The Teaching of the Early Church on the Use of Wine and Strong Drink. New York: Columbia University Press, 1927.

Reinhardt, K.F. Germany 2000 Years. New York: Ungar, 1962.

Renouard, Y. Vignobles, vignes et vins de France au moyen âge (Vineyards, vines and wines of France in the Middle Ages). Le moyen âge, 3:337-349, 1960.

REFERENCES

Restif de la Bretonne, N. La vie de mon père (Life of My Father). (1779.) Edited by J. and R. Wittman. Paris: Plon, 1933.

Restif de la Bretonne, N. Les nuits de Paris. (1788.) Translated by Linda Asher and Ellen Fertig. Les nuits de Paris or The Nocturnal Spectator. New York: Random House, 1964.

Revel, J. A capital city's privileges: Food supplies in early-modern Rome. In: Forster, R., and Ranum, O., eds. Food and Drink in History. Baltimore: Johns Hopkins Press, 1979. pp. 37-49.

Rice, E. The Foundations of Early Modern Europe, 1460-1559. New York: W.W. Norton, 1970.

Rice, T. The Scythians. New York: Praeger, 1957.

Richard of Devizes. Chronicle. (1190-1192.) London: H.G. Bohn, 1848.

Rix, K. James Boswell. Journal of Alcoholism, 10:73-77, 1975.

Robb, M.J. Scotch Whisky. New York: Dutton, 1951.

Roberts, S.K. Alehouses, brewing and government under the early Stuarts. Southern History, 2:45-71, 1980.

Robinson, G. Rural Russia under the Old Regime. London: Longmans, Green and Co., 1932.

Robiquet, J. Daily Life under the French Revolution. Translated by J. Kirkup. New York: Macmillan, 1964.

Roehl, R. Patterns and structure of demand. In: Fontana Economic History of Europe. Vol. 1: The Middle Ages. New York: Barnes and Nobles, 1976. pp. 107-143.

Roger de Hoveden. The Annals. (c. 1200.) Edited and translated by H.T. Riley. 2 vols. London: H.G. Bohn, 1853. Reprint New York, AMS Press, 1968.

Rolleston, J.D. Voltaire and Alcoholism. British Journal of Inebriety. 22(4):184-185, 1925.

Rolleston, J.D. Alcoholism in classical antiquity. British Journal of Inebriety, 24:101-120, 1927.

Rolleston, J.D. Alcoholism: A review of the literature, II. Bulletin of Hygiene, 4:453-460, 1929.

Rolleston, J.D. Alcoholism in mediaeval England. British Journal of Inebriety, 31:33-50, 1933.

Rolleston, J.D. Alcoholism in Saint Simon's Memoires. British Journal of Inebriety, 38:118-121, 1941.

Rolleston, J.D. Pepys and alcoholism. British Journal of Inebriety 41:70-77, 1944.

Römer, A. Luther und die Trinksitten (Luther and drinking customs). Die Alcoholfrage, 13/14:100-114, 1916.

Rooney, P.; Buchanan, W.; and MacNeil, A. Robert Ferguson: Poet and patient (1750-1774). The Practicioner, 219:402-407, 1977.

Rorabaugh, W.J. The Alcoholic Republic: An American Tradition. New York: Oxford University Press, 1979.

Ross, J. Whisky. London: Routledge and Kegan Paul, 1970.

Ross, J.B., and McLaughlin, M.M. The Portable Renaissance Reader. New York: Viking Press, 1953.

Roueché, B. The Neutral Spirit: A Portrait of Alcohol. Boston: Little, Brown and Co., 1960.

Roueché, B. Alcohol in human culture. In: Lucia, S., ed. Alcohol and Civilization. New York: McGraw-Hill, 1963. pp. 167-183.

Rudé, G. The London "mob" of the eighteenth century. Historical Journal, 2:1-8, 1959a.

Rudé, G. "Mother Gin" and the London riots of 1736. Guildhall Miscellany, 10:53-63, 1959b.

Rudé, G. Hanoverian London, 1714-1808. Berkeley and Los Angeles: University of California Press, 1971.

Rudwin, M. The Origin of the German Carnival Comedy. New York: G.E. Stechert and Co., 1920.

Rush, B. An Inquiry into the Effects of Arden Spirits upon the Human Body and Mind. (1784.) 8th ed. Brookfield, Mass., 1814. Reprint in: Grob, G., ed. Nineteenth Century Medical Attitudes toward Alcohol Addiction. New York: Arno Press, 1981.

Sagarra, E. A Social History of Germany, 1648-1914. New York: Holmes and Meier, 1977.

Saint-Evremond, C. The Works of Monsieur de St.-Evremond. Translated by M. des Maizeaux. 2nd ed. 3 vols. London, 1728.

Saint-Germain, J. La Reynie et la police au grand siècle (La Reynie and the Police in the 17th Century). Paris: Hachette, 1962.

Saint-Germain, J. La vie quotidienne en France à la fin du grand siècle (Daily Life in France at the End of the 17th Century). Paris: Hachette, 1965.

Saint-Simon, duc de. Historical Memoirs. (1695-1751.) Abridged, edited and translated by L. Norton. 3 vols. New York: McGraw Hill Book Co., 1967-1972.

Salasa, M. Emile Zola and the concept of alcoholism in 19th century France. British Journal on Alcohol and Alcoholism 12:14-22, 1977.

Salimbene degli Adami. Chronicle. (1282-1287.) Partially translated by P. Hermann. In: XIII Century Chronicles. With introduction and notes by M.T. Laureilhe. Chicago: Franciscan Herald Press, 1961.

Samuelson, J. History of Drink. London: Trubner and Co., 1878.

Sandmaier, M. Bacchantic maidens and temperance daughters: A historical perspective on female alcohol use. Alcohol Health and Research World, 4(4):41-51, Summer 1980.

Schölzer, A.L. Revolutioner der Diät von Europa seit 300 Jahren (The revolution in the diet of Europe over the past 300 years). Inhalts, 44:93-120, 1781.

Scottish Law Review, 59:85-87, 1943. Early Scots liquor laws.

Sée, H. Economic and Social Conditions in France in the Eighteenth Century. Translated by E.H. Zeydel. New York: Crofts and Co., 1927.

Seltman, C. Wine in the Ancient World. London: Routledge and Kegan Paul, 1957.

Seneca. Moral Letters. Vols. 2 and 3. Translated by R. Gummere. Loeb Classical Library, 1925.

Seward, D. Monks and Wine. London: Mitchell Beazley Publisher, 1979.

Seymour, T.D. Life in the Homeric Age. New ed. New York: Macmillan, 1914.

Scriptores Historiae Augustae. Translated by D. Magie. 3 vols. Loeb Classical Library, 1921-1932.

Shennan, J.H. Philippe, Duke of Orleans: Regent of France, 1715-1723. London: Thames and Hudson, 1979.

Shipkey, R. Problems in alcoholic production and controls in early 19th century Ireland. Historical Journal, 16(2):291-302, 1973.

Shurtleff, N.B., ed. Records of the Governor and Company of the Massachusetts Bay in New England. 5 vols. Boston, 1853.

Sidonius. Poems and Letters. Translated by W. Anderson. 2 vols. Loeb Classical Library. London, 1936.

Sigerist, H. A History of Medicine. New York: Oxford University Press, 1951.

Simon, A. History of the Wine Trade in England. 3 vols. London: Wyman and Sons, 1906, 1907, 1909.

Simon, A. The Book of the Grape: The Wine Trade Text Book. London: Duckworth and Co., 1920.

REFERENCES

Simon, A. Bottlescrew Days; Wine Drinking in England during the 18th Century. London: Duckworth and Co., 1926.

Simon, A. Drink. London: Burke Publishing, 1948.

Sivéry, G. Les comtes de Hainaut et le commerce du vin au XIVe siècle et au début du XVe siècle (The Counts of Hainant and Wine Commerce in the 14th Century and the Beginning of the 15th Century). Lille: University of Lille, 1969.

Slicher van Bath, B.H. The Agrarian History of Western Europe, A.D. 500-1850. Translated by O. Ordish. London: Edward Arnold, 1963.

Smith, A. An Inquiry into the Nature and Causes of the Wealth of Nations. (1776.) Edited by E. Cannon. 6th ed. London: Methuen and Co., 1950.

Smollett, T. The Expedition of Henry Clinker. (1771.) New York: Modern Library, 1929.

Smollett, T. The History of England, from the Revolution of 1688 to the Death of George II. Vols. 3 and 4. London: Printed for J. Wallis, 1805.

Smollett, T. Travels through France and Italy. (1763-1765.) Edited by F. Felsenstein. New York: Oxford University Press, 1979.

Sombart, W. Luxury and Capitalism. Ann Arbor: University of Michigan Press, 1967.

The Spectator. Edited by G.G. Smith. New York: E.P. Dutton, 1907.

Spiller, B. The story of beer. Geographical Magazine, 28:86-94, 143-154, 169-181, 1955.

Spring, J.A., and Buss, D.A. Three centuries of alcohol in the British diet. Nature, 270:567-572, December 15, 1977.

Stacpoole, H. François Villon. New York: G.P. Putnam's Sons, 1916.

Staley, E. The Guilds of Florence. London: Metheun and Co., 1906. Reprint New York: Benjamin Blom, 1967.

Stanihurst, R. Description of Ireland. In: Holinshed, R. Chronicles of England, Scotland, and Ireland. (2nd ed., 1587.) Vol. 6: Ireland. London, 1808. pp. 1-69.

Stivers, R. A Hair of the Dog: Irish Drinking and American Stereotypes. University Park: Pennsylvania State University Press, 1976.

Stolleis, M. "Vom dem grewlicken Laster der Trunckenheit" -- Trinkverbote im 16. und 17. Jahrhundert (On the vicious vice of drunkenness. Drinking prohibitions in the 16th and 17th century). In: Volger, G., ed. Rausch und Realität: Drogen im Kulturvergleich. Vol. 1. Cologne: Rautenstrauch-Joest-Muesums, 1981. pp. 98-105.

REFERENCES

Stone, L. The Crisis of the Aristocracy. Abridged ed. London: Oxford University Press, 1967.

Stone, L. The new eighteenth century. New York Review of Books, 31(5):42-48, March 29, 1984.

Strabo. The Geography of Strabo. Translated by H. Jones. 8 vols. Loeb Classical Library, 1917-1932.

Strauss, G. Nuremberg in the Sixteenth Century. Bloomington: Indiana University Press, 1979.

Stubbes, P. The Anatommie of Abuses. London, 1583. Reprint New York: Da Capo Press, 1972.

Suetonius. History of the Twelve Caesars. Translated by R. Graves. London: Penguin Classics, 1957.

Sumberg, S.L. The Nuremberg Schembart Carnival. New York: Columbia University Press, 1941.

Sutherland, D. Raise Your Glasses: A Lighthearted History of Drinking. London: Macdonald, 1969.

Swart, K.W. The Miracle of the Dutch Republic as Seen in the 17th Century. London: H.K. Lewis, 1967.

Swift, J. A Project for the Advancement of Religion and Reformation of Manners. (1709.) In: The Works of Jonathan Swift. Arranged by T. Sheridan. Edited by J. Nicolas. Vol. 3. London: J. Johnson, 1808.

Symonds, J. Wine, Women and Song. Portland, Maine: Thomas B. Mosher, 1899.

Szasz, T. Ceremonial Chemistry: The Ritual Persecution of Drugs, Addicts, and Pushers. Garden City, New York: Doubleday, 1974.

Tacitus. Histories and Annals. Translated by C. Moore (Histories) and J. Jackson (Annals). 4 vols. Loeb Classical Library, 1935-1937.

Tacitus. Germania. Translated by H. Mattingly. In: On Brittania and Germany. Baltimore: Penguin Books, 1948.

Taylor, J. Drinke and Welcome. London: Anne Griffin, 1637.

Temple, W. Observations on the United Provinces of the Netherlands. (London, 1673.) In: The Works of Sir William Temple. Vol. 1. London: S. Hamilton, 1814.

Theophrastus. Inquiry into Plants. Translated by A. Hort. 2 vols. Loeb Classical Library, 1916.

Thomann, G. Colonial Liquor Laws. New York: U.S. Brewers' Association, 1887.

REFERENCES

Thomas, K. Work and leisure in pre-industrial society: Conference paper and discussion. Past and Present, 29:50-66, 1964.

Thomas, K. Religion and the Decline of Magic. London: Weidenfeld and Nicolson, 1971.

Thompson, C. The Mystery and Art of the Apothecary. Philadelphia: J.B. Lippincott Co., 1929.

Thompson, E.P. The Making of the English Working Class. New York: Vantage Books, 1963.

Thompson, E.P. Patrician society, plebian culture. Journal of Social History, 7:382-405, 1974.

Tibullus. Catullus, Tibullus, and Perviqilium Veneris. Revised and translated by J.P. Postgate (Tibullus). Loeb Classical Library, 1950.

Tilly, C. The Vendée. London: Edward Arnold, 1964.

Townsend, J. A Dissertation on the Poor Laws. (1786.) Berkeley and Los Angeles: University of California Press, 1971.

Trevelyan, G.M. English Social History. London: Longman Group Limited, 1978.

Trotter, T. An Essay, Medical, Philosophical, and Chemical on Drunkenness. (1804.) Philadelphia: Finley, 1813. Reprint New York: Arno Press, 1981.

Tucker, J. An impartial Inquiry into the Benefit and Damage arising to one Nation from the Present very Great Use of Low-priced Spirituous Liquors. London: T. Trege, 1751.

Tuohy, W. "Even the bad times are good" at Old Bushmills. Los Angeles Times, January 4, 1981, part I, p. 7.

Tuohy, W. Scotch: Few toasts in slow season. Los Angeles Times, August 21, 1982, pp. 1, 31.

Turner, B. The government of the body: Medical regimen and the rationalization of diet. British Journal of Sociology, 33:254-269, 1982.

Turner, T. The Diary of a Georgian Shopkeeper. (c. 1755.) Abridged and Edited by G.H. Jenings. Oxford: Oxford University Press, 1979.

Tyrrell, I. Sobering Up: From Temperance to Prohibition in Antebellum America, 1800-1860. Westport, Conn.: Greenwood Press, 1979.

Ukers, W.H. All About Coffee. New York: Tea and Coffee Trade Journal Co., 1935.

Uribe, C.A. Brown Gold: The Amazing Story of Coffee. New York: Random House, 1954.

Vadé, J.-J. La pipe casée. (c. 1750.) In: Oeuvres. Paris: Les marchands de nouveautes, 1831.

Van Laun, H. History of French Literature. New York: G.P. Putnam's Sons, 1909.

Vauban, S. Project d'un dîme royale. (1707.) Translated as: A Project for a Royal Tythe. London: J. Matthews, 1708.

Villon, F.. The Testament. (c. 1462.) In: Complete Works. Translated by Anthony Bonner. New York: David McKay Co., 1960.

Vincent, J.M. European blue laws. Annual Report of the American Historical Association, 1898, pp. 357-372.

Vincent, J.M. Costume and Conduct in the Laws of Basel, Bern, and Zurich, 1370-1800. Baltimore: Johns Hopkins Press, 1935.

Vogt, I. "Upper class and working class drinking cultures in 19th-century Germany." Paper presented at the Conference on the Social History of Alcohol, sponsored by the Alcohol Research Group, Berkeley, California, January 1984.

Voltaire. Philosophical Dictionary. (1764.) Translated by P. Gay. 2 vols. New York: Basic Books, 1962.

Waddell, H. The Wandering Scholars. 7th ed. London: Constable and Co., 1949.

Wain, J. Samuel Johnson. 2nd ed. London: Macmillan, 1980.

Walter of Henley. Husbandry. Translated by E. Lamond. London: Longman, Green and Co., 1890.

Walzer, M. The Revolution of Saints. New York: Atheneum, 1971.

Warmington, E., trans. Remains of Old Latin. 4 vols. Loeb Classical Library, 1935.

Warner, R.H., and Rosett, H.L. The effects of drinking on offspring: An historical survey of the American and British literature. Journal of Studies on Alcohol, 36:1395-1420, 1975.

Wasson, R. Soma: Divine Mushroom of Immortality. New York: Harcourt Brace Jovanovich, [1968].

Wasson, R.; Gordon, A.; and Carl, A. The Road of Eleusis: Unveiling the Secret of the Mysteries. New York: Harcourt Brace Jovanovich, 1978.

Watney, J. Beer is Best: A History of Beer. London: Peter Owen, 1974.

Watney, J. Mother's Ruin: The History of Gin. London: Peter Owen, 1976.

Webb, S., and Webb, B. History of Liquor Licensing in England Principally from 1700 to 1830. London: Longmans, Green and Co., 1903.

Weber, M. The Protestant Ethic and the Spirit of Capitalism. Translated by Talcott Parsons. New York: Charles Scribner's Sons, 1958.

REFERENCES

Weinberg, F. The Wine and the Will: Rabelais's Bacchic Christianity. Detroit: Wayne State University Press, 1972.

Wellman, F. Coffee. New York: Interscience Publishers, 1961.

Werveke, H. van. Le commerce des vins Français au moyen âge (The French wine trade in the Middle Ages). Revue belge philologique, 12:1096-1101, 1933.

Wesley, J. Primitive Physic. (1747.) London: Epworth Press, 1960.

Whitener, D. Prohibition in North Carolina, 1715-1945. Chapel Hill: University of North Carolina Press, 1945.

Wiley, W. The Gentleman of Renaissance France. Cambridge: Harvard University Press, 1954.

Williams, A. The Police of Paris 1718-1789. Baton Rouge, 1979.

Williams, B. The Whig Supremacy, 1714-1760. 2nd edition. Oxford: Clarendon Press, 1962.

Williams, C. "Taverns, Tapsters and Topers." 2 Vols. Ph.D. dissertation, Louisiana State University, 1969.

Williams, G.P., and Brake, G.T. Drink in Great Britain, 1900-1979. London: B. Edsall and Co., 1980.

Wilson, C.A. Food and Drink in Britain from the Stone Age to Recent Times. London: Anchor Press, 1973.

Wilson, C.H. The Dutch Republic and the Civilization of the Seventeenth Century. London: Weidenfeld and Nicolson, 1968.

Wilson, G.B. Alcohol and the Nation: A Contribution to the Study of the Liquor Problem in Great Britain from 1800 to 1935. London: Nicholson and Watson, 1940.

Wilson, G.B. The liquor problem in England and Wales: A survey from 1860 to 1935. British Journal of Inebriety, 38:141-165, 1941.

Wilson, R. Scotch: The Formative Years. London: Constable, 1970.

Wilson, R. Scotch: Its History and Romance. Newton Abbot, England: David and Charles, 1973.

Winkler, A. Drinking on the American frontier. Quarterly Journal on Studies in Alcohol, 29:413-445, 1968.

Winnington-Ingram, R. Euripides and Dionysus. Cambridge: Cambridge University Press, 1948.

Wright, L. Middle-Class Culture in Elizabethan England. Chapel Hill: University of North Carolina Press, 1935.

REFERENCES

Wright-St. Clair, R.E. Beer in therapeutics: A historical annotation. New Zealand Medical Journal, 61:512-513, 1962.

Wrightson, K. Alehouses, order and reformation in rural England, 1590-1660. In: Yeo, E., and Yeo, S., eds. Popular Culture and Class Conflict 1590-1914: Explanations in the History of Labour and Leisure. New Jersey: Humanities Press, 1981. pp. 1-27.

Xenophon. Anabasis. Translated by C. Brownson. In: Xenophon. Vols. 2-3. Loeb Classical Library, 1922.

Xenophon. Constitution of the Lacedaemonians. Translated by E. Marchant. In: Xenophon. Scripta minora. Loeb Classical Library, 1925.

Xenophon. Cyropaedia. 2 vols. Translated by W. Miller. Loeb Classical Library, 1914.

Xenophon. Hiero. Translated by E. Marchant. In: Xenophon. Scripta Minora. Loeb Classical Library, 1925.

Xenophon. Memorabilia and Oeconomicus. Translated by E. Marchant. Loeb Classical Library, 1938.

Xenophon. Symposium. Translated by O.J. Todd. In: Xenophon. Vol. 3. Loeb Classical Library, 1922.

Young, A. A Tour in Ireland, 1776-1779. (1780.) London: Casscle and Co., 1887.

Young, A. Travels in France during the Years 1787, 1788 and 1789. (London, 1794.) Edited by C. Maxwell. Cambridge: Cambridge University Press, 1929.

Young, A. Travels in France during the Years 1787, 1788 and 1789. (London, 1794.) Edited by J. Kaplow. Garden City, New York: Doubleday, Anchor, 1969.

Younge, R. [R. Junius]. The Drunkards Character. London: R. Badger, 1638.

Younger, W. Gods, Men, and Wine. London: Food and Wine Society, 1966.

Zirker, M.R., Jr. Fielding's Social Pamphlets: A Study of "An Enquiry into the Causes of the Late Increase of Robbers" and "A Proposal for Making an Effectual Provision for the Poor". Berkeley and Los Angeles: University of California Press, 1966.

Zumthor, P. Daily Life in Rembrandt's Holland. Translated by S. Taylor. London: Weidenfeld and Nicholson, 1962.

Indexes

Index references are made, not to page numbers, but to geographic and chronological headings where corresponding information can be located within the text. The text is arranged chronologically by date and subdivided by geographic or cultural headings. This order has been reversed in the indexes to facilitate geographic access. Each index term refers first to geographic or cultural headings (listed alphabetically in capital letters), and then, chronologically, to dates under which the geographic headings are located within the text. B.C. dates are indicated in the index by a preceeding dash. (For example, –450 refers to the year 450 B.C.)

Name Index

The Name Index contains references to all historical individuals mentioned in the text and to all authors of works cited as primary sources within the text. Titles of works with anonymous authorship are listed alphabetically by title in a separate section at the end of the index.

ANONYMOUS

Peoples and Places Index

The Peoples and Places Index contains references to all countries, nations, groups of people, and specific geographic locations throughout the text. This index is particularly useful in tracking peoples and places in the text that are located under other geographic headings of primary interest. It also identifies those major sections of the text that are actually located under corresponding headings. Where possible, country names are given preference over other geographic, national, or religious designators.

Subject Index

The Subject Index contains references to topics covered throughout the text, including groups of people that are not geographically designated. In most cases, the specific word used in the text is assigned as the index term with cross-references to related terms. Beverages are frequently indexed under both general and specific terms.

About the Authors

Gregory A. Austin is Research Historian and Information Specialist at the Southern California Research Institute, Los Angeles, California. He serves as principal investigator on two federally-funded research projects relating to the history of drug and alcohol use and is the author of *Perspectives on the History of Psychoactive Substance Use,* part of the National Institute on Drug Abuse Research Issues Series of which he also was the former Chief Editor. With Michael L. Prendergast, he is the compiler of the two-volume *Drug Use and Abuse: A Guide to Research Findings* (ABC-Clio, 1984). He received his Ph.D. in history from the University of California, Los Angeles.

Robin Room is Scientific Director of the Alcohol Research Group, Institute of Epidemiology and Behavioral Medicine, Medical Research Institute of San Francisco, and Adjunct Associate Professor with the Department of Social and Administrative Health Sciences, School of Public Health, University of California, Berkeley. He has authored and coauthored numerous alcohol and drug research studies under the sponsorship of such notable organizations as The National Institute of Drug Abuse, the Finnish Foundation of Alcohol Studies, and the Toronto Addiction Research Foundation. In 1983, Room was corecipient of the Jellinek Memorial Award for "outstanding contribution to knowledge in the field of alcohol studies."